Sturm-Liouville Theory

Mathematical
Surveys
and
Monographs

Volume 121

Sturm-Liouville Theory

Anton Zettl

American Mathematical Society

EDITORIAL COMMITTEE

Jerry L. Bona Peter S. Landweber
Michael G. Eastwood Michael P. Loss

J. T. Stafford, Chair

2000 *Mathematics Subject Classification.* Primary 34B20, 34B24; Secondary 47B25.

For additional information and updates on this book, visit
www.ams.org/bookpages/surv-121

Library of Congress Cataloging-in-Publication Data
Zettl, Anton.
 Sturm-Liouville theory / Anton Zettl.
  p. cm. — (Mathematical surveys and monographs ; v. 121)
 Includes bibliographical references and index.
 ISBN 0-8218-3905-5 (alk. paper)
 1. Sturm-Liouville equation. I. Title. II. Mathematical surveys and monographs ; no. 121.

QA379.Z48 2005
515′.35–dc22
                                                                                    2005048214

Copying and reprinting. Individual readers of this publication, and nonprofit libraries acting for them, are permitted to make fair use of the material, such as to copy a chapter for use in teaching or research. Permission is granted to quote brief passages from this publication in reviews, provided the customary acknowledgment of the source is given.

 Republication, systematic copying, or multiple reproduction of any material in this publication is permitted only under license from the American Mathematical Society. Requests for such permission should be addressed to the Acquisitions Department, American Mathematical Society, 201 Charles Street, Providence, Rhode Island 02904-2294, USA. Requests can also be made by e-mail to reprint-permission@ams.org.

  © 2005 by the American Mathematical Society. All rights reserved.
   The American Mathematical Society retains all rights
   except those granted to the United States Government.
     Printed in the United States of America.
  ∞ The paper used in this book is acid-free and falls within the guidelines
     established to ensure permanence and durability.
     Visit the AMS home page at http://www.ams.org/
      10 9 8 7 6 5 4 3 2 1   10 09 08 07 06 05

Contents

Preface	ix
Part 1. Existence and Uniqueness Problems	1
Chapter 1. First Order Systems	3
1. Introduction	3
2. Existence and Uniqueness of Solutions	3
3. Variation of Parameters	8
4. The Gronwall Inequality	8
5. Bounds and Extensions to the Endpoints	10
6. Continuous Dependence of Solutions on the Problem	12
7. Differentiable Dependence of Solutions on the Data	14
8. Adjoint Systems	19
9. Inverse Initial Value Problems	20
10. Comments	21
Chapter 2. Scalar Initial Value Problems	25
1. Introduction	25
2. Existence and Uniqueness	25
3. Continuous Extensions to the Endpoints	26
4. Continuous Dependence of Solutions on the Problem	27
5. Differentiable Dependence of Solutions on the Data	28
6. Sturm Separation and Comparison Theorems	32
7. Periodic Coefficients	35
8. Comments	38
Part 2. Regular Boundary Value Problems	41
Chapter 3. Two-Point Regular Boundary Value Problems	43
1. Introduction	43
2. Transcendental Characterization of the Eigenvalues	43
3. The Fourier Equation	47
4. The Space of Regular Boundary Value Problems	53
5. Continuity of Eigenvalues and Eigenfunctions	54
6. Differentiability of Eigenvalues	56
7. Finite Spectrum	57
8. Green's Function	61
9. Comments	65
Chapter 4. Regular Self-Adjoint Problems	69

1.	Introduction	69
2.	Canonical Forms of Self-Adjoint Boundary Conditions	71
3.	Existence of Eigenvalues	72
4.	Dependence of Eigenvalues on the Problem	76
5.	The Prüfer Transformation	81
6.	Separated Boundary Conditions	86
7.	Coupled Boundary Conditions	90
8.	An Elementary Existence Proof for Coupled Boundary Conditions	91
9.	Monotonicity of Eigenvalues	95
10.	Multiplicity of Eigenvalues	96
11.	Green's Function	96
12.	Finite Real Spectrum	98
13.	Comments	104

Chapter 5. Regular Left-Definite and Indefinite Problems   107
1. Introduction   107
2. Definition and Characterization of Left-Definite Problems   108
3. Existence of Eigenvalues   112
4. Continuous Dependence of Eigenvalues on the Problem   114
5. Eigenvalue Inequalities   116
6. Differentiability of Eigenvalues   118
7. T-Left-Definite Problems   121
8. Indefinite Problems and Complex Eigenvalues   122
9. Comments   125

Part 3. Oscillation and Singular Existence Problems   129

Chapter 6. Oscillation   131
1. Introduction   131
2. Principal and Non-Principal Solutions   131
3. Oscillation Criteria   133
4. Oscillatory Characterizations   137
5. Comments   140

Chapter 7. The Limit-Point, Limit-Circle Dichotomy   143
1. Introduction   143
2. System Regularization of the Scalar Equation   143
3. Endpoint Classifications: R, LC, LP, O, NO, LCNO, LCO   145
4. LP and LC Conditions   146
5. Comments   151

Chapter 8. Singular Initial Value Problems   155
1. Introduction   155
2. Scalar Regularization with Regularizing Functions   155
3. Factorization of Solutions near an LCNO Endpoint   156
4. Limit-Circle "Initial Value Problems"   158
5. Comments   159

Part 4. Singular Boundary Value Problems   161

Chapter 9. Two-Point Singular Boundary Value Problems — 163
1. Introduction — 163
2. The Minimal and Maximal Domains and Lagrange Form — 163
3. Transcendental Characterization of Eigenvalues — 165
4. Green's Function — 167
5. Comments — 169

Chapter 10. Singular Self-Adjoint Problems — 171
1. Introduction — 171
2. The Lagrange Form — 171
3. The Minimal and Maximal Domains and Self-Adjoint Operators — 173
4. Operator Theory Characterization and Self-Adjoint Boundary Conditions — 174
5. The Friedrichs Extension — 193
6. Nonoscillatory Endpoints — 194
7. Oscillatory Endpoints — 200
8. Behavior of Eigenvalues near a Singular Boundary — 201
9. Approximating a Singular Problem with Regular Problems — 204
10. Green's Function — 205
11. Multiplicity of Eigenvalues — 206
12. Summary of Spectral Properties — 208
13. Comments — 211

Chapter 11. Singular Indefinite Problems — 215
1. Introduction and Results — 215
2. Krein Spaces — 219
3. Self-Adjoint Operators in Krein Spaces — 220
4. A Construction of Left-Definite Krein Spaces — 223
5. Proof of Theorems 11.1.1 and 11.1.3 — 225
6. Comments — 227

Chapter 12. Singular Left-Definite Problems — 229
1. Introduction — 229
2. An Associated One Parameter Family of Right Definite Operators — 232
3. Existence of Eigenvalues — 235
4. Lemmas and Proofs — 240
5. LC Non-Oscillatory Problems — 243
6. Further Eigenvalue Properties in the LCNO Case — 248
7. Approximating a Singular Problem with Regular Problems — 249
8. Floquet Theory of Left-Definite Problems — 257
9. Comments — 259

Part 5. Examples and other Topics — 263

Chapter 13. Two Intervals — 265
1. Introduction — 265
2. Notation and Basic Assumptions — 266
3. Characterization of all Self-Adjoint Extensions — 266
4. Comments — 275

Chapter 14. Examples	277
Chapter 15. Notation	293
Chapter 16. Comments on Some Topics not Covered	295
Chapter 17. Open Problems	299
Bibliography	303

Preface

In 1836-1837 Sturm and Liouville published a series of papers on second order linear ordinary differential equations including boundary value problems. The influence of their work was such that this subject became known as Sturm-Liouville theory. The impact of these papers went well beyond their subject matter to general linear and nonlinear differential equations and to analysis generally, including functional analysis. Prior to this time the study of differential equations was largely limited to the search for solutions as analytic expressions. Sturm and Liouville were among the first to realize the limitations of this approach and to see the need for finding properties of solutions directly from the equation even when no analytic expressions for solutions are available.

Many thousands of papers, by Mathematicians, Physicists, Engineers and others, have been written since then. Yet, remarkably, this subject is an intensely active field of research today. Dozens of papers are published on Sturm-Liouville Problems (SLP) every year.

In 1910 Hermann Weyl published one of the most widely quoted papers in analysis [**607**]. Just as the 1836-37 papers of Sturm and Liouville started the study of *regular* SLP, the 1910 paper of Weyl initiated the investigation of *singular* SLP. The development of quantum mechanics in the 1920's and 1930's, the proof of the general spectral theorem for unbounded self-adjoint operators in Hilbert space by von Neumann and Stone, and the fundamental work of Titchmarsh [**573**] provided some of the motivation for further investigations into the spectral theory of Sturm-Liouville operators.

The purpose of this monograph is twofold: (i) to give a modern survey of some of the basic properties of the Sturm-Liouville equation and (ii) to bring the reader to the forefront of knowledge on some aspects of SLP.

On numerous occasions I have been asked: Where can I find a readable introduction to Sturm-Liouville problems? Although the subject matter of SLP is briefly discussed in many books, these discussions tend to be sketchy, particularly in the singular case. Even for the regular case, a general discussion of separated and coupled self-adjoint boundary conditions is not easy to find in the existing literature. We hope that this monograph will serve as a readable introduction to SLP and, at the same time, provide an up to date account of some parts of this fascinating subject.

A major stimulus for the writing of this monograph was the authors' collaboration with Paul Bailey and Norrie Everitt on the development of the fortran code SLEIGN2 for the numerical computation of eigenvalues and eigenfunctions of *regular and singular, separated and coupled, self-adjoint* SLP. All nine files of the SLEIGN2 software package as well as numerous related papers can be downloaded from: http:/www.math.niu.edu/~zettl/SL2/.

The code is designed to be used by novice and expert alike. It comes with a user friendly interface and an interactive help tutorial. When used with some theoretical results in papers, some of which are available from the web page just mentioned, SLEIGN2 can also be used to approximate parts of the essential (continuous) spectrum, e.g. the starting point of the essential spectrum, the first few spectral bands or gaps, etc.

Although the subject of Sturm-Liouville problems is over 170 years old, a surprising number of the results surveyed here are of recent origin, some were published within the last couple of years and a few are not in print at the time of this writing.

The book is divided into five parts. Part I deals with existence and uniqueness questions for initial value problems including the continuous and differentiable dependence of solutions on all the parameters of the problem. Regular two-point boundary value problems are discussed in Part II, non-self-adjoint problems in Chapter 3, classical (right-definite) self-adjoint problems in Chapter 4 and left-definite and indefinite problems in Chapter 5. Oscillation, the limit-point/limit-circle dichotomy and singular initial value problems are covered in Part III. Singular self-adjoint, non-self-adjoint, right-definite, left-definite and indefinite problems are studied in Part IV. Part V contains chapters on notation, topics not covered, the two-interval theory of boundary value problems and a chapter on examples. These examples have been chosen to illustrate the depth and diversity of Sturm-Liouville theory.

When I started this project it was my intention to provide detailed proofs of all results and to give an elementary proof whenever possible. But I soon realized that this task was beyond my energy level. So I have compromised by providing some detailed proofs, as elementary as possible, and readable references to all proofs not given. Many of these references can be found on the web address mentioned above.

I have been privileged to work with many mathematicians and am grateful to all of them. Special thanks go to my co-authors: F.V. Atkinson, P.B. Bailey, J. Billingham, B.M. Brown, X. Cao, E.A. Coddington, R. J. Cooper, H.I. Dwyer, M.S.P. Eastham, W. Eberhard, W.D. Evans, W.N. Everitt, Z.M. Franco, G. Freiling, H. Frentzen, M. Giertz, J. Goldstein, J. Gunson, K. Haertzen, D.B. Hinton, H.G. Kaper, R.M. Kauffman, A.C. King, L. Kong, Q. Kong, M.K. Kwong, A.M. Krall, G. Leaf, H. Lekkerkerker, Q. Lin, L. Littlejohn, M. Marletta, D.K.R. McCormack, M. Möller, H.-D. Niessen, D. Race, B.S. Garbow, T.T. Read, J. Ridenhour, C. Shubin, G. Stolz, H. Volkmer, J. Weidmann, J.S.W. Wong, and H. Wu. In particular I thank Paul Bailey for introducing me to the wonderful world of computing, for his seemingly infinite patience in writing, debugging and improving the SLEIGN2 code.

I am greatly indebted to my colleagues and friends Qingkai Kong, Man Kam Kwong, and Hongyou Wu for many hundreds of hours of enjoyable and productive collaboration.

I am especially grateful to my friend and collaborator for more than a quarter century, W.N. Everitt. Norrie's characteristically careful and exacting criticisms have resulted in numerous improvements not only of the contents but also the presentation of this monograph. Moreover, I like to think that some of his infinite enthusiasm for Mathematics and his masterful writing style have rubbed off on me.

Special thanks go to the two anonymous referees for their thorough reading of the manuscript. Their corrections, suggestions and criticisms have resulted in numerous improvements.

Last, but certainly not least, I thank my wife Sandra for her help with the database for the references and, especially, for helping with the hardware and software problems that arose during the typing of this manuscript with Scientific Workplace (SWP). Also for her tolerance and understanding during this and many other Mathematics projects.

The world of Mathematics is full of wonders and of mysteries, at least as much so as the physical world.

Part 1

Existence and Uniqueness Problems

da Vinci, Leonardo (1452-1519)
> He who loves practice without theory is like the sailor who boards ship without a rudder and compass and never knows where he may cast.
>
> No human investigation can be called real science if it cannot be demonstrated mathematically.

Stewart, Ian
> The successes of the differential equation paradigm were impressive and extensive. Many problems, including basic and important ones, led to equations that could be solved. A process of self-selection set in, whereby equations that could not be solved were automatically of less interest than those that could.

Does God Play Dice? The Mathematics of Chaos, Blackwell, Cambridge, MA, 1989.

CHAPTER 1

First Order Systems

H. S. Wall
> The Mathematician is an artist whose medium is the mind and whose creations are ideas.

1. Introduction

This chapter is devoted to the study of basic properties of first order systems of general dimension n. Although our primary interest is in the case $n = 2$ we include the higher order case since it can be studied with basically the same methods.

Notation. An open interval is denoted by (a,b) with $-\infty \leq a < b \leq \infty$; $[a,b]$ denotes the closed interval which includes the left endpoint a and the right endpoint b, regardless of whether these are finite or infinite, $\mathbb{R}$ denotes the reals, $\mathbb{C}$ the complex numbers, and
$$\mathbb{N} = \{1, 2, 3, \ldots\}, \ \mathbb{N}_0 = \{0, 1, 2, \ldots\}, \ \mathbb{Z} = \{\ldots, -2, -1, 0, 1, 2, \ldots\}.$$
For any interval J of the real line, open, closed, half open, bounded or unbounded, by $L(J, \mathbb{C})$ we denote the linear manifold of complex valued Lebesgue measurable functions y defined on J for which
$$\int_a^b |y(t)|\,dt \equiv \int_J |y(t)|\,dt \equiv \int_J |y| < \infty.$$

The notation $L_{loc}(J, \mathbb{C})$ is used to denote the linear manifold of functions y satisfying $y \in L([\alpha, \beta], \mathbb{C})$ for all compact intervals $[\alpha, \beta] \subseteq J$. If $J = [a, b]$ and both of a and b are finite, then $L_{loc}(J, \mathbb{C}) = L(J, \mathbb{C})$. Also, we denote by $AC_{loc}(J)$ the collection of complex-valued functions y which are absolutely continuous on all compact intervals $[\alpha, \beta] \subseteq J$. The symbols $L(J, \mathbb{R})$ and $L_{loc}(J, \mathbb{R})$ are defined similarly.

For a given set S, $M_{n,m}(S)$ denotes the set of $n \times m$ matrices with entries from S. If $n = m$ we write $M_n(S) = M_{n,n}(S)$; also if $m = 1$ we sometimes write S^n for $M_{n,1}(S)$. The norm of a constant matrix as well as the norm of a matrix function P is denoted by $|P|$. This may be taken as
$$|P| = \sum |p_{ij}|.$$

2. Existence and Uniqueness of Solutions

DEFINITION 1.2.1 (Solution). *Let J be any interval, open, closed, half open, bounded or unbounded; let $n, m \in \mathbb{N}$, let $P : J \to M_n(\mathbb{C})$, $F : J \to M_{n,m}(\mathbb{C})$. By a solution of the equation*
$$Y' = PY + F \text{ on } J$$

we mean a function Y from J into $M_{n,m}(\mathbb{C})$ which is absolutely continuous on all compact subintervals of J and satisfies the equation a.e. on J. A matrix function is absolutely continuous if each of its components is absolutely continuous.

THEOREM 1.2.1 (Existence and uniqueness). *Let J be any interval, open, closed, half open, bounded or unbounded; let $n, m \in \mathbb{N}$. If*

$$P \in M_n(L_{loc}(J, \mathbb{C})) \tag{1.2.1}$$

and

$$F \in M_{n,m}(L_{loc}(J, \mathbb{C})) \tag{1.2.2}$$

then every initial value problem (IVP)

$$Y' = PY + F \tag{1.2.3}$$

$$Y(u) = C, \ u \in J, \ C \in M_{n,m}(\mathbb{C}) \tag{1.2.4}$$

has a unique solution defined on all of J. Furthermore, if C, P, F, are all real-valued, then there is a unique real valued solution.

PROOF. We give two proofs of this important theorem; the second one is the standard successive approximations proof. As we will see later the analytic dependence of solutions on the spectral parameter λ follows more readily from the second proof than the first.

For both proofs we note that if Y is a solution of the IVP (1.2.3), (1.2.4) then an integration yields

$$Y(t) = C + \int_u^t (PY + F), \ t \in J. \tag{1.2.5}$$

Conversely, every solution of the integral equation (1.2.5) is also a solution of the IVP (1.2.3), (1.2.4).

Choose c in J, $c \neq u$. We show that (1.2.5) has a unique solution on $[u, c]$ if $c > u$ and on $[c, u]$ if $c < u$. Assume $c > u$. Let

$$B = \{Y : [u, c] \to M_{n,m}(\mathbb{C}), \ Y \ continuous\}.$$

Following Bielecki [**67**] we define the norm of any function $Y \in B$ to be

$$\|Y\| = \sup \left\{ \exp\left(-K \int_u^t |P(s)| \, ds\right) |Y(t)|, \ t \in [u, c] \right\}, \tag{1.2.6}$$

where K is a fixed positive constant $K > 1$. It is easy to see that with this norm B is a Banach space. Let the operator $T : B \to B$ be defined by

$$(TY)(t) = C + \int_u^t (PY + F)(s) \, ds, \ t \in [u, c], \ Y \in B. \tag{1.2.7}$$

Then for $Y, Z \in B$ we have

$$|(TY)(t) - (TZ)(t)| \leq \int_u^t |P(s)||Y(s) - Z(s)| ds$$

and hence
$$\exp\left(-K\int_u^t |P(s)|\,ds\right)|(TY)(t)-(TZ)(t)|$$
$$\leq \|Y-Z\|\int_u^t |P(s)|\exp\left(-K\int_s^t |P(r)|\,dr\right)ds$$
$$\leq \frac{1}{K}\|Y-Z\|.$$

Therefore
$$\|TY-TZ\| \leq \frac{1}{K}\|Y-Z\|.$$

From the contraction mapping principle in Banach space it follows that the map T has a unique fixed point and therefore the IVP (1.2.3), (1.2.4) has a unique solution on $[u,c]$. The proof for the case $c<u$ is similar; in this case the norm of B is modified to
$$\|Y\| = \sup\{\exp\left(K\int_t^u |P(s)|\,ds\right)|Y(t)|,\ t\in[c,u]\}.$$

Since there is a unique solution on every compact subinterval $[u,c]$ and $[c,u]$ for $c\in J$, $c\neq u$ it follows that there is a unique solution on J. To establish the furthermore part take the Banach space of real-valued functions and proceed similarly. This completes the first proof.

For the second proof we construct a solution of (1.2.5) by successive approximations. Define
$$Y_0(t) = C,\ Y_{n+1}(t) = C + \int_u^t (PY_n + F),\ t\in J,\ n=0,1,2,\ldots \qquad (1.2.8)$$

Then Y_n is a continuous function on J for each $n\in N_0$. We show that the sequence $\{Y_n : n\in N_0\}$ converges to a function Y uniformly on each compact subinterval of J and that the limit function Y is the unique solution of the integral equation (1.2.5) and hence also of the IVP (1.2.3), (1.2.4). Choose $b\in J$, $b>u$ and define
$$p(t) = \int_u^t |P(s)|\,ds,\ t\in J;\ B_n(t) = \max_{u\leq s\leq t}|Y_{n+1}(s)-Y_n(s)|,\ u\leq t\leq b. \qquad (1.2.9)$$

Then
$$Y_{n+1}(t) - Y_n(t) = \int_u^t P(s)\,[Y_n(s)-Y_{n-1}(s)]\,ds,\ t\in J,\ n\in\mathbb{N}. \qquad (1.2.10)$$

From this we get
$$|Y_2(t)-Y_1(t)| \leq B_0(t)\int_u^t |P(s)|\,ds = B_0(t)\,p(t) \leq B_0(b)\,p(b),\ u\leq t\leq b. \qquad (1.2.11)$$

$$|Y_3(t)-Y_2(t)| \leq \int_u^t |P(s)|\,|Y_2(s)-Y_1(s)|\,ds \leq \int_u^t |P(s)|\,B_0(s)\,p(s)\,ds$$
$$\leq B_0(t)\int_u^t |P(s)|\,p(s)\,ds \leq B_0(b)\,\frac{p^2(t)}{2!}$$
$$\leq B_0(b)\,\frac{p^2(b)}{2!},\ u\leq t\leq b.$$

From this and mathematical induction we get

$$|Y_{n+1}(t) - Y_n(t)| \leq B_0(b) \frac{p^n(b)}{n!}, \quad u \leq t \leq b.$$

Hence for any $k \in N$

$$|Y_{n+k+1}(t) - Y_n(t)| \leq |Y_{n+k+1}(t) - Y_{n+k}(t)| + |Y_{n+k}(t) - Y_{n+k-1}(t)| +$$
$$\ldots + |Y_{n+1}(t) - Y_n(t)|$$
$$\leq B_0(b) \frac{p^n(b)}{n!}[1 + \frac{p(b)}{n+1} + \frac{p^2(b)}{(n+2)(n+1)} + \ldots].$$

Choose m large enough so that $p(b)/(n+1) \leq 1/2$, then $p^2(b)/((n+2)(n+1)) \leq 1/4$, etc. when $n > m$ and the term in brackets is bounded above by 2. It follows that the sequence $\{Y_n : n \in N_0\}$ converges uniformly, say to Y, on $[u, b]$. From this it follows that Y satisfies the integral equation (1.2.5) and hence also the IVP (1.2.3), (1.2.4) on $[u, b]$.

To show that Y is the unique solution assume Z is another one; then Z is continuous and therefore $|Y - Z|$ is bounded, say by $M > 0$ on $[u, b]$. Then

$$|Y(t) - Z(t)| = \left|\int_u^t P(s)[Y(s) - Z(s)]\,ds\right| \leq M \int_u^t |P(s)|\,ds \leq M\,p(t), \quad u \leq t \leq b.$$

Now proceeding as above we get

$$|Y(t) - Z(t)| \leq M \frac{p^n(t)}{n!} \leq M \frac{p^n(b)}{n!}, \quad u \leq t \leq b, \; n \in \mathbb{N}.$$

Therefore $Y = Z$ on $[u, b]$. There is a similar proof for the case when $b < u$. This completes the second proof. □

It is interesting to observe that the initial approximation $Y_0(t) = C$ can be replaced with $Y_0(t) = G(t)$ for any continuous function G without any essential change in the proof.

To study the dependence of the unique solution on the parameters of the problem we introduce a convenient notation. Let J be an interval. For each $P \in M_n(L_{loc}(J, \mathbb{C}))$, each $F \in M_{n,m}(L_{loc}(J, \mathbb{C}))$, each $u \in J$ and each $C \in M_{n,m}(\mathbb{C})$ there is, according to Theorem 1.2.1, a unique $Y \in M_{n,m}(AC_{loc}(J))$ such that $Y' = PY + F$, $Y(u) = C$. We use the notation

$$Y = Y(\cdot, u, C, P, F) \tag{1.2.12}$$

to indicate the dependence of the unique solution Y on these quantities. Below, if the variation of Y with respect to some of the variables u, C, P, F is studied while the others remain fixed we abbreviate the notation (1.2.12) by dropping those quantities which remain fixed. Thus we may use $Y(t)$ for the value of the solution at $t \in J$ when u, C, P, F are fixed or $Y(\cdot, u)$ to study the variation of the solution function Y with respect to u, $Y(\cdot, P)$ to study Y as a function of P, etc.

THEOREM 1.2.2 (Rank invariance). *Let* $J = (a, b)$, *and assume that* $P \in M_n(L_{loc}(J, \mathbb{C}))$. *If* Y *is an* $n \times m$ *matrix solution of*

$$Y' = PY \text{ on } J, \tag{1.2.13}$$

then we have

$$\text{rank}\,Y(t) = \text{rank}\,Y(u), \quad t, u \in J. \tag{1.2.14}$$

2. EXISTENCE AND UNIQUENESS OF SOLUTIONS

Moreover, if $m = n$, then for any $u, t \in J$, we have

$$(\det Y)(t) = (\det Y)(u) \exp\left(\int_u^t \operatorname{trace} P(s)\, ds\right). \qquad (1.2.15)$$

PROOF. The formula (1.2.15) follows from the fact that $y = \det Y$ satisfies the first order scalar equation $y' - py = 0$ where $p = \operatorname{trace} P$. To prove the general case, let $Y(u) = C$ and let rank $C = r$. If $r = 0$, then $Y(t) = 0$ for all t by Theorem 1.2.1. For $r > 0$ let C_i, $i = 1, \ldots r$ be linearly independent columns of C and construct a nonsingular $n \times n$ matrix D by adding $n - r$ appropriate constant vectors to C_i, $i = 1, \ldots, r$. Denote by Z the solution of (1.2.13) satisfying the initial condition $Z(u) = D$. Then by (1.2.15) rank $Z(t) = n$, for $t \in J$. Hence the first r columns of $Z(t)$, $Z_1(t), \ldots, Z_r(t)$ are linearly independent. From this and the uniqueness part of Theorem 1.2.1 the (constant) n-vectors $Y_1(t), Y_2(t), \ldots, Y_r(t)$ are linearly independent since $Z_j = Y_j$ on J. Hence rank $Y(t) \geq r$, for $t \in J$. Now suppose that rank $Y(c) > r$ for some c in J. Then by repeating the above argument with u replaced by c we reach the conclusion that rank $Y(t) > r$ for all $t \in J$. But this contradicts rank $Y(u) = r$ and concludes the proof. □

Formula (1.2.15) is sometimes called Abel's formula. It follows from this formula that if a solution is nonsingular at some point $u \in J$ then it is nonsingular at every point of J. The rank invariance of solutions given by (1.2.14) extends this result to the case when the matrix Y is not square but without a formula corresponding to (1.2.15).

THEOREM 1.2.3 (Everitt and Race). *Let $P : J \to M_n(C)$ and $F : J \to M_{n,1}(C)$, $J = (a, b)$, $-\infty \leq a < b \leq \infty$. If, for any $u \in J$ and any linearly independent constant vectors $C_1, \ldots, C_n$, each initial value problem*

$$Y' = PY + F, \quad Y(u) = C_i, \quad i = 1, \ldots, n,$$

has a unique solution Y_i on J, then $P \in M_n(L_{loc}(J, \mathbb{C}))$ and $F \in M_n(L_{loc}(J, \mathbb{C}))$. Furthermore, if each Y_i is a C^1 solution, then there exist such P and F which are continuous.

PROOF. We first prove the special case when $F = 0$ on J. Let Y_i be a vector solution satisfying $Y_i(u) = C_i$ and let Y be the matrix whose $i-th$ column is Y_i, $i = 1, \ldots, n$. Then the matrix solution Y is nonsingular in some neighborhood N_u of u. Choose

$$P = Y' Y^{-1} \quad \text{on} \quad N_u.$$

Let K be a compact subinterval of J. Since the open cover $\{N_u : u \in J\}$ of K has a finite subcover, we can conclude that Y is uniquely defined and invertible on K. Since Y is continuous and invertible on K, it follows that Y^{-1} is continuous and hence bounded on K. Also, Y' is integrable on K since Y is absolutely continuous on K by virtue of the fact that it is a solution on J. Therefore $P \in M_n(L_{loc}(J, \mathbb{C}))$.

To establish the case when F is not identically zero on J, let Y be a vector solution of $Y' = PY + F$ satisfying $Y(u) = 0$ and choose a solution Z of this equation such that $Z(u) = C$ and let $V = Z - Y$. Then $V' = PV$ and $V(u) = C$. Since this holds for arbitrary C we may conclude from the special case established above that $P \in M_n(L_{loc}(J, \mathbb{C}))$. Hence $F = V' - PV \in M_n(L_{loc}(J, \mathbb{C}))$. The furthermore statement is clear from the proof. □

3. Variation of Parameters

Let $P \in M_n(L_{loc}(J))$. From Theorem 1.2.1 we know that for each point u of J there is exactly one matrix solution X of (1.2.13) satisfying $X(u) = I_n$ where I_n denotes the $n \times n$ identity matrix.

DEFINITION 1.3.1 (Primary fundamental matrix Φ). *For each fixed $u \in J$ let $\Phi(\cdot, u)$ be the fundamental matrix of (1.2.13) satisfying*

$$\Phi(u, u) = I_n \, .$$

Note that for each fixed u in J, $\Phi(\cdot, u)$ belongs to $M_n(AC_{loc}(J))$. Furthermore, if J is compact and $P \in M_n(L(J, \mathbb{C}))$, then u can be an endpoint of J and $\Phi(\cdot, u)$ belongs to $M_n(AC(J))$. (This is clear from the proof of Theorem 1.2.1, or see Theorem 1.5.2 below.) By Theorem 1.2.2, $\Phi(t, u)$ is invertible for each $t, u \in J$ and we note that

$$\Phi(t, u) = Y(t) Y^{-1}(u) \tag{1.3.1}$$

for any fundamental matrix Y of (1.2.13).

We call Φ the primary fundamental matrix of (1.2.13) and also write

$$\Phi = \Phi(P) = (\Phi_{rs})_{r,s=1}^n \, , \quad \Phi(P)(t, u) = \Phi(t, u, P). \tag{1.3.2}$$

Observe that for any constant $n \times m$ matrix C, ΦC is also a solution of (1.2.13). If C is a constant nonsingular $n \times n$ matrix then ΦC is a fundamental matrix solution and every fundamental matrix solution has this form.

The next result is called the variation of parameters formula and is fundamental in the theory of linear differential equations.

THEOREM 1.3.1 (Variation of Parameters Formula). *Let J be any interval, let $P \in M_n(L_{loc}(J, \mathbb{C}))$ and let $\Phi = \Phi(\cdot, \cdot, P)$ be the primary fundamental matrix of (1.2.13) defined above. Let $F \in M_{n,m}(L_{loc}(J, \mathbb{C}))$, $u \in J$ and $C \in M_{n,m}(\mathbb{C})$. Then*

$$Y(t) = \Phi(t, u, P) C + \int_u^t \Phi(t, s, P) F(s) \, ds, \quad t \in J \tag{1.3.4}$$

is the solution of (1.2.3), (1.2.4). Note that if J is compact and $P \in M_n(L(J))$, $F \in M_{n,m}(L(J))$, then $Y \in M_{n,m}(AC(J))$, and u can be an endpoint or an interior point of J.

PROOF. Clearly $Y(u) = C$. Differentiate (1.3.4) and substitute into the equation (1.2.3). □

4. The Gronwall Inequality

Since we need a Gronwall inequality which is slightly more general than the one usually found in the literature we state and proof it here.

THEOREM 1.4.1 (The Gronwall Inequality). *(i) (The "right" Gronwall inequality) Let $J = [a, b]$. Assume g in $L(J, \mathbb{R})$ with $g \geq 0$ a.e., f real valued and continuous on J. If y is continuous, real valued, and satisfies*

$$y(t) \leq f(t) + \int_a^t g(s) \, y(s) \, ds \, , \quad a \leq t \leq b, \tag{1.4.1}$$

then
$$y(t) \leq f(t) + \left(\int_a^t f(s)\,g(s)\,\exp\left(\int_s^t g(u)\,du\right)ds\right), \ a \leq t \leq b. \quad (1.4.2)$$

For the special case when $f(t) = c$, a constant, we get
$$y(t) \leq c\,\exp\left(\int_a^t g(s)\,ds\right), \quad t \in J. \quad (1.4.3)$$

For the special case when f is nondecreasing on $[a, b]$ we get
$$y(t) \leq f(t)\,\exp\left(\int_a^t g(s)\,ds\right), \ a \leq t \leq b. \quad (1.4.4)$$

(ii) *(The "left" Gronwall inequality) Let $J = [a, b]$. Assume g is in $L(J, \mathbb{R})$, $g \geq 0$ a.e. and f real valued and continuous on J. If y is continuous, real valued, and satisfies*
$$y(t) \leq f(t) + \int_t^b g(s)\,y(s)\,ds, \ a \leq t \leq b, \quad (1.4.5)$$

then
$$y(t) \leq f(t) + \left(\int_t^b f(s)\,g(s)\,\exp\left(\int_t^s g(u)\,du\right)ds\right), \ a \leq t \leq b. \quad (1.4.6)$$

For the special case when $f(t) = c$, a constant, we get
$$y(t) \leq c\,\exp\left(\int_t^b g(s)\,ds\right), \ a \leq t \leq b. \quad (1.4.7)$$

For the special case when f is nondecreasing on $[a, b]$ we have
$$y(t) \leq f(t)\,\exp\left(\int_t^b g(s)\,ds\right), \ a \leq t \leq b. \quad (1.4.8)$$

PROOF. For part (i) let $z(t) = \int_a^t g\,y$, $t \in J$ and note that
$$z' = g\,y \leq g\,f + g\,z.; \ z' - g z \leq g f \text{ a.e.}$$

Hence we have
$$\exp\left(-\int_a^s g(u)du\right)[z'(s) - g(s)z(s)] = \left[\exp\left(-\int_a^s g(u)du\right)z(s)\right]'$$
$$\leq g(s)\,f(s)\,\exp\left(-\int_a^s g(u)du\right), \ a \leq s \leq b.$$

Integrating from a to t we get
$$\exp\left(-\int_a^t g(u)du\right)z(t) \leq \int_a^t g(s)\,f(s)\,\exp\left(-\int_a^s g(u)du\right)ds, \ a \leq t \leq b.$$

From (1.4.1) and the above line we obtain
$$y(t) \leq f(t) + z(t) \leq f(t) + \exp\left(\int_a^t g(u)du\right)\int_a^t g(s)\,f(s)\,\exp\left(-\int_a^s g(u)du\right)ds$$
$$= f(t) + \int_a^t g(s)\,f(s)\,\exp\left(\int_s^t g(u)du\right)ds, \ a \leq t \leq b.$$

This concludes the proof of (1.4.2). The two special cases follow from (1.4.2). For part (ii) let $z(t) = \int_t^b g\, y$ and note that

$$z' + gz \geq -g\, f,$$

then proceed as in part (i). □

5. Bounds and Extensions to the Endpoints

In this section we investigate bounds for solutions and the continuous extension of solutions to the endpoints of the underlying interval.

THEOREM 1.5.1. *Let $J = (a,b)$, $-\infty \leq a < b \leq \infty$, let $n,m \in \mathbb{N}$. Suppose that $P \in M_n(L(J,\mathbb{C}))$; $F \in M_{n,m}(L(J,\mathbb{C}))$. Assume that for some $u \in J$, $C \in M_{n,m}(\mathbb{C})$, we have*

$$Y' = PY + F \quad \text{on } J, \quad Y(u) = C. \tag{1.5.1}$$

Then

$$|Y(t)| \leq \left(|C| + \int_a^b |F|\right) \exp\left(\int_a^b |P|\right), \quad a < t < b. \tag{1.5.2}$$

PROOF. Note that (1.5.1) is equivalent to

$$Y(t) = C + \int_u^t \left(P(s)\,Y(s) + F(s)\right) ds, \; a < t < b. \tag{1.5.3}$$

Case 1. $u \leq t < b$. From (1.5.3) we get

$$|Y(t)| \leq |C| + \left|\int_u^t (PY + F)\right| \leq |C| + \int_u^t (|P|\,|Y| + |F|)$$
$$\leq \left(|C| + \int_u^b |F|\right) + \int_u^t (|P|\,|Y|), \; u \leq t < b.$$

From this and Gronwall's inequality we obtain

$$|Y(t)| \leq \left(|C| + \int_u^b |F|\right) \exp\left(\int_u^t |P|\right) \leq \left(|C| + \int_u^b |F|\right)\left(\int_u^b |P|\right), \; u \leq t < b.$$

Case 2. $a < t \leq u$. From (1.5.3)

$$|Y(t)| \leq |C| + \left|\int_u^t (PY + F)\right| \leq |C| + \int_t^u (|P|\,|Y| + |F|)$$
$$\leq \left(|C| + \int_a^u |F|\right) + \int_t^u (|P|\,|Y|), \; a < t \leq u.$$

From this and the "left" Gronwall inequality we get

$$|Y(t)| \leq \left(|C| + \int_a^u |F|\right) \exp\left(\int_t^u |P|\right)$$
$$\leq \left(|C| + \int_a^u |F|\right) \exp\left(\int_a^u |P|\right), \; a < t \leq u.$$

Combining the two cases we conclude that (1.5.2) holds. □

Below we will show that, under the conditions of Theorem 1.5.1, the inequality $a < t < b$ can be replaced with $a \leq t \leq b$ in (1.5.2). For this $Y(a)$ and $Y(b)$ are defined as limits. This holds for both finite and infinite endpoints a, b.

THEOREM 1.5.2 (Continuous extensions to endpoints). *Let* $J = (a, b)$, $-\infty \leq a < b \leq \infty$. *Assume that*

$$P \in M_n(L_{loc}(a,b), \mathbb{C}); \quad F \in M_{n,m}(L_{loc}(a,b), \mathbb{C}). \tag{1.5.4}$$

i) Suppose, in addition to (1.5.4), that

$$P \in M_n(L(a,c), \mathbb{C}); \quad F \in M_{n,m}(L(a,c), \mathbb{C}) \tag{1.5.5}$$

for some $c \in (a, b)$. *For some* $u \in J$ *and* $C \in M_{n,m}(\mathbb{C})$, *let* Y *be the solution of the IVP (1.2.3), (1.2.4) on* J. *Then*

$$Y(a) = \lim_{t \to a^+} Y(t) \tag{1.5.6}$$

exists and is finite.

ii) Suppose that, in addition to (1.5.4), P, F satisfy

$$P \in M_n(L(c,b), \mathbb{C}); \quad F \in M_{n,m}(L(c,b), \mathbb{C}) \tag{1.5.7}$$

for some $c \in (a, b)$. *For some* $u \in J$ *and* $C \in M_{n,m}(\mathbb{C})$, *let* Y *be the solution of the IVP (1.2.3), (1.2.4) on* J. *Then*

$$Y(b) = \lim_{t \to b^-} Y(t) \tag{1.5.8}$$

exists and is finite.

PROOF. We establish Theorem 1.5.2 for b; the proof for the endpoint a is similar and hence omitted. It follows from (1.5.2) that $|Y|$ is bounded on $[c, b)$ for $c \in J$, say by B. Let $\{b_i\}$ be any strictly increasing sequence converging to b. Then for $j > i$ we have

$$|Y(b_j) - Y(b_i)| = \left| \int_{b_i}^{b_j} PY \right| \leq B \int_{b_i}^{b_j} |P|.$$

From this and the absolute continuity of the Lebesgue integral it follows that $\{Y(b_i) : i \in N\}$ is a Cauchy sequence and hence converges to a finite limit. $\square$

The next result establishes the rank invariance of solutions of homogeneous systems at the endpoints of the underlying interval and establishes the existence and uniqueness of solutions of initial value problems when the initial condition is specified at an endpoint.

THEOREM 1.5.3 (Rank invariance at endpoints). *Let* $J = (a, b)$, $-\infty \leq a < b \leq \infty$. *Assume that*

$$P \in M_n(L_{loc}(a,b), \mathbb{C}). \tag{1.5.9}$$

i) Suppose, in addition to (1.5.9), that

$$P \in M_n(L(a,c), \mathbb{C}) \tag{1.5.10}$$

for some $c \in (a, b)$. *Let, for some* $u \in J$ *and* $C \in M_{n,m}(\mathbb{C})$, Y *be the solution of the IVP (1.2.3), (1.2.4) with $F = 0$ on J. Then*

$$\operatorname{rank} Y(a) = \operatorname{rank} Y(u) \tag{1.5.11}$$

where $Y(a)$ is given by (1.5.6). Moreover, given any $C \in M_{n,m}(\mathbb{C})$ there exists a unique solution Y of the "endpoint" value problem:

$$Y' = PY, \quad Y(a) = C. \tag{1.5.12}$$

ii) Suppose, in addition to (1.5.9), that

$$P \in M_n(L(c,b), \mathbb{C}) \tag{1.5.13}$$

for some $c \in (a, b)$. Let, for some $u \in J$ and $C \in M_{n,m}(\mathbb{C})$, Y be the solution of the IVP (1.2.3), (1.2.4) with $F = 0$ on J. Then

$$\operatorname{rank} Y(b) = \operatorname{rank} Y(u) \tag{1.5.14}$$

where $Y(b)$ is given by (1.5.8). Moreover, given any $C \in M_{n,m}(\mathbb{C})$ there exists a unique solution Y of the "endpoint" value problem:

$$Y' = PY, \quad Y(b) = C. \tag{1.5.15}$$

PROOF. Note that (1.5.11) and (1.5.14) do not follow directly from (1.2.14) and (1.5.6) or (1.5.8) since the rank of a matrix is not a continuous function of the matrix. We argue as follows: Let $Y(u) = C$, $\operatorname{rank} C = r$. If $r = 0$, then $Y(t) = 0$ for all $t \in J$ and $Y(b) = 0$ by (1.5.8). If $r > 0$, let $C_1, \ldots, C_r$ be linearly independent columns of $Y(u)$ and construct a nonsingular $n \times n$ matrix D by adding $n - r$ appropriate columns to C_j, $j = 1, \ldots, r$. Let Z denote the solution of (1.2.13) determined by the initial condition $Z(u) = D$. It follows from (1.2.15) that $Z(t)$ is nonsingular for each $t \in J$ and hence $Z(b)$ is nonsingular by (1.5.8) and (1.2.15). Therefore $Z_1(b), \ldots, Z_r(b)$ are linearly independent. By Theorem 1.2.1 $Y_r(t) = Z_r(t)$ for $t \in J$ and hence also for $t = b$ by (1.5.8); thus $\operatorname{rank} Y(b) \geq r$. If $\operatorname{rank} Y(b) = k > r$ then $\sum_1^k c_j Y_j(t) = 0$ for $t \in J$ and hence also for $t = b$, contradicting $k > r$. The proof for the endpoint a is similar. This establishes (1.5.11) and (1.5.14). To prove the moreover parts of the Theorem consider the primary fundamental matrix Φ, choose $u \in J$ and determine the solution Y of (1.2.13) by the initial condition $Y(u) = \Phi(b, u) C$. Then $Y(b) = C$. Note that $\Phi(b, u)$ exists by (1.5.8). The proof of (1.5.12) is similar. □

6. Continuous Dependence of Solutions on the Problem

The next result establishes bounds for solutions of initial value problems; these are then used in Theorem 1.6.2 to show that solutions of initial value problems depend continuously on all parameters of the problem.

THEOREM 1.6.1. Let $u, v \in J = (a, b)$, $-\infty \leq a < b \leq \infty$, $C, D \in M_{n,m}(\mathbb{C})$, $P, Q \in M_n(L(J, \mathbb{C}))$, $F, G \in M_{n,m}(L(J, \mathbb{C}))$. Assume

$$Y' = PY + F \text{ on } J, \ Y(u) = C; \quad Z' = QZ + G \text{ on } J, \ Z(v) = D. \tag{1.6.1}$$

Then

$$|Y(t) - Z(t)| \leq K \exp\left(\int_a^b |Q|\right), \quad a \leq t \leq b, \tag{1.6.2}$$

where

$$K = |C - D| + \left|\int_u^v |F|\right| + M\left|\int_u^v |P|\right| + \int_a^b |F - G| + M\int_a^b |P - Q|, \tag{1.6.3}$$

and
$$M = \left(|C| + \int_a^b |F|\right) \exp\left(\int_a^b |P|\right). \tag{1.6.4}$$

PROOF. For $a < t < b$ this follows from the Gronwall inequality as in the proof of Theorem 1.5.1. The case $t = a$ and $t = b$ then follows from Theorem 1.5.2. □

THEOREM 1.6.2 (Continuous dependence). *Let $J = (a, b)$, $-\infty \leq a < b \leq \infty$, $u \in J$, $C \in M_{n,m}(\mathbb{C})$, $P \in M_n(L(J, \mathbb{C}))$, and $F \in M_{n,m}(L(J, \mathbb{C}))$. Let $Y = Y(\cdot, u, C, P, F)$ be the solution of (1.2.3), (1.2.4) on J. Then Y is a continuous function of all its variables u, C, P, F uniformly on the closure of J; more precisely, for fixed P, F, u, C; given any $\epsilon > 0$ there is a $\delta > 0$ such that if $v \in J$, $D \in M_{n,m}(\mathbb{C})$, $Q \in M_n(L(J, \mathbb{C}))$, and $G \in M_{n,m}(L(J, \mathbb{C}))$ satisfy*
$$|u - v| + |C - D| + \int_a^b |P - Q| + \int_a^b |F - G| < \delta, \tag{1.6.5}$$
then
$$|Y(t, u, C, P, F) - Y(t, v, D, Q, G)| < \epsilon, \quad a \leq t \leq b. \tag{1.6.6}$$

Note that $Y(t, u, C, P, F)$ is jointly continuous in u, C, P, F, uniformly for t in the closure of J.

PROOF. The absolute continuity of the Lebesgue integral and (1.6.5) imply that the constant K in (1.6.3) can be made arbitrarily small. The conclusion then follows from (1.6.2). □

THEOREM 1.6.3. *Let $J = (a, b)$, $-\infty \leq a < b \leq \infty$, let $P_k \in M_{n,n}(L_{loc}(J, \mathbb{C}))$, $F_k \in M_{n,m}(L_{loc}(J, \mathbb{C}))$, $C_k \in M_{n,m}$, $u_k \in J$, $k \in \mathbb{N}_0 = \{0, 1, 2, \ldots\}$. Assume*
(i) $P_k \to P_0$ as $k \to \infty$
locally in $L_{loc}(J, \mathbb{C})$ in the sense that for each compact subinterval K of J we have
$$\int_K |P_k - P_0| \to 0 \quad as \quad k \to \infty;$$
(ii) $F_k \to F_0$ as $k \to \infty$
locally in $L_{loc}(J)$ in the sense that for each compact subinterval K of J we have
$$\int_K |F_k - F_0| \to 0 \quad as \quad k \to \infty;$$
(iii) $C_k \to C_0 \in C$ as $k \to \infty$;
(iv) $u_k \to u_0 \in J$ as $k \to \infty$.
Then
$$Y(t, u_k, C_k, P_k, F_k) \to Y(t, u_0, C_0, P_0, F_0) \quad as \quad k \to \infty$$
locally uniformly on J, i.e., uniformly in t on each compact subinterval of J.

Moreover, if $P_k \in M_n(L(J), \mathbb{C})$, $F_k \in M_{n,m}(L(J), \mathbb{C})$, and (i), (ii) hold in $L(J, \mathbb{C})$, i.e. with K replaced by J and (iii), (iv) hold, then
$$Y(t, u_k, C_k, P_k, F_k) \to Y(t, u_0, C_0, P_0, F_0) \quad as \quad k \to \infty$$
uniformly on the closure of J.

PROOF. This follows from Theorem 1.6.2. □

7. Differentiable Dependence of Solutions on the Data

Theorem 1.6.2 shows that the solution of the initial value problem (1.2.3), (1.2.4) with $P \in M_{n,n}(L_{loc}(J,\mathbb{C}))$, $F \in M_{n,m}(L_{loc}(J,\mathbb{C}))$, $C \in M_{n,m}(\mathbb{C})$ depends continuously on the problem. In this section we show that this dependence is differentiable with respect to each parameter of the problem.

DEFINITION 1.7.1. *A map T from a Banach space X into a Banach space Z, $T : X \to Z$, is differentiable at a point $x \in X$ if there exists a bounded linear map $T'(x) : X \to Z$ such that*

$$|T(x+h) - T(x) - T'(x)h| = o(|h|), \text{ as } h \to 0 \text{ in } X.$$

That is, for each $\varepsilon > 0$ there is a $\delta > 0$ such that

$$|T(x+h) - T(x) - T'(x)h| \leq \varepsilon(|h|) \text{ for all } h \in X \text{ with } |h| < \delta.$$

If such a map $T'(x)$ exists, it is unique and is called the Frechet derivative of T at x. A map T is differentiable on a set $S \subset X$ if it is differentiable at each point of S. In this case the derivative is a map : $x \to T'(x)$ from S into the Banach space $L(X,Z)$ of all bounded linear operators from X into Z denoted by T'. To say that T' is continuously differentiable on S or T is C^1 on S means that the map T' is continuous in the operator topology of the Banach space $L(X,Z)$.

The differentiability of the solution

$$Y = Y(t, u, C, P, F)$$

with respect to t follows from the definition of solution. The differentiability of Y with respect to u is established in the next lemma.

LEMMA 1.7.1. *Let the hypotheses and notation of Theorem 1.2.1 hold. Fix t, C, P, F and consider Y as a function of u. Then $Y \in AC_{loc}(J)$.*

PROOF. It follows from the representation (1.3.1) that the primary fundamental matrix $\Phi(t,u)$ is differentiable with respect to u, since the inverse of a differentiable matrix is differentiable. The differentiability of Y with respect to u then follows from the variation of parameters representation

$$Y(t,u) = \Phi(t,u)C + \int_u^t \Phi(t,s) F(s) \, ds.$$

This concludes the proof. $\square$

For fixed t, u, Y is a function of C, P, F mapping

$$M_{n,m}(\mathbb{C}) \times M_n(L(J,\mathbb{C})) \times M_{n,m}(L(J,\mathbb{C}))$$

into $M_{n,m}(L(J,\mathbb{C}))$. By Theorem 1.6.2, Y is continuous in C, P, F. Is it differentiable in C? in P? in F?

THEOREM 1.7.1. *For fixed $t, u \in J$, $P \in M_n(L(J,\mathbb{C}))$ and $F \in M_{n,m}(L(J,\mathbb{C}))$, the solution $Y = Y(t, u, C, P, F)$ of (1.2.3), (1.2.4) is differentiable in C; its derivative is given by*

$$Y'(C) = \frac{\partial Y}{\partial C}(t, u, C, P, F) = \Phi(t, u, P). \tag{1.7.1}$$

Thus we have

$$Y'(C) H = \Phi(t, u, P) H, \quad H \in M_{n,m}(\mathbb{C}). \tag{1.7.2}$$

The derivative $Y'(C)$ is constant in C and in F.

PROOF. This follows directly from the variation of parameters formula and the definition of derivative. □

THEOREM 1.7.2. *Let $J = [a, b]$. Fix $t, u \in J$, $C \in M_{n,m}(\mathbb{C})$, $P \in M_n(L(J, \mathbb{C}))$, and $F \in M_{n,m}(L(J, \mathbb{C}))$; let $Y = Y(t, u, C, P, F)$. We have*

$$Y'(F)(H) = \frac{\partial Y}{\partial F}(t, u, C, P, F)(H) = \int_u^t \Phi(t, s) H(s) \, ds, \ H \in M_{n,m}(L(J)). \quad (1.7.3)$$

Here the right side of equation (1.7.3) defines a bounded linear operator on the space $M_{n,m}(L(J, \mathbb{C}))$. The derivative $Y'(F)$ is constant in F.

PROOF. From the variation of parameters formula (1.3.4) we get

$$Y(t, u, C, P, F + H) - Y(t, u, C, P, F) = \int_u^t \Phi(t, s, P) H(s) \, ds.$$

The conclusion follows from this equation and the definition of derivative. □

Before stating the next Theorem we give two lemmas. These may be of independent interest.

LEMMA 1.7.2. *Let $J = (a, b)$, $-\infty \leq a < b \leq \infty$, let $P \in M_n(L_{loc}(J, \mathbb{C}))$, let $u \in J$. Then for any $t \in J$ we have*

$$\Phi(t, u, P) = I + \int_u^t P + \int_u^t P(r) \int_u^r P(s) \, ds dr$$
$$+ \int_u^t P(r) \int_u^r P(s) \int_u^s P(x) \, dx ds dr + \cdots. \quad (1.7.4)$$

PROOF. This follows directly from the successive approximations proof of the existence-uniqueness Theorem : Start with the first approximation $\Phi_0 = I$; then $\Phi_1 = I + \int_u^t P$, $\Phi_1 = I + \int_u^t P + \int_u^t P(r) \int_u^r P(s) \, ds dt$, etc. □

The next Lemma establishes a product formula for fundamental solutions. This can be viewed as an extension of the exponential law; see Lemma 1.7.3, Remark 1.7.1 and Corollary 1.7.1 below.

LEMMA 1.7.3 (Product formula). *Let $P, H \in M_n(L_{loc}(J, \mathbb{C}))$. Then for any $t, u \in J$ we have*

$$\Phi(t, u, P + H) = \Phi(t, u, P) \, \Phi(t, u, S), \quad (1.7.5)$$

where

$$S = \Phi^{-1}(\cdot, u, P) \, H \, \Phi(\cdot, u, P). \quad (1.7.6)$$

PROOF. The proof consists in showing that both sides satisfy the same initial value problem and then using the existence-uniqueness Theorem. □

LEMMA 1.7.4 (Exponential law). *Let $P, H \in M_n(L_{loc}(J, \mathbb{C}))$. If P commutes with the integral of H in the sense that*

$$P(t) \left(\int_u^s H \right) = \left(\int_u^s H \right) P(t), \quad s, t, u \in J, \quad (1.7.7)$$

then the exponential law holds:

$$\Phi(t, u, P + H) = \Phi(t, u, P) \, \Phi(t, u, H). \quad (1.7.8)$$

PROOF. It follows from Lemmas 1.7.2, 1.7.3 and hypothesis (1.7.7) that
$$\Phi(\cdot, u, P)\, H = H\, \Phi(\cdot, u, P)$$
and hence $S = H$ in (1.7.6). □

THEOREM 1.7.3. *Let J be a compact interval $[a, b]$. Fix $t, u \in J$, $C \in M_{n,m}(\mathbb{C})$, $F \in M_{n,m}(L(J, \mathbb{C}))$. For $P \in M_n(L(J, \mathbb{C}))$ let $Y = Y(t, u, C, P, F)$ be the unique solution of the initial value problem (1.2.3), (1.2.4). Then the map $P \to Y(t, u, C, P, F)$ from the Banach space $M_n(L(J, \mathbb{C}))$ to the (finite dimensional) Banach space $M_{n,m}(\mathbb{C})$ is differentiable and its derivative*

$$Y'(P) = \frac{\partial Y}{\partial P}(t, u, C, P, F) \tag{1.7.9}$$

is the bounded linear transformation from the Banach space $M_n(L(J, \mathbb{C}))$ to the Banach space $M_{n,m}(\mathbb{C})$ given, for any $H \in M_n(L(J, \mathbb{C}))$ by

$$\begin{aligned} Y'(P)\, H &= \Phi(t, u, P) \left(\int_u^t \Phi^{-1}(r, u, P) H(r) \Phi(r, u, P)\, dr \right) C \\ &\quad + \int_u^t \Phi(t, r, P) \left(\int_r^t \Phi^{-1}(s, u, P) H(s) \Phi(s, u, P)\, ds \right) F(r)\, dr. \end{aligned} \tag{1.7.10}$$

PROOF. Fix t, u, C, F and let $Y(t, P) = Y(t, u, C, P, F)$. From the variation of parameters formula it follows that for $H \in M_{n,n}(L(J, \mathbb{C}))$ and S defined by (1.7.6) we have

$$Y(t, P + H) - Y(t, P)$$
$$= \Phi(t, P + H)\, C + \int_u^t \Phi(t, s, P + H)\, F(s)\, ds - \Phi(t, P)\, C - \int_u^t \Phi(t, s, P)\, F(s)\, ds$$
$$= \Phi(t, u, P)\, [\Phi(t, u, S) - I]\, C + \int_u^t \Phi(t, s, P)[\Phi(t, r, S) - I] F(r) dr$$
$$= \Phi(t, u, P) \left[\int_u^t S + \int_u^t S(x) \int_u^x S(y) dy dx \right.$$
$$\left. + \int_u^t S(x) \int_u^x S(y) \int_u^y S(z) dz dy dx + \ldots \right] C$$
$$+ \int_u^t \Phi(t, r, P) \left[\int_r^t S + \int_r^t S(x) \int_r^x S(y) dy dx \right.$$
$$\left. + \int_r^t S(x) \int_r^x S(y) \int_r^y S(z) dz dy dx + \ldots \right] F(r) dr.$$

Hence

$$Y(t, u, P + H) - Y(t, u, P) - \Phi(t, u, P) \left(\int_u^t S(r) dr \right) C$$
$$- \int_u^t \Phi(t, r, P) \left(\int_r^t S(x) dx \right) F(r) dr$$
$$= \Phi(t, u, P) \left[\int_u^t S(x) \int_u^x S(y) dy dx + \int_u^t S(x) \int_u^x S(y) \int_u^y S(z) dz dy dx + \ldots \right] C$$

$$+ \int_u^t \Phi(t,r,P) \left[\int_r^t S(x) \int_r^x S(y) dy dx \right.$$
$$\left. + \int_r^t S(x) \int_r^x S(y) \int_r^y S(z) dz dy dx + \ldots \right] F(r) dr$$
$$= E(H).$$

Noting that $|S|(b-a)| \leq |kH|$ for some $k \in \mathbb{R}$ and that the functions $|\Phi(t,u,P)|$ and $|\Phi^{-1}(t,u,P)|$ are bounded on J, there exists an $M > 0$ such that
$$|E(H)| \leq M|C| \left[|S(b-a)|^2 + |S(b-a)|^3 \ldots \right] + M|F| \left[|S(b-a)|^2 + |S(b-a)|^3 \ldots \right]$$
$$\leq M|C||kH| \left[|kH| + |kH|^2 + \ldots \right] + M|F||kH| \left[|kH| + |kH|^2 + \ldots \right].$$

From this it follows that
$$\frac{|E(H)|}{|H|} \to 0 \text{ as } |H| \to 0 \text{ in } M_n(L(J)).$$

This completes the proof. □

THEOREM 1.7.4. *Let the hypotheses and notation of Theorem 1.7.3 hold and assume, in addition, that the commutativity hypothesis (1.7.7) is satisfied. Then*

(1)
$$H(t) \, \Phi(t,u,P) = \Phi(t,u,P) \, H(t), \quad t, u \in J. \quad (1.7.11)$$

(2) *The exponential law holds, i.e.*
$$\Phi(t,u,P+H) = \Phi(t,u,P) \, \Phi(t,u,H), \quad t, u \in J. \quad (1.7.12)$$

(3) *Formula (1.7.10) reduces to*
$$Y'(P)(H) = \Phi(t,u,P) \left(\int_u^t H(s) \, ds \right) C$$
$$+ \int_u^t \Phi(t,r,P) \left(\int_r^t H(s) \, ds \right) F(r) \, dr, \quad t, u \in J. \quad (1.7.13)$$

Note however that $Y'(P)$ is not the operator defined by the right-hand side of (1.7.13) since H cannot be restricted to satisfy the commutativity hypothesis (1.7.7) due to the definition of the derivative $Y'(P)$.

PROOF. This follows from Theorem 1.7.3 and Lemma 1.7.4. □

REMARK 1.7.1. In the special case when P and H are constant matrices we have
$$Y'(P)(H) = e^{(t-u)P} \left(\int_u^t e^{(u-r)P} H e^{(r-u)P} \, dr \right) C$$
$$+ \int_u^t e^{(t-r)P} \left(\int_r^t e^{(u-s)P} H e^{(s-u)P} ds \right) F(r) \, dr. \quad (1.7.14)$$

Note that if P and H are constant and commute, then (1.7.14) reduces to
$$Y'(P)(H) = (t-u)e^{(t-u)P} H C + \int_u^t (t-r)e^{(t-r)P} H F(r) \, dr.$$

But this reduction does not hold, in general, for constant matrices which do not commute.

COROLLARY 1.7.1. *Consider the exponential map of matrices:*
$$E(A) = e^A, \ A \in M_n(\mathbb{C}).$$
The Frechet derivative of E is the bounded linear operator from $M_n(\mathbb{C})$ into $M_n(\mathbb{C})$ given by
$$E'(A) H = e^A \int_0^1 e^{-rA} H e^{rA} \, dr, \ H \in M_n(\mathbb{C}). \tag{1.7.15}$$

PROOF. This is the special case of Theorem 1.7.4 when $a = 0 = u$, $b = 1$, $P(t) = A$ for all $t \in [0,1]$, $F \equiv 0$, $C = I$. □

REMARK 1.7.2. Note that (1.7.15) reduces to the more familiar formula $E'(A) = E(A)$ for all $A \in M_n(\mathbb{C})$ only in the one dimensional case $n = 1$. When $n > 1$ (1.7.15) reduces to $E'(A) = E(A)$ only for constant multiples $A = cI_n$, $c \in \mathbb{C}$, of the identity matrix I_n, since only multiples of the identity satisfy the commutativity condition with respect to all matrices in $M_n(\mathbb{C})$.

REMARK 1.7.3. In Corollary 1.7.1 $M_n(\mathbb{C})$ can be replaced by $M_n(\mathbb{R})$; in fact $M_n(\mathbb{C})$ can be replaced by an arbitrary Banach algebra. See [**393**].

For fixed t, u, C, F replace P by $P + zW$ and fix P and W. It is well known that the solution
$$Y = Y(t, u, C, P + zW, F)$$
of the intial value problem (1.2.3), (1.2.4) with P replaced by $P + zW$ is an entire function of z. What is its derivative:
$$Y'(z) = \frac{\partial Y}{\partial z}?$$
This question is answered by

THEOREM 1.7.5. *Let $J = (a,b)$, $-\infty \leq a < b \leq \infty$, $t, u \in J$, $C \in M_{n,m}$, $P, W \in M_n(L(J, \mathbb{C}))$, $F \in M_{n,m}(L(J, \mathbb{C}))$; let $Y = Y(t, u, C, P + zW, F)$ denote the unique solution of (1.2.3), (1.2.4) for each $z \in C$. Then Y is an entire function of z and*
$$Y'(z) = \Phi(t, u, P + zW) \left(\int_u^t \Phi^{-1}(r, u, P + zW) W(r) \Phi(r, u, P + zW) \, dr \right) C$$
$$+ \int_u^t \Phi(t, r, P + zW)$$
$$\cdot \left(\int_r^t \Phi^{-1}(s, u, P + zW) W(s) \Phi(s, u, P + zW) \, ds \right) F(r) \, dr. \tag{1.7.16}$$

PROOF.
$$[Y(t, u, C, P + (z+h)W, F) - Y(t, u, C, P + zW, F)]$$
$$= [\Phi(t, u, P + (z+h)W) - \Phi(t, u, P + zW)] C$$
$$+ \int_u^t [\Phi(t, r, P + (z+h)W) - \Phi(t, r, P + zW)] F(r) \, dr.$$
Let
$$S(z) = \Phi^{-1}(\cdot, u, P + zW) W(\cdot) \Phi(\cdot, u, P + zW).$$

Proceeding similarly to the proof of Theorem 1.7.3 we get

$$\Phi(t, u, P + (z+h)W) - \Phi(t, u, P + zW)$$
$$= \Phi(t, u, P + (z+h)W)[\Phi(t, u, hS(z)) - I]$$
$$= \Phi(t, u, P + zW)[h \int_u^t S(z) + o(h)]$$
$$= h\Phi(t, u, P + zW) \int_u^t S(z) + o(h).$$

Combining these two identities we get

$$[Y(t, u, C, P + (z+h)W, F) - Y(t, u, C, P + zW, F)]$$
$$= h \left[\Phi(t, u, P + zW) \left(\int_u^t S(z) \right) C + \int_u^t \Phi(t, r, P + zW) \left(\int_r^t S(z) \right) F(r) \, dr \right]$$
$$+ o(h).$$

And the result follows. □

THEOREM 1.7.6. *Let $J = (a, b)$, $-\infty \leq a < b \leq \infty$, $t, u \in J$, $C \in M_{n,m}(\mathbb{C})$, $P \in M_n(L(J, \mathbb{C}))$, $F \in M_{n,m}(L(J, \mathbb{C}))$; and for each $z \in C$ let $Y = Y(t, u, C, P + zW, F)$ denote the unique solution of (1.2.3), (1.2.4) for each $W \in L(J)$. Then Y is a differentiable function of W and*

$$Y'(W) H = z \, \Phi(t, u, P + zW) \left(\int_u^t \Phi^{-1}(r, u, P + zW) H(r) \, \Phi(r, u, P + zW) \, dr \right) C$$
$$+ z \int_u^t \Phi(t, r, P + zW)$$
$$\cdot \left(\int_r^t \Phi^{-1}(s, u, P + zW) H(s) \, \Phi(s, u, P + zW) \, ds \right) F(r) \, dr, \qquad (1.7.17)$$

for $H \in L(J, \mathbb{C})$.

PROOF. The proof is similar to that of Theorem 1.7.3 and hence omitted. □

8. Adjoint Systems

The concept of "adjointness" plays an important role in the study of boundary value problems just as it does in matrix theory. The results in this section will be used later.

LEMMA 1.8.1. *Let P, Q be any $k \times k$ complex matrix functions on J. Let F, G be $k \times m$ complex matrix functions on J. If $Y' = PY + F$ on J and $Z' = QZ + G$ on J and C is a constant $k \times k$ complex matrix, then*

$$(Z^*CY)' = Z^*(Q^*C + CP)Y + Z^*CF + G^*CY \quad on \quad J. \qquad (1.8.1)$$

PROOF. This follows from a straightforward computation and is therefore omitted. □

COROLLARY 1.8.1. *Let the assumptions and notation be as in Lemma 1.8.1. If, in addition, C is invertible and $Q = -C^{-1^*}P^*C^*$, then*

$$(Z^*CY)' = Z^*CF + G^*CY. \qquad (1.8.2)$$

PROOF. This follows from (1.8.1). □

The fundamental matrices of adjoint systems are closely related to each other. The next result gives this relationship. It plays an important role in the theory of adjoint and, in particular, self-adjoint boundary value problems.

THEOREM 1.8.1 (The Adjointness Identity). *Let $P \in M_n(L_{loc}(J, \mathbb{C}))$, let $E \in M_n(\mathbb{C})$. Assume*

$$E^{-1}E^* = I \text{ or } E^{-1}E^* = -I \tag{1.8.3}$$

and define

$$P^+ = -E^{-1} P^* E. \tag{1.8.4}$$

Then

$$\Phi(t, s, P) = E^{-1} \Phi^*(s, t, P^+) E, \quad s, t \in J. \tag{1.8.5}$$

PROOF. Fix $s \in J$ and let

$$Z(t) = E^{-1*}\Phi^*(t, s, P) E^* \Phi(t, s, P^+), \quad t \in J.$$

Note that $Z(s) = I$ and

$$Z'(t) = E^{-1*}[P(t)\Phi(t,s,P)]^* E^* \Phi(t,s,P^+) + E^{-1*}\Phi^*(t,s,P) E^* P^+(t) \Phi(t,s,P^+)$$
$$= E^{-1*}\Phi^*(t,s,P) E E^{-1}P^*(t) E E^{-1} E^* \Phi(t,s,P^+)$$
$$\quad + E^{-1*}\Phi^*(t,s,P) E^* P^+(t) \Phi(t,s,P^+)$$
$$= -E^{-1*}\Phi^*(t,s,P) E^* P^+(t) \Phi(t,s,P^+)$$
$$\quad + E^{-1*}\Phi^*(t,s,P) E^* P^+(t) \Phi(t,s,P^+)$$
$$= 0, \quad t \in J,$$

using (1.8.3) and (1.8.4). Hence $Z(t) = I$, for $t \in J$. That this is equivalent to (1.8.5) follows from the representation $\Phi(t, s, P^+) = Y(t) Y^{-1}(s)$, $s, t \in J$, for any fundamental matrix Y of $Y' = P^+ Y$. $\square$

9. Inverse Initial Value Problems

Notation. Given d n-dimensional vectors $Y_1, Y_2, \ldots, Y_d$ we denote the $n \times d$ matrix whose $i-th$ column is Y_i, $i = 1, \ldots, d$, by

$$Y = [Y_1, Y_2, \ldots, Y_d]. \tag{1.9.1}$$

Above we started with a coefficient matrix P and, possibly, a nonhomogeneous term F and then studied the existence of solutions and their properties. Here we reverse this. Given a number of functions, under what conditions are they solutions of a first order linear system? For the sake of completeness we state the theorem for both the direct and the inverse problems.

THEOREM 1.9.1. *Let $1 \leq d \leq n$, $P \in M_n(L_{loc}(J, \mathbb{C}))$. Assume that Y_i, $i = 1, \ldots, d$ are vector solutions of*

$$Y' = PY. \tag{1.9.2}$$

If

$$\text{rank}\,[Y_1, Y_2, \ldots, Y_d](t) = d \tag{1.9.3}$$

for some t in J, then this is true for every t in J.

Conversely, let $Y_i \in M_{n,1}(AC_{loc}(J))$, $i = 1, \ldots, d$, $1 \leq d \leq n$ and assume that

$$\text{rank}[Y_1, \ldots, Y_d](t) = d, \quad for \quad t \in J. \tag{1.9.4}$$

Then there exists an $n \times n$ matrix $P \in M_n(L_{loc}(J, \mathbb{C}))$ such that Y_i, $i = 1, \ldots, d$, are solutions of (1.9.2).

Furthermore, if $Y_i \in M_{n,1}(C^1(J), \mathbb{C})$, $i = 1, \ldots, d$ then there exists a continuous such P.

PROOF. The first part is contained in Theorem 1.2.2 so we only prove the second part. If $d = n$ take $P = Y'Y^{-1}$. If $d < n$ we construct an $n \times n$ matrix

$$M = [Y_1, Y_2, \ldots, Y_d, Y_{d+1}, \ldots, Y_n]$$

as follows. For each $t_1 \in J$ there is a $d \times d$ nonsingular submatrix of the $n \times d$ matrix $[Y_1, \ldots, Y_d](t_1)$. Let its rows be numbered by $r_1, \ldots, r_d$. To the right of the first row which in not one of these place the first row of the $(n-d) \times (n-d)$ identity matrix; to the right of the second row which is not one of these place the second row of the $(n-d) \times (n-d)$ identity matrix, and so on. Thus each of Y_i for $i > d$ is a constant matrix with all components zero except one which is the number 1. For each $t_1 \in J$ the matrix M so constructed is nonsingular at t_1 and by continuity $\det M(t) \neq 0$ for all t in some neighborhood N_{t_1} of t_1. Take

$$P(t) = M'(t) M^{-1}(t), \text{ for } t \in N_{t_1}.$$

Any compact subinterval of J can be covered by a finite number of such neighborhoods N_{t_1} and hence P can be defined on J. On points which are covered by more than one such neighborhood, P is multiply defined, we just choose one definition, say the one determined by the lowest numbered neighborhood. Clearly $P \in M_n(L_{loc}(J, \mathbb{C}))$ and Y_i $i = 1, \ldots, d$ are solutions. This completes the proof of the first part of (ii).

To prove the furthermore part we note that the constructed matrix P is piecewise continuous by construction. Thus to get a continuous P we remove the multiply defined aspect of the above construction as follows: On a subinterval which is covered by two or more of the neighborhoods N_t discard all definitions of M used above - just on this subinterval - then connect the two remaining pieces together in such a way as to keep M nonsingular on J. Then construct a new P from the new M as above for all $t \in J$. This results in a continuous P and completes the proof. □

10. Comments

Much of Chapter 1 is based on the paper [**396**] by Kong and Zettl. Below we comment on each section separately.

(1) The notation for matrix functions such as $M_n(L(J, \mathbb{C}))$ is taken from [**476**].
(2) The sufficiency of the local integrability conditions of Theorem 1.2.1 are well known - see [**487**] or [**600**]; the necessity given by Theorem 1.2.3 is due to Everitt and Race - see [**231**]. Except for the use of the Bielecki norm the first proof of Theorem 1.2.1 is the standard successive approximations argument, although it is dressed in the clothes of the Contraction Mapping Theorem in Banach space here. The advantage of the Bielecki norm is that it yields a global proof; the sup norm would only give a local proof and then one has to patch together the intervals of existence. The second proof is a minor variant of the usual successive approximations argument.

The constancy of the rank of solutions given by Theorem 1.2.2 is known - see [**319**] or [**515**] but we haven't seen it stated under these general conditions. It is surprising how many authors, including the two just mentioned, assume continuity of the coefficients when local Lebesgue integrability suffices. This is of some consequence both theoretically and numerically when coefficients are approximated by piece-wise constants, piece-wise linear functions, etc.

(3) The variation of parameters formula given by Theorem 1.3.1 is standard, but our notation is not. We use a notation which shows the dependence of the primary fundamental matrix on the coefficient matrix P. This is useful for the differentiation results that follow.

(4) A detailed discussion of the Gronwall inequality is given here because it is a very useful tool and we do not want any extraneous assumptions on f and g. The Gronwall inequality has many extensions: see - [**56**].

(5) Theorem 1.5.1 is elementary but we have not seen it stated in this generality. The continuous extensions of solutions given by Theorem 1.5.2 are a special case of much more powerful results e.g. Levinson's asymptotic theorem - [**123**]. Often the existence of limits of solutions are stated only for infinite endpoints. *We want to emphasize here that the relevant consideration is not whether the endpoint is finite or infinite but whether the coefficient matrix P and the inhomogeneous term F are integrable or not all the way to the endpoint.* Theorem 1.5.3 may be new in [**396**].

(6) Theorems 1.6.1, 1.6.2 and 1.6.3 illustrate clearly that the natural space in which to study solutions of linear ode's is $L^1_{loc}(J, \mathbb{C})$ in the singular case and $L^1(J, \mathbb{C})$ in the regular case.

(7) Sections 7 and 9 were influenced to some extend by the treatment of the inverse spectral theory for regular Sturm-Liouville problems by Poeschel and Trubowitz in [**522**]. Theorems 1.7.1 and 1.7.2 are trivial consequences of the variation of parameters formula; Theorems 1.7.3, 1.7.5 and 1.7.6 are not so trivial consequences of the Variation of Parameters Formula and may be new in [**396**].

Formula (1.7.15) of Corollary 1.7.1 for the derivative of the matrix exponential function is included here, although it is a completely straightforward consequence of Theorem 1.7.4, for the following reasons: (i) we couldn't find it in any of the books on matrices we looked at and (ii) a different and rather strange and obscure formula was published in the Monthly [**576**]. So Q. Kong and the author submitted a brief note to the Monthly containing formula (1.7.15) with a short proof based on the proof of Theorem 1.7.3 but specialized to this very special case. The letter from the editor stated that our note was rejected by "the matrix mafia". This in spite of the fact that, as far as we can ascertain and as hard as it is to believe, formula (1.7.15) seems to be new. It was subsequently published in [**393**]. (To add another strange twist to this bizarre saga, in this paper, the authors themselves state that formula 1.7.15 is not new; this statement was based on the referee's claim which was later found to be inaccurate.) We also note that, as mentioned above, this formula is valid in any Banach Algebra.

(8) Adjoint systems of this type were used by Atkinson [**21**]. They will be used in the next chapter to provide an elegant proof of a very general

Lagrange identity due to Everitt and Neuman; see [**230**]. Theorem 1.8.1, the Adjointness Identity, is due to Zettl; see [**620**], [**622**], [**633**].

(9) These kinds of inverse problems are discussed by Hartman [**319**] and Petrovski [**515**] but not quite in this generality.

CHAPTER 2

Scalar Initial Value Problems

R. L. Moore
   That student is taught the best who is told the least.

1. Introduction

In this chapter we apply the results of Chapter 1 to the study of initial value problems (IVP) consisting of the second order scalar equation

$$-(py')' + qy = f \text{ on } J \qquad (2.1.1)$$

together with initial conditions

$$y(c) = h, \ (py')(c) = k, \ c \in J, \ h, k \in \mathbb{C} \qquad (2.1.2)$$

where

$$J = (a,b), \ -\infty \le a < b \le \infty, \ \frac{1}{p}, q, f : J \to \mathbb{C}. \qquad (2.1.3)$$

We show that the solutions of these IVP depend not only continuously but differentiably on all parameters of the problem.

2. Existence and Uniqueness

DEFINITION 2.2.1 (Solution). *By a solution of equation (2.1.1) we mean a function $y : J \to \mathbb{C}$ such that y and $y^{[1]} = py'$ are absolutely continuous on each compact subinterval of J and the equation is satisfied a.e. on J. Given a solution y we refer to $y^{[1]}$ as its quasi-derivative to distinguish it from the classical derivative y'.*

Note that the classical derivative of a solution y, in general, exists only almost everywhere but the quasi-derivative $y^{[1]} = (py')$ is absolutely continuous on all compact subintervals of J and thus exists and is continuous at each point of the open interval J.

THEOREM 2.2.1. *Assume that*

$$1/p, q, f \in L_{loc}(J, \mathbb{C}). \qquad (2.2.1)$$

Then every initial value problem (IVP) (2.1.1), (2.1.2), (2.1.3) has a solution defined on all of J and this solution is unique. Moreover, if all the data p, q, f, h, k is real, then there is a unique real-valued solution on J.

PROOF. Let

$$P = \begin{bmatrix} 0 & 1/p \\ q & 0 \end{bmatrix}, \ F = \begin{bmatrix} 0 \\ f \end{bmatrix}, \ Y = \begin{bmatrix} y \\ py' \end{bmatrix}. \qquad (2.2.2)$$

Then the equation (2.1.1) is equivalent to the first order system
$$Y' = PY + F \text{ on } J, \qquad (2.2.3)$$
in the sense that, given any scalar solution y of (2.1.1) the vector Y defined by (2.2.2) is a solution of the system (2.2.3) and conversely, given any vector solution Y of system (2.2.3) its top component y is a solution of (2.1.1). Theorem 2.2.1 follows from this system representation and Theorem 1.2.1 of Chapter 1. □

The next theorem shows that the conditions (2.2.1) are minimal for all IVP to have unique solutions.

THEOREM 2.2.2. *If the initial value problem (2.1.1), (2.1.2), (2.1.3) has a unique solution in a neighborhood of u for each $u \in J$ and each pair $h, k \in \mathbb{C}$, then (2.2.1) holds.*

PROOF. This follows from Theorem 1.2.3 of Chapter 1 - the Everitt and Race Theorem. □

In order to study the continuous and differentiable dependence of the solutions of IVP we introduce a convenient notation.

NOTATION 2.2.1. *Given (2.2.1), the unique solution y of (2.1.1) and its quasi-derivative $y^{[1]}$ are denoted by*
$$y = y(\cdot, c, h, k, 1/p, q, f), \quad y^{[1]} = y^{[1]}(\cdot, c, h, k, 1/p, q, f), \qquad (2.2.4)$$
to highlight their dependence on these quantities. Note that solutions depend on $1/p$, not on p. This is clear from the representation (2.2.2), (2.2.3).

In the theory of boundary value problems the spectral parameter λ and the weight function w play important roles; thus we also study the equation
$$-(py')' + qy = \lambda w y \text{ on } J, \ \lambda \in \mathbb{C}, \qquad (2.2.5)$$
where
$$1/p, q, w \in L_{loc}(J, \mathbb{C}), \ J = (a, b), \ -\infty \leq a < b \leq \infty. \qquad (2.2.6)$$
For this case we use the notation
$$y = y(\cdot, c, h, k, 1/p, q, w, \lambda), \ y^{[1]} = y^{[1]}(\cdot, c, h, k, 1/p, q, w, \lambda) \qquad (2.2.7)$$
to highlight the dependence of y and $y^{[1]}$ on these quantities. Theorems 2.2.1 and 2.2.2 can be applied to this case simply by taking $f = 0$ and replacing q by $q - \lambda w$ in equation (2.1.1).

Below, when we study the dependence of y and $y^{[1]}$ on one of these quantities with all the others fixed we further abbreviate this notation by simply omitting all the fixed variables. Thus we write $y = y(\cdot, c)$ when we wish to study the unique solution as a function of c, $y = y(\cdot, q)$ to study the dependence of y on q, etc. Similarly for the quasi-derivative $y^{[1]}$.

3. Continuous Extensions to the Endpoints

The nature of the coefficients near the endpoints determines the behavior of the solutions there. What is "regular" behavior of solutions near an endpoint and when does it occur? These are the sort of questions we pursue in this section.

DEFINITION 2.3.1 (Regular and Singular Endpoints). *Let $J = (a, b), -\infty \leq a < b \leq \infty$, and consider the equation (2.1.1) with conditions (2.2.1).*

The endpoint a is said to be *regular (or equation (2.1.1) is regular at a)* if

$$1/p, q, f \in L((a,d), \mathbb{C}) \tag{2.3.1}$$

for some $d \in J$; otherwise it is called *singular*. Similarly, the endpoint b is said to be *regular* if

$$1/p, q, f \in L((d,b), \mathbb{C}) \tag{2.3.2}$$

for some $d \in J$; otherwise it is called *singular*. Note that, given condition (2.2.1), if (2.3.1) or (2.3.2) hold for some $d \in J$ then they hold for any $d \in J$.

REMARK 2.3.1. **In much of the literature an infinite endpoint is automatically classified as singular in contrast with Definition 2.3.1. We propose this definition in view of the fact that, given (2.2.6), (2.3.2) is necessary and sufficient for all solutions y of (2.1.1) and their quasi-derivatives $y^{[1]}$ to have a finite limit at b. See Theorem 2.3.1 below.**

It is not the finite or infinite nature of the endpoint b but condition (2.3.2) which determines whether or not all solutions of (2.1.1) and their quasi-derivatives have finite limits at b. We contend that this is a natural definition of "regular" behavior at b. Similar remarks apply at the endpoint a.

Next we take up the question of the continuous extension to the endpoints of solutions and their quasi-derivatives of the scalar SL equation.

THEOREM 2.3.1. *Let (2.2.1) hold. Then*

- *the limits*

$$y(a) = \lim_{t \to a^+} y(t), \quad y^{[1]}(a) = \lim_{t \to a^+} y^{[1]}(t) \tag{2.3.3}$$

both exist and are finite for every solution y and its quasi-derivative $y^{[1]}$ of (2.1.1) if and only if (2.3.1) holds;

- *the limits*

$$y(b) = \lim_{t \to b^+} y(t), \quad y^{[1]}(b) = \lim_{t \to b^+} y^{[1]}(t) \tag{2.3.4}$$

both exist and are finite for the solution y and its quasi-derivative $y^{[1]}$ of (2.1.1) if and only if (2.3.2) holds.

PROOF. This follows from Theorems 1.5.2 and 2.2.2. □

4. Continuous Dependence of Solutions on the Problem

Next we take up the question of the continuous dependence of solutions and their quasi-derivatives of initial value problems for the second order scalar equation, on all parameters of the problem.

THEOREM 2.4.1 (Continuous dependence of solutions on the problem). *Let (2.2.1) hold. Let y denote the unique solution of the initial value problem (2.1.1), (2.1.2) according to Theorem 2.2.1. Then each of y and its quasi-derivative $y^{[1]}$ is a jointly continuous function of all its variables, uniformly on compact subintervals of J. More precisely, given $c_j \in J$, $h_j, k_j \in \mathbb{C}$, $1/p_j, q_j, f_j \in L_{loc}(J, \mathbb{C})$, $j = 1, 2,$*

and given $\epsilon > 0$ and a compact subinterval $K = [a_1, b_1]$ of J containing c_1 and c_2, there exists a number $\delta > 0$ such that if

$$|c_1-c_2|+|h_1-h_2|+|k_1-k_2|+\int_K (|1/p_1 - 1/p_2| + |q_1 - q_2| + |f_1 - f_2|) < \delta, \quad (2.4.1)$$

then

$$|y(t, c_1, h_1, k_1, 1/p_1, q_1, f_1) - |y(t, c_2, h_2, k_2, 1/p_2, q_2, f_2)| < \epsilon \quad (2.4.2)$$

and

$$|y^{[1]}(t, c_1, h_1, k_1, 1/p_1, q_1, f_1) - y^{[1]}(t, c_2, h_2, k_2, 1/p_2, q_2, f_2)| < \epsilon \quad (2.4.3)$$

both for all $t \in K$.

Furthermore, if $1/p, q, f \in L(J, \mathbb{C})$, J is a compact interval, and (2.4.1) holds with $K = J$, then (2.4.2) and (2.4.3) hold on J, i.e. for $a \le t \le b$.

PROOF. This is a consequence of Theorem 1.6.2. □

5. Differentiable Dependence of Solutions on the Data

The differentiability of the solution

$$y(t, c, h, k, 1/p, q, w, \lambda)$$

of (2.2.5), (2.2.6) and its quasi-derivative

$$y^{[1]} = (py')(t, c, h, k, 1/p, q, w, \lambda)$$

with respect to t follows from the definition of solution; the differentiability of y and $y^{[1]}$ with respect to c is a consequence Lemma 1.7.1. The differentiability of y and of $y^{[1]}$ with respect to the other variables is studied in this section.

THEOREM 2.5.1 (Differentiable dependence on initial conditions). *Let (2.2.5), (2.2.6) hold. Let u, v be solutions of (2.2.5) determined by the initial conditions*

$$u(c) = 0, \ (pu')(c) = 1; \ v(c) = 1, \ (pv')(c) = 0, \ c \in J.$$

Using the notation (2.2.7) with the convention that all fixed variables are omitted, we have that each of the following maps from $\mathbb{C}$ to $\mathbb{C}$:

$$h \to y(t, c, h, k, 1/p, q, w, \lambda), \ h \to y^{[1]}(t, c, h, k, 1/p, q, w, \lambda),$$
$$k \to y(t, c, h, k, 1/p, q, w, \lambda), \ k \to y^{[1]}(t, c, h, k, 1/p, q, w, \lambda)$$

is differentiable and the derivatives are given by:

$$y'(h) = v(t) h, \ h \in \mathbb{C},$$
$$(y^{[1]})'(h) = (pv')(t) h, \ h \in \mathbb{C},$$
$$y'(k) = u(t) k, \ k \in \mathbb{C},$$
$$(y^{[1]})'(k) = (pu')(t) k, k \in \mathbb{C},$$

respectively. Note that here $y'(h)$ denotes the derivative of y with respect to h, with all other variables, $t, c, k, 1/p, q, w, \lambda$ fixed; and in $(y^{[1]})'(h) = (py')'(h)$ the outside prime denotes the derivative of the quasi-derivative (py') with respect to h. Thus the two primes in $(py')'$ have different meanings in this formula - the outside one is for differentiation with respect to h, the inside one for differentiation with respect to t - but since t is fixed here this should not cause confusion. Similar remarks apply to the formulas for differentiation with respect to k.

Let $K = [a_1, b_1]$ be a compact subinterval of J. Each of the following maps from $\mathbb{C}$ to the Banach space $C(K)$:

$$h \to y(\cdot, c, h, k, 1/p, q, w, \lambda), \ h \to y^{[1]}(\cdot, c, h, k, 1/p, q, w, \lambda),$$
$$k \to y(\cdot, c, h, k, 1/p, q, w, \lambda), \ k \to y^{[1]}(\cdot, c, h, k, 1/p, q, w, \lambda)$$

is differentiable and its Frechet derivative is given by

$$y'(k)(g) = v\, g, \ g \in C(K),$$
$$y^{[1]'}(k)(g) = (pv')\, g, \ g \in C(K),$$
$$y'(k)(g) = u\, g, \ g \in C(K),$$
$$y^{[1]'}(k)(g) = (pu')\, g, \ g \in C(K),$$

respectively.

PROOF. This is a straightforward consequence of the variation of parameters formula and the definition of the Frechet derivative. See the above remarks about notation. □

To compute the derivatives of y and $y^{[1]}$ with respect to $1/p$, q, w and λ we use the primary fundamental matrix of the system representation of equation (2.2.5): Let

$$P = \begin{bmatrix} 0 & 1/p \\ q & 0 \end{bmatrix}, \ W = \begin{bmatrix} 0 & 0 \\ w & 0 \end{bmatrix}$$

and let $\Phi(\cdot, \cdot, P, w, \lambda) = (\phi_{ij})$ to be the primary fundamental matrix of the system $Y' = (P - \lambda W)Y$. See Definition 1.3.1 for Φ.

THEOREM 2.5.2 (Differentiable dependence on the equation). *Let conditions (2.2.5), (2.2.6) hold, and let K be a compact subinterval of J. Fix t, $c \in K$, h, $k \in \mathbb{C}$. Then the maps $q \to y(t, q)$, $q \to y^{[1]}(t, q)$, $w \to y(t, w)$, $w \to y^{[1]}(t, w)$, $1/p \to y(t, 1/p)$, $1/p \to y^{[1]}(t, 1/p)$ from $L(K, \mathbb{C})$ to $\mathbb{C}$ as well as the maps $\lambda \to y(t, \lambda)$, $\lambda \to y^{[1]}(t, \lambda)$ from $\mathbb{C} \to \mathbb{C}$ are differentiable and their derivatives are given by*

$$y'(t, q)(r) = -\int_c^t \phi_{1,2}(t, s)\, y(t, s)\, r(s)\, ds, \ r \in L(K, \mathbb{C}),$$

$$(py')'(t, q)(r) = -\int_c^t \phi_{2,2}(t, s)\, y(t, s)\, r(s)\, ds, \ r \in L(K, \mathbb{C}),$$

$$y'(t, 1/p)(r) = \int_c^t \phi_{1,2}(t, s)\, q(s) \left(\int_c^s (py')(x)\, r(x)\, dx \right) ds$$
$$+ \int_c^t (py')(x)\, r(x)\, dx, \ r \in L(K, \mathbb{C}),$$

$$(py')'(t, 1/p)(r) = \int_c^t \phi_{2,2}(t, s)\, q(s) \left(\int_c^s (py')(x)\, r(x)\, dx \right) ds, \ r \in L(K, \mathbb{C}),$$

$$y'(t, w)(r) = \lambda \int_c^t \phi_{1,2}(t, s, w)\, y(s, w)\, r(s)\, ds, \ r \in L(K, \mathbb{C});$$

$$(py')'(t, w)(r) = \lambda \int_c^t \phi_{2,2}(t, s, w)\, y(s, w)\, r(s)\, ds, \ r \in L(K, \mathbb{C});$$

$$y'(t,\lambda) = \int_c^t \phi_{1,2}(t,s,\lambda)\, w(s)\, y(s,\lambda)\, ds, \ \lambda \in \mathbb{C};$$

$$(y^{[1]})'(t,\lambda) = \int_c^t \phi_{2,2}(t,s,\lambda)\, w(s)\, y(s,\lambda)\, ds, \ \lambda \in \mathbb{C}.$$

PROOF. We prove some of these formulas, the proofs of the others are similar and hence omitted. Let

$$-(py')' + qy = 0, \ y(c) = h, \ (py')(c) = k;$$
$$-(pz')' + (q+r)z = 0, \ z(c) = h, \ (pz')(c) = k.$$

Let $x = z - y$. Then

$$-(px')' + qx = -r\,z, \ x(c) = 0, \ (px')(c) = 0.$$

From the variation of parameters formula it follows that

$$x(t) = \int_c^t \phi_{1,2}(t,s)(-r(s))\, z(s)\, ds.$$

Letting $z = y + (z - y)$ we get

$$z(t) - y(t) + \int_c^t \phi_{1,2}(t,s)(r(s))\, y(s)\, ds = -\int_c^t \phi_{1,2}(t,s)[z(s) - y(s)]\, r(s)\, ds$$
$$= o(r) \text{ as } r \to 0 \text{ in } L(J,\mathbb{C}).$$

The last equality follows from the fact that $\phi_{1,2}$ is bounded on $K \times K$ and $z \to y$ uniformly on K by the furthermore part of Theorem 2.4.1. Similarly we get

$$(pz' - py')(t) = (px')(t) = \int_c^t \phi_{2,2}(t,s)(-r(s)\, z(s)\, ds.$$

From these equations, proceeding as above, we obtain the formulas for $y(t,q)(r)$ and for $(y^{[1]})'(t,q)(r)$.

To derive the formulas for the derivatives with respect to $1/p$ we proceed as follows. Let

$$\frac{1}{p_r} = \frac{1}{p} + r, \ r \in L(J,\mathbb{C})$$

and let $y = y(t, 1/p)$, $z = z(t, 1/p_r)$. Set

$$x(t) = \int_c^t \frac{1}{p}(py' - p_r z').$$

Then

$$-(px')' + qx = f, \ x(c) = 0, \ (px')(c) = 0, \ f(t) = -q(t)\int_c^t (p_r z')\, r.$$

From the variation of parameters formula we get

$$x(t) = -\int_c^t \phi_{1,2}(t,s)\, q(s) \left(\int_c^s (p_r z')(u)\, r(u)du\right) ds,$$

$$(px')(t) = -\int_c^t \phi_{2,2}(t,s)\, q(s) \left(\int_c^s (p_r z')(u)\, r(u)du\right) ds.$$

Note that
$$z(t) - y(t) = \int_c^t [\frac{1}{p_r}(p_r z') - \frac{1}{p}(py')]$$
$$= \int_c^t [\frac{1}{p}(p_r z') - \frac{1}{p}(py')] + \int_c^t (p_r z') r$$
$$= -x(t) + \int_c^t (p_r z') r$$
$$= \int_c^t \phi_{1,2}(t,s) q(s) \left(\int_c^s (p_r z')(u) r(u) du \right) ds + \int_c^t (p_r z') r.$$

Setting
$$p_r z' = py' + [p_r z' - py']$$
we obtain
$$z(t) - y(t) - \int_c^t \phi_{1,2}(t,s) q(s) \left(\int_c^s (py')(u) r(u) du \right) ds + \int_c^t (py') r$$
$$= \int_c^t \phi_{1,2}(t,s) q(s) \left(\int_c^s [(p_r z') - py'](u) r(u) du \right) ds + \int_c^t [(p_r z') - py'] r$$
$$= o(r) \text{ as } r \to 0 \text{ in } L(J).$$

The last equality follows from the boundedness of $\phi_{1,2}$ on $K \times K$, from $q \in L(K, \mathbb{C})$, and from the fact that, by Theorem 2.4.1, $(p_r z') \to py'$ uniformly on K as $r \to 0$ in $L(J, \mathbb{C})$. □

REMARK 2.5.1. There is an interesting and subtle point involved in the proof of Theorem 2.5.2: $1/p + r$ may be identically zero on a subinterval of J. As noted above, below (2.2.4), the solutions y and their quasi-derivatives $y^{[1]}$ depend on $1/p$, not on p. Therefore $1/p$ may be identically zero on a subinterval of J or even on all of J. This is allowed by the existence-uniqueness Theorem 1.2.1 and subsequent theorems. However, the equation (2.2.5) has to be interpreted properly in this case as Atkinson [21] has pointed out. See Sections 3.7, 4.8 and 4.12 below for further elaboration of this point. Atkinson uses the notation
$$-(\frac{1}{p}y')' + qy = \lambda w y$$
but this notation has not been widely adopted.

For regular equations each solution y and its quasi-derivative $y^{[1]}$ are not only entire functions of λ but have order at most $1/2$.

THEOREM 2.5.3. *Assume that*
$$1/p, q, w \in L(J, \mathbb{C}), \quad J = (a,b), \quad -\infty \leq a < b \leq \infty. \tag{2.5.1}$$
Then every nontrivial solution y of (2.2.5) and its quasi-derivative $y^{[1]}$ are entire functions of λ of order at most $1/2$. More precisely, there exist positive constants M, B, δ such that
$$|y(t, \lambda)| \leq B e^{M \sqrt{|\lambda|}}, \ a \leq t \leq b, \ |\lambda| \geq \delta,$$
$$|(py')(t, \lambda)| \leq B e^{M \sqrt{|\lambda|}}, \ a \leq t \leq b, \ |\lambda| \geq \delta.$$

PROOF. Let $v = py'$ then $v' = (q - \lambda w) y$. Fix λ and let prime " $'$ " denote differentiation with respect to t. Then

$$[|\lambda| |y|^2 + |v|^2]' = [|\lambda| \bar{y} y + \bar{v} v]'$$
$$= |\lambda| (\frac{1}{\bar{p}} y \bar{v} + \frac{1}{p} v \bar{y}) + \bar{v} (q - \lambda w) y + v (\bar{q} - \bar{\lambda} \bar{w}) \bar{y}.$$

From this and the elementary inequality

$$2 |a b| \leq \frac{|\lambda| |a|^2 + |b|^2}{\sqrt{|\lambda|}}, \; |\lambda| \neq 0$$

we get

$$[|\lambda| |y|^2 + |v|^2]' \leq \frac{|\lambda| |y|^2 + |v|^2}{\sqrt{|\lambda|}} \left(|\lambda| \frac{1}{|p|} + |q| + |\lambda| |w| \right)$$

and hence

$$[\log (|\lambda| |y|^2 + |v|^2)]' \leq \sqrt{|\lambda|} \frac{1}{|p|} + \frac{1}{\sqrt{|\lambda|}} |q| + \sqrt{|\lambda|} |w|.$$

An integration yields

$$|\lambda| |y(t,\lambda)|^2 + |v(t,\lambda)|^2 \leq C \, e^{\sqrt{|\lambda|} \int_a^t (\frac{1}{|p|} + |w|) + \frac{1}{\sqrt{|\lambda|}} \int_a^t |q|}$$

$$\leq B \, e^{M \sqrt{|\lambda|}}, \; 0 < M = \int_a^b (\frac{1}{|p|} + |w|) < \infty, \; e^{\left(\frac{1}{\sqrt{|\lambda|}} \int_a^b |q| \right)} < B < \infty.$$

This completes the proof. □

6. Sturm Separation and Comparison Theorems

Two of the most celebrated results in the theory of second order linear differential equations are the Sturm Separation and Comparison Theorems. These are established in this section. But first, in order to make them meaningful we show that the zeros of nontrivial solutions are isolated at all regular points.

THEOREM 2.6.1. *Let (2.2.5), (2.2.6) hold with $\mathbb{C} = \mathbb{R}$ and assume that $p > 0$ a.e. on J and λ is real. Then the zeros of every nontrivial solution y of (2.2.5) are isolated in the interior of J and also at regular endpoints of J. If a nontrivial solution y has a zero at a regular endpoint of J then there is an appropriate one sided neighborhood of this endpoint in which y has no other zero. Thus only a singular endpoint of J can be an accumulation point of zeros of any nontrivial solution y of (2.2.5).*

PROOF. First we show that if y has consecutive zeros at $c, d \in (a, b)$, $c < d$, then $(py')(h) = 0$ for some $h \in (c, d)$. We have

$$0 = y(d) - y(c) = \int_c^d y' = \int_c^d \frac{1}{p} (py') = (py')(h) \int_c^d \frac{1}{p}$$

by the Mean Value Theorem for the Lebesgue integral. (Recall that (py') is continuous on J.) Hence either $(py')(h) = 0$ or $\int_c^d \frac{1}{p} = 0$, but the latter would imply that $\frac{1}{p} = 0$ a.e. in (c, d) in contradiction to the hypothesis that $p > 0$ a.e. in (a, b).

Now to prove the Theorem suppose there exists a sequence $\{t_n \in (a, b) : n \in N_0\}$ such that $t_n \to t_0$ and $y(t_n) = 0$, $n \in N_0$. Then $y(t_0) = 0$ since y is continuous and from the first part of the proof we get a sequence $\{h_n : n \in N\}$ with $h_n \to t_0$ such that $(py')(h_n) = 0$. Since $y^{[1]}$ is continuous in (a,b) it follows that $y^{[1]}(t_0) = 0$. But $y(t_0) = 0$ and $y^{[1]}(t_0) = 0$ implies that y is identically zero on J by the uniqueness of initial value problems. This contradiction completes the proof. $\square$

THEOREM 2.6.2 (Sturm Separation Theorem). *Let (2.2.5), (2.2.6) hold with $\mathbb{C} = \mathbb{R}$ and assume that $p > 0$ a.e. on J and λ is real. Suppose that y and z are linearly independent solutions of (2.2.5). Then z has a zero strictly between any two zeros of y.*

PROOF. Suppose $y(c) = 0 = y(d)$. Since y and z are linearly independent neither is the trivial solution and they have no common zero. By Theorem 2.6.1 we may assume that c and d are consecutive zeros of y and that $c < d$. Further, replacing y by $-y$ if necessary, we may assume that $y > 0$ on the open interval (c,d). Then
$$0 < y(t) - y(c) = \int_c^t \frac{1}{p}(py').$$
This implies that $(py')(c) > 0$. Otherwise, since py' is continuous and $(py')(c)$ is not zero, the assumption that $(py')(c) < 0$ implies that it is negative in some right neighborhood of c and this contradicts the above equation. Similarly we get that $(py')(d) < 0$.

Now multiplying the equation for y by z and the equation for z by y and subtracting we get
$$0 = z(py')' - y(pz')' = [z(py') - y(pz')]'.$$
An integration yields
$$z(d)(py')(d) - z(c)(py')(c) = 0.$$
Assume that z has no zero in (c,d). If $z > 0$ on (c,d) then the left hand side of the above equation is positive giving a contradiction. A similar contradiction is reached if $z < 0$ completing the proof. $\square$

The next example shows that the hypothesis $p > 0$ a.e. on J cannot be omitted in the Sturm Separation Theorem.

EXAMPLE 2.6.1. *The equation*
$$-(py')' = 0 \quad \text{on} \quad (0,1), \quad \frac{1}{p(t)} = t^{-2}\sin(\frac{1}{t}), \quad 0 < t < 1$$
has solutions $u(t) = 1$, $v(t) = \cos(\frac{1}{t})$, on $(0,1)$. The solution u has no zero on the interval $(0,1)$ while the solution v is oscillatory at 0 since it has an infinite number of zeros in any right neighborhood of 0. The endpoint 0 is singular. Consider this equation on an interval $J = (\varepsilon, 1)$, $0 < \varepsilon < 1$. Then the solution u has no zero on J and for any positive integer n we can choose ε sufficiently small so that the solution v has n zeros in J.

DEFINITION 2.6.1. *Given a singular endpoint of equation*
$$(py')' + qy = 0 \quad \text{on} \quad J, \quad 1/p, q \in L_{loc}(J, \mathbb{R}), \quad p > 0 \text{ a.e.} \quad \text{on} \quad J = (a,b) \quad (2.6.1)$$
it follows from the Sturm Separation Theorem that either all nontrivial solutions are oscillatory at this endpoint or none of them is. Thus oscillation at a given singular endpoint is a property of the equation itself, not just of a particular solution. We

say that the equation (2.6.1) is oscillatory (O) at a if for every $c \in J$ there is a nontrivial solution which has an infinite number of zeros in the interval (a, c). Similarly for the endpoint b.

REMARK 2.6.1. This definition leads to the classification of all equations (2.6.1) into two mutually inclusive and disjoint classes which we denote by O (oscillatory) and NO (nonoscillatory). This classification holds at each singular endpoint and, of course, may be different at the two endpoints a, b of the underlying interval J. The problem of finding necessary and sufficient conditions on p and q, *which can be verified for each equation*, for oscillation to hold is still today an active research problem. Numerous verifiable sufficient conditions and also necessary conditions are known but these do not cover all equations. There are even necessary and sufficient conditions known, see chapter 3 of [**368**], but these are not verifiable for each equation. A major difficulty in obtaining verifiable necessary and sufficient conditions is due to the fact that "natural" conditions for oscillation are of "interval" type in contrast to NO conditions which have to be essentially pointwise. By "interval" type conditions we mean conditions on p and q which are required to hold only on a sequence of intervals with only very weak restrictions on the complements of these intervals. To produce a zero of every nontrivial solution on a given interval it is sufficient to give conditions on p and q only on a subinterval. For example by comparing with a constant coefficient equation on a subinterval one can use the Sturm Comparison Theorem to guarantee a zero for every solution on this subinterval and therefore on the whole interval. See the paper [**413**] by Kwong and Zettl for a method called "the telescoping method" for obtaining "interval" type oscillation conditions.

Next we discuss the Comparison Theorem which is an important tool in obtaining sufficient conditions for oscillation.

THEOREM 2.6.3 (Sturm Comparison Theorem). *Let (2.6.1) hold. Consider a comparison equation*

$$(Pz')' + Qz = 0 \quad \text{on} \quad J = (a, b). \tag{2.6.2}$$

Assume $1/P, Q \in L_{loc}(J, \mathbb{R})$ *and satisfy*

$$Q \geq q, \quad 0 < P \leq p \quad \text{on} \quad J. \tag{2.6.3}$$

Suppose y is a non-trivial solution of (2.6.1) satisfying $y(c) = 0 = y(d)$ for some $c, d \in J$, $c < d$. Then every solution of (2.6.2) has a zero in the closed interval $[c, d]$.

PROOF. Since the zeros of y are isolated by Theorem 2.6.1 we may assume that c and d are consecutive zeros of y. Replacing y by $-y$, if necessary, we may also assume that $y > 0$ on (c, d). Suppose that z is a nontrivial solution of (2.6.2) which has no zero in $[c, d]$. Then a direct, albeit tedious, computation yields the Picone identity:

$$[\frac{y}{z}(py'z - Pz'y)]' = (Q - q)y^2 + (p - P)y'^2 + P\frac{(yz' - zy')^2}{z^2}. \tag{2.6.4}$$

An integration yields

$$\int_c^d (Q - q)y^2 + \int_c^d (p - P)y'^2 = -\int_c^d P\frac{(yz' - zy')^2}{z^2} \tag{2.6.5}$$

The hypothesis (2.6.3) implies that $yz' - zy' = 0$ a.e. on $[c, d]$. Hence $f = y/z = 0$ and, consequently, $y = 0$ on $[c, d]$. By the existence-uniqueness theorem this implies that y is the trivial solution on J and this contradiction completes the proof. □

We end this section with an example to show that when p changes sign the behavior of the classical and quasi-derivatives of a solution can be quite different.

EXAMPLE 2.6.2. *Let*

$$p(t) = \frac{1}{\cos(\ln(t))}, \ 0 < t \leq 1.$$

Then $1/p \in L(0, 1)$ and the equation

$$-(py')' = 0 \text{ on } (0, 1)$$

is regular at 0 and at 1 and has

$$y = \int \cos(\ln(t)) \, dt, \ v(t) = 1$$

as solutions. Note that the point 0 is an accumulation point of zeros of y' since $y'(t_k) = 0$ where

$$t_k = e^{-k\pi/2} \to 0, \text{ as } k \to \infty,$$

and $(py')(t) = 1$ for $t \in [0, 1]$. The Wronskian

$$W(y, v)(t) = \begin{vmatrix} y & v \\ py' & pv' \end{vmatrix}(t) = -1, \ 0 \leq t \leq 1,$$

but the classical Wronskian

$$\begin{vmatrix} y & v \\ y' & v' \end{vmatrix}(t) = -\cos(\ln(t)), \ 0 < t \leq 1$$

is nonconstant with an accumulation point of zeros at 0.

7. Periodic Coefficients

In this section we specialize the SL equation to the periodic coefficient case. This case sometimes goes under the heading of Floquet theory. Throughout this section we assume that the coefficients of the equation

$$(py')' + qy = \lambda wy, \ \lambda \in \mathbb{C}, \text{ on } J \tag{2.7.1}$$

with $J = \mathbb{R}$ or $J = (a, \infty)$ for some a, $-\infty < a < \infty$, are periodic with the same period and that h is the smallest positive number such that

$$r(t + h) = r(t), \ q(t + h) = q(t), \ w(t + h) = w(t), \ \text{for a.a. } t \in \mathbb{R}, \tag{2.7.2}$$
$$r = 1/p, \ q, \ w \in L_{loc}(J, \mathbb{R}).$$

Do equations with h-periodic coefficients have nontrivial h-periodic solutions? In general the answer is no but they only miss by a multiplicative factor.

THEOREM 2.7.1. *Let (2.7.1), (2.7.2) hold. There exists a $\rho \in \mathbb{C}$ and a nontrivial solution z such that*

$$z(t + h) = \rho \, z(t), \ t \in J. \tag{2.7.3}$$

PROOF. Determine solutions u, v with the initial conditions
$$u(0) = 1 = (pv')(0), \ v(0) = 1 = (pu')(0). \tag{2.7.4}$$

Let $y(t) = u(t+h)$. Then y is a solution:
$$\begin{bmatrix} y \\ y^{[1]} \end{bmatrix}'(t) = \begin{bmatrix} u \\ u^{[1]} \end{bmatrix}'(t+h) = \begin{bmatrix} 0 & r \\ q - \lambda w & 0 \end{bmatrix}(t+h) \begin{bmatrix} u \\ u^{[1]} \end{bmatrix}(t+h)$$
$$= \begin{bmatrix} 0 & r \\ q - \lambda w & 0 \end{bmatrix}(t) \begin{bmatrix} y \\ y^{[1]} \end{bmatrix}(t).$$

Similarly, $v(t+h)$ is a solution of (2.7.1). Hence there exist constants A_{ij} such that
$$u(t+h) = A_{11}u(t) + A_{12}v(t), \ v(t+h) = A_{21}u(t) + A_{22}v(t), \ t \in \mathbb{R}.$$

Let
$$A = \begin{bmatrix} A_{11} & A_{12} \\ A_{21} & A_{22} \end{bmatrix}.$$

Then A is nonsingular, since $\det(A) = 0$ would imply that $u(t+h)$, $v(t+h)$ are linearly dependent and this would lead to the contradiction that $u(t), v(t)$ are linearly dependent. Now (2.7.3) is equivalent with each of the following equations:
$$c\,u(t+h) + d\,v(t+h) = \rho\,[cu(t) + dv(t)],$$
$$c[A_{11}u(t) + A_{12}v(t)] + d[A_{21}u(t) + A_{22}v(t)] = \rho\,[cu(t) + dv(t)],$$
$$[c(A_{11} - \rho) + dA_{21}]u(t) + [cA_{12} + d(A_{22} - \rho)]v(t) = 0, \ t \in \mathbb{R}.$$

Since u, v are linearly independent (2.7.3) holds for a nontrivial solution z if and only if the algebraic system
$$c(A_{11} - \rho) + dA_{21} = 0,$$
$$cA_{12} + d(A_{22} - \rho) = 0,$$

has a nontrivial solution for (c, d); and this so if and only if
$$\det \begin{bmatrix} A_{11} - \rho & A_{21} \\ A_{12} & A_{22} - \rho \end{bmatrix} = \rho^2 - (trace A)\rho + \det(A) = 0;$$

in other words, if and only if ρ is an eigenvalue of A^t, the transpose of A.

Since $\det(A) \neq 0$, A has an eigenvalue $\rho \neq 0$ in $\mathbb{C}$. From this quadratic equation for ρ it follows that its solutions ρ_1, ρ_2 satisfy:
$$\rho_1 \rho_2 = \det(A) \neq 0.$$

From the definition of u, v and A we have
$$A = \begin{bmatrix} u(h) & u^{[1]}(h) \\ v(h) & v^{[1]}(h) \end{bmatrix}.$$

So $\det A = \det A^t = W(u, v)(h) = 1$, since the Wronskian is constant and $W(u, v)(0) = 1$. Hence
$$\det A = 1 = \rho_1 \rho_2; \ \rho_1 = 1/\rho_2; \ trace A = u(h) + v^{[1]}(h) = D.$$
$\square$

Theorem 2.7.1 holds for each fixed λ. To study the dependence on λ we let
$$D = D(\lambda) = u(h, \lambda) + v^{[1]}(h, \lambda).$$

7. PERIODIC COEFFICIENTS

DEFINITION 2.7.1. *Let (2.7.1), (2.7.2) hold. Let ρ_1, ρ_2 denote the roots of the quadratic equation*

$$\rho^2 - (\operatorname{trace} A)\rho + \det(A) = 0$$

with A defined as in the proof of Theorem 2.7.1. We call ρ_1, ρ_2 the characteristic multipliers of equation (2.7.1):

$$\rho_j = \rho_j(r, q, w, \lambda),\ j = 1, 2.$$

DEFINITION 2.7.2. *Let (2.7.1), (2.7.2) hold. Let $\rho_1 = e^{h m_1}$, $\rho_2 = e^{h m_2}$. Then m_1 and m_2 are called characteristic exponents of equation (2.7.1): $m_j = m_j(r, q, w),\ j = 1, 2$. Note that m_1 and m_2 are not uniquely determined but their real parts are unique.*

REMARK 2.7.1. *If $\rho_1(\lambda) = 1$, then $\rho_2(\lambda) = 1$ and there is an h-periodic solution, i.e this value of λ is a periodic eigenvalue on the interval $[0, h]$. Note that λ is not, in general, a geometrically double eigenvalue since not all solutions need be periodic. If $\rho_1(\lambda) = -1$ then $\rho_2(\lambda) = -1$ and there is a semi-periodic solution; this value of λ is a semi-periodic eigenvalue on the interval $[0, h]$ and a periodic eigenvalue on the interval $[0, 2h]$.*

THEOREM 2.7.2. *Let (2.7.1), (2.7.2) hold. There are linearly independent solutions z_1, z_2 such that for all $t \in J$, either*

(1) $z_1(t) = e^{m_1 t} k_1(t),\ z_2(t) = e^{m_2 t} k_2(t)$ *for some $m_1, m_2 \in \mathbb{C}$, not necessarily distinct, and such that k_j is h-periodic, $j = 1, 2$; or*
(2) $z_1(t) = e^{m t} k_1(t),\ z_2(t) = e^{m t}[t k_1(t) + k_2(t)]$ *where $m \in \mathbb{C}$ and k_j is h-periodic, $j = 1, 2$.*

PROOF.

CASE 1. *$\rho_1 \neq \rho_2$. Let z_1, z_2 be the solutions satisfying (2.7.3) for $\rho = \rho_1$, $\rho = \rho_2$ respectively, according to Theorem 2.7.1. Let $\rho_1 = e^{h m_1}$, $\rho_2 = e^{h m_2}$, $k_j(t) = e^{-m_j t} z_j(t),\ j = 1, 2$. Then*

$$k_j(t+h) = e^{-m_j t} e^{-m_j h} z_j(t+h) = e^{-m_j t} e^{-m_j h} e^{m_j h} z_j(t) = k_j(t),\ j = 1, 2;\ t \in \mathbb{R}.$$

CASE 2. *$\rho_1 = \rho_2 = \rho$. Proceeding as in the previous case we get a solution $z_1(t) = e^{m_1 t} k_1(t)$ with k h-periodic. Choose a solution z_2 such that z_1, z_2 are linearly independent. Consider*

$$z_2(t+h) = c\, z_1(t) + d\, z_2(t).$$

Then

$$W(z_1, z_2)(t+h) = [z_1 z_2^{[1]} - z_2 z_1^{[1]}](t+h) = [\rho z_1(t)(c\, z_1^{[1]}(t) + d\, z_2^{[1]}(t))$$
$$-\rho\, z_1^{[1]}(t)(c\, z_1(t) + d\, z_2(t))] = \rho\, d\, W(z_1, z_2)(t).$$

Since the Wronskian is constant it follows that $\rho d = 1$. From $\rho_1 \rho_2 = 1$, it follows that $d = \rho$. Thus $z_2(t+h) = c\, z_1(t) + \rho\, z_2(t)$.

CASE 3. *If $c = 0$, this reduces to the first case.*

Assume $c \neq 0$. Let $k_1(t) = e^{-m\,t} z_1(t)$ and $k_2(t) = e^{-m\,t} z_2(t) - \frac{c}{\rho h} t\, k_1(t)$. Then k_1 is h-periodic and

$$k_2(t+h) = e^{-m\,(t+h)} z_2(t+h) - \frac{c}{\rho h}(t+h)\, k_1(t)$$

$$= e^{-m\,t} e^{-m\,h} [c\, z_1(t) + \rho\, z_2(t)] - \frac{c}{\rho h}(t+h)\, k_1(t)$$

$$= e^{-m\,t} e^{-m\,h} c\, z_1(t) + e^{-m\,t} z_2(t) - \frac{c}{\rho h}(t+h)\, k_1(t)$$

$$= e^{-m\,t} z_2(t) - \frac{c}{\rho h} t\, k_1(t) = k_2(t).$$

So Case 2 holds with z_2 replaced by $(\rho h/c) z_2$.

$\square$

We next state as Corollaries two results which are consequences of the proof of Theorem 2.7.2.

COROLLARY 2.7.1. *Let the hypotheses and notation of Theorem 2.7.2 hold. The first part of Theorem 2.7.2 holds whenever there are two linearly independent solutions of*

$$z(t+h) = \rho\, z(t)$$

with either the same or different values of ρ, while the second part holds when there is only one such linearly independent solution.

COROLLARY 2.7.2. *Let the hypotheses and notation of Theorem 2.7.2 hold. The first part of Theorem 2.7.2 holds if and only if the matrix A^t and therefore A has two linearly independent eigenvectors (this always holds if the eigenvalues of A are distinct but may also hold if the eigenvalues of A are equal). The second part holds when A has only one linearly independent eigenvector. In particular when $\rho_1 = \rho_2$ the first part holds when the rank of the matrix $(A - \rho I)$ is 0 and the second part holds when the rank of $(A - \rho I) = 1$. Recall that*

$$A = A(\lambda) = \begin{bmatrix} u(h,\lambda) & u^{[1]}(h,\lambda) \\ v(h,\lambda) & v^{[1]}(h,\lambda) \end{bmatrix}.$$

REMARK 2.7.2. It is worth noting that Theorems 2.7.1 and 2.7.2 are established under the assumption (2.7.2). In particular no continuity assumptions and no sign restrictions have been placed on the coefficients $r = 1/p, q, w$. Thus these coefficients can change sign, be step functions or be identically zero on subintervals of the fundamental interval $[0, h]$. See the comments for this section below for further remarks and references.

8. Comments

These are made for each section separately.

(1) Here we just introduce the initial value problems to be studied.
(2) In Theorem 2.2.1 the sufficiency of the local integrability condition (2.2.1) is well known, the necessity established in Theorem 2.2.2 is due to Everitt and Race [**231**].
(3) The continuous extensions of solutions and their quasi-derivatives given by Theorem 2.3.1 are surely not new, but we don't know of a reference where they can be found in this generality.

(4) Again, other than [**396**], we don't know of a reference where the continuous dependence of solutions of initial value problems on $1/p, q, w$ *in the L^1 norm* is established. The continuous dependence of solutions on initial conditions i.e. on c, h, k is discussed in the book by Hille [**326**].

(5) The differentiable dependence of solutions and their quasi-derivatives on each parameter as well as the formulas for the derivatives is based on Kong and Zettl [**396**].

Is y jointly differentiable in all its variables: $c, h, k, 1/p, q, w$? Ditto for py'. What are the derivatives?

Theorem 2.5.3 is adapted from Atkinson [**21**]. It shows that the solutions of any initial value problem for the equation (2.2.5) with real or complex coefficients satisfying condition (2.5.1) are entire functions of order *at most* $1/2$.

It follows from the asymptotic form of the eigenvalues of regular self-adjoint SLP that for p, q, w real-valued and $1/p > 0$, $w > 0$ the non-trivial solutions - as functions of λ - are of order exactly $1/2$. Under what more general conditions on p, q, w are the non-trivial solutions of exact order $1/2$? It follows from results established in Chapter 4 that the order is exactly $1/2$ if $p \geq 0, w \geq 0$ and p and w are both positive on a common subinterval of J.

Under what general conditions (other than the trivial case $p = q = w = 0$) are the solutions of the SL equation of order zero as functions of λ? In Chapters 3 and 4 we will construct solutions which are polynomial functions of λ and hence of order 0. For these problems $1/p$ and w (and q) are alternately identically zero on adjacent subintervals.

Recently Binding and Volkmer [**81**] have found classes of SLP with solutions which, as functions of λ, are of order r, for any r satisfying $0 < r < 1/2$.

(6) The Sturm Separation and Comparison Theorems are classical; indeed they are among the classic theorems of analysis. For some reason many authors of papers and books establish these results only for continuous coefficients. This is unnecessarily limiting for both theoretical as well as numerical reasons. So we give these classic results for locally integrable coefficients.

(7) The SL equation with periodic coefficients is often called Hill's equation, [**341**], [**88**], [**150**], [**149**], [**146**]. The study of this equation is also referred to as Floquet theory. This fascinating theory has a voluminous literature, see the books by Eastham [**153**] and Weidmann [**600**] and the references therein. In [**153**] strong smoothness and positivity restrictions are placed on the coefficients, however many, but not all, of the proofs given there are valid under much less severe restrictions on the coefficients. A couple of proofs using a boundedness assumption on one or more of the coefficients are an exception. The proofs of Theorems 2.7.1 and 2.7.2 are taken from [**153**] and are valid, as pointed out in Remark 2.7.1, without any continuity or sign restrictions on the coefficients $r = 1/p, q, w$. Note in particular that each of these coefficients can be a step function, can change sign, can be identically zero on one or more subintervals of the fundamental interval $[0, h]$. In the most extreme case when all three of these coefficients

are identically zero on the whole interval $[0, h]$ solutions and their quasi-derivatives are constant and the matrix A is the identity matrix. When w is identically zero there is no λ dependence. Note that $r = 1/p$ identically zero is also allowed, i.e. p may be infinite at each point of a subinterval or on the entire interval. Taking coefficients identically zero on appropriate subintervals reduces the differential SL equation to a difference equation. Thus, remarkably, the theory of differential equations includes a class of difference equations!

When p and w are positive and q is real-valued, Floquet theory leads to a characterization of the spectrum of self-adjoint realizations of the SL equation in the Hilbert space $L^2(J, w)$. This spectrum consists of an infinite number of spectral "bands" separated by "gaps"; any or all of the gaps may be missing in a particular case. The endpoints of the intervals forming these spectral bands and gaps are all determined by the periodic and semi-periodic eigenvalues of the SL equation on the fundamental interval $[0, h]$. These endpoints, and therefore the spectral bands and gaps, can be computed with SLEIGN2.

Is there a similar result when p or w change sign? For the start of a "left-definite" Floquet theory when $p > 0$ and w changes sign; see Section 12.7 and the recent paper by Marletta and Zettl [**464**]. See Chapter 5 for an introduction to left-definite problems. Is there a Floquet theory in the Krein space $L^2(J, w)$ when w changes sign analogous to the Floquet theory in the Hilbert space $L^2(J, w)$ when $w > 0$?

For an extension of the "bands and gaps" result to problems with a complex-valued coefficient; see Rofe-Beketov [**551**].

Part 2

Regular Boundary Value Problems

Sylvester, J.J. (1814-1897)
> ...there is no study in the world which brings into more harmonious action all the faculties of the mind than [mathematics], ... or, like this, seems to raise them, by successive steps of initiation, to higher and higher states of conscious intellectual being....

Presidential Address to British Association, 1869.

> So long as a man remains a gregarious and sociable being, he cannot cut himself off from the gratification of the instinct of imparting what he is learning, of propagating through others the ideas and impressions seething in his own brain, without stunting and atrophying his moral nature and drying up the surest sources of his future intellectual replenishment.
>
> The world of ideas which it [mathematics] discloses or illuminates, the contemplation of divine beauty and order which it induces, the harmonious connexion of its parts, the infinite hierarchy and absolute evidence of the truths with which it is concerned, these, and such like, are the surest grounds of the title of mathematics to human regard, and would remain unimpeached and unimpaired were the plan of the universe unrolled like a map at our feet, and the mind of man qualified to take in the whole scheme of creation at a glance.

Presidential Address to British Association, 1869.

CHAPTER 3

Two-Point Regular Boundary Value Problems

Hardy, Godfrey H. (1877-1947)
> I believe that mathematical reality lies outside us, that our function is to discover or observe it, and that the theorems which we prove, and which we describe grandiloquently as our "creations," are simply the notes of our observations.

A Mathematician's Apology, London, Cambridge University Press, 1941.

> Archimedes will be remembered when Aeschylus is forgotten, because languages die and mathematical ideas do not. "Immortality" may be a silly word, but probably a mathematician has the best chance of whatever it may mean.

A Mathematician's Apology, London, Cambridge University Press, 1941.

1. Introduction

In this chapter we study regular Sturm-Liouville problems (SLP) with general, not necessarily self-adjoint, two point boundary conditions (BC). The main tool for this study is the theory of analytic functions of a complex variable.

2. Transcendental Characterization of the Eigenvalues

A regular two point SLP consists of the equation

$$-(py')' + qy = \lambda w y \text{ on } J = (a,b), \ -\infty \leq a < b \leq \infty, \quad (3.2.1)$$

where

$$r = 1/p, \ q, \ w \in L(J, \mathbb{C}), \ \lambda \in \mathbb{C}, \quad (3.2.2)$$

together with boundary conditions

$$AY(a) + BY(b) = 0, \ Y = \begin{bmatrix} y \\ py' \end{bmatrix}, \ A, B \in M_2(\mathbb{C}). \quad (3.2.3)$$

Here $M_2(\mathbb{C})$ denotes the 2×2 matrices with complex entries. From Section 1.6 we know that $Y(a)$, $Y(b)$ exist as finite limits (for a, b finite or infinite) so that (3.2.3) is well defined. Let

$$P = \begin{bmatrix} 0 & 1/p \\ q & 0 \end{bmatrix}, \ W = \begin{bmatrix} 0 & 0 \\ w & 0 \end{bmatrix}. \quad (3.2.4)$$

Then, as shown in Chapter 2, the scalar equation (3.2.1) is equivalent to the first order system

$$Y' = (P - \lambda W)Y = \begin{bmatrix} 0 & 1/p \\ q - \lambda w & 0 \end{bmatrix} Y, \ Y = \begin{bmatrix} y \\ py' \end{bmatrix}. \quad (3.2.5)$$

Let $\Phi(\cdot, u, P, w, \lambda)$ be the primary fundamental matrix of (3.2.5) and recall that

$$\Phi' = (P - \lambda W)\Phi \quad \text{on} \quad J, \ \Phi(u, u, \lambda) = I, \ a \le u \le b, \ \lambda \in \mathbb{C}. \tag{3.2.6}$$

Define the characteristic function δ by

$$\delta(\lambda) = \delta(a, b, A, B, P, w, \lambda) = \det[A + B\,\Phi(b, a, P, w, \lambda)], \ \lambda \in \mathbb{C}. \tag{3.2.7}$$

This function δ is the transcendental function of the title of this section, it is also called a characteristic function. We will show below that its zeros are precisely the eigenvalues of the problem.

DEFINITION 3.2.1. *By a trivial solution of equation (3.2.1) on some interval I we mean a solution y which is identically zero on I and whose quasi-derivative $z = py'$ is also identically zero on I. (I may be a subinterval of J or it may be the whole interval J.) Note that, under the general hypotheses (3.2.2), a solution y may be identically zero on I but its quasi-derivative py' is not necessarily zero on I.*

DEFINITION 3.2.2. *Let (3.2.2) hold. A complex number λ is called an eigenvalue of the boundary value problem consisting of (3.2.1), (3.2.3) if the equation (3.2.1) has a nontrivial solution on J satisfying the boundary conditions (3.2.3). Such a solution is called an eigenfunction of λ. Any multiple of an eigenfunction is also an eigenfunction. If there are two linearly independent eigenfunctions for the same λ then we say that λ has geometric multiplicity two. If there is only one linearly independent eigenfunction of λ we say that λ is a simple eigenvalue or that λ has geometric multiplicity one. Since for each $\lambda \in \mathbb{C}$ the equation (3.2.1) has exactly two linearly independent solutions, each eigenvalue has geometric multiplicity either one or two.*

REMARK 3.2.1. Condition (3.2.2) does not restrict the coefficients to be real valued, and if they are real valued, it does not restrict the sign of any of the coefficients r, q, w. Also, each of r, q, w is allowed to be identically zero on one or more subintervals of J. If r is identically zero on a subinterval I, then all solutions y are constant on I. Note that if this constant is zero for some solution y then its quasi-derivative $z = py'$ may be a nonzero constant on I. Similarly, if both q and w are identically zero on a subinterval I, then py' is constant on I for any solution y. This constant may be nonzero even when y is identically zero on I. These statements can be clearly seen, and are best interpreted, from the system formulation of equation (3.2.1):

$$y' = r\,z, \quad z' = (q - \lambda w)\,y \quad \text{on } J, \ z = py', \ r = \frac{1}{p}.$$

An interval of zeros of a nontrivial solution y is counted as a single zero in the results on the numbers of zeros of solutions, in particular of eigenfunctions.

REMARK 3.2.2. Recall from Section 2.3 that condition (3.2.2) implies that y and (py') exist as finite limits at each (finite or infinite) endpoint a, b. Hence the boundary condition (3.2.3) is well defined.

LEMMA 3.2.1. *Let (3.2.1), (3.2.2), (3.2.3) hold. Then the characteristic function δ is well defined and is an entire function of λ for fixed (a, b, A, B, P, w).*

PROOF. It follows from Theorem 1.5.2 that for fixed λ, P, w, the primary fundamental matrix $\Phi(b, a, \lambda, P, w)$ exists and is continuous at a and b. The entire dependence on λ follows from Theorem 2.5.3. □

LEMMA 3.2.2. *Let (3.2.2) hold. Then*
(1) *A complex number λ is an eigenvalue of the BVP (3.2.1), (3.2.3) if and only if $\delta(\lambda) = 0$.*
(2) *The geometric multiplicity of an eigenvalue λ is equal to the number of linearly independent vector solutions $C = Y(a)$ of the linear algebra system*

$$[A + B \; \Phi(b, a, \lambda)] \, C = 0. \tag{3.2.8}$$

PROOF. Suppose $\delta(\lambda) = 0$. Then (3.2.8) has a nontrivial vector solution for C. Solve the IVP
$$Y' = (P - \lambda W)Y \text{ on } J, \; Y(a) = C.$$
Then
$$Y(b) = \Phi(b, a, \lambda) \, Y(a) \text{ and } [A + B \, \Phi(b, a, \lambda)] \, Y(a) = 0.$$

From this it follows that the top component of Y, say, y is an eigenfunction of the BVP (3.2.1), (3.2.3); this means λ is an eigenvalue of this BVP. Conversely, if λ is an eigenvalue and y an eigenvector of λ, then $Y = \begin{bmatrix} y \\ py' \end{bmatrix}$ satisfies $Y(b) = \Phi(b, a, \lambda) \, Y(a)$ and consequently $[A + B \, \Phi(b, a, \lambda)] \, Y(a) = 0$. Since $Y(a) = 0$ would imply that y is the trivial solution in contradiction to it being an eigenfunction, we have that $\det[A + B \, \Phi(b, a, \lambda)] = 0$. If (3.2.8) has two linearly independent solutions for C, say C_1, C_2, then solve the IVP with the initial conditions $Y(a) = C_1, Y(a) = C_2$ to obtain solutions Y_1, Y_2. Then Y_1, Y_2 are linearly independent vector solutions of (3.2.5) and their top components y_1, y_2 are linearly independent solutions of (3.2.1). Conversely, if y_1, y_2 are linearly dependent solutions of (3.2.1) we can reverse the steps above to obtain two linearly independent vector solutions of the algebraic system (3.2.8). □

The next result shows that any given complex number is an eigenvalue of geometric multiplicity two for precisely one boundary condition.

LEMMA 3.2.3. *Let (3.2.1), (3.2.2), (3.2.3) hold with $B = -I$. A given number $\lambda \in \mathbb{C}$ is an eigenvalue of geometric multiplicity two if and only if*
$$A = \Phi(b, a, \lambda).$$

PROOF. This follows from Lemma 3.2.2 and its proof. □

LEMMA 3.2.4. *For the BVP (3.2.1), (3.2.2), (3.2.3) exactly one of the following four cases holds:*
(1) *There are no eigenvalues in $\mathbb{C}$.*
(2) *Every complex number is an eigenvalue.*
(3) *There are exactly n eigenvalues in $\mathbb{C}$ for some $n \in \mathbb{N} = \{1, 2, 3, ...\}$.*
(4) *There are an infinite but countable number of eigenvalues in $\mathbb{C}$ and these have no finite accumulation point in $\mathbb{C}$.*

PROOF. This follows directly from Lemma 3.2.1 and Lemma 3.2.2 and the well known fact that the zeros of an entire function are isolated and therefore have no accumulation point in the finite complex plane $\mathbb{C}$. □

REMARK 3.2.3. Every self-adjoint regular SLP (see Chapter 4 for a definition of self-adjoint regular problems) with positive weight function w falls into category 4 of Lemma 3.2.4. Section 3.3 below contains simple examples illustrating cases 1 and 2. Are there examples for case 3? Atkinson in his book [**21**] hints at the existence of such examples but does not construct any. Such examples are constructed in Section 3.7 for every $n \in \mathbb{N}$; see also Sections 4.8 and 4.12.

It is convenient to classify the boundary conditions (BC) (3.2.3) into two mutually exclusive classes: separated and coupled. Note that, since the BC are homogeneous, multiplication by a nonzero constant or a nonsingular matrix leads to equivalent boundary conditions.

LEMMA 3.2.5 (Separated Boundary Conditions). *Let (3.2.1), (3.2.2), (3.2.3) hold. Fix P, W, J and assume*

$$A = \begin{bmatrix} A_1 & A_2 \\ 0 & 0 \end{bmatrix}, \; B = \begin{bmatrix} 0 & 0 \\ B_1 & B_2 \end{bmatrix}. \qquad (3.2.9)$$

Then

$$\delta(\lambda) = -A_2 B_1 \phi_{11}(b,a,\lambda) - A_2 B_2 \phi_{21}(b,a,\lambda)$$
$$+ A_1 B_1 \phi_{12}(b,a,\lambda) + A_1 B_2 \phi_{22}(b,a,\lambda)$$

for $\lambda \in \mathbb{C}$.

PROOF. This follows from the definition of δ and a direct computation. □

The characterization of the eigenvalues as zeros of an entire function given by Lemma 3.2.2 reduces to a simpler and more informative form when the boundary conditions are self-adjoint and coupled. This reduction is given by the next lemma.

LEMMA 3.2.6 (Coupled Self-Adjoint Boundary Conditions). *Let (3.2.1), (3.2.2), (3.2.3) hold and let $\Phi = (\phi_{ij})$ be the primary fundamental matrix of the system (3.2.5). Fix P, W, J and assume that*

$$B = -I, \; A = e^{i\gamma} K, \; -\pi < \gamma \leq \pi, \; K \in SL_2(\mathbb{R}), \qquad (3.2.10)$$

i.e. K is a real 2×2 matrix with determinant 1. Let $K = (k_{ij})$ and define

$$D(\lambda, K) = k_{11} \, \phi_{22}(b,a,\lambda) - k_{12} \, \phi_{21}(b,a,\lambda)$$
$$- k_{21} \, \phi_{12}(b,a,\lambda) + k_{22} \, \phi_{11}(b,a,\lambda) \qquad (3.2.11)$$

for $\lambda \in \mathbb{C}$. Note that $D(\lambda, K)$ does not depend on γ. Then

(1) *The complex number λ is an eigenvalue of BVP (3.2.1), (3.2.2), (3.2.3), (3.2.10) if and only if*

$$D(\lambda, K) = 2 \cos \gamma, \; -\pi < \gamma \leq \pi. \qquad (3.2.12)$$

(2) *If p, q, w are real valued and λ is an eigenvalue for $A = e^{i\gamma} K$, $B = -I$, $0 < \gamma < \pi$, with eigenfunction u, then λ is also an eigenvalue for $A = e^{-i\gamma} K$, $B = -I$, but with eigenfunction $\bar{u}$.*

PROOF. By (1.2.15) we have $\det \Phi(b,a,\lambda) = 1$. We abbreviate $(\phi_{ij}(b,a,\lambda))$ to ϕ_{ij}. By (3.2.7), (3.2.11) and noting that $\det K = 1$ we get

$$\delta(\lambda) = \det(e^{i\gamma} K - \Phi) = \begin{vmatrix} e^{i\gamma} k_{11} - \phi_{11} & e^{i\gamma} k_{12} - \phi_{12} \\ e^{i\gamma} k_{21} - \phi_{21} & e^{i\gamma} k_{22} - \phi_{22} \end{vmatrix}$$
$$= 1 + e^{2i\gamma} - e^{i\gamma} D(\lambda).$$

Hence $\delta(\lambda) = 0$ if and only if (3.2.12) holds. Part (2) follows from (3.2.12) and by taking conjugates of equation (3.2.1). □

REMARK 3.2.4. Although the matrices (3.2.10) determine self-adjoint boundary conditions (these are the canonical form of all coupled self-adjoint BC, see Chapters 4 and 10 below), no conditions other than (3.2.2) are assumed on p, q, w in Lemma 3.2.6 part (1). In particular no symmetry (formal self-adjointness) or definiteness assumption is made on equation (3.2.1). Thus the characterization of the eigenvalues given by (3.2.12) applies not only to so–called left-definite, right-definite and indefinite SLP but the coefficients p, q and the weight function w can be complex valued. Furthermore one or more of $1/p, q, w$ can be identically zero on one or more subintervals of J.

3. The Fourier Equation

Thomson, [Lord Kelvin] William (1824-1907)

Fourier is a mathematical poem.

We now pause to consider the simplest SLP for a number of reasons: (i) to illustrate some results of the previous section, (ii) to indicate some of the coming attractions, (iii) because a remarkable number of properties of SLP for the simplest SL equation have natural extensions to the general case.

Consider the equation

$$-y'' = \lambda y \text{ on } (a,b), \ -\infty \le a < b \le \infty, \ \lambda \in \mathbb{C}. \tag{3.3.1}$$

We include the case when one or both endpoints are infinite here, even though this is a singular problem, ($p = 1 = w$ are not in $L(J, \mathbb{R})$ if J is unbounded) and singular boundary value problems are discussed in Part 4, to highlight the interplay between regular and singular problems.

We claim that to fully understand regular SLP requires a perspective which includes the singular case.

Each infinite endpoint is in the LP case since the constant 1 is a non $L^2(J, \mathbb{R})$ solution for $\lambda = 0$ and J unbounded. Thus no boundary conditions are required or allowed at the LP endpoints $-\infty$ and $+\infty$ and there is one and only one self-adjoint realization, say S, of the equation (3.3.1) in the space $L^2(-\infty, \infty)$. The spectrum of S, $\sigma(S)$, contains no eigenvalues and coincides with the essential (continuous) spectrum $\sigma_e(S)$; we have

$$\sigma(S) = \sigma_e(S) = [0, \infty).$$

For definitions of the terms used in the above paragraph and for proofs of the stated facts the reader is referred to Part 4 and to ([**487**]) and ([**600**]).

In this case

$$P = \begin{bmatrix} 0, & 1 \\ 0, & 0 \end{bmatrix}, \ W = \begin{bmatrix} 0, & 0 \\ 1, & 0 \end{bmatrix}.$$

Since these are fixed for this example we will omit them in the notation for Φ in this section. This primary fundamental matrix $\Phi = \Phi(t, u, P, W, \lambda) = \Phi(t, u, \lambda)$ is determined as the unique solution of the initial value problem

$$\Phi' = (P - \lambda W) \Phi, \ \Phi(u) = I, \ t, u \in \mathbb{R}, \ \lambda \in \mathbb{C}.$$

To compute Φ we choose an analytic branch of the square root function $\sqrt{z}$ as follows:
$$\mu = \sqrt{\lambda} = r^{1/2} e^{\frac{1}{2} i \theta}, \text{ for } \lambda = r e^{i\theta}, \text{ with } -\pi < \theta \leq \pi,$$
and obtain
$$\Phi(t, u, \lambda) = \begin{bmatrix} \cosh(i\mu(t-u)), & \frac{1}{i\mu} \sinh(i\mu(t-u)) \\ i\mu \sinh(i\mu(t-u)), & \cosh(i\mu(t-u)) \end{bmatrix}, \quad (3.3.2)$$
$$t, u \in \mathbb{R}, \ \lambda \in \mathbb{C}, \ \lambda \neq 0, \ \mu = \sqrt{\lambda},$$
and
$$\Phi(t, u, 0) = \begin{bmatrix} 1, & t-u \\ 0, & 1 \end{bmatrix}, \ t, u \in \mathbb{R}. \quad (3.3.3)$$

Note that for fixed $t, u \in \mathbb{R}$ the fundamental matrix $\Phi(t, u, \lambda)$ is analytic at λ for each $\lambda \in \mathbb{C}$ including $\lambda = 0$. This is a consequence of the second proof of Theorem 1.2.1. (It can be confirmed directly from the series expansions of the hyperbolic *sinh* and *cosh* functions.)

For the convenience of the reader we now recall some definitions and properties of hyperbolic functions which will be used below; see Abramowitz and Stegun [2] for more details.

(1) $\sinh z = e^z - e^{-z}$ $z \in \mathbb{C}$, $2 \cosh z = e^z + e^{-z}$, $z \in \mathbb{C}$, $\tanh z = \sinh z / \cosh z$.
(2) $\sinh z = -i \sin iz$, $\cosh z = \cos iz$, $\tanh z = -i \tan iz$.
(3) $\sinh(z + 2k\pi i) = \sinh z$, $\cosh(z + 2k\pi i) = \cosh z$.
(4) $(\sinh z)' = \cosh z$, $(\cosh z)' = \sinh z$.
(5) $\sin z = 0$ if and only if $z = k\pi$, $k \in \mathbb{Z}$; $\cos z = 0$ if and only if $z = (2k+1)\pi/2$, $k \in \mathbb{Z}$.
(6) The general solutions of the equations $\sin x = z$, $\cos x = z$, $\tan x = z$ are given by, respectively:
(7) $x = (-1)^k \arcsin z + k\pi$, $k \in \mathbb{Z}$.
(8) $x = \pm \arccos z + 2k\pi$, $k \in \mathbb{Z}$.
(9) $x = \arctan z + k\pi$, $k \in \mathbb{Z}$.
(10) For $-1 \leq t \leq 1$, $\arcsin t$, $\arccos t$ are real and $-\pi/2 \leq \arcsin t \leq \pi/2$, $0 \leq \arccos t \leq \pi$.

Now let $-\infty < a < b < \infty$ and consider the two point boundary condition
$$AY(a) + BY(b) = 0, \quad A, B \in M_2(\mathbb{C}). \quad (3.3.4)$$
We discuss coupled and separated boundary conditions separately.
- Coupled BC. Let
$$A = \begin{bmatrix} 1, & 0 \\ 0, & 1 \end{bmatrix}, \ B = \begin{bmatrix} c, & 0 \\ 0, & d \end{bmatrix}. \quad (3.3.5)$$
For (3.3.5) the characteristic function is given by
$$\delta(\lambda) = 1 + cd + (c+d) \cosh(i\mu(b-a)), \ \mu = \sqrt{\lambda} \neq 0. \quad (3.3.6)$$

In all of the examples below the case $\lambda = 0$ needs to be checked separately since $\lambda = 0$ plays a special role in these formulas.

We now consider a number of special cases of (3.3.5).
(1) $c = -d$. Then $\delta(\lambda) = 1 - c^2$, a constant independent of λ. If this constant is zero then every complex number is an eigenvalue; if this constant is not zero, then no complex number is an eigenvalue. In particular we have

(a) For $c = 1$, $d = -1$ every complex number is an eigenvalue.
(b) For $c = -1$, $d = 1$ every complex number is an eigenvalue.
(c) For $c \in \mathbb{C}$, $c \neq 1$, $c \neq -1$, and $d = -c$ no complex number is an eigenvalue.

(2) $c \neq -d$. The characteristic equation for the eigenvalues is:
$$\cosh((i\mu(b-a))) = \cos(-\mu(b-a)) = -\frac{1+cd}{c+d} = h.$$

The roots of this equation are given by
$$-\mu(b-a) = \pm \arccos h + 2k\pi, \ k \in \mathbb{Z},$$
$$\arccos z = \int_z^1 \frac{dt}{(1-t^2)^{1/2}} = \pi/2 - \arcsin z,$$
$$\arcsin z = \int_0^z \frac{dt}{(1-t^2)^{1/2}},$$

and both of these complex integrals must be taken along a path which does not cross the real axis.

When h is real and $-1 \leq h \leq 1$ then the roots for $\mu(b-a)$ are real and we get
$$\mu(b-a) = \mp \arccos h - 2k\pi, \ k \in \mathbb{Z}.$$

From this and $\mu = s > 0$ we get
$$\mu(b-a) = \arccos h + 2k\pi, \ k \in N_0.$$

We now consider some special cases of this case:

(a) $c = i = d$. Here $h = 0$,
$$\mu(b-a) = \pi/2 + k\pi, \ k \in N_0$$
and the eigenvalues are
$$\lambda_n = \frac{(\pi/2 + n\pi)^2}{(b-a)^2}, \ n \in N_0.$$

(b) $c = -1 = d$. This is the self-adjoint periodic (P) case with $h = 1$. First note that $\lambda = 0$ is an eigenvalue for this case. Our search for the other eigenvalues leads to $\mu(b-a) = 2k\pi$, $k \in N$ and consequently the eigenvalues $\{\lambda_n^P : n \in N_0\}$ are given by:
$$0, \ \frac{(2\pi)^2}{(b-a)^2}, \ \frac{(2\pi)^2}{(b-a)^2}, \ \frac{(4\pi)^2}{(b-a)^2}, \ \frac{(4\pi)^2}{(b-a)^2}, \ \frac{(6\pi)^2}{(b-a)^2}, \ \frac{(6\pi)^2}{(b-a)^2}, \ \frac{(8\pi)^2}{(b-a)^2}, \ \cdots$$
with eigenfunctions u_n given by:
$$1, \ \sin(2\pi(t-a)), \ \cos(2\pi(t-a)), \ \sin(4\pi(t-a)), \ \cos(4\pi(t-a)),$$
$$\sin(6\pi(t-a)), \ \cos(6\pi(t-a)), \ \sin(8\pi(t-a)), \ \ldots$$

Note that $\lambda_0 = 0$ is simple and all other eigenvalues are double.

(c) $c = 1 = d$. This is the self-adjoint semi-periodic (S) case and gives $h = -1$. We note that $\lambda = 0$ is not an eigenvalue in this case. Proceeding as above we get
$$\mu(b-a) = \pi + 2k\pi = (2k+1)\pi, \ k \in N_0.$$

Hence the eigenvalues $\{\lambda_n^S : n \in N_0\}$ in this case are given by
$$\frac{(\pi)^2}{(b-a)^2}, \ \frac{(\pi)^2}{(b-a)^2}, \ \frac{(3\pi)^2}{(b-a)^2}, \ \frac{(3\pi)^2}{(b-a)^2}, \ \frac{(5\pi)^2}{(b-a)^2}, \ \frac{(5\pi)^2}{(b-a)^2}, \ \frac{(7\pi)^2}{(b-a)^2}, \ \cdots$$

with eigenfunctions u_n:

$$\sin(\pi(t-a)),\ \cos(\pi(t-a)),\ \sin(3\pi(t-a)),\ \cos(3\pi(t-a)),$$
$$\sin(5\pi(t-a)),\ \cos(5\pi(t-a)),\ \sin(7\pi(t-a)),\ \ldots$$

In this case all eigenvalues, including λ_0, are double.

(d) $c = 1/d$, $c \in \mathbb{R}$, $c \neq 0$. This gives $h = -(1+1)/(c+1/c)$. Since $c + \frac{1}{c} > 2$ if $c > 0$ and $c \neq 1$; $c + \frac{1}{c} < -2$ if $c < 0$ and $c \neq -1$, we have $-1 < h < 1$. Let

$$\pm t_0(c) = \arccos(h),\ 0 < t_0(c) < \pi.$$

Then the roots are given by

$$(b-a)\mu = \pm t_0(c) + 2k\pi,\ k \in N_0$$

and so the eigenvalues are

$$\lambda_n(c) = \frac{(\pm t_0(c) + 2n\pi)^2}{(b-a)^2},\ n \in N_0$$

(e) $c = e^{i\alpha} = d$, $0 < \alpha < \pi$. Here

$$h = -\frac{1+e^{2i\alpha}}{2e^{i\alpha}} = -\cos\alpha,\ t_0(\alpha) = \arccos(-\cos\alpha) = \pi - \alpha \in (0,\pi).$$

So

$$(b-a)\mu = \pi - \alpha + 2k\pi,\ k \in N_0,$$

and therefore

$$\lambda_n(\alpha) = \frac{(\pi - \alpha + 2n\pi)^2}{(b-a)^2},\ n \in N_0.$$

- Separated BC. Here we take

$$A = \begin{bmatrix} A_1, & A_2 \\ 0, & 0 \end{bmatrix},\ B = \begin{bmatrix} 0, & 0 \\ B_1, & B_2 \end{bmatrix},\ A_j, B_j \in \mathbb{C},\ j = 1, 2.$$

A direct calculation gives

$$-\delta(\lambda) = A_1 B_1 \Phi_{12}(b,a,\lambda) + A_1 B_2 \Phi_{22}(b,a,\lambda) - A_2 B_1 \Phi_{11}(b,a,\lambda) - A_2 B_2 \Phi_{21}(b,a,\lambda)$$
$$= (A_1 B_2 - A_2 B_1)\cosh(i\mu(b-a)) + (\frac{1}{i\mu} A_1 B_1 - i\mu A_2 B_2)\sinh(i\mu(b-a)).$$

Note that $\delta(\mu)$ is periodic with fundamental period $2k\pi i$, so if there is one eigenvalue then there is a countable infinity of them. For each eigenvalue there is only one linearly independent eigenfunction. We now consider some special cases. For each case we list the characteristic equation, its roots for μ and the corresponding eigenvalues. As in the earlier cases $\lambda = 0$ has to be checked independently.

(1) $A_1 B_2 - A_2 B_1 = 0$. Note that this case includes both the Dirichlet (D) and Neumann (N) boundary conditions. Here we have

$$\delta(\lambda) = (\frac{1}{i\mu} A_1 B_1 - i\mu A_2 B_2)\sinh((i\mu(b-a)).$$

To find all eigenvalues we proceed as follows: (i) we find all eigenvalues produced by the roots of the sinh factor, (ii) check to see if any of these roots yield a non-zero root of the first factor, and (iii) we check separately to see if $\lambda = 0$ is an

3. THE FOURIER EQUATION

eigenvalue. Clearly the first factor can produce at most one eigenvalue and that only in exceptional cases,

$$\sinh((i\mu(b-a))) = -i\sin((-\mu(b-a))) = 0,$$
$$-\mu(b-a) = (-1)^k \arcsin(0) + k\pi = k\pi, \ k \in \mathbb{Z}.$$

Since $\mu = s > 0$ we get the following eigenvalues from the periodic factor:

$$\lambda_n = \frac{(n\pi)^2}{(b-a)^2}, \ n \in \mathbb{N}.$$

(2) $A_1 = 1 = B_1$, $A_2 = 0 = B_2$. For these Dirichlet (D) BC $\lambda = 0$ is not an eigenvalue and also the first factor does not produce an eigenvalue; hence all the eigenvalues are given by

$$\lambda_n^D = \frac{((n+1)\pi)^2}{(b-a)^2}, \ n \in \mathbb{N}_0.$$

(3) $A_1 = 0 = B_1$, $A_2 = 1 = B_2$. These are the Neuman (N) BC:

$$-i\mu \sinh(i\mu(b-a)) = \mu \sin(-\mu(b-a)) = 0,$$
$$\mu(b-a) = k\pi, \ k \in \mathbb{Z}.$$

Since $\lambda = 0$ is also an eigenvalue in this case and the first factor does not produce an eigenvalue we get that

$$\lambda_n^N = \frac{(n\pi)^2}{(b-a)^2}, \ n \in \mathbb{N}_0.$$

Note that $\lambda_0^N(a,b) = 0$ for all a, b, $-\infty < a < b < \infty$.
(4) $A_1 = 1 = B_2$, $A_2 = 0 = B_1$.

$$\cosh(i\mu(b-a)) = \cos(-\mu(b-a)) = 0,$$
$$-\mu(b-a) = \pm \arccos(0) + 2k\pi, \ k \in \mathbb{Z},$$
$$\lambda_n^{DN} = \frac{(\pi/2 + 2n\pi)^2}{(b-a)^2}, \ n \in \mathbb{N}_0.$$

(5) $A_1 = 0 = B_2$, $A_2 = 1 = B_1$.

$$-\cosh(i\mu(b-a)) = -\cos((-\mu)(b-a)) = 0.$$

Here we have the same roots and hence the same eigenvalues as in the previous case

$$\lambda_n^{ND} = \frac{(\pi/2 + 2n\pi)^2}{(b-a)^2}, \ n \in \mathbb{N}_0.$$

(6) $A_1 B_2 - A_2 B_1 \neq 0$.

$$\coth(i\mu(b-a)) = \frac{(\frac{1}{\mu} A_1 B_1 - i\mu A_2 B_2)}{A_1 B_2 - A_2 B_1}.$$

The roots of this equation are not so easy to find explicitly since the unknown μ appears on both sides. However, numerical approximations can be obtained from a root finder code.

Some observations

(1) Let $\{a_k : k \in N\}$ be a decreasing sequence to $-\infty$; $\{b_k : k \in N\}$ an increasing sequence to $+\infty$ and let
$$E = \{\lambda_n^D(a_k, b_k) : n \in \mathbb{N}_0, \ k \in \mathbb{N}.\}$$
Then E is dense in $[0, \infty)$. Thus every point of the essential spectrum of the self-adjoint realization S of the Fourier equation on $(-\infty, \infty)$ is the limit of a sequence of eigenvalues of regular problems on the intervals (a_k, b_k), $k \in \mathbb{N}$. This illustrates a general result, to be discussed in Chapter 10, about the approximation of the spectrum of singular SLP by eigenvalues of a sequence of regular SLP. The spectrum of a singular problem cannot, in general, be approximated by any one regular problem, but we will see in Chapter 10 that quite often the spectrum of a given singular problem can be approximated quite well by the spectra *of a sequence of regular problems*.

(2) For any $n \in \mathbb{N}_0$ the Dirichlet eigenvalue λ_n^D is greater than or equal to λ_n for any other self-adjoint boundary condition. This also illustrates a general fact which is valid for arbitrary regular self-adjoint SLP and for singular problems with LCNO endpoints. See Section 4.8 below.

(3) For any fixed self-adjoint boundary condition (see Chapter 4 for a definition of self-adjoint boundary conditions) there is a smallest eigenvalue but not a largest; in fact for any self-adjoint boundary condition λ_n goes to ∞ asymptotically like n^2 i.e.
$$\frac{\lambda_n}{n^2} \to c \in \mathbb{R}, \quad \text{as} \quad n \to \infty.$$

(4) $\lambda_n^D(a, b) \to 0$ as $(b - a) \to \infty$ for each $n \in \mathbb{N}_0$.

(5) $\lambda_n^D(a, b) \to \infty$ as $(b - a) \to 0$ for each $n \in \mathbb{N}_0$.

(6) $\lambda_0^D(a, b, q) \to \infty$ as $(b - a) \to 0$ for $q(t) = c$, a constant. But

(7) $\lambda_0^N(a, b, q) \to c$ as $(b - a) \to 0$ for $q(t) = c$, a constant. Thus, as the length of the interval shrinks to zero the difference between the Dirichlet and Neuman eigenvalues goes to infinity. This is a general phenomenon of SLP found recently by Kong and Zettl, see [**392**], [**394**], [**135**], [**136**].

(8) When $p > 0$, $w > 0$ and the boundary conditions are separated and self-adjoint (see Chapter 4 below for a definition of self-adjoint boundary conditions, separated and coupled) the $n-th$ eigenfunction u_n is unique up to constant multiples and has exactly n zeros in the open interval (a, b) for each $n \in \mathbb{N}_0$.

(9) For coupled self-adjoint boundary conditions there may be geometrically double eigenvalues i.e. eigenvalues for which there are two linearly dependent eigenfunctions. In such cases, although each eigenvalue λ_n is uniquely defined by the ordering scheme $-\infty < \lambda_0 \leq \lambda_1 \leq \lambda_2 \leq \lambda_3 \leq \ldots$, there is some arbitrariness in the indexing of eigenfunctions for double eigenvalues. Consequently the exact number of zeros of u_n for general SLP with coupled boundary conditions cannot be specified exactly as in the separated case but see Section 4.8.

(10) The lowest eigenvalue of the Fourier equation with periodic boundary conditions is simple, all others are double; for semi-periodic boundary conditions all eigenvalues, including the first one are double. We will see below that this too is a general phenomenon : For general SLP with coupled self-adjoint boundary conditions determined by a matrix $K \in SL(2, \mathbb{R})$ either $\lambda_0(K)$ or $\lambda_0(-K)$ is simple. However, the fact that all other eigenvalues are double in the Fourier case with

$K = I$ is not typical. In fact it is extreme. In general double eigenvalues are not common.

(11) Example 5.9 on page 27 of [**285**] shows that the algebraic and geometric multiplicities for *non-self-adjoint* boundary conditions are not the same, in general, in contrast with the self-adjoint case.

4. The Space of Regular Boundary Value Problems

If two SLP are "close" to each other then are their eigenvalues and eigenfunctions also "close" to each other? To study the "closeness" of two BVP we introduce a " boundary value problem space" with a metric Let

$$J = (a', b'), \ -\infty \le a' < b' \le \infty,$$
$$\Omega = \{\omega = (a, b, A, B, 1/p, q, w)\} \quad (3.4.1)$$

such that

$$-\infty \le a' < a < b < b' \le \infty, \ A, B \in M_2(\mathbb{C}), \ 1/p, q, w \in L_{loc}(J, \mathbb{C}).$$

By an eigenvalue of $\omega \in \Omega$ we mean an eigenvalue of the BVP determined by ω. Note that, in this section, we have changed the notation for the endpoints of the interval J from a to a' and b to b'; this is so that we can use a and b to denote endpoints of varying subintervals of J.

Do the eigenvalues and eigenfunctions depend continuously on the problem? Does a small change of the problem result in a small change of each eigenvalue and each eigenfunction? To study these questions we want to compare the spectrum of different problems which may be defined on different intervals. Each $\omega \in \Omega$ determines a unique SLP: a, b the interval, A, B the boundary condition, and the restrictions of p, q, w to $[a, b]$ the equation. Observe that the values of p, q, w outside the interval $[a, b]$, i.e. in $(a', b') \setminus [a, b]$, do not affect the spectrum of the problem determined by ω. To account for this and to facilitate comparisons between eigenvalues of problems defined on different intervals we let

$$\widetilde{\Omega} = \{\widetilde{\omega} = (a, b, A, B, \widetilde{1/p}, \widetilde{q}, \widetilde{w})\} \quad (3.4.2)$$

where

$$\widetilde{q} = \begin{cases} q & \text{on } [a, b], \\ 0 & \text{otherwise,} \end{cases} \quad (3.4.3)$$

and $\widetilde{1/p}$, $\widetilde{w}$ are defined similarly.

For the topology of Ω we use a metric d defined as follows: (Note that Ω is not a vector space.) For

$$\omega = (a, b, A, B, 1/p, q, w) \in \Omega, \ \omega_0 = (a_0, b_0, A_0, B_0, 1/p_0, q_0, w_0) \in \Omega,$$

define

$$d(\omega, \omega_0) = |a - a_0| + |b - b_0| + ||A - A_0|| + ||B - B_0|| \quad (3.4.4)$$
$$+ \int_{a'}^{b'} \left(|\widetilde{\tfrac{1}{p}} - \widetilde{\tfrac{1}{p_0}}| + |\widetilde{q} - \widetilde{q_0}| + |\widetilde{w} - \widetilde{w_0}| \right)$$

where $||\cdot||$ denotes any matrix norm. Note that d is a metric: For any $\omega, \omega_0, \omega_1 \in \Omega$ we have

(1) $d(\omega, \omega_0) \ge 0$;
(2) $d(\omega, \omega_0) = 0$ if and only if $\omega = \omega_0$;

(3) $d(\omega, \omega_0) = d(\omega_0, \omega)$;
(4) $d(\omega, \omega_0) \leq d(\omega, \omega_1) + d(\omega_1, \omega_0)$.

5. Continuity of Eigenvalues and Eigenfunctions

The zeros of a polynomial are continuous functions of the coefficients. Analytic functions can be considered as polynomials of infinite degree since they have a power series expansion. Do the zeros of an analytic function depend continuously on its coefficients? Since there are, in general, an infinite number of coefficients we have to introduce a topology before a meaningful answer to this question can be given. Fortunately it turns out that a metric topology is sufficient to give an answer of yes. This is the content of the next lemma.

LEMMA 3.5.1 (Continuity of the zeros of an analytic function). *Let A be an open set in the complex plane $\mathbb{C}$, F a metric space, f a continuous complex valued function on $A \times F$ such that for each $\alpha \in F$, the map $z \to f(z, \alpha)$ is an analytic function on A. Let B be an open subset of A whose closure $\bar{B}$ in $\mathbb{C}$ is compact and contained in A, and let $\alpha_0 \in F$ be such that no zero of $f(z, \alpha_0)$ is on the boundary of B. Then there exists a neighborhood W of α_0 in F such that :*

(1) *For any $\alpha \in W$, $f(z, \alpha)$ has no zero on the boundary of B;*
(2) *for any $\alpha \in W$, the sum of the orders of the zeros of $f(z, \alpha)$ contained in B is independent of α.*

PROOF. See page 248 in Dieudonné [**138**]. □

The next result is an application of Lemma 3.5.1 to Sturm-Liouville problems.

THEOREM 3.5.1. *Let $\omega_0 = (a_0, b_0, A_0, B_0, 1/p_0, q_0, w_0) \in \Omega$. Assume that $\lambda(\omega_0)$ is an isolated eigenvalue of the problem (3.2.1), (3.2.2), (3.2.3). Then, given any $\epsilon > 0$, there exists a $\delta > 0$ such that if $\omega = \omega(a, b, A, B, 1/p, q, w) \in \Omega$ satisfies*

$$d(\omega, \omega_0) < \delta \tag{3.5.1}$$

then the SLP ω has an eigenvalue $\lambda(\omega)$ satisfying

$$|\lambda(\omega) - \lambda(\omega_0)| < \epsilon. \tag{3.5.2}$$

Furthermore, if $\lambda(\omega_0)$ is algebraically simple then there is exactly one algebraically simple $\lambda(\omega)$ satisfying (3.5.2); if $\lambda(\omega_0)$ is an algebraically double eigenvalue, then for ω satisfying (3.5.1) either (3.5.2) holds for an algebraically double eigenvalue of ω or for exactly two algebraically simple ones.

PROOF. See Kong and Zettl [**394**], Theorem 3.1. □

REMARK 3.5.1. The algebraic multiplicity of an eigenvalue of $\omega \in \Omega$ is the order of its zero as a root of the characteristic function discussed in Section 3.2. In Section 4.8 below it is shown that the geometric and algebraic multiplicities of eigenvalues of regular self-adjoint SLP are the same. Thus the word "algebraically" can be replaced by "geometrically" or both can be omitted entirely in Theorem 3.5.1 for self-adjoint problems with positive weight function and positive leading coefficient.

REMARK 3.5.2. Note that there is no index on λ in (3.5.2). In the general non-self-adjoint case the meaning of λ_n is far from clear; even in the self-adjoint case when the eigenvalues are real but unbounded above and below the meaning of λ_n is not clear. We will show in the next chapter that, in the self-adjoint case when the eigenvalues are bounded below and λ_n is defined in the usual way it is, in general, not a continuous function of the problem. In particular, for self-adjoint problems with p and w positive, λ_n does depend continuously on the coefficients and on the endpoints, but not on the boundary conditions. Thus in inequality (3.5.2) the index on the two eigenvalues may be different. Nevertheless in all cases by "jumping" the index (if there is an "index"), if necessary, each isolated eigenvalue can be embedded in a "continuous eigenvalue branch". In the next section we will see that, whether these continuous eigenvalue branches are defined by a constant index or not, each continuous eigenvalue branch is differentiable.

REMARK 3.5.3. Besides establishing the continuity of the eigenvalues within continuous eigenvalue branches for self-adjoint and non-self-adjoint SLP, Theorem 3.5.1 is also an existence theorem for eigenvalues. Given any self-adjoint SLP ω_0 with a countably infinite number of isolated eigenvalues, we can conclude from Theorem 3.5.1 that every SLP ω which is sufficiently close to ω_0, *whether it is self-adjoint or not*, must have n eigenvalues close to n eigenvalues of $\lambda(\omega_0)$, for any positive integer n. Note that it does not follow directly from Theorem 3.5.1 that ω must have an *infinite* number of eigenvalues close to those of ω_0.

REMARK 3.5.4. **Below when we speak of an isolated eigenvalue being embedded in a 'continuous eigenvalue branch' we will mean a continuous branch in the sense of Theorem 3.5.1.**

By a normalized eigenfunction of an SLP with a positive weight function w we mean an eigenfunction u that satisfies

$$\int_a^b |u|^2 w = 1. \tag{3.5.3}$$

Next we state a result for normalized eigenfunctions. Note that these are not uniquely determined. In the case of a simple eigenvalue with only real-valued eigenfunctions they are unique up to sign, but for a double eigenvalue there are pairs of linearly independent normalized eigenfunctions.

Below when considering SLP on two subintervals of J we extend the solutions and their quasi-derivatives continuously to the whole of the open interval J. This can be done by the existence-uniqueness theorem and is done to facilitate comparisons between solutions and their quasi-derivatives on different intervals.

THEOREM 3.5.2. *Let the notation and hypotheses of Theorem 3.5.1 hold and assume that $w_0 > 0$ on J.*

(i) Assume the eigenvalue $\lambda(\omega_0)$ is simple for some $\omega_0 \in \Omega$ and let $u(\cdot, \omega_0)$ denote a normalized eigenfunction of $\lambda(\omega_0)$. Then there is a neighborhood M of ω_0 in Ω such that $\lambda(\omega)$ is simple for every $\omega \in M$ with a positive weight function w, and there exist normalized eigenfunctions $u(\cdot, \omega)$ of $\lambda(\omega)$ for $\omega \in M$ such that

$$u(\cdot, \omega) \to u(\cdot, \omega_0), \quad (pu')(\cdot, \omega) \to (pu')(\cdot, \omega_0), \quad as \ \omega \to \omega_0 \ in \ \Omega, \tag{3.5.4}$$

both uniformly on any compact subinterval K of (a', b').

(ii) Assume $\lambda(\omega)$ is a double eigenvalue for all ω with positive weight functions in some neighborhood M of ω_0 in Ω. Let $u(\cdot, \omega_0)$ be any normalized eigenfunction of $\lambda(\omega_0)$. Then there exist normalized eigenfunctions $u = (\cdot, \omega)$ of $\lambda(\omega)$ such that (3.5.4) hold, both uniformly on any compact subinterval K of (a', b'). Note that

in this case, given two linearly independent normalized eigenfunctions $u_j(\cdot, \omega_0)$ of $\lambda(\omega_0)$ there exist a pair $u_j(\cdot, \omega)$ of linearly independent normalized eigenfunctions of $\lambda(\omega)$ such that $u_j(\cdot, \omega) \to u_j(\cdot, \omega_0)$ as $\omega \to \omega_0$ in Ω for $j = 1, 2$.

PROOF. See [394], Theorem 3.2. □

6. Differentiability of Eigenvalues

Hugo Rossi

> In the fall of 1972 President Nixon announced that the rate of increase of inflation was decreasing. This was the first time a sitting president used the third derivative to advance his case for reelection.

Mathematics Is an Edifice, Not a Toolbox, Notices of the AMS, v. 43, no. 10, October.

In this section we show that the continuous eigenvalue branches guaranteed by Theorem 3.5.1 are differentiable, even when they are not defined by a fixed index, and we find formulas for the derivatives

THEOREM 3.6.1. *Let* $\omega = (a, b, A, B, 1/p, q, w) \in \Omega$ *with* $w > 0$ *on* J. *Fix* a, b, A, B *and assume that* A, B *satisfy the self-adjointness conditions:*

$$A E A^* = B E B^*, \quad \text{rank}(A|B) = 2, \quad E = \begin{bmatrix} 0 & -1 \\ 1 & 0 \end{bmatrix}.$$

Suppose that $\lambda(\omega)$ *is an isolated simple eigenvalue of* ω. *Then there is a simple closed curve* Γ *in* $\mathbb{C}$ *with* $\lambda(\omega)$ *in its interior and a neighborhood* U *of* ω *in* Ω *such that for any* ρ *in* U *the BVP identified with* ρ *has exactly one eigenvalue inside* Γ *and this eigenvalue is simple. This map* $\lambda : U \to \mathbb{C}$ *is differentiable with respect to*

(1) q *with* p, w *fixed and its Frechet derivative is the bounded linear transformation given by*

$$\lambda'(q) h = \int_a^b |u(\cdot, q)|^2 h, \quad h \in L^1((a, b), \mathbb{C}) \tag{3.6.1}$$

where $u = u(\cdot, q)$ *is a normalized eigenfunction of* $\lambda(q)$;

(2) $1/p$ *with* q, w *fixed and its Frechet derivative is the bounded linear transformation given by*

$$\lambda'(1/p) h = -\int_a^b |(pu')(\cdot, 1/p)|^2 h, \quad h \in L^1((a, b), \mathbb{C}), \tag{3.6.2}$$

where $u = u(\cdot, 1/p)$ *is a normalized eigenfunction of* $\lambda(1/p)$;

(3) w *with* $1/p, q$ *fixed and its Frechet derivative is the bounded linear transformation given by*

$$\lambda'(w) h = -\lambda(w) \int_a^b |u(\cdot, w)|^2 h, \quad h \in L^1((a, b), \mathbb{C}), \tag{3.6.3}$$

where $u = u(\cdot, w)$ *is a normalized eigenfunction of* $\lambda(w)$.

PROOF. See [475]. □

THEOREM 3.6.2. *Fix $a, b \in J$, $a < b$. Let $\omega = (A, B, 1/p, q, w) \in \Omega = (M_2(C))^2 \times (L^1((a,b), \mathbb{C}))^3$ Assume that $\lambda(\omega)$ is an isolated simple eigenvalue of the SLP ω. Then there is a simple closed curve Γ in $\mathbb{C}$ with $\lambda(\omega)$ in its interior and a neighborhood U of ω in Ω such that for each $\varphi \in U$ the BVP identified with φ has exactly one eigenvalue $\lambda(\varphi)$ inside Γ and this eigenvalue is simple. This map $\lambda : U \to C$ is differentiable at ω and its derivative is the bounded linear transformation given by:*

$$\lambda'(\omega)\rho = \int_a^b \{-py'\bar{p}\bar{z}'(1/r) + [g - \lambda(\rho)v]y\bar{z}\} + d^*[CY(a) + DY(b)], \quad (3.6.4)$$

where $\rho = (C, D, 1/r, g, v) \in \Omega$, y, z are biorthogonal solutions of the given and its adjoint boundary value problems at $\lambda(\omega)$, i.e.

$$-(py')' + qy = \lambda(\omega) \, w \, y, \ AY(a) + BY(b) = 0, \ Y = \begin{bmatrix} y \\ py' \end{bmatrix} \quad (3.6.5)$$

$$-(\bar{p}z')' + \bar{q}z = \bar{\lambda}(\omega)\bar{w}z, \ Z = \begin{bmatrix} z \\ \bar{p}z' \end{bmatrix}, \ \int_a^b y\,\bar{z}\,w = 1, \quad (3.6.6)$$

and where $d \in \mathbb{C}^2$ is such that

$$Z(a) = EA^*d, \ Z(b) = -EB^*d, \ E = \begin{bmatrix} 0 & -1 \\ 1 & 0 \end{bmatrix}. \quad (3.6.7)$$

PROOF. See [**475**]. □

REMARK 3.6.1. Note that for Theorem 3.6.2 no self-adjointness hypothesis is needed: the coefficients p, q and the weight function w may change sign or even be complex-valued; the boundary conditions need not be self-adjoint. The existence of infinitely many eigenvalues for non-self-adjoint SLP under some strong additional hypotheses on the coefficients is well known for so called Birkhoff regular and Stone regular SLP; see Mennicken and Möller [**471**], and Eberhard and Freiling [**159**], Eberhard, Freiling and Zettl [**160**].

7. Finite Spectrum

In this section we construct SLP with exactly n eigenvalues for each non-negative integer n. We also show that these n eigenvalues can be arbitrarily distributed throughout the complex plane $\mathbb{C}$. More precisely : Given any k disjoint open sets Ω_i in $\mathbb{C}$ and any k integers n_i there exists a SLP with exactly n_k eigenvalues in Ω_i for $i = 1, 2, ..., k$. In the self-adjoint case much more is known, see Section 4.12 below. This construction involves coefficients $r = 1/p$ and q, w which are identically zero on adjacent subintervals. For such problems of 'Atkinson type' it is convenient to formulate equation (3.2.1) as the equivalent system:

$$y' = r\,z, \quad z' = (q - \lambda w)\,y \quad \text{on } J, \ z = py', \ r = \frac{1}{p}. \quad (3.7.1)$$

We consider two point boundary conditions (BC) of the form:

$$AY(a) + BY(b) = 0, \quad Y = \begin{bmatrix} y \\ py' \end{bmatrix}, \ A, B \in M_2(\mathbb{C}), \quad (3.7.2)$$

where $M_2(\mathbb{C})$ denotes the set of square matrices of order 2 over $\mathbb{C}$.

LEMMA 3.7.1. *Let (3.2.2) hold and let* $\Phi(t,\lambda) = [\phi_{ij}](t,\lambda)$ *denote the fundamental matrix of the system (3.7.1) determined by the initial condition* $\Phi(a,\lambda) = I$, $\lambda \in \mathbb{C}$. *Then (see Lemma 3.2.2) a complex number* λ *is an eigenvalue of the SLP (3.2.1), (3.7.2) (or (3.7.1) and (3.7.2)) if and only if*

$$\delta(\lambda) = \det[A + B\,\Phi(b,\lambda)] = 0.$$

Let

$$C = \begin{bmatrix} c_{11} & c_{12} \\ c_{21} & c_{22} \end{bmatrix} := \begin{bmatrix} a_{22}b_{11} - a_{12}b_{21} & a_{11}b_{21} - a_{21}b_{11} \\ a_{22}b_{12} - a_{12}b_{22} & a_{11}b_{22} - a_{21}b_{12} \end{bmatrix}. \tag{3.7.3}$$

Then for $\lambda \in \mathbb{C}$ *we have*

$$\begin{aligned}\delta(\lambda) = \det(A) + \det(B) &+ c_{11}\phi_{11}(b,\lambda) + c_{12}\phi_{12}(b,\lambda) \\ &+ c_{21}\phi_{21}(b,\lambda) + c_{22}\phi_{22}(b,\lambda).\end{aligned} \tag{3.7.4}$$

PROOF. The first statement is contained in Lemma 3.2.2; formula (3.7.4) can be verified with a straight-forward computation. □

The SLP (3.2.1), (3.7.2), or equivalently (3.7.1), (3.7.2), is said to be degenerate if in (3.7.4) either $\delta(\lambda) \equiv 0$ for all $\lambda \in \mathbb{C}$ or $\delta(\lambda) \neq 0$ for every $\lambda \in \mathbb{C}$. In the former case every complex number is an eigenvalue, in the latter there is no eigenvalue.

To generate SLP with finite spectrum we construct a characteristic function $\delta(\lambda)$ which is a polynomial in λ. This requires a detailed study of the structure of the principal fundamental matrix which we now embark upon.

We recall the basic condition (3.2.2)

$$r = 1/p,\ q,\ w \in L(J, \mathbb{C}),\ \lambda \in \mathbb{C}, \tag{3.7.5}$$

and partition the interval $J = (a,b)$ as follows

$$a = a_0 < a_1 < ... < a_{n-1} < a_n = b \tag{3.7.6}$$

for some odd positive integer $n = 2m+1$. Suppose that

$$\begin{aligned} r = 0 \text{ on } (a_{2i}, a_{2i+1}),\ q_{2i} = \int_{a_{2i}}^{a_{2i+1}} q,\ w_{2i} = \int_{a_{2i}}^{a_{2i+1}} w \neq 0,\quad & i \in \mathbb{N}_0,\ 2i+1 \leq n; \\ w = q = 0 \text{ on } (a_{2i+1}, a_{2i+2}),\ r_{2i+1} = \int_{a_{2i+1}}^{a_{2i+2}} r \neq 0,\quad & i \in \mathbb{N}_0,\ 2i+2 \leq n. \end{aligned} \tag{3.7.7}$$

We determine the structure of the principal fundamental matrix of system (3.7.1), under the conditions (3.7.6), (3.7.7).

LEMMA 3.7.2. *Let (3.7.5), (3.7.6), (3.7.7) hold. Let* $\Phi(t,\lambda) = [\phi_{ij}](t,\lambda)$ *be the fundamental matrix solution of the system (3.7.1) determined by the initial condition* $\Phi(a,\lambda) = I$ *for each* $\lambda \in \mathbb{C}$. *Then we have that*

$$\Phi(a_1, \lambda) = \begin{bmatrix} 1 & 0 \\ q_0 - \lambda w_0 & 1 \end{bmatrix}, \tag{3.7.8}$$

$$\Phi(a_3, \lambda) = \begin{bmatrix} 1 + (q_0 - \lambda w_0)r_1 & r_1 \\ (q_0 - \lambda w_0) + (q_2 - \lambda w_2) + (q_0 - \lambda w_0)(q_2 - \lambda w_2)r_1 & 1 + (q_2 - \lambda w_2)r_1 \end{bmatrix}, \tag{3.7.9}$$

and in general, for $i \leq m$,

$$\Phi(a_{2i+1}, \lambda) = \begin{bmatrix} 1 & r_{2i-1} \\ q_{2i} - \lambda w_{2i} & 1 + (q_{2i} - \lambda w_{2i})r_{2i-1} \end{bmatrix} \Phi(a_{2i-1}, \lambda). \quad (3.7.10)$$

PROOF. Observe from (3.7.7) that y is constant on each subinterval where r is identically zero and z is constant on every subinterval where both q and w are identically zero. The result follows from repeated applications of (3.7.7). □

Let $R = \Pi_{i=0}^{m-1} r_{2i+1}$, and note that $b = a_{2m+1}$. Then this structure of Φ and mathematical induction yield the following:

COROLLARY 3.7.1. *Let the hypotheses and notation of Lemma 3.7.2 hold. For the fundamental matrix Φ we have that*

$$\phi_{11}(b, \lambda) = R \, \Pi_{i=0}^{m-1}(q_{2i} - \lambda w_{2i}) + \tilde{\phi}_{11}(\lambda), \quad (3.7.11)$$

$$\phi_{12}(b, \lambda) = R \, \Pi_{i=1}^{m-1}(q_{2i} - \lambda w_{2i}) + \tilde{\phi}_{12}(\lambda), \quad (3.7.12)$$

$$\phi_{21}(b, \lambda) = R \, \Pi_{i=0}^{m}(q_{2i} - \lambda w_{2i}) + \tilde{\phi}_{21}(\lambda), \quad (3.7.13)$$

$$\phi_{22}(b, \lambda) = R \, \Pi_{i=1}^{m}(q_{2i} - \lambda w_{2i}) + \tilde{\phi}_{22}(\lambda). \quad (3.7.14)$$

where $\tilde{\phi}_{ij} = o(R)$ as $\min\{r_{2i+1} : i = 0, \ldots, m-1\} \to \infty$ for fixed q, w, and λ, $i, j = 1, 2$.

Now we construct regular SLP with general self-adjoint and non-self-adjoint BC which have exactly m eigenvalues for each $m \in \mathbb{N}$.

THEOREM 3.7.1. *Let $n = 2m + 1$ for $m \in \mathbb{N}$, and let (3.7.5), (3.7.6), (3.7.7) hold. Let $C = (c_{ij})$ be the matrix given by (3.7.3). Then:*
  (1) *If $c_{21} \neq 0$, then the SLP (3.7.1), (3.7.2) has exactly $m + 1$ eigenvalues λ_j, $j = 0, 1, \ldots, m$.*
  (2) *If $c_{21} = 0$ and $c_{11}w_0 + c_{22}w_{2m} \neq 0$, then the SLP (3.7.1), (3.7.2) has exactly m eigenvalues λ_j, $j = 0, 1, \ldots, m - 1$.*
  (3) *If $c_{21} = c_{11} = c_{22} = 0$, but $c_{12} \neq 0$, then the SLP (3.7.1), (3.7.2) has exactly $m - 1$ eigenvalues λ_j, $j = 0, 1, \ldots, m - 2$.*
  (4) *If neither of the above conditions holds, then the SLP (3.7.1), (3.7.2) either has l eigenvalues for $l \in \{1, \ldots, m-1\}$, or is degenerate.*

PROOF. We note from (3.7.7) that the degrees of $\phi_{11}(b, \lambda)$, $\phi_{12}(b, \lambda)$, $\phi_{21}(b, \lambda)$, and $\phi_{22}(b, \lambda)$ in λ are $m, m-1, m+1$, and m, respectively. Then the results follow from Lemma 3.7.1 and Corollary 3.7.1. □

As special cases of Theorem 3.7.1 we consider the SLP with the following boundary conditions:
1. Separated BC:

$$\begin{array}{l} A_1 y(a) + A_2 (py')(a) = 0, \quad A_1, A_2 \in \mathbb{C}, \quad (A_1, A_2) \neq (0, 0), \\ B_1 y(b) + B_2 (py')(b) = 0, \quad B_1, B_2 \in \mathbb{C}, \quad (B_1, B_2) \neq (0, 0). \end{array} \quad (3.7.15)$$

2. Coupled BC:

$$Y(b) = AY(a), \quad Y = \begin{bmatrix} y \\ py' \end{bmatrix}, \quad A \in M_2(\mathbb{C}). \tag{3.7.16}$$

The next corollary is an immediate consequences of Theorem 3.7.1.

COROLLARY 3.7.2. *Let the assumptions and notation of Theorem 3.7.1 hold.*
(i) The separated BVP (3.7.1), (3.7.15) has exactly $m+1$ eigenvalues if $A_2 B_2 \neq 0$, exactly m if $A_2 B_2 = 0$ and $(A_2, B_2) \neq (0,0)$, and exactly $m-1$ if $A_2 = B_2 = 0$.
(ii) The coupled BVP (3.7.1), (3.7.16) has exactly $m+1$ eigenvalues if $a_{12} \neq 0$, exactly m if $a_{12} = 0$ and $a_{22}w_0 + a_{11}w_{2m} \neq 0$, exactly $m-1$ if $a_{12} = a_{22} = a_{11} = 0$ and $a_{21} \neq 0$.

PROOF. It is easy to see that for the separated BC (3.7.15)

$$C = \begin{bmatrix} -A_2 B_1 & A_1 B_1 \\ -A_2 B_2 & A_1 B_2 \end{bmatrix},$$

and for the coupled BC (3.7.16)

$$C = \begin{bmatrix} a_{22} & -a_{21} \\ -a_{12} & a_{11} \end{bmatrix}.$$

Hence the conclusions follow from Theorem 3.7.1. $\square$

From Theorem 3.7.1 we know that for each positive integer m there are SLP with exactly m eigenvalues. The next theorem will show that these m eigenvalues can be located anywhere in the complex plane.

THEOREM 3.7.2. *Let the assumptions and notation of Theorem 3.7.1 hold. Given any k disjoint open sets $\mathcal{N}_i$ in $\mathbb{C}$ and any k integers n_i there exists a SLP with exactly n_i eigenvalues in $\mathcal{N}_i$, for $i = 1, 2, ..., k$.*

PROOF. Let $m = \sum_{i=1}^{k} n_i$ and $n = 2m + 1$. Construct a SLP in the form of (3.7.1) with separated BC (3.7.15) under the assumptions (3.7.5), (3.7.6), (3.7.7), and $A_1 = B_2 = 0$, $A_2 = B_1 = 1$. Then the characteristic function defined by (3.7.4) becomes

$$\delta(\lambda) = \phi_{11}(b, \lambda) = R \prod_{i=0}^{m-1}(q_{2i} - \lambda w_{2i}) + \tilde{\phi}_{11}(\lambda)$$

where $\tilde{\phi}_{11} = o(R)$ as $\min\{r_0, \ldots, r_{2m-1}\} \to \infty$ for fixed q, w, and λ. Since q and w can be chosen arbitrarily, we can choose them such that $\tilde{\delta}(\lambda) := \prod_{i=0}^{m-1}(q_{2i} - \lambda w_{2i})$ has exactly n_i roots in $\mathcal{N}_i$, and none on the boundary of $\mathcal{N}_i$, $i = 1, \ldots, k$. Choose $r_{2i+1}, i = 0, \ldots, m-1$ so large that

$$\left| \tilde{\phi}_{11}(\lambda) \right| < R \prod_{i=0}^{m-1} |q_{2i} - \lambda w_{2i}|.$$

Then it follows from Rouche's theorem that $\delta(\lambda)$ has exactly n_i roots in $\mathcal{N}_i$, $i = 1, \ldots, k$. $\square$

COROLLARY 3.7.3. *Let the assumptions and notation of Theorem 3.7.1 hold. For every $m \in \mathbb{N}$ there exists a SLP in the form of (3.7.1) and (3.7.2) with piecewise constant coefficients which has exactly m eigenvalues. Furthermore, given any k disjoint open sets $\mathcal{N}_i$ in $\mathbb{C}$ and any k integers n_i there exists a SLP with piecewise constant coefficients having exactly n_i eigenvalues in $\mathcal{N}_i$, for $i = 1, 2, ..., k$.*

PROOF. This follows from the preceding results and the observation that for hypothesis (3.7.7) to hold, p_i, q_i, w_i can be chosen to be constant in the i-th subinterval. □

We end this section with an example illustrating what happens in the extreme cases when either $1/p$ or q and w are zero on the whole interval J.

EXAMPLE 3.7.1. *Recall that the Sturm-Liouville equation (3.2.1) is equivalent to the system (3.7.1).*

(i) Suppose $r = 1/p \equiv 0$ on (a,b). Then every $\lambda \in \mathbb{C}$ is an eigenvalue of (3.7.1) with the Dirichlet boundary condition:
$$y(a) = 0 = y(b).$$
In fact, for any $\lambda \in \mathbb{C}$, the unique solution determined by the initial condition $y(a, \lambda) = 0$, $z(a, \lambda) = 1$ is a nontrivial solution satisfying these Dirichlet boundary conditions and hence an eigenfunction.

(ii) Suppose $w \equiv q \equiv 0$ on (a,b). Then every $\lambda \in \mathbb{C}$ is an eigenvalue of (3.7.1) with the Neumann boundary condition:
$$z(a) = 0 = z(b).$$
In fact, for any $\lambda \in \mathbb{C}$, the unique solution of (3.7.1) determined by the initial condition $y(a, \lambda) = 1$, $z(a, \lambda) = 0$ is a nontrivial solution satisfying the Neumann boundary condition and hence an eigenfunction.

8. Green's Function

In this section we construct Green's function for two point boundary value problems and discuss adjoint problems. Just as the concept of "adjointness" is important in matrix theory so also is it important in the theory of boundary value problems.

The construction of the Green's function given here is not the standard one found in most books or papers. The standard construction introduces, a priori, a recipe involving a jump discontinuity of the quasi-derivative along the diagonal. This seems artificial at first glance and can be avoided by constructing a matrix Green's function for the equivalent first order system first, then extracting the top right component to obtain a Green's function for the scalar problem [**491**]. The jump discontinuity of the quasi-derivative appears naturally in this construction as, of course, it must since the Green's function is unique.

Let
$$\frac{1}{p}, q, w, f \in L^1(J, \mathbb{C}), \quad J = (a, b), \quad -\infty \le a < b \le \infty, \quad \lambda \in \mathbb{C}, \tag{3.8.1}$$

$$P = \begin{bmatrix} 0 & 1/p \\ q & 0 \end{bmatrix}, \; W = \begin{bmatrix} 0 & 0 \\ w & 0 \end{bmatrix}, \; F = \begin{bmatrix} 0 \\ f \end{bmatrix}, \; Y = \begin{bmatrix} y \\ py' \end{bmatrix}, \; A, B \in M_2(\mathbb{C}). \tag{3.8.2}$$

The scalar boundary value problem
$$-(py')' + qy = \lambda w y + f, \quad AY(a) + BY(b) = 0 \tag{3.8.3}$$
is equivalent to the system boundary value problem
$$Y' = (P - \lambda W)Y + F, \quad AY(a) + BY(b) = 0. \tag{3.8.4}$$

Let $\Phi = \Phi(\cdot,\cdot,\lambda)$ be the primary fundamental matrix of the homogeneous system

$$Y' = (P - \lambda W)Y. \tag{3.8.5}$$

Note that $\Phi(t,u,\lambda) = \Phi(t,a,\lambda)\,\Phi(a,u,\lambda)$ for $a \le t, u \le b$. This follows from the representation $\Phi(t,u,\lambda) = Y(t)Y^{-1}(u)$ for any fundamental matrix solution Y of (3.8.5).

We suppress the dependence of Φ on P, W since these are fixed here. Recall from Section 1.5 that $\Phi(b,a,\lambda)$ exists.

THEOREM 3.8.1. *Let (3.8.1) to (3.8.5) hold. Let $\lambda \in \mathbb{C}$. Then the following three statements are equivalent:*

(1) *When $f = 0$ on J, the boundary value problem (3.8.3) (and consequently also (3.8.4)) has only the trivial solution.*
(2) *The matrix $[A + B\,\Phi(b,a,\lambda)]$ has an inverse.*
(3) *For every $f \in L^1(J,\mathbb{C})$ each of the problems (3.8.3), (3.8.4) has a unique solution.*

Furthermore, if $[A+B\,\Phi(b,a,\lambda)]^{-1}$ exists, and the matrix function K is defined by

$$K(t,u,\lambda) = \begin{cases} \Phi(t,a,\lambda)\,U(\lambda)\,\Phi(b,u,\lambda), & a \le t < u \le b, \\ \Phi(t,a,\lambda)U(\lambda)\,\Phi(b,u,\lambda) + \Phi(t,u,\lambda), & a \le u \le t \le b, \end{cases} \tag{3.8.6}$$

where

$$U(\lambda) = -[A + B\,\Phi(b,a,\lambda)]^{-1}B. \tag{3.8.7}$$

THEOREM 3.8.2. *Then for any $f \in L^1(J,\mathbb{C})$ the unique solution y of (3.8.3) and the unique solution Y of (3.8.4), respectively, are given by*

$$y(t) = \int_a^b K_{12}(t,u,\lambda)f(u)du, \quad a \le t \le b, \tag{3.8.8}$$

$$Y(t) = \int_a^b K(t,u,\lambda)F(u)du, \quad a \le t \le b. \tag{3.8.9}$$

Moreover, the Green's function K_{12} is continuous on $[a,b] \times [a,b]$ and is the unique continuous function on $[a,b] \times [a,b]$ with property (3.8.8), i.e. if (3.8.8) holds for every $f \in L^1(J,\mathbb{C})$ with K_{12} replaced by a continuous G_{12}, then $K_{12} = G_{12}$. We use the notation $K(\cdot,\cdot,\lambda,A,B,P,W)$ to indicate the dependence of K on $A, B, P, W,$ and λ.

PROOF. Recall from Chapters 1 and 2 that Y is a solution of

$$Y' = (P - \lambda W)Y + F \text{ on } J \tag{3.8.10}$$

if and only if y is a solution of

$$-(py')' + qy = \lambda wy + f \text{ on } J \tag{3.8.11}$$

where

$$Y = \begin{bmatrix} y \\ py' \end{bmatrix}.$$

For $C = \begin{bmatrix} c_1 \\ c_2 \end{bmatrix}$, $c_j \in \mathbb{C}$, determine a solution Y of (3.8.10) on J by the initial condition

$$Y(a,\lambda) = C.$$

8. GREEN'S FUNCTION

Then y is a solution of (3.8.11) determined by the initial conditions $y(a, \lambda) = c_1$, $(py')(a, \lambda) = c_2$. Note that
$$Y(t, \lambda) = \Phi(t, a, \lambda) C, \quad a \leq t \leq b,$$
and from the variation of parameters formula (1.3.4) we have
$$Y(t, \lambda) = \Phi(t, a, \lambda) C + \int_a^t \Phi(t, s, \lambda) F(s) ds, \quad a \leq t \leq b. \tag{3.8.12}$$
In particular,
$$Y(b, \lambda) = \Phi(b, a, \lambda) C + \int_a^b \Phi(b, s, \lambda) F(s) ds.$$
Note that the integral in (3.8.12) exists since $f \in L^1(J, \mathbb{C})$ implies that the solutions have finite limits at the endpoints by Theorem 1.5.2 and are therefore bounded in a neighborhood of each endpoint. Let $D(\lambda) = [A + B \, \Phi(b, a, \lambda)]$ and observe that
$$AY(a, \lambda) + BY(b, \lambda) = D(\lambda) C + B \int_a^b \Phi(b, s, \lambda) F(s) ds. \tag{3.8.13}$$

Recall from Chapter 1 that, when $f = 0$ on J, Y and y are nontrivial solutions if and only if C is not the zero vector. We see from (3.8.13) that, when $f = 0$ on J, (and therefore $F = 0$ on J) there is a nontrivial solution Y (and a nontrivial solution y of (3.8.3)) satisfying the boundary condition $AY(a, \lambda) + BY(b, \lambda) = 0$ if and only if $D(\lambda)$ is singular. It also follows from (3.8.13) that there is a unique solution Y satisfying the boundary condition $AY(a) + BY(b) = 0$ *for every* $f \in L^1(J, \mathbb{C})$ if and only if $D(\lambda)$ is nonsingular. Similarly there is a unique solution y satisfying the boundary condition $AY(a) + BY(b) = 0$ *for every* $f \in L^1(J, \mathbb{C})$ if and only if $D(\lambda)$ is nonsingular.

We now construct the Green's function. Assume that $D(\lambda)$ is nonsingular. Let
$$C = D^{-1}(\lambda)(-B) \int_a^b \Phi(b, s, \lambda) F(s) ds.$$
Then $AY(a, \lambda) + BY(b, \lambda) = 0$ and we get from (3.8.12)
$$Y(t, \lambda) = \Phi(t, a, \lambda)[D^{-1}(\lambda)(-B) \int_a^b \Phi(b, s, \lambda) F(s) ds] + \int_a^t \Phi(t, s, \lambda) F(s) ds$$
$$= \int_a^b \Phi(t, a, \lambda)[D^{-1}(\lambda)(-B) \Phi(b, s, \lambda) F(s) ds] + \int_a^t \Phi(t, s, \lambda) F(s) ds$$
$$= \int_a^b K(t, s, \lambda) F(s) ds, \quad a \leq t \leq b.$$

In the last step we used the property that $\Phi(b, s, \lambda) = \Phi(b, a, \lambda) \Phi(a, s, \lambda)$ and the definition of K.

Note that (3.8.8) follows from this identity by taking the upper right component.

To prove the moreover statement note that the Green's function K_{12} is continuous on $[a, b] \times [a, b]$ and assume that there is a continuous G_{12} on $[a, b] \times [a, b]$ such that (3.8.8) holds for every $f \in L^1(J, \mathbb{C})$. Then
$$0 = \int_a^b [K_{12}(t, s, \lambda) - G_{12}(t, s, \lambda)] f(s) ds, \quad a \leq t \leq b.$$

Assume that there is a point (h, k) in $[a, b] \times [a, b]$ such that $K_{12}(h, k, \lambda) \neq G_{12}(h, k, \lambda)$. By continuity and a change in notation if necessary we may assume that $K_{12} - G_{12}$ is positive on a square $[c, d] \times [u, v]$. By choosing a function f which has compact support $[c, d]$ and is positive we get a contradiction.

For additional information see Neuberger [**491**], and Zettl [**620**], [**622**], [**622**]. The latter three references contain extensions to higher order two point boundary value as well as to problems with multi-point boundary conditions. $\square$

REMARK 3.8.1. The matrix Green's function $K(t, u, \lambda)$ has a jump discontinuity when $t = u$. But since this jump is the identity matrix I, the Green's function $K_{12}(t, u, \lambda)$ of the scalar problem (and also $K_{21}(t, u, \lambda)$) is continuous on $[a, b] \times [a, b]$ even along the diagonal where $t = u$.

Next we prove some lemmas which will be used to establish the Green's function identity of adjoint problems

LEMMA 3.8.1. *Let P and W be defined by (3.8.1), (3.8.2) and E by (3.8.19). Then for any $\lambda \in \mathbb{C}$ we have*

$$\overline{P} - \overline{\lambda W} = E(P - \lambda W)^* E. \qquad (3.8.14)$$

PROOF. This can be verified by a direct computation. $\square$

LEMMA 3.8.2 (The Adjoint Identity). *Let (3.8.1) hold with $f = 0$ on J; let P, W, Y be defined by (3.8.2), let $\Phi(\cdot, \cdot, \lambda, P, W)$ be the primary fundamental matrix of (3.8.5) and let E be given by (3.8.19). Then for any $\lambda \in \mathbb{C}$ we have*

$$\Phi(t, u, \lambda, P, W) = -E\, \Phi^*(u, t, \overline{\lambda}, \overline{P}, \overline{W})\, E, \quad a \leq t, u \leq b. \qquad (3.8.15)$$

PROOF. This follows from Theorem 1.8.1, the 'Adjoint Identity'. $\square$

LEMMA 3.8.3. *Let the hypotheses and notation of Theorem 3.8.1 hold. Then for all t, u, with $a \leq t, u \leq b$ we have*

$$K(t, u, \lambda, A, B, P, W) - E K^*(u, t, \overline{\lambda}, C, D, \overline{P}, \overline{W}) E$$
$$= \Phi(t, a, \lambda, P, W)\, H(\lambda)\, \Phi(a, u, \lambda, P, W), \qquad (3.8.16)$$

where

$$H(\lambda) = U(\lambda) - EV^*(\lambda)E - I, \qquad (3.8.17)$$
$$V(\lambda) = -[C + D\, \Phi(b, a, \overline{\lambda}, \overline{P}, \overline{W})]^{-1} D\, \Phi(b, a, \overline{\lambda}, \overline{P}, \overline{W}). \qquad (3.8.18)$$

PROOF. This follows from Lemma 3.8.2 and the definition of the matrix Green's functions, see formula (3.8.6). $\square$

LEMMA 3.8.4. *Let the hypotheses and notation of Lemma 3.8.3 hold. Then $H(\lambda) = 0$ if and only if*

$$AEC^* = BED^*, \quad E = \begin{bmatrix} 0 & -1 \\ 1 & 0 \end{bmatrix}. \qquad (3.8.19)$$

PROOF. Note that $E^{-1} = -E = E^*$. Now $H(\lambda) = 0$ is equivalent to

$$I = U(\lambda) - EV^*(\lambda)E$$
$$= -[A + B\, \Phi(b, a, \lambda, P, W)]^{-1} B\, \Phi(b, a, \lambda, P, W)$$
$$+ E\{[C + D\, \Phi(b, a, \overline{\lambda}, \overline{P}, \overline{W})]^{-1} D\, \Phi(b, a, \overline{\lambda}, \overline{P}, \overline{W})\}^* E.$$

Multiplying on the left by $A + B\,\Phi(b,a,\lambda,P,W)$ and on the right by $E\,[C + D\,\Phi(b,a,\overline{\lambda},\overline{P},\overline{W})]^*$ we obtain the equivalent identity

$$B\,\Phi(b,a,\lambda,P,W)E[C + D\,\Phi(b,a,\overline{\lambda},\overline{P},\overline{W})]^*$$
$$- [A + B\,\Phi(b,a,\lambda,P,W)]E\,\Phi^*(b,a,\overline{\lambda},\overline{P},\overline{W})D^*$$
$$= [A + B\Phi(b,a,\lambda,P,W)]E[C + D\,\Phi(b,a,\overline{\lambda},\overline{P},\overline{W})]^*$$

Using the Adjoint Identity, see Theorem 1.8.1, the property $\Phi(b,a)\Phi(a,b) = \Phi(b,b) = I$ and canceling like terms the last identity reduces to the first part of (3.8.19). □

THEOREM 3.8.3 (Adjoint Matrix Green's Functions). *Let (3.8.1) to (3.8.5) hold, let $C, D \in M_2(\mathbb{C})$. Assume that*

$$[A + B\,\Phi(b,a,P,W,\lambda)]^{-1} \text{ and } [C + D\,\Phi(b,a,\overline{P},\overline{W},\overline{\lambda})]^{-1}$$

both exist for some $\lambda \in \mathbb{C}$. Then

$$K(t,u,\lambda,A,B,P,W) = EK^*(u,t,\overline{\lambda},C,D,\overline{P},\overline{W})E, \quad a \leq t,u \leq b, \qquad (3.8.20)$$

if and only if

$$AEC^* = BED^*, \quad E = \begin{bmatrix} 0 & -1 \\ 1 & 0 \end{bmatrix}.$$

PROOF. This follows directly from Lemmas 3.8.3 and 3.8.4. □

THEOREM 3.8.4 (Adjoint Green's Functions). *Let (3.8.1) to (3.8.5) hold, let $C, D \in M_2(\mathbb{C})$. Assume that $[A+B\,\Phi(b,a,P,W,\lambda)]^{-1}$ and $[C+D\,\Phi(b,a,\overline{P},\overline{W},\overline{\lambda})]^{-1}$ both exist for some $\lambda \in \mathbb{C}$. Then*

$$K_{12}(t,u,\lambda,A,B,P,W) = \overline{K}_{12}(u,t,\overline{\lambda},C,D,\overline{P},\overline{W}), \quad a \leq t,u \leq b. \qquad (3.8.21)$$

THEOREM 3.8.5. *If and only if*

$$AEC^* = BED^*, \quad E = \begin{pmatrix} 0 & -1 \\ 1 & 0 \end{pmatrix}.$$

PROOF. Note that the upper right component of identity (3.8.20) gives (3.8.21). Thus the sufficiency of condition (3.8.11) follows directly from Theorem 3.8.2. To prove the necessity assume that (3.8.21) holds for all t, u in the given range. Then $H(\lambda) = 0$ follows since for fixed u, $\phi_{11}(t,a)$ and $\phi_{12}(t,a)$ are linearly independent functions of t and, on the other hand, $\phi_{12}(a,u)$ and $\phi_{22}(a,u)$ are linearly independent functions of u. □

9. Comments

These are made separately for each section.

- (1) Much less is known about non-self-adjoint problems than self-adjoint ones. In the self-adjoint case the well developed and beautiful theory of self-adjoint operators in Hilbert space can be applied. The theory of non-self-adjoint operators, particularly the spectral theory, is rather fragmentary compared to the self-adjoint case.

(2) The transcendental characterization of the eigenvalues given by Lemma 3.2.2 is elementary. The characterization of Lemma 3.2.6 for coupled self-adjoint boundary conditions but a general non-symmetric (not formally self-adjoint) equation is also elementary. The latter is given in terms of the canonical representation: $Y(b) = e^{i\gamma} K Y(a)$, $K \in SL_2(\mathbb{R})$. This representation, although known (it is mentioned without a proof or reference to a proof in the introduction of the book on inverse spectral theory by Pöschel and Trubowitz ([**522**]), has not been widely exploited. Since the boundary conditions are invariant under multiplication by a nonsingular matrix, when studying the continuity of the eigenvalues as functions of the boundary conditions it is useful to have a canonical representation for these.

In principle the characterization of the eigenvalues as roots of a transcendental equation can be used to compute them numerically. Use an ode solver to compute the transcendental functions $\delta(\lambda)$ or $D(\lambda)$ and then use a root finder. There are two major difficulties associated with this scheme: (i) there are, in general e.g. in the self-adjoint case, infinitely many eigenvalues and therefore infinitely many roots. Which one should the root finder search for? If it succeeds in finding one which one is it? In the self-adjoint case when p and w are positive this question is answered by the indexing scheme - the eigenvalues can be indexed uniquely by $\mathbb{N}_0$. How should the eigenvalues be indexed in the non-self-adjoint case? Much more research is needed before this question has a "natural" answer.

Canonical forms for non-self-adjoint boundary conditions have recently been introduced by Haertzen, Kong, Wu and Zettl in [**285**], [**286**]. These authors also introduce a geometric and topological structure on the space of boundary conditions and apply methods of algebraic geometry to the study of SLP. This seems to be a fresh approach to this very old subject.

(3) It is remarkable how many properties of the simplest SLP carry over to the general case. So we have included an extensive, but not comprehensive, discussion of the Fourier equation.

(4) As mentioned in Chapter 2 we believe that the L^1 norm is the natural norm to use for the study of regular SLP, so it seems to us that the metric space space Ω introduced here provides a natural setting for the investigation of the dependence of eigenvalues and eigenfunctions on all parameters of the problem. In ([**522**]) the authors study regular problems with $p = 1 = w$ and assume that $q \in L^2(J, \mathbb{R})$ for J a compact interval. Can the hypothesis $q \in L^2(J, \mathbb{R})$ be replaced by $q \in L^1(J, \mathbb{R})$?

(5) The continuity results here are due to Haertzen, Kong, Wu and Zettl [**384**], [**383**], [**285**], [**286**]. As already mentioned in this section the continuity result for eigenvalues is also an existence result of perturbation type: Two problems which are close to each other must have (at least a finite number) of eigenvalues close to each other. So if one problem is known to have eigenvalues so must the other. In particular, if one problem is self-adjoint with an infinite number of real eigenvalues then all nearby problems must have at least a finite number of eigenvalues close to the eigenvalues of the given self-adjoint problem. These eigenvalues may not be real unless the

nearby problem is also self-adjoint. Can this be extended to an infinite number of eigenvalues? If a non-self-adjoint problem is sufficiently close to a self-adjoint one must it have an eigenvalue close to every one of the self-adjoint eigenvalues?

(6) The differentiability results discussed here are due to Möller and Zettl [**475**] and to Kong and Zettl [**392**], [**394**]. The differentiability results with respect to the endpoints were inspired by earlier such results of Dauge and Helffer [**135**]. For extensions of all these results to higher order ode's see Kong, Wu and Zettl [**383**]. The paper of Möller and Zettl [**475**] shows that simple eigenvalues of Fredholm operators in Banach spaces are differentiable functions of the operator. This has wide applications to e.g. matrix theory, ordinary differential equations, partial differential equations, etc.

(7) The construction of self-adjoint and non-self-adjoint Sturm-Liouville problems with discrete spectrum consisting of exactly n eigenvalues for each positive integer n is due to Kong, Wu and Zettl [**391**] . Although Atkinson [**21**] hints at the existence of regular Sturm-Liouville problems with only a finite number of eigenvalues he gives no example. As far as we know the construction given here provides the first such examples. This construction yields non-self-adjoint and self-adjoint SLP with exactly n eigenvalues for each nonnegative integer n. Furthermore Theorem 3.7.2 shows that these n eigenvalues can be arbitrarily distributed throughout the complex plane (the real line in the self-adjoint case). In the self-adjoint case we know how - for a fixed equation - the eigenvalues change when the boundary conditions vary. See Section 4.12 for further results in the self-adjoint case, particularly regarding inequalities among eigenvalues for varying boundary conditions. Little seems to be known about effect on the eigenvalues when the boundary conditions are varied in the non-self-adjoint case but see Haertzen, Kong, Wu and Zettl [**285**]. Even less is known about the effect on eigenvalues for fixed boundary conditions when the coefficients are varied, but see [**522**]. In the self-adjoint case it is known that the $n-th$ eigenvalue depends continuously on the coefficients, see Chapter 4 below. Also see Chapter 4 for results on the, in general discontinuous, dependence of the $n-th$ eigenvalue on the boundary conditions.

The following questions remain open:
- (i) Given any n distinct points in the complex plane, is there a SLP whose eigenvalues are exactly these points?
- (ii) Given any n points in the complex plane no one of which is repeated more than once, is there a SLP whose eigenvalues are exactly these points?
- (iii) Given any n distinct points on the real line, is there a self-adjoint SLP whose eigenvalues are exactly these points? (See the Comments Section at the end of Chapter 5 for an answer to this question.) Does inverse spectral theory shed any light on this?
- (iv) Given any n points on the real line, no one of which is repeated more than once, is there a self-adjoint SLP whose eigenvalues are

exactly these points? Does inverse spectral theory shed any light on this?

(8) As already mentioned the construction of the Green's function given here is not the standard one, [**491**]. We believe it is more natural and instructive than the standard construction given e.g. in Coddington and Levinson [**123**]. This construction extends naturally to multi-point problems [**620**], [**622**] and, as we will see in Part 4, also to the singular case.

Eberhard and Freiling [**159**], Eberhard, Freiling and Zettl [**160**], Mennicken and Möller [**471**], have established the existence of infinitely many eigenvalues for the symmetric equation (3.2.1) but with non-self-adjoint BC. (Eberhard, Freiling and Eberhard, Freiling and Zettl also allow complex-valued coefficients.) We have

THEOREM 3.9.1 (Mennicken-Möller). *Assume that*

$$p, q, w \in L^1(J, \mathbb{R}), \quad J = [a, b], \quad -\infty < a < b < \infty, \quad p > 0, \ w > 0,$$
$$p = w \in AC(J), \ p'/p, q/w \in L^r(J, \mathbb{R}), \ 1 < r \leq \infty.$$

Then the BVP (3.2.1), (3.2.9) with $(A_1, A_2) \neq (0,0) \neq (B_1, B_2)$ has a countably infinite number of eigenvalues and the BVP (3.2.1) together with the coupled BC,

$$Y(b) = A Y(a)$$

has a countably infinite number of eigenvalue for any $A \in M_2(\mathbb{C})$ satisfying either $a_{12} \neq 0$ or $a_{12} = 0$ and $a_{11} + a_{22} \neq 0$.

PROOF. See [**471**], [**159**]. □

These authors also establish an expansion theorem for these cases. Does this result hold under less restrictive conditions on the real-valued coefficients? Is there a similar result for complex valued coefficients?

CHAPTER 4

Regular Self-Adjoint Problems

Ellis, Havelock
> It is here [in mathematics] that the artist has the fullest scope of his imagination.

The Dance of Life.

1. Introduction

In this chapter we specialize to the self-adjoint case. This case has been studied intensively since 1836-1837 when the seminal papers of Sturm and Liouville appeared. Much more is known about self-adjoint problems than non-self-adjoint ones. We try to strike a balance between discussing basic facts and recent developments.

A self-adjoint SLP consists of the symmetric differential equation

$$-(py')' + qy = \lambda w y \quad \text{on} \quad J = (a,b), \ -\infty \leq a < b \leq \infty, \quad (4.1.1)$$

with coefficients satisfying:

$$r = 1/p,\ q,\ w \in L^1(J, \mathbb{R}); \quad (4.1.2)$$

and boundary conditions

$$AY(a) + BY(b) = 0,\ Y = \begin{bmatrix} y \\ py' \end{bmatrix}, \quad (4.1.3)$$

where

$$A, B \in M_2(\mathbb{C}),\ A E A^* = B E B^*, \ rank\,(A:B) = 2,\ E = \begin{bmatrix} 0 & -1 \\ 1 & 0 \end{bmatrix}. \quad (4.1.4)$$

REMARK 4.1.1. For the convenience of the readers who may not be interested in non-self-adjoint problems and thus may not have read the first three chapters in detail, we repeat the remarks made earlier that condition (4.1.2) does not restrict the sign of any of the coefficients r, q, w. Also, each of $r = 1/p$, q, w is allowed to be identically zero on subintervals of J. If r is identically zero on a subinterval I, then there exist solutions y which are identically zero on I but whose quasi-derivative $z = py'$ is a nonzero constant on I. A maximal such interval of zeros is counted as a single zero in the results on the numbers of zeros of eigenfunctions in this chapter.

DEFINITION 4.1.1. *By a trivial solution of equation (4.1.1) on some interval I we mean a solution y which is identically zero on I and whose quasi-derivative $z = py'$ is also identically zero on I.*

DEFINITION 4.1.2. *A real valued measurable function f is said to change sign on an interval I if it assumes positive values on a set of positive Lebesgue measure and also assumes negative values on a set of positive Lebesgue measure.*

REMARK 4.1.2. Recall from Chapter 2 that condition (4.1.2) guarantees that the boundary condition (4.1.3) is well defined since all solutions and their quasi-derivatives can be continuously extended to the endpoints. This for bounded and unbounded intervals J.

REMARK 4.1.3. The terms "symmetric" differential equation and "self-adjoint" boundary conditions are taken from the theory of differential operators in the Hilbert space $H = L^2(J, w)$, see Chapter 10, or [**487**], [**600**]. This is the beautiful 'GKN' theory, named after Glazman, Krein and Naimark. The assumption that $w > 0$ on J ensures that H is a Hilbert space. Although we will use weaker assumptions on w in this chapter we nevertheless adopt this terminology for convenience. It may be a little misleading when w changes sign, since Richardson [**546**], see also Mingarelli [**472**], Binding and Volkmer [**80**], has pointed out that then there are problems (4.1.1) to (4.1.4) with nonreal eigenvalues.

The next theorem shows that eigenfunctions u of nonreal eigenvalues cannot be normalized by the requirement that
$$\int_a^b |u|^2 w = 1.$$

THEOREM 4.1.1. *Let (4.1.1) to (4.1.4) hold. If λ is a nonreal eigenvalue of (4.1.1) to (4.1.4) and u is an eigenfunction of λ, then*
$$\int_a^b |u|^2 w = 0. \tag{4.1.5}$$

PROOF. The self-adjoint boundary conditions (4.1.3), (4.1.4) are either separated or coupled and, if coupled, can be put into the canonical form (see Section 4.2 below)
$$B = -I, \quad A = e^{i\gamma}K, \quad K \in SL_2(\mathbb{R}), \quad -\pi < \gamma \leq \pi.$$
Assume that $\lambda \in \mathbb{C}$ is a nonreal eigenvalue for a coupled boundary condition in this canonical form. Multiply the equation of (4.1.1) by $\overline{y}$ and its conjugate equation by y, and subtract to get
$$(\lambda - \overline{\lambda})y\overline{y}w = -\overline{y}(py')' + y(p\overline{y}')' = [y(p\overline{y}') - \overline{y}(py')]' = [y, \overline{y}]'.$$
Noting that $K^*EK = E$ by (4.1.4) we get
$$(\lambda - \overline{\lambda})\int_a^b |y|^2 w = [y, \overline{y}](b) - [y, \overline{y}](a) = (EY(b), Y(b))) - (EY(a), Y(a))$$
$$= (Ee^{i\gamma}KY(a), e^{i\gamma}KY(a)) - (EY(a), Y(a))$$
$$= (K^*EKY(a), Y(a)) - (EY(a), Y(a)) = 0.$$

Hence (4.1.5) holds since λ is nonreal. For the case of separated boundary conditions the proof is similar; for this case we can see that the right-hand side of the equation is zero by an even more direct substitution. □

REMARK 4.1.4. Clearly (4.1.5) may hold when w changes sign but cannot hold under the classical assumption that $w > 0$ a.e. on J. This positivity condition on w can be weakened considerably to allow $w \geq 0$, see Theorems 4.3.2, 4.3.3. 4.3.4 below. Also (4.1.5) cannot hold when w changes sign but the problem is left-definite; in this case the eigenvalues are all real as we will see in Chapter 5 below.

2. Canonical Forms of Self-Adjoint Boundary Conditions

The boundary condition (4.1.3) is clearly invariant under multiplication by a nonsingular matrix. In preparation for the investigation of how eigenvalues change when the boundary condition is changed we discuss canonical forms of self-adjoint boundary conditions in this section.

For our purposes here it is convenient to divide the self-adjoint boundary conditions into three mutually exclusive subclasses and to use the following canonical representations of these subclasses:

(1) **Separated self-adjoint BC.** These are

$$A_1 y(a) + A_2 (py')(a) = 0, \; A_1, A_2 \in \mathbb{R}, \; (A_1, A_2) \neq (0,0), \qquad (4.2.1)$$

$$B_1 y(b) + B_2 (py')(b) = 0, \; B_1, B_2 \in \mathbb{R}, \; (B_1, B_2) \neq (0,0). \qquad (4.2.2)$$

These separated conditions can be parameterized as follows:

$$\cos\alpha \, y(a) - \sin\alpha \, (py')(a) = 0, \; 0 \le \alpha < \pi, \qquad (4.2.3)$$

$$\cos\beta \, y(b) - \sin\beta \, (py')(b) = 0, \; 0 < \beta \le \pi. \qquad (4.2.4)$$

Note the different normalization in (4.2.4) for β than that used for α in (4.2.3). This is for convenience in stating some of the results below. It is also the normalization used by the code SLEIGN2.

(2) **All *real* coupled self-adjoint BC.** These can be formulated as follows:

$$Y(b) = K Y(a), \quad Y = \begin{bmatrix} y \\ (py') \end{bmatrix} \qquad (4.2.5)$$

where $K \in SL_2(\mathbb{R})$, i.e. K satisfies

$$K = \begin{bmatrix} k_{11} & k_{12} \\ k_{21} & k_{22} \end{bmatrix}, \; k_{ij} \in \mathbb{R}, \; \det K = 1. \qquad (4.2.6)$$

(3) **All *complex* coupled self-adjoint BC.** These are:

$$Y(b) = e^{i\gamma} K Y(a), \quad Y = \begin{bmatrix} y \\ (py') \end{bmatrix} \qquad (4.2.7)$$

where K satisfies (4.2.6) and $-\pi < \gamma < 0$, or $0 < \gamma < \pi$.

LEMMA 4.2.1. *Given a boundary condition (4.1.3) with matrices A, B satisfying (4.1.4), it is equivalent to exactly one of the separated, real coupled, or complex coupled boundary conditions defined above.*

PROOF. This is postponed to Theorem 10.4.3. The Hilbert space operator theory characterization (the GKN theory) of self-adjoint realizations of general Sturm-Liouville equations is given in Section 10.4 and this theory is then applied to both regular and singular problems. In the regular case considered here it shows that conditions (4.1.4) characterize *all* self-adjoint realizations of equation (4.1.1) under condition (4.1.2) with $w > 0$. Thus explaining where the self-adjointness conditions (4.1.4) 'come from'. The regular canonical representations stated above are then proven and singular canonical forms are also established. The GKN theory gives a precise meaning to the phrase 'self-adjoint Sturm-Liouville problem'. □

3. Existence of Eigenvalues

In this section we focus on the existence of eigenvalues of self-adjoint SLP; these consist of equation (4.1.1) and boundary conditions (4.1.3) but with some restrictions on the coefficients in addition to (4.1.2) - no additional conditions are required, besides (4.1.4), on the boundary conditions. The reasons for needing more requirements on the coefficients are not only to avoid pathological cases such as $r = 1/p$, or w being identically zero on the underlying interval J but also to avoid nonreal eigenvalues. As mentioned in Section 4.1 Richardson pointed out in 1918 that even for the classical case when $p > 0$, but w changes sign on J, the SLP (4.1.1), (4.1.2) may have nonreal eigenvalues even for the Dirichlet boundary condition. In Chapter 5 we give an upper bound on the number of nonreal eigenvalues and find sufficient conditions for all eigenvalues to be real even when w changes sign.

We start with the basic existence theorem for the case when (4.1.2) holds and $w > 0$ on J. The conditions on w and p are weakened in subsequent theorems. Note that q may be positive, negative or change sign in these theorems.

THEOREM 4.3.1. *Consider the self-adjoint SLP consisting of equation (4.1.1) with coefficients satisfying (4.1.2), and the boundary conditions (4.1.3), (4.1.4). Assume that p is real-valued on J and*

$$w > 0 \text{ a.e. on } J.$$

(The boundary conditions may be separated or coupled, and in the latter case either real coupled or complex coupled. Note that no sign restriction is placed on p or q). Then:

(1) *All eigenvalues are real, isolated with no finite accumulation point, and there are an infinite but countable number of them.*
(2) *If p changes sign on J, then the eigenvalues are unbounded below and above and can be ordered to satisfy*

$$\ldots \leq \lambda_{-2} \leq \lambda_{-1} \leq \lambda_0 \leq \lambda_1 \leq \lambda_2 \leq \ldots \qquad (4.3.1)$$

with $\lambda_n \to +\infty$, $\lambda_{-n} \to -\infty$, as $n \to \infty$. Each eigenvalue may be geometrically simple or double but there cannot be two consecutive equalities in (4.3.1) since, for any value of λ, equation (4.1.1) has exactly two linearly independent solutions.
(3) *If p changes sign on J and the boundary conditions are separated then strict inequality holds everywhere in (4.3.1).*
(4) *If $p > 0$ on J and the boundary conditions are real and coupled, then the eigenvalues are bounded below and can be ordered to satisfy*

$$-\infty < \lambda_0 \leq \lambda_1 \leq \lambda_2 \leq \ldots; \quad \lambda_n \to +\infty, \text{ as } n \to \infty. \qquad (4.3.2)$$

Each eigenvalue may be simple or double but there cannot be two consecutive equalities in (4.3.2) since, for any value of λ, equation (4.1.1) has exactly two linearly independent solutions. Note that λ_n is well defined for each $n \in \mathbb{N}_0$ but there is some arbitrariness in the indexing of the eigenfunctions corresponding to a double eigenvalue since every nontrivial solution of the equation for such an eigenvalue is an eigenfunction. Given such an indexing scheme, let u_n be a real-valued eigenfunction of $\lambda_n(K)$ for real coupled boundary conditions, then the number of zeros of u_n in the open interval J is 0 or 1, if $n = 0$, and $n - 1$ or n or $n + 1$ if $n \geq 1$.

3. EXISTENCE OF EIGENVALUES

(5) *If $p > 0$ and the boundary conditions are complex coupled then all eigenvalues are simple and strict inequality holds everywhere in (4.3.2). Moreover, if u_n is an eigenfunction of $\lambda_n(\gamma, K)$, $0 < \gamma < \pi$, $-\pi < \gamma < 0$ then the number of zeros of $\operatorname{Re} u_n$ on $[a, b]$ is 0 or 1 if $n = 0$, and $n - 1$ or n or $n + 1$ if $n \geq 1$. The same conclusion holds for $\operatorname{Im} u_n$. Furthermore, u_n has no zero in $[a, b]$, $n \in \mathbb{N}_0$.*

(6) *If $p > 0$ and the boundary conditions are separated then strict inequality holds everywhere in (4.3.2). Furthermore, if u_n is an eigenfunction of λ_n, then u_n is unique up to constant multiples and has exactly n zeros in the open interval J, $n \in \mathbb{N}_0$.*

(7) *If $p > 0$, then for any self-adjoint boundary condition, separated, real coupled or complex coupled, the following asymptotic formula holds:*

$$\frac{\lambda_n}{n^2} \to c = \pi^2 \left(\int_a^b \sqrt{\frac{w}{p}} \right)^{-2}, \quad \text{as } n \to \infty. \tag{4.3.3}$$

(8) *If p changes sign and $p^+(t) = p(t)$, if $p(t) > 0$ and 0 otherwise; $p^-(t) = -p(t)$, if $p(t) < 0$ and 0 otherwise, then (4.3.3) takes the form:*

$$\frac{\lambda_n}{n^2} \to c = \pi^2 \left(\int_a^b \sqrt{\frac{w}{p^+}} \right)^{-2}, \quad \frac{\lambda_n}{n^2} \to c = -\pi^2 \left(\int_a^b \sqrt{\frac{w}{p^-}} \right)^{-2}, \tag{4.3.4}$$

as $n \to \infty$ and as $n \to -\infty$, respectively.

PROOF. The "standard" Hilbert space proof that all eigenvalues are real and that there are an infinite number of them can be found in [**123**]. This proof "works" when w is positive on the underlying interval J, so that $L^2(J, w)$ is a Hilbert space. Although stronger assumptions on the coefficients are used in [**123**], the proof given extends readily to our integrability conditions here, including the case when p changes sign. Also this proof applies to unbounded intervals. See Section 11 of this chapter for some details. This proof is non-elementary in the sense that it depends on the theory of self-adjoint operators in Hilbert space.

The fact that the eigenvalues are isolated and have no finite accumulation point follows from their characterization as the zeros of the entire characteristic function $\delta(\lambda)$, see Lemmas 3.2.1 and 3.2.2.

Möller in [**474**] showed that the spectrum of a SLP is not bounded below whenever p assumes negative values on a set of positive Lebesgue measure, even when this set contains no interval. Similarly the spectrum is not bounded above whenever p assumes positive values on a set of positive Lebesgue measure. It is well known that the eigenvalues are bounded below when p and w are both positive on J; see [**474**]. The ordering in part (2) follows again from the characterization of the eigenvalues as zeros of the entire function $\delta(\lambda)$. Part (3) is elementary - for separated boundary conditions there can be at most one linearly independent solution for any value of λ. Part (4) follows from the earlier parts except for the last sentence regarding the zero properties of the eigenfunctions; a proof of this can be found in [**384**].

The fact that all eigenvalues of a complex coupled self-adjoint boundary condition are simple follows from the characterization

$$\Phi(b, a, \lambda_n) = e^{i\gamma} K$$

of the double eigenvalues in Section 3.2. In this self-adjoint case both $\Phi(b,a,\lambda_n)$ and K are real and hence $\gamma = 0$ or $\gamma = \pi$. The zero properties of the eigenfunctions is proven in Theorem 4.8 of [**384**].

Part (6) is a classic result in Sturm-Liouville theory if not Analysis, see [**21**], [**123**], and Section 5 below.

For the asymptotic formulas in parts (7) and (8) see Atkinson and Mingarelli [**29**].

An elementary proof of the existence of eigenvalues for the case of separated BC when $p > 0$ and $w > 0$ on J is well known and given in Section 5 of this chapter. It is based on the Prüfer transformation. The Prüfer transformation proof is elementary in the sense that it does not invoke operator theory from Hilbert space, but elementary here does not mean easy. Recently an elementary proof for the general coupled case when $p > 0$ and $w > 0$ on J was found by Eastham, Kong, Wu and Zettl in [**145**]. This will be discussed in Section 8 and in the Comments Section at the end of this chapter. □

The conditions on p and w in Theorem 4.3.1 can be considerably weakened. The hypothesis that p is real-valued maps $J \to \mathbb{R}$, and $1/p \in L^1(J,\mathbb{R})$ allows p to have "mild" zeros on J but does not allow $r = 1/p$ to be identically zero on a subinterval of J. Some results below allow $r = 1/p$ and w to be identically zero on subintervals of J. Although p is allowed to change sign, for all results in this section we require $w > 0$. The sign of the coefficient q is immaterial in the existence results of this section. In Chapter 5 we investigate the case when $p > 0$ but w changes sign. In general, these problems are 'indefinite' and may have nonreal eigenvalues but there is an important subclass of these problems, the so called left-definite (LD) problems. The spectrum of these LD problems is real and can be characterized in terms of the spectra of a one parameter family of (right-definite) problems with weight function $|w|$ as we will see in Chapter 5. The sign of q, directly or indirectly, does play a role in the LD case.

To avoid needless repetitions, since we are dealing with Lebesgue integrable functions here, we just write $f > 0$ on J instead of $f > 0$ a.e. on J; similarly for $f \geq 0$.

THEOREM 4.3.2 (Everitt, Kwong, Zettl). *Assume that*

$$r = 1/p,\ q,\ w \in L^1(J,\mathbb{R}),\ J = (a,b),\ -\infty \leq a < b \leq \infty,$$
$$r > 0,\ w \geq 0,\ on\ J,\ \int_a^b w > 0. \tag{4.3.5}$$

Then the SLP (4.1.1), with separated boundary conditions (4.2.3), (4.2.4) has only real eigenvalues, there are an infinite but countable number of them, and they can be ordered to satisfy

$$-\infty < \lambda_0 < \lambda_1 < \lambda_2 < \ldots;\ and\ \lambda_n \to +\infty\ as\ n \to \infty. \tag{4.3.6}$$

If u_n is an eigenfunction of λ_n, then u_n is unique up to constant multiples. Let z_n denote the number of zeros of u_n in the open interval (a,b), $n \in \mathbb{N}_0 = \{0,1,2,\ldots\}$. Then

$$z_{n+1} = z_n + 1,\ n \in \mathbb{N}_0.$$

For any integer $m \geq 0$ there exists a SLP with separated boundary conditions such that $z_0 = m$. A sufficient but not necessary condition that $z_0 = 0$ is that $w > 0$ a.e. on J.

PROOF. See [**214**]. □

REMARK 4.3.1. Theorem 4.3.2 seems to be the most general result available to establish the existence of an infinite number of isolated real eigenvalues for the case when $r > 0$, and $w \geq 0$, on J. The integral condition on w eliminates the case when w is identically zero on J, which would mean that λ has no effect on the boundary conditions. In the next theorem the hypothesis $r > 0$, on J is weakened to $r \geq 0$ on J, but at the expense of considerably more restrictions on w. Note the subtle but important difference between the hypotheses $r > 0$ *on J and $r \geq 0$ on J.*

THEOREM 4.3.3 (Atkinson). *Assume that*

$$r = 1/p,\ q,\ w \in L^1(J, \mathbb{R}),\ J = (a,b),\ -\infty \leq a < b \leq \infty,$$

$$r \geq 0,\ w \geq 0,\ \text{on } J,\ \int_a^t w > 0,\ \int_t^b w > 0,\ \int_a^b r > 0,\ \text{for any } t \in J,$$

and $w = 0$ on $(c,d) \subset J$ implies $q = 0$ on (c,d). (4.3.7)

Then the SLP (4.1.1) with separated boundary conditions (4.2.3), (4.2.4) has only real and simple eigenvalues, they are bounded below and can be ordered to form a finite or infinite sequence satisfying

$$-\infty < \lambda_0 < \lambda_1 < \lambda_2 < \ldots$$

If u_n is an eigenfunction of λ_n, then u_n is unique up to constant multiples and has exactly n zeros in the open interval J. (Recall that under these conditions a "zero" may be a whole subinterval.)

PROOF. See Theorem 8.4.5 in [**21**]. □

REMARK 4.3.2. The phrase "a finite or infinite sequence" is a quote from F. V. Atkinson's book cited in the proof. It suggests - to this writer - that Atkinson was aware that under the conditions of Theorem 4.3.3 there may only be a finite number of eigenvalues. But he gives no example. We construct such examples in Section 4.12. Theorems 4.3.3 and 4.3.4 are stated in [**21**] only for bounded intervals J. However, Atkinson mentions that this is only for "convenience"; in other words the results and their proofs extend readily to unbounded intervals J.

THEOREM 4.3.4 (Atkinson). *Let the hypotheses and notation of Theorem 4.3.3 hold and, in addition, suppose that there exists an infinite increasing sequence $\{c_i : i \in \mathbb{N}\}$ of points in J such that*

$$\int_{c_{2i}}^{c_{2i+1}} w > 0,\ \int_{c_{2i+1}}^{c_{2i+2}} r > 0. \tag{4.3.8}$$

Then the SLP (4.1.1), with separated boundary conditions (4.2.3), (4.2.4) has only real eigenvalues, there are an infinite but countable number of them, and they can be ordered to satisfy

$$-\infty < \lambda_0 < \lambda_1 < \lambda_2 < \ldots;\ \text{and}\ \lambda_n \to +\infty\ \text{as}\ n \to \infty. \tag{4.3.9}$$

If u_n is an eigenfunction of λ_n, then u_n is unique up to constant multiples and has exactly n zeros in the open interval J. (Recall that under these conditions a "zero" may be a whole subinterval.)

PROOF. See Theorem 8.4.6 in [**21**]. □

COROLLARY 4.3.1. *Let the hypotheses and notation of Theorem 4.3.3 hold. If $r = 1/p$ and w are both positive on a common subinterval of J, then the conclusions of Theorem 4.3.4 hold; in particular there are an infinite number of eigenvalues.*

PROOF. Just choose distinct points c_i in the common subinterval where r and w are positive and (4.3.4) holds. □

REMARK 4.3.3. In Section 4.12 we construct SLP with only a finite number of eigenvalues. By Corollary 4.3.1 w and $r = 1/p$ cannot be positive on a common subinterval. In this construction the interval J is partitioned into subintervals such that w and $r = 1/p$ are alternatively identically zero on adjacent subintervals.

NOTATION 4.3.1. *To study the variation of the eigenvalues with respect to the parameters of the problem we use the following notations:*

$$\lambda_n = \lambda_n(a, b, \alpha, \beta, r, q, w); \quad u_n = u_n(\cdot, a, b, \alpha, \beta, r, q, w), \quad (4.3.10)$$

for the separated boundary conditions (4.2.3), (4.2.4);

$$\lambda_n = \lambda_n(a, b, K, r, q, w); \quad u_n = u_n(\cdot, a, b, K, r, q, w), \quad (4.3.11)$$

for real coupled boundary conditions (4.2.5); and

$$\lambda_n = \lambda_n(a, b, \gamma, K, r, q, w); \quad u_n = u_n(\cdot, a, b, \gamma, K, r, q, w), \quad (4.3.12)$$

for complex coupled boundary conditions (4.2.7).

NOTATION 4.3.2. *Note that in all cases above when the eigenvalues are bounded below we use the index set $\mathbb{N}_0$ and order the eigenvalues as in (4.3.9). This indexing scheme defines λ_n uniquely. On the other hand, when the eigenvalues are unbounded below and above we use $\mathbb{Z}$ for the indexing set. Observe that this does not define λ_n uniquely, in particular any specific eigenvalue can be λ_0. The code SLEIGN2 (for singular problems with a limit-circle oscillatory endpoint) defines λ_0 as the smallest nonnegative eigenvalue, λ_1 the next smallest etc. and λ_{-1} as the largest negative eigenvalue etc. (If $\lambda = 0$ is a double eigenvalue, then $\lambda_1 = \lambda_0 = 0$.) This is one of infinitely many ways to define λ_n uniquely when the eigenvalues are unbounded in both directions. Binding and Volkmer [**80**] have established a different and natural indexing scheme for the case when p changes sign. This is based on a Prüfer type characterization of the eigenvalues and is given in Theorem 4.6.5 below.*

4. Dependence of Eigenvalues on the Problem

It was shown in Section 3.5 that the eigenvalues of self-adjoint and non-self-adjoint SLP depend continuously on the problem in the sense that each isolated eigenvalue can be embedded in a continuous eigenvalue branch as one or more of the parameters of the problem varies. But this continuous branch is not necessarily determined by a fixed index. In those cases when λ_n is well defined it is not, in general, a continuous function of the problem. In particular, Everitt, Möller and Zettl [**227**], [**228**] have shown that λ_n does not depend continuously on the separated boundary conditions. Eastham, Kong, Wu and Zettl have shown that

λ_n is also not a continuous function of the coupled boundary conditions and have characterized the discontinuities. On the other hand Kong and Zettl [**392**] have shown that λ_n is a continuous function of the endpoints a, b and recently in [**384**], Kong, Wu and Zettl have shown that λ_n is a continuous function of each of the coefficients r, q, w, with $r = 1/p$. Thus the continuity and differentiability results of Chapter 3 for the general non-self-adjoint case can be made more explicit in the self-adjoint case. To study whether the eigenvalues do or do not change only "a little" when the problem is changed "slightly" we need a metric. So we introduce spaces of self-adjoint SLP with a metric.

The space of self-adjoint SLP used in this section is denoted by Ω_{s-a} and defined by

$$\Omega_{s-a} = \{\omega = (a, b, A, B, 1/p, q, w)\} \tag{4.4.1}$$

where we assume that

$$-\infty \leq a' < a < b < b' \leq \infty,\ A, B \in M_2(\mathbb{C}),$$
$$A E A^* = B E B^*,\ rank\,(A : B) = 2,\ E = \begin{bmatrix} 0, & -1 \\ 1, & 0 \end{bmatrix},$$
$$1/p,\ q,\ w \in L_{loc}(J, \mathbb{R}),\ w > 0\ \text{a.e. on}\ J = (a', b'). \tag{4.4.2}$$

In (4.4.2) we use the notation $J = (a', b')$, $-\infty \leq a' < a < b < b' \leq \infty$, because we want to study the variation of the eigenvalues with respect to the endpoints a, b as they vary within the interval J. Note that, although the problem may be singular on (a', b'), it is regular on each interval (a, b) with $-\infty \leq a' < a < b < b' \leq \infty$ by (4.4.2).

For the canonical form (4.2.5), (4.2.6) of the special case of separated self-adjoint BC we use the notation

$$\Omega_{ss} = \{\omega = (a, b, \alpha, \beta, 1/p, q, w)\}; \tag{4.4.3}$$

for the complex self-adjoint coupled case we let

$$\Omega_{cc} = \{\omega = (a, b, \gamma, K, 1/p, q, w),\ -\pi < \gamma < 0,\ \text{or}\ 0 < \gamma < \pi\}; \tag{4.4.4}$$

and for the real coupled case i.e. when $\gamma = 0$ or $\gamma = \pi$ we use the notation

$$\Omega_{rc} = \{\omega = (a, b, K, 1/p, q, w)\}. \tag{4.4.5}$$

REMARK 4.4.1. **Note that no sign restriction is placed on p.** The reason for restricting w to be positive is so that the well developed theory of self-adjoint differential operators in the weighted Hilbert space $L^2((a,b), w)$ can be applied. This is the so called "right definite" (RD) case. The condition $w > 0$ can be relaxed to $w \geq 0$ but care must be taken when w is identically zero on a subinterval. See Section 3 above. Note that from (4.4.2) it follows that p may have "mild" zeros in J, p can change sign but cannot be identically zero on a subinterval of J. In particular, p can be a step function with positive and negative values.

There is also an extensive literature for the case when w changes sign. Some of these problems can be studied in a different Hilbert space - not $L^2((a,b), w)$ - a "left-definite" Hilbert space and so these problems are called left-definite (LD). There are at least two different approaches to the study of LD problems, we will discuss one of them in Chapter 5 and make some general remarks and give references for the other approach in the Comments Section of Chapter 5. In the general case, problems with a weight function w that changes sign can be studied using Krein or

Pontryagin space theory but, although these abstract theories are well developed, there are few concrete and explicit applications to SLP; applications which provide information not available from other methods. We will mention a few open problems at the end of Chapter 5.

THEOREM 4.4.1. *Let A, B be fixed matrices satisfying the self-adjoint conditions (4.1.4), let (4.4.2) hold and assume that $p > 0$ on $J = (a', b')$, $-\infty \leq a' < b' \leq \infty$. Consider problems $\omega = \omega(a, b, A, B, 1/p, q, w) \in \Omega_{s-a}$ with $a' < a < b < b'$. For each $n \in \mathbb{N}_0$, let $\lambda_n = \lambda_n(\omega)$ be the $n-th$ eigenvalue of ω as given by (4.3.2) of Theorem 4.3.1. Then for each $n \in \mathbb{N}_0$, λ_n is a continuous function of the equation. In particular, for each $n \in \mathbb{N}_0$, $\lambda_n(1/p)$ is a continuous function of $1/p \in L^1(J, \mathbb{R})$, $p > 0$; $\lambda_n(q)$ is a continuous function of $q \in L^1(J, \mathbb{R})$; $\lambda_n(w)$ is a continuous function of $w \in L^1(J, \mathbb{R})$, $w > 0$; $\lambda_n(a)$ is a continuous function of a, $\lambda_n(b)$ is a continuous function of b.*

PROOF. See Section 2 of Kong, Wu and Zettl [**384**]. □

Next we study the dependence of the eigenvalues on the boundary conditions for a fixed equation, i.e. for fixed $a, b, 1/p, q, w$. In this connection we first introduce the "jump set $\mathbb{J}$" of separated and coupled boundary conditions on which, as we will see below, λ_n has a jump discontinuity.

DEFINITION 4.4.1 (Jump set of boundary conditions). *The "jump set of boundary conditions $\mathbb{J}$" is the union of*

(1) *the (real and complex) coupled conditions*

$$Y(b) = e^{i\gamma} K Y(a), \quad Y = \begin{bmatrix} y \\ (py') \end{bmatrix}, \quad -\pi < \gamma \leq \pi,$$

where the 2×2 matrix $K = (k_{ij}) \in SL(2, \mathbb{R})$ satisfies $k_{12} = 0$, and
(2) *the separated boundary conditions*

$$A_1 y(a) + A_2 (py')(a) = 0, \ A_1, A_2 \in \mathbb{R}, \ (A_1, A_2) \neq (0, 0),$$
$$B_1 y(b) + B_2 (py')(b) = 0, \ B_1, B_2 \in \mathbb{R}, \ (B_1, B_2) \neq (0, 0),$$

satisfying $A_2 B_2 = 0$.

THEOREM 4.4.2. *Let A, B be matrices satisfying the self-adjoint conditions (4.1.4), let (4.4.2) hold and assume that $p > 0$ on $J = (a', b')$, $-\infty \leq a' < b' \leq \infty$. Consider problems $\omega = \omega(a, b, A, B, 1/p, q, w)$ with $a' < a < b < b'$. For each $n \in \mathbb{N}_0$, let $\lambda_n = \lambda_n(\omega)$ be the $n-th$ eigenvalue of ω as given by (4.3.2) of Theorem 4.3.1; let $\mathbb{J}$ be given by Definition 4.4.1. Fix the equation i.e. fix $a, b, 1/p, q, w$ and consider λ_n as a function of the boundary conditions. Then:*

(1) *If the boundary condition is not on the jump set $\mathbb{J}$, then λ_n is a continuous function of the boundary condition for each $n \in \mathbb{N}_0$.*
(2) *Let $n \in \mathbb{N}$, and assume that the coupled real boundary condition K is in $\mathbb{J}$ and $\lambda_n = \lambda_{n-1}$. Then λ_n is continuous at K.*
(3) *The lowest eigenvalue λ_0 has an infinite jump discontinuity at each, separated or (real or complex) coupled, boundary condition in $\mathbb{J}$.*
(4) *Let $n \in \mathbb{N}$. If the boundary condition is in $\mathbb{J}$ and λ_n is simple, then λ_n has a finite jump discontinuity at this boundary condition.*

PROOF. See Section 3 in [**384**]. □

The nature of the jump discontinuities given by Theorem 4.4.2 can be specified explicitly. This is done in the next theorem for separated BC; for the coupled case the reader is referred to Theorems 3.39, 3.73, 3.76, and Propositions 3.71, 3.72 in [**384**].

THEOREM 4.4.3. *Let the hypotheses and notation of Theorem 4.4.1 hold. Fix a, b, p, q, w. Consider the separated boundary conditions:*

$$A_1 y(a) + A_2 (py')(a) = 0, \ A_1, A_2 \in \mathbb{R}, \ (A_1, A_2) \neq (0,0),$$
$$B_1 y(b) + B_2 (py')(b) = 0, \ B_1, B_2 \in \mathbb{R}, \ (B_1, B_2) \neq (0,0).$$

- *Fix B_1, B_2 and let $A_1 = 1$. Consider $\lambda_n = \lambda_n(A_2)$ as a function of $A_2 \in \mathbb{R}$. Then for each $n \in \mathbb{N}_0$, $\lambda_n(A_2)$ is continuous at A_2 for $A_2 > 0$ and $A_2 < 0$ but has a jump discontinuity at $A_2 = 0$. More precisely we have:*
  (1) $\lambda_n(A_2) \to \lambda_n(0)$ as $A_2 \to 0^-$, $n \in \mathbb{N}_0$.
  (2) $\lambda_0(A_2) \to -\infty$ as $A_2 \to 0^+$.
  (3) $\lambda_{n+1}(A_2) \to \lambda_n(0)$ as $A_2 \to 0^+$.
- *Fix A_1, A_2 and let $B_1 = 1$. Consider $\lambda_n = \lambda_n(B_2)$ as a function of $B_2 \in \mathbb{R}$. Then for each $n \in \mathbb{N}_0$, $\lambda_n(B_2)$ is continuous at B_2 for $B_2 > 0$ and $B_2 < 0$ but has a jump discontinuity at $B_2 = 0$. More precisely we have:*
  (1) $\lambda_n(B_2) \to \lambda_n(0)$ as $B_2 \to 0^+$, $n \in \mathbb{N}_0$.
  (2) $\lambda_0(B_2) \to -\infty$ as $B_2 \to 0^-$.
  (3) $\lambda_{n+1}(B_2) \to \lambda_n(0)$ as $B_2 \to 0^-$.

PROOF. See Everitt, Möller and Zettl [**227**]. □

REMARK 4.4.2. Note that $\lambda_0(A_2)$ has an infinite jump discontinuity at $A_2 = 0$, but for all $n \geq 1$, $\lambda_n(A_2)$ has a finite jump discontinuity at $A_2 = 0$, $\lambda_n(A_2)$ is left but not right continuous at 0. Similarly, $\lambda_0(B_2)$ has an infinite jump discontinuity at $B_2 = 0$, but for all $n \geq 1$, $\lambda_n(B_2)$ has a finite jump discontinuity at $B_2 = 0$; $\lambda_n(B_2)$ is right but not left continuous at 0. In all cases $\lambda_n(0)$ is embedded in a continuous branch of eigenvalues as A_2 or B_2 passes through zero but this branch is not given by a fixed index n; in order to preserve continuity the index "jumps" from n to $n+1$ as A_2 or B_2 pass through zero from the appropriate direction.

REMARK 4.4.3. This forced "index jumping" in order to stay on a continuous branch of eigenvalues plays an important role in some of the algorithms and their numerical implementations used in the code SLEIGN2 [**43**] for the numerical approximation of the spectrum of regular and singular SLP.

REMARK 4.4.4. Kong and Zettl [**394**] have shown that each continuous eigenvalue branch is in fact differentiable everywhere including the point $A_2 = 0$ (or $B_2 = 0$) where the index jumps. This also follows from Möller and Zettl [**475**].

Now that the continuities and discontinuities of λ_n have been completely characterized it is natural to investigate the differentiability of λ_n as a function of all parameters of the problem. This we embark upon next.

THEOREM 4.4.4. *Let $\omega = (a, b, A, B, 1/p, q, w) \in \Omega_{s-a}$, where Ω_{s-a} is defined in (4.4.1) through (4.4.2) and assume that $p > 0$. Let λ_n be defined as in Theorem 4.3.1 and let u_n be a normalized eigenfunction of λ_n.*

(1) *Fix all the components of ω except the left endpoint a and consider $\lambda_n = \lambda_n(a)$ as a function of a. Assume that for some a, $\lambda_n(a)$ is a simple eigenvalue of $\omega(a)$. Then there is a neighborhood U of a and a neighborhood V of $\lambda_n(a)$ such that for every c in U the SLP $\omega(c)$ has exactly one eigenvalue in V and it is simple. The map $\lambda_n : U \to V$ is continuous on U and differentiable almost everywhere in U and we have*

$$\lambda_n'(a) = \frac{1}{p(a)}|pu_n'|^2(a) - |u_n|^2(a)[q(a) - \lambda_n(a)w(a)] \quad \text{a.e. in } U. \qquad (4.4.6)$$

Furthermore, if p, q, w are continuous at a and $p(a) \neq 0$, then (4.4.6) holds at the point a.

(2) *Fix all the components of ω except b. Assume that for some b, $\lambda_n(b)$ is a simple eigenvalue of $\omega(b)$. Then there is a neighborhood U of b and a neighborhood V of $\lambda_n(b)$ such that for every c in U the SLP $\omega(c)$ has exactly one eigenvalue in V and it is simple. The map $\lambda_n : U \to V$ is continuous on U and differentiable almost everywhere in U and we have*

$$\lambda_n'(b) = -\frac{1}{p(b)}|pu_n'|^2(b) + |u_n|^2(b)\left[q(b) - \lambda_n(b)w(b)\right] \quad \text{a.e. in } U. \qquad (4.4.7)$$

Furthermore, if p, q, w are continuous at b and $p(b) \neq 0$, then (4.4.7) holds at the point b.

PROOF. See [**394**]. $\square$

REMARK 4.4.5. In his well known monograph on variational methods for eigenvalue problems [**604**] Hans Weinberger states, without proof or reference to a proof, that the Dirichlet eigenvalues are decreasing functions of the length of the interval but that this is not true for the Neumann eigenvalues. (The Dirichlet case follows from the variational characterization.) Formulas (4.4.6) and (4.4.7) shed a great deal of light on this phenomenon.

Assume that $p > 0$. For Dirichlet boundary conditions we have: $u(a) = u(a, a, 0, \pi) = 0$ and $u(b, b, 0, \pi) = 0$. Hence the second term in (4.4.6) and in (4.4.7) is zero; thus it is clear from these formulas that the Dirichlet eigenvalues are increasing functions of the left endpoint a and decreasing functions of the right endpoint b. It is also clear from these formulas that this is not true, in general, *for any other separated boundary conditions*. However if q/w is bounded above, say by C, then *for any separated boundary condition* all eigenvalues greater than C are increasing functions of the left endpoint a and decreasing functions of the right endpoint b. Since for any fixed regular SLP the eigenvalues $\lambda_n \to \infty$ asymptotically as n^2 it is clear that if q/w is bounded above only the eigenvalues of low index can fail to be monotonic functions of the length of the interval.

THEOREM 4.4.5. *Let the hypotheses and notation of Theorem 4.4.4 hold. Fix a, b, p, q, w. (Recall that A, B are replaced by α, β for the canonical form of separated BC; by γ, K for general coupled BC and by K for the canonical form of real coupled boundary conditions.)*

(1) *Fix all components of ω except γ and let $\lambda_n = \lambda_n(\gamma)$ and let $u_n = u_n(\cdot, \gamma)$ be a normalized eigenfunction of λ_n. Then λ_n is differentiable at γ for any γ satisfying $-\pi < \gamma < 0$ or $0 < \gamma < \pi$ and*

$$\lambda_n'(\gamma) = -2 \operatorname{Im}[u_n(b)\,(pu_n')(b)], \qquad (4.4.8)$$

where $\operatorname{Im}[z]$ denotes the imaginary part of z.

(2) Fix all components of ω except K. Assume that $\lambda_n = \lambda_n(K)$ is a simple eigenvalue and $u_n = u_n(\cdot, K)$ a normalized eigenfunction of $\lambda_n(K)$. Then there exists a neighborhood U of K and a neighborhood V of $\lambda_n(K)$ such that for every G in U satisfying $G \in SL_2(\mathbb{R})$, the BVP $\omega(G)$ has exactly one eigenvalue in V and it is simple. The map $\lambda_n : U \to V$ is differentiable at K and its Frechet derivative is given by the bounded linear transformation defined by

$$\lambda_n'(K) H = [p\overline{u_n}'(b), -\overline{u}_n(b)] H K^{-1} \begin{bmatrix} u_n(b) \\ (pu_n')(b) \end{bmatrix}, \quad H \in M_{2,2}(\mathbb{C}). \quad (4.4.9)$$

PROOF. See [**394**] for (1) and for the self-adjoint case of (2) when $K + H$ is restricted to lie in $SL_2(\mathbb{R})$, and see [**475**] for the general case of (2). □

THEOREM 4.4.6. *Let the hypotheses and notation of Theorem 4.4.4 hold. Fix a, b, A, B. Assume that λ_n is a simple eigenvalue of ω for some $n \in \mathbb{N}_0$ and u_n is a normalized eigenfunction of λ_n. Then there is a simple closed curve Γ in $\mathbb{C}$ with $\lambda_n(\omega)$ in its interior and a neighborhood U of ω in Ω_{s-a} such that for any ρ in U, the SLP ρ has exactly one eigenvalue in the interior of Γ and this eigenvalue is simple. This map $\lambda_n : U \to \mathbb{C}$ is differentiable with respect to*

(1) *q with p, w fixed and its Frechet derivative is the bounded linear transformation given by*

$$\lambda_n'(q) h = \int_a^b |u_n(\cdot, q)|^2 h, \quad h \in L^1((a,b), \mathbb{R}); \quad (4.4.10)$$

(2) *$1/p$ with q, w fixed and its Frechet derivative is the bounded linear transformation given by*

$$\lambda_n'(1/p) h = -\int_a^b |u_n^{[1]}(\cdot, 1/p)|^2 h, \quad h \in L^1((a,b), \mathbb{R}); \quad (4.4.11)$$

(3) *w with $1/p, q$ fixed and its Frechet derivative is the bounded linear transformation given by*

$$\lambda_n'(w) h = -\lambda_n(w) \int_a^b |u_n(\cdot, w)|^2 h, \quad h \in L^1((a,b), \mathbb{R}). \quad (4.4.12)$$

PROOF. This result is proved in [**475**] for continuous eigenvalue branches of general non-self-adjoint SLP. By Theorem 4.4.1 λ_n is a continuous function of $1/p$, q and w in the self-adjoint case, this justifies the use of the index n in formulas (4.4.10), (4.4.11), (4.4.12). □

5. The Prüfer Transformation

The Prüfer transformation provides a "polar coordinate" factorization $y = \rho \sin \theta$, $(py') = \rho \cos \theta$ of all non-trivial solutions y and their quasi-derivatives py' of the second order linear Sturm-Liouville equation such that ρ has no zero and θ satisfies a first order nonlinear differential equation which is independent of ρ. This equation is studied in this section. It is called the Prüfer equation. In the next section we will derive a number of consequences for SLP. One of these consequences is the existence and characterization of the eigenvalues for all self-adjoint separated boundary conditions; another is a proof of the Sturm Comparison Theorem without Picone's identity. We start with the Comparison Theorem and continuous dependence of solutions of initial value problems for the Prüfer equation. Consider

$$\theta'_j = r_j \cos^2 \theta_j + g_j \sin^2 \theta_j \quad \text{on} \quad J = [a,b], \quad -\infty \le a < b \le \infty, \quad j = 1, 2; \quad (4.5.1)$$

$$\theta_j(c_j) = \alpha_j, \quad c_j \in J, \quad \alpha_j \in \mathbb{R}, \quad j = 1, 2; \quad (4.5.2)$$

$$r_j = 1/p_j, \quad g_j \in L^1(J, \mathbb{R}), \quad j = 1, 2. \quad (4.5.3)$$

THEOREM 4.5.1. *Let (4.5.1) to (4.5.3) hold. Each initial value problem has a unique real valued solution which is defined on all of J. Given any $\varepsilon > 0$ there exists a $\delta > 0$ such that if*

$$|c_1 - c_2| + |\alpha_1 - \alpha_2| + \int_a^b |r_1 - r_2| + \int_a^b |g_1 - g_2| < \delta, \quad (4.5.4)$$

then

$$|\theta(t, c_1, \alpha_1, r_1, g_1) - \theta(t, c_2, \alpha_2, r_2, g_2)| < \varepsilon \quad \text{for all} \quad t \in J. \quad (4.5.5)$$

REMARK 4.5.1. Here we follow our notational convention and indicate the dependence of θ on the parameters of the problem. Note that no sign restriction is placed on r_j or q_j. This theorem shows not only that $\theta(\cdot, c, \alpha, r, g)$ is jointly continuous in (c, α, r, g) uniformly for $t \in J$ but at any fixed point $(r, g) \in L^1(J, \mathbb{R}) \times L^1(J, \mathbb{R})$ this continuity is uniform in $\alpha \in \mathbb{R}$ and for $c, t \in J$.

PROOF. Let $\phi = \theta_2 - \theta_1$ with $\theta_j = \theta(t, c_j, \alpha_j, r_j, g_j)$, $j = 1, 2$. Then

$$\phi' = [r_2 - r_1]\cos^2\theta_2 + [g_2 - g_1]\sin^2\theta_2 + r_1(\cos^2\theta_2 - \cos^2\theta_1) + g_1(\sin^2\theta_2 - \sin^2\theta_1). \quad (4.5.6)$$

Define

$$u(t) = \begin{cases} \dfrac{\sin\theta_2(t) - \sin\theta_1(t)}{\theta_2(t) - \theta_1(t)}, & \text{when} \quad \theta_2(t) \ne \theta_1(t), \\ \cos\theta_2(t), & \text{when} \quad \theta_2(t) = \theta_1(t); \end{cases} \quad (4.5.7)$$

$$k = (g_1 - r_1)[\sin\theta_2 + \sin\theta_1]\, u \quad \text{on} \quad J; \quad (4.5.8)$$

$$h = (r_2 - r_1)\cos^2\theta_2 + [g_2 - g_1]\sin^2\theta_2 \quad \text{on} \quad J. \quad (4.5.9)$$

Then $u \in C(J, \mathbb{R})$, $k, h \in L^1(J, \mathbb{R})$ and

$$\phi' = k\phi + h \quad \text{on} \quad J.$$

Solving this first order linear equation we get

$$\phi(t) = \phi(c) \exp\left(\int_c^t k\right) + \int_c^t \exp\left(\int_s^t k\right) h(s)\, ds, \quad s, t \in J. \quad (4.5.10)$$

Hence

$$|\phi(t)| \le |\phi(c)|\, M + M \int_a^b |h|, \quad M = \exp\left(\int_a^b |k|\right), \quad t \in J. \quad (4.5.11)$$

Note that $h \in L^1(J, \mathbb{R})$ since $|h| \le |(r_2 - r_1)| + |g_2 - g_1|$. The special case when $c_1 = c_2 = c$ follows from formula (4.5.11).

Assume that $c_1 < c_2$ and $g_1 = g_2 = g$, $p_1 = p_2 = p$. Then $h = 0$ and (4.5.11) reduces to
$$|\phi(t)| \leq |\phi(c)| M. \tag{4.5.12}$$
Define
$$f(t, \theta) = r(t) \cos^2 \theta + g(t) \sin^2 \theta, \quad for \quad t \in J, \quad \theta \in \mathbb{R}.$$
Let $\theta_j = \theta(\cdot, c_j, \alpha_j, p, g)$, $j = 1, 2$. Then
$$\theta_j(t) = \alpha_j + \int_{c_1}^{t} f(s, \theta_j(s))\, ds, \quad t \in J, \quad j = 1, 2,$$
and
$$\theta_2(c_2) - \theta_1(c_2) = \alpha_2 - \alpha_1 - \int_{c_1}^{c_2} f(s, \theta_1(s))\, ds. \tag{4.5.13}$$

From (4.5.12) and the continuity of the integral it follows that $|\phi(c_2)|$ can be made arbitrarily small when α_2 is sufficiently close to α_1 and when c_2 is sufficiently close to c_1. It then follows from (4.5.13) that $|\phi(t)|$ can be made arbitrarily small when c_1, c_2 and α_1, α_2 are sufficiently close together since
$$M = \exp\left(\int_a^b |k|\right) \leq \exp\left(2 \int_a^b |r - g|\right).$$

This completes the proof for the special case when $c_1 < c_2$. The proof of the case $c_1 > c_2$ is similar.

To complete the proof we write
$$\theta(t, c_1, \alpha_1, r_1, g_1) - \theta(t, c_2, \alpha_2, r_2, g_2)$$
$$= \theta(t, c_1, \alpha_1, r_1, g_1) - \theta(t, c_2, \alpha_2, r_1, g_1) + \theta(t, c_2, \alpha_2, r_1, g_1) - \theta(t, c_2, \alpha_2, r_2, g_2)$$
and use the established special cases. □

THEOREM 4.5.2. *Let (4.5.1) to (4.5.3) hold and assume, in addition, that*
$$p_2 \leq p_1, \quad g_1 \leq g_2 \quad a.e. \ on \quad J. \tag{4.5.14}$$
Let the solutions θ_j be determined by the initial conditions
$$\theta_j(a) = \alpha, \quad 0 \leq \alpha < \pi, \quad j = 1, 2.$$

(1) *If $\theta_2(c) > \theta_1(c)$ for some $c \in [a, b]$, then $\theta_2(t) > \theta_1(t)$ for $t \in [c, b]$.*
(2) *If $\theta_2(c) \geq \theta_1(c)$ for some $c \in [a, b]$, then $\theta_2(t) \geq \theta_1(t)$ for $t \in [c, b]$.*
(3) *If $\theta_2(c) < \theta_1(c)$ for some $c \in [a, b]$, then $\theta_2(t) < \theta_1(t)$ for $t \in [a, c]$.*
(4) *If $\theta_2(c) \leq \theta_1(c)$ for some $c \in [a, b]$, then $\theta_2(t) \leq \theta_1(t)$ for $t \in [a, c]$.*
(5) *If $g_2 > g_1$ on $(c, d) \subset [a, b]$, then $\theta_2(c) \geq \theta_1(c)$ implies that $\theta_2(t) > \theta_1(t)$ for $t \in [c, b]$.*
(6) *If $g_2 > g_1$ on $(c, d) \subset [a, b]$, then $\theta_2(c) \leq \theta_1(c)$ implies that $\theta_2(t) < \theta_1(t)$ for $t \in [a, c]$.*

PROOF. Let $\phi = \theta_2 - \theta_1$, $r_j = 1/p_j$, $j = 1, 2$. Proceeding similarly to the proof of Theorem 4.5.1 we obtain
$$\phi' = f\phi + h$$

where
$$f = (g_2 - r_1)(\sin\theta_2 + \sin\theta_1)\frac{\sin\theta_2 - \sin\theta_1}{\theta_2 - \theta_1},$$
$$h = (r_2 - r_1)\cos^2\theta_2 + [g_2 - g_1]\sin^2\theta_2,$$

satisfy
$$|f| \leq 2(|g_2| + r_1), \quad 0 \leq h(t) \leq (r_2 - r_1) + [g_2 - g_1].$$

In particular, $f, h \in L^1(J, \mathbb{R})$. Let
$$k(t) = \exp\left(-\int_c^t f\right), \quad t \in J.$$

Then $k > 0$ on J and
$$(k\phi)' = k(\phi' - f\phi) = kh \geq 0 \quad \text{on} \quad [c, b].$$

Hence $k\phi$ is nondecreasing on $[c, b]$. Thus $\phi(c) > 0$ implies $0 \leq k(t)\phi(t)$ for $c < t$ and $\phi(t) > 0$ for $t \in [c, b]$. Similarly, $\phi(c) \geq 0$ implies that $\phi(t) \geq 0$ for $t \in [c, b]$. This completes the proofs of parts (1) and (2); parts (3) and (4) can be proven similarly.

To prove part (5) assume there exists $u \in (c, b]$ such that $\phi(u) = 0$. By the previous cases this implies that $\phi(t) = 0$ for $t \in (c, u]$. Hence on the interval $(c, u]$ we have: $\phi' = 0$ and consequently $h = 0$. But this implies $\sin^2\theta_1 = 0$ which in turn leads to $\theta_1 = 0 \pmod{\pi}$ and thus we get $0 = \theta_1' = r_1$. This contradiction completes the proof of part (5); and part (6) can be proven similarly. □

REMARK 4.5.2. If $p_2^{-1} > p_1^{-1}$ on $[c, d] \subset [a, b]$ and $\phi(c) \geq 0$ then $\theta_2(t) > \theta_1(t)$ for $t \in (c, b]$ except for the case when $g_2 = g_1$ a.e. on $[c, v] \subset [c, d]$ and $\theta_1(c) = \theta_2(c) = \pi/2 \pmod{\pi}$.

To develop the properties of θ needed below we consider
$$r = p^{-1}, q, w \in L^1(J, \mathbb{R}), \quad J = [a, b], \quad p > 0, \quad w > 0 \quad \text{a.e. on} \quad J; \quad (4.5.15)$$
$$\theta(a, \lambda) = \alpha, \quad 0 \leq \alpha < \pi, \quad \lambda \in \mathbb{R}; \quad (4.5.16)$$
$$\theta' = r\cos^2\theta + (\lambda w - q)\sin^2\theta \quad \text{on} \quad J = [a, b]. \quad (4.5.17)$$

THEOREM 4.5.3. *Let (4.5.15) to (4.5.17) hold. Then the unique solution $\theta(t, \lambda)$ is defined on J and has the following properties:*

(1) $\theta(b, \lambda)$ *is continuous and strictly increasing in λ;*
(2) *if $\theta(c, \lambda) = k\pi$ for some $c \in (a, b)$, some $\lambda \in \mathbb{R}$ and some $k \in \mathbb{N}$, then $\theta(t, \lambda) > k\pi$ for $c < t \leq b$;*
(3) $\theta(b, \lambda) \to \infty$ *as $\lambda \to \infty$;*
(4) $\theta(b, \lambda) \to 0$ *as $\lambda \to -\infty$.*

PROOF. Part (1) follows from Theorem 4.5.1, part (2) from Theorem 4.5.2. To prove part (3) let $\tan\phi = \lambda^{1/2}\tan\theta$ for $\lambda > 0$ and determine ϕ uniquely by requiring $|\theta - \phi| < \pi/2$. Then
$$\phi' = \lambda^{1/2} r\cos^2\phi + \lambda^{1/2} w\sin^2\phi - \lambda^{-1/2} q\sin^2\phi \quad \text{on} \quad J = [a, b], \quad \text{for} \quad \lambda > 0.$$

Note that
$$\lambda^{-1/2}\int_a^b q\sin^2\phi \leq \lambda^{-1/2}\int_a^b |q| \to 0 \quad as \quad \lambda \to \infty.$$

This observation and an integration yield

$$\phi(b,\lambda) \geq \phi(a,\lambda) + \lambda^{1/2} \int_a^b \min(r,w) - \lambda^{-1/2} \int_a^b |q|.$$

Now note that

$$\int_a^b \min(r,w) > 0$$

and hence $\phi(b,\lambda) \to \infty$ as $\lambda \to \infty$. Therefore $\theta(b,\lambda) \to \infty$ as $\lambda \to \infty$.

To prove (4) let

$$\theta_{-\infty}(t) = \lim_{\lambda \to -\infty} \theta(t,\lambda), \quad t \in (a,b].$$

This limit exists since $\theta(t,\lambda)$ is strictly increasing in λ and $0 \leq \theta(b,\lambda) \leq \theta(0,\lambda)$ for $\lambda < 0$. We want to show that $\theta_{-\infty}(t) = 0$ for all $t \in (a,b]$. Integrating (4.5.17) we get

$$\theta(b,\lambda) = \alpha + \int_a^b r(s)\cos^2\theta(s,\lambda)ds + \int_a^b [\lambda w(s) - q(s)]\sin^2\theta(s,\lambda)ds,$$

$$\lambda \int_a^b w(s)\sin^2\theta(s,\lambda)ds = \theta(b,\lambda) - \alpha - \int_a^b r(s)\cos^2\theta(s,\lambda)ds + \int_a^b q(s)\sin^2\theta(s,\lambda)ds.$$

Each term on the right-hand side is bounded for $\lambda < 0$. Hence

$$\int_a^b w(s)\sin^2\theta(s,\lambda)\,ds \to 0 \quad as \quad \lambda \to -\infty.$$

Let $\{\lambda_n : n \in \mathbb{N}\} \to -\infty$ and define, for almost all $s \in [a,b]$, $f_n(s) = w(s)\sin^2\theta(s,\lambda_n)$. Then

$$|f_n(s)| \leq w(s) \quad a.e., \quad 0 < \int_a^b w(s)ds < \infty,$$

and

$$f_n(s) \to w(s)\sin^2\theta_{-\infty}(s) \quad as \quad n \to \infty, \quad \text{pointwise} \quad a.e. \quad \text{on} \quad J.$$

By the Lebesgue Dominated Convergence Theorem we get

$$\int_a^b f_n(s)ds \to \int_a^b w(s)\sin^2\theta_{-\infty}(s)\,ds \quad as \quad n \to \infty.$$

Hence

$$\int_a^b w(s)\sin^2\theta_{-\infty}(s)ds = 0,$$

and thus $w(s)\sin^2\theta_{-\infty}(s)ds = 0$, a.e. and $\theta_{-\infty}(s) = 0 (\mathrm{mod}\,\pi)$.

To show that $\theta_{-\infty}(t) = 0$ for all $t \in (a,b]$ we first show that $\theta_{-\infty}(t) = 0$ for all $t \in (a,c]$ for some $c \in (a,b]$:

For $\lambda < 0$, $s < t$, $s, t \in [a,b]$,

$$\theta(t,\lambda) - \theta(s,\lambda) = \int_s^t \frac{1}{p} + \lambda \int_s^t w(x)\sin^2\theta(x,\lambda)dx - \int_s^t (q(x) + r(x))\sin^2\theta(x,\lambda)dx$$

$$\leq \int_s^t r(x)dx - \int_s^t q(x)\sin^2\theta(x,\lambda)dx.$$

Let $\lambda \to -\infty$, then $\theta(t,\lambda) \to 0 (\bmod\, \pi)$ from above and
$$\theta_{-\infty}(t) - \theta_{-\infty}(s) \le \int_s^t r.$$
In particular, $\theta_{-\infty}(t) \le \alpha + \int_s^t r < \pi$ for all $t \in [a,c]$ for some $c \in (a,b]$. Hence $\theta_{-\infty}(t) = 0$ for all $t \in (a,c]$. Let $c = lub\{d \in (a,b]$ such that $\theta_{-\infty}(t) = 0$ for all $t \in (a,d]\}$. We want to show that $c = b$. If $c < b$, take $s = c < t \le b$. Then
$$\theta_{-\infty}(t) \le \theta_{-\infty}(c) + \int_s^t r = \int_s^t r < \pi,$$
for all $t \in [c,h]$ for some h, $c < h < b$. This contradicts the definition of c. Therefore $\theta_{-\infty}(t) = 0$ for all $t \in (a,b]$ and $\theta(t,\lambda) \to 0^+$ as $\lambda \to -\infty$ for all $t \in (a,b]$. This concludes the proof of Theorem 4.5.3. □

6. Separated Boundary Conditions

In this section we study properties of eigenvalues for separated boundary conditions. An elementary proof for their existence can be based on the Prüfer transformation and the Prüfer equation from the previous section. The phase function θ also yields a characterization of the eigenvalues which is useful both theoretically and numerically.

In order to discuss the relationship between the SL equation and the equations arising from the Prüfer transformation we consider the equations

$$-(py')' + qy = \lambda w y \quad \text{on} \quad J, \tag{4.6.1}$$

$$\theta' = p^{-1} \cos^2 \theta + (\lambda w - q) \sin^2 \theta \quad \text{on} \quad J, \tag{4.6.2}$$

$$\rho' = [(p^{-1} + q - \lambda w) \sin \theta \cos \theta] \rho \quad \text{on} \quad J, \tag{4.6.3}$$

where

$$1/p, q, w \in L^1(J, \mathbb{R}), \quad \lambda \in \mathbb{R}, \quad p > 0 \quad \text{a.e.} \quad \text{on} \quad J = [a,b], \quad -\infty < a < b < \infty. \tag{4.6.4}$$

THEOREM 4.6.1. *Let (4.6.1) to (4.6.4) hold.*

(1) *Then every initial value problem for equation (4.6.2) has a unique real-valued solution and this solution is defined on all of J.*

(2) *Suppose θ and ρ are solutions of (4.6.2) and (4.6.3), respectively. Then $y = \rho \sin \theta$ is a solution of (4.6.1) on J and $py' = \rho \cos \theta$.*

(3) *Suppose y is a non-trivial solution of (4.6.1). Then there exists a solution θ of (4.6.2) and a solution ρ of (4.6.3) satisfying $\rho(t) \ne 0$ for $t \in J$, such that $y = \rho \sin \theta$ and $py' = \rho \cos \theta$.*

PROOF. Part (1) is known from the theory of first order non-linear differential equations; it can also be proved directly using the successive approximations method employed in the proof of Theorem 1.2.1, the linear existence-uniqueness theorem. Part (2) can be verified by a direct computation. For (3) determine ρ from
$$\rho^2 = y^2 + (py')^2$$
and determine θ from $\tan \theta(t) = y(t)/(py')(t)$, $\theta(a) = \alpha \in [0,\pi)$ if $(py')(t) \ne 0$ and from $\cot \theta(t) = (py')(t)/y(t)$, $\theta(a) = \alpha \subset [0,\pi)$ if $(py')(t) = 0$ and show that for both cases the same equation (4.6.2) is satisfied in a neighborhood of t and hence (4.6.2) holds on J. □

Consider the SLP consisting of the equation
$$-(py')' + qy = \lambda wy, \text{ on } (a,b), \ -\infty < a < b < \infty, \qquad (4.6.5)$$
together with separated boundary conditions
$$A_1 y(a) + A_2 (py')(a) = 0, \ (A_1, A_2) \neq (0,0), \ A_1, A_2 \in \mathbb{R}, \qquad (4.6.6)$$

$$B_1 y(b) + B_2 (py')(b) = 0, \ (B_1, B_2) \neq (0,0), \ B_1, B_2 \in \mathbb{R}, \qquad (4.6.7)$$
and coefficients satisfying
$$p, q, w : (a,b) \to \mathbb{R}, \ 1/p, q, w \in L(a,b), \ p > 0, \ w > 0, \text{ a.e. on } (a,b). \qquad (4.6.8)$$

THEOREM 4.6.2. *Let (4.6.5) to (4.6.8) hold. Then*
(1) *all eigenvalues are real and simple;*
(2) *there are an infinite but countable number of eigenvalues $\{\lambda_n : n \in \mathbb{N}_0\}$, they are bounded below, can be ordered to satisfy the inequalities*
$$-\infty < \lambda_0 < \lambda_1 < \lambda_2 < \lambda_3 < \cdots$$
and $\lambda_n \to \infty$ as $n \to \infty$.
(3) *If $u_n = u_n(\cdot, \lambda_n)$ is an eigenfunction of λ_n, then u_n has exactly n zeros in the open interval (a,b).*
(4) *Choose $\alpha \in [0, \pi)$ such that*
$$\tan \alpha = \frac{-A_2}{A_1}, \quad \text{if} \quad A_1 \neq 0, \quad \text{and} \quad \alpha = \pi/2 \quad \text{if} \quad A_1 = 0;$$
similarly, choose $\beta \in (0, \pi]$ such that
$$\tan \beta = \frac{-B_2}{B_1}, \quad \text{if} \quad B_1 \neq 0, \quad \text{and} \quad \beta = \pi/2 \quad \text{if} \quad B_1 = 0.$$
Then each eigenvalue λ_n is the unique solution $\lambda = \lambda_n$ of the equation
$$\theta(b, \lambda) = \beta + n\pi, \quad n \in \mathbb{N}_0, \qquad (4.6.9)$$
where θ is the solution (4.6.2) determined by the initial condition $\theta(a, \lambda) = \alpha$ for each $\lambda \in \mathbb{R}$.
(5) *The sequence of eigenfunctions $\{u_n = u_n(\cdot, \lambda_n) : n \in \mathbb{N}_0\}$ can be normalized to be an orthonormal sequence in the Hilbert space $H = L^2(J, w)$ i.e.*
$$\int_a^b u_n \overline{u}_m w = \begin{cases} 0, & \text{if} \quad n \neq m \\ 1, & \text{if} \quad n = m. \end{cases}$$
Furthermore, the orthonormal sequence $\{u_n = u_n(\cdot, \lambda_n) : n \in \mathbb{N}_0\}$ is complete in H i.e. for any $f \in H$ we have
$$f = \sum_0^\infty c_n u_n, \quad c_n = \int_a^b f \, u_n \, w.$$

Here the left equality is to be interpreted as meaning that the partial sums of the series on the right side of the equation converge to f in the norm of H.

PROOF. This is well known. Some details for parts (1) to (4) as well as the orthogonality of the eigenfunctions will be given below; for a completeness proof see [**123**], although the proof given there has stronger hypotheses, it can readily be adapted to yield a proof of part (5). □

THEOREM 4.6.3. *Let $\omega = (a, b, 0, \pi, 1/p, q, w) \in \Omega_{ss}$ and fix a, p, q, w. Assume that*

$$p > 0 \text{ a.e. and } q^2/w \in L_{loc}((a', b'), \mathbb{R}), \quad -\infty \leq a' < a < b < b' \leq \infty.$$

Then for any $n \in \mathbb{N}_0$, $\lambda_n^D(b)$ is strictly decreasing in b for $b \in (a, b')$ and

$$\lambda_n^D(b) \to \infty \text{ as } b \to a^+.$$

PROOF. See [**392**]. □

THEOREM 4.6.4. *Let $\omega = (a, b, \alpha, \pi/2, 1/p, q, w) \in \Omega_{ss}$ (i.e. we have an arbitrary separated self-adjoint BC at a but a Neumann condition at b.) Assume that*

$$Q = q/w \in AC_{loc}[a, b'), \; p(b) \geq \delta > 0 \text{ for } b \in (a, b').$$

Then

(1) $\lambda_0(b) \to Q(a) = q(a)/w(a)$ as $b \to a^+$.
(2) $\lambda_n(b) \to \infty$ as $b \to a^+$ for $n = 1, 2, 3, \ldots$.
(3) *If Q is decreasing in (a, b') then $\lambda_n(b)$ is decreasing in (a, b') and $\lambda_n(b) \geq Q(b)$ for each $n \in \mathbb{N}_0$.*
(4) *If Q is increasing in (a, b') and $Q(b) \to \infty$ as $b \to b'$ the $\lambda_0(b)$ is increasing in (a, b') and $\lambda_0(b) \leq Q(b)$; for $n \in \mathbb{N}$, $\lambda_n(b)$ has a unique extremum in (a, b') and this extremum is a strict minimum.*
(5) *If Q has a unique extremum in (a, b') and this extremum is a strict minimum and $Q(b) \to \infty$ as $b \to b'$ then for any $n \in \mathbb{N}_0$, $\lambda_n(b)$ has a unique extremum in (a, b') and this extremum is a strict minimum.*

PROOF. See Theorem 4.4 in [**392**]. □

Hinton and Lewis [**333**] introduced property BD of the spectrum: the spectrum is bounded below and discrete. The next example shows that property BD does not depend continuously on the coefficient $1/p$. This is just one of many illustrations of the delicate dependence of the spectrum on the problem.

EXAMPLE 4.6.1. *Consider the BVP with Dirichlet BC and the equation*

$$-(p_\varepsilon y')' = \lambda y \text{ on } (0, 1),$$

where $\varepsilon \in [0, 1]$ and

$$p_\varepsilon(t) = \begin{cases} -1, & \text{if } 0 \leq t \leq \varepsilon, \\ 1, & \text{if } \varepsilon < t \leq 1. \end{cases}$$

Then for $\varepsilon = 0$ the spectrum is bounded below but for each $\varepsilon > 0$ the spectrum is unbounded below. Note that $1/p_\varepsilon \to 1/p_0$ in $L^1((0,1), \mathbb{R})$.

The characterization (4.6.9) of Theorem 4.6.2 is established under the hypotheses $p > 0$ and $w > 0$. These assumptions on p and w guarantee that the spectrum is bounded below and (4.6.9) holds for each λ_n, $n \in \mathbb{N}_0$. This characterization of λ_n is interesting from a theoretical and a numerical perspective. Numerically it can be used to compute each eigenvalue independently of all other eigenvalues; this is done in the code SLEIGN2. Theoretically it can be used to study the dependence of λ_n on the problem. Also it follows directly from (4.6.9) that each eigenfunction of λ_n has exactly n zeros in the open interval J. When p changes sign the spectrum is unbounded above and below. Does (4.6.9) hold in this case for all the positive and negative eigenvalues? The next theorem gives an affirmative answer to this question.

THEOREM 4.6.5 (Binding and Volkmer). *Consider the SLP consisting of the equation*
$$-(py')' + qy = \lambda w y \quad on \quad J = (a,b),$$
together with separated boundary conditions
$$A_1 y(a) + A_2(py')(a) = 0, \ (A_1, A_2) \neq (0,0), \ A_1, A_2 \in \mathbb{R},$$
$$B_1 y(b) + B_2(py')(b) = 0, \ (B_1, B_2) \neq (0,0), \ B_1, B_2 \in \mathbb{R},$$
and coefficients satisfying
$$1/p, q, w \in L(J, \mathbb{R}), \ w > 0 \quad a.e. \quad on \quad J = (a,b), \ -\infty < a < b < \infty.$$

Assume that p changes sign in J. Then this SLP has only real and simple eigenvalues, there are an infinite but countable number of them, they are unbounded below and above and can be indexed and ordered to satisfy
$$.. < \lambda_{-3} < \lambda_{-2} < \lambda_{-1} < \lambda_0 < \lambda_1 < \lambda_2 < \lambda_3 < ... \quad (4.6.10)$$

Let θ be defined as in Theorem 4.6.1 above. Then for each integer $n \in \mathbb{Z}$ there is exactly one eigenvalue λ_n and it is the unique solution of the equation
$$\theta(b, \lambda_n) = n\pi + \beta. \quad (4.6.11)$$

There are no other eigenvalues. Here β is defined as in Theorem 4.6.2 and equation (4.6.9).

PROOF. See Binding and Volkmer [80] and the next remark. □

REMARK 4.6.1. The fact that all eigenvalues are real and that there are an infinite but countable number of them follows from the "standard" Hilbert space proof - using the Hilbert space $L^2(J, w)$ – and the self-adjoint operator realization of this SLP, see [123], [487]. Möller [474] showed that these eigenvalues are unbounded above and below. From the characterization of the eigenvalues as zeros of the characteristic function, see Lemmas 3.2.1 and 3.2.2, the eigenvalues are isolated with no finite accumulation point. The simplicity of the eigenvalues is clear from the separated boundary conditions. The indexing of the eigenvalues so that (4.6.10) is not unique, in fact rather arbitrary. It can be made more definite as follows: If $\lambda = 0$ is an eigenvalue denote it by λ_0 and let λ_1 denote the smallest positive eigenvalue. This exists: Let
$$\lambda_1 = \inf\{\lambda_n : \lambda_n > 0\},$$
then λ_1 is an eigenvalue by the continuity of the characteristic function $\delta(\lambda)$ and it is positive since $\lambda = 0$ is isolated. Similarly $\lambda_2 = \inf\{\lambda_n > \lambda_1\}$ is an eigenvalue greater than λ_1, etc. The same argument can be used when $\lambda = 0$ is not an eigenvalue. This is the indexing scheme used by the code SLEIGN2 for the numerical computation of the eigenvalues in the singular limit-circle oscillatory case. As already mentioned it is rather arbitrary, one can replace $\lambda = 0$ in this scheme by any real λ and use it for a "pivot". The characterization of the eigenvalues in terms of the Prüfer angle θ given by equation (4.6.11) is definite and explicit. It is interesting from a theoretical and numerical perspective. For instance, if one computes an eigenvalue as a root of the characteristic equation $\delta(\lambda) = 0$, the question arises: Which eigenvalue is it? The characterization (4.6.11) gives a definite answer to this question for this class of SLP. In general indexing the eigenvalues in some definite and explicit manner is a difficult open problem for general SLP; even in the case when the coefficients are

real-valued and the boundary conditions are self-adjoint, e.g. when both p and w change sign.

REMARK 4.6.2. When comparing the characterizations (4.6.9) of Theorem 4.6.2 with (4.6.11) of Theorem 4.6.6 it is important to realize that from the former (when $p > 0$) it follows that each eigenfunction u_n of λ_n in Theorem 4.6.2 has exactly n zeros in the open interval J, but that, when p changes sign, there is no analogous result in spite of the beautiful characterization of the eigenvalues λ_n by equation (4.6.11) of Theorem 4.6.6. Binding and Volkmer show with theorems and examples that the zero properties of eigenfunctions under the conditions of Theorem 4.6.3 can be quite "strange", for instance eigenfunctions can have an infinite number of zeros in the interval J. We give one of their examples:

EXAMPLE 4.6.2 (Binding and Volkmer). Let $J = (0,1)$, $w = 1$, $q \in L^1(J, \mathbb{R})$, and define p by
$$\frac{1}{p(t)} = 2t\cos(1/t) + \sin(1/t),\ 0 < t < 1.$$
Consider the SLP
$$-(py')' + qy = \lambda y \text{ on } J,$$
with a boundary condition
$$y(0) = 0,\ B_1 y(1) + B_2 (py')(1) = 0,\ B_1, B_2 \in \mathbb{R},\ (B_1, B_2) \neq (0,0).$$

Every eigenfunction has an infinite number of zeros accumulating at the left endpoint 0.

Note that this is a regular problem on J; in particular 0 is a regular endpoint. Compare this with Theorem 2.6.1.

7. Coupled Boundary Conditions

In this section we specialize some of the results of Chapter 3 for non-self-adjoint problems to the self-adjoint case and obtain some additional information for the latter case.

A self-adjoint regular SLP consists of the equation

$$-(py')' + qy = \lambda wy \text{ on } J,\ J = (a,b),\ -\infty \leq a < b \leq \infty,\ 1/p, q, w \in L^1(J, \mathbb{R}), \tag{4.7.1}$$

together with boundary conditions
$$AY(a) + BY(b) = 0,\quad A, B \in M_2(\mathbb{C}),\ \operatorname{rank}(A:B) = 2,$$
$$AEA^* = BEB^*,\ E = \begin{bmatrix} 0 & -1 \\ 1 & 0 \end{bmatrix}. \tag{4.7.2}$$

The *coupled* regular self-adjoint boundary conditions have the canonical form:
$$Y(b) = e^{i\gamma} K Y(a),\ -\pi < \gamma \leq \pi,\ K \in SL_2(\mathbb{R}),\ Y = \begin{bmatrix} y \\ py' \end{bmatrix}. \tag{4.7.3}$$

LEMMA 4.7.1. *Let (4.7.1) and (4.7.3) hold and let $D(\lambda)$ be defined by (3.2.11). Then a number λ is an eigenvalue of the SLP (4.7.1), (4.7.3) if and only if*
$$D(\lambda) = 2\cos\gamma. \tag{4.7.4}$$

PROOF. This is a special case of Lemma 3.2.6. □

LEMMA 4.7.2. *Let Φ be the primary fundamental matrix of the system representation of equation (4.7.1). A number λ is an eigenvalue of geometric multiplicity two of the SLP (4.7.1), (4.7.3) if and only if*

$$\Phi(b,\lambda) = e^{i\gamma}K. \tag{4.7.5}$$

PROOF. This is a special case of Lemma 3.2.3. □

LEMMA 4.7.3. *Let (4.7.1), (4.7.3) hold. Assume that $-\pi < \gamma < 0$ or $0 < \gamma < \pi$, and suppose $w > 0$. Then every eigenvalue is simple.*

PROOF. With the hypotheses on p and w all eigenvalues are real. The conclusion then follows from (4.7.5) since $\Phi(b,\lambda)$ and K are both real-valued. □

LEMMA 4.7.4. *For the SLP (4.7.1), (4.7.3) with $p > 0$, $w > 0$ we have*

$$\lambda_n(e^{i\gamma}K) = \lambda_n(e^{-i\gamma}K) \tag{4.7.6}$$

for any $\gamma \in (-\pi, \pi]$ and any $n \in \mathbb{N}_0$. Furthermore, if z is an eigenfunction of $\lambda_n(e^{i\gamma}K)$, then its complex conjugate $\bar{z}$ is an eigenfunction of $\lambda_n(e^{-i\gamma}K)$.

PROOF. The result (4.7.5) follows from the characterization (4.7.4). The furthermore statement can be verified by a direct substitution. □

THEOREM 4.7.1. *Let (4.7.1), (4.7.3) hold. Fix J, p, q, w and assume $p > 0$, $w > 0$. Consider $\lambda_n(e^{i\gamma}K)$ as a function of $K = (k_{ij}) \in SL_2(\mathbb{R})$. Then:*
- *If $k_{12} \neq 0$, the $\lambda_n(e^{i\gamma}K)$ is a continuous function of K for each $n \in \mathbb{N}_0$.*
- *If $k_{12} = 0$ and λ_n is simple, then λ_n is not continuous at K for each $n \in \mathbb{N}_0$.*

PROOF. This is a special case of Theorem 4.3.3; also see Sections 2, 3, and 4 of [384]. □

8. An Elementary Existence Proof for Coupled Boundary Conditions

In this section we study the variation of the eigenvalues with respect to the boundary conditions for a fixed equation. Both separated and coupled BC are considered and inequalities established for each class and between these classes.

For $K \in SL_2(\mathbb{R})$, $K = \begin{bmatrix} k_{11} & k_{12} \\ k_{21} & k_{22} \end{bmatrix}$, denote by $\mu_n = \mu_n(K)$ and $\nu_n = \nu_n(K)$, $n \in \mathbb{N}_0$, the eigenvalues for the separated boundary conditions

$$y(a) = 0, \quad k_{22}y(b) - k_{12}y^{[1]}(b) = 0, \tag{4.8.1}$$

$$y^{[1]}(a) = 0, \quad k_{21}y(b) - k_{11}y^{[1]}(b) = 0, \tag{4.8.2}$$

respectively. Note that $(k_{22}, k_{12}) \neq (0,0) \neq (k_{21}, k_{11})$ since $\det K = 1$. Therefore each of these is a self-adjoint separated boundary condition with a countably infinite number of only real eigenvalues.

THEOREM 4.8.1. *Let (4.1.1), (4.1.2) hold and assume that $p > 0$, $w > 0$ on J. Let μ_n and ν_n, $n \in \mathbb{N}_0$ be the eigenvalues for (4.8.1), and (4.8.2), respectively. Let $\omega = (a, b, A, B, 1/p, q, w) \in \Omega_{s-a}$. Fix a, b, p, q, w and let $A = e^{i\gamma}K$, $K = (k_{ij})$, $B = -I$, where $K \in SL_2(\mathbb{R})$ i.e. we have the BC*

$$Y(b) = e^{i\gamma}KY(a), \quad -\pi < \gamma \leq \pi. \tag{4.8.3}$$

Let μ_n and ν_n, $n \in \mathbb{N}_0$ be the eigenvalues for (4.8.1), and (4.8.2), respectively.

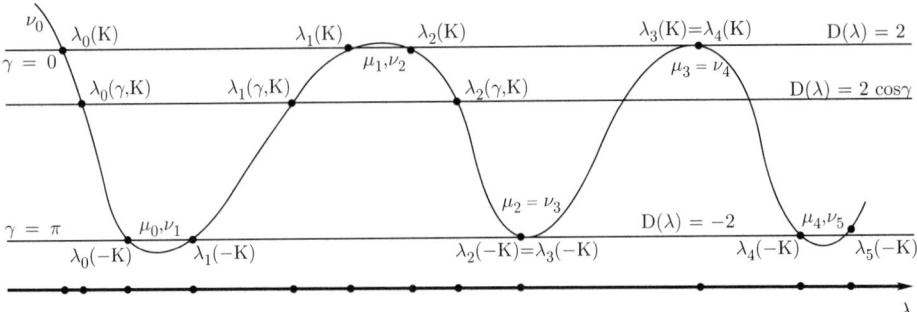

FIGURE 1. The graph of $D(\lambda) = 2\cos(\gamma)$ illustrating the inequalities (4.8.5). For (4.8.4) the position of ν_n moves to ν_{n+1}.

Denote the eigenvalues for (4.8.3) by $\lambda_n(\gamma, K)$, abbreviated to $\lambda_n(K)$ when $\gamma = 0$, for $n \in N_0$.

- Suppose that $k_{12} < 0$ and $k_{11} \leq 0$. Then
(1) $\lambda_0(K)$ is simple;
(2) $\lambda_0(K) < \lambda_0(-K)$ and
(3) the following inequalities hold for $-\pi < \gamma < 0$ and $0 < \gamma < \pi$:

$$-\infty < \lambda_0(K) < \lambda_0(\gamma, K) < \lambda_0(-K) \leq \{\mu_0, \nu_0\}$$
$$\leq \lambda_1(-K) < \lambda_1(\gamma, K) < \lambda_1(K) \leq \{\mu_1, \nu_1\}$$
$$\leq \lambda_2(K) < \lambda_2(\gamma, K) < \lambda_2(-K) \leq \{\mu_2, \nu_2\}$$
$$\leq \lambda_3(-K) < \lambda_3(\gamma, K) < \lambda_3(-K) \leq \{\mu_3, \nu_3\} \leq \cdots \quad (4.8.4)$$

- Suppose that $k_{12} \leq 0$ and $k_{11} > 0$. Then
(1) $\lambda_0(K)$ is simple;
(2) $\lambda_0(K) < \lambda_0(-K)$ and
(3) the following inequalities hold for $-\pi < \gamma < 0$ and $0 < \gamma < \pi$:

$$\nu_0 \leq \lambda_0(K) < \lambda_0(\gamma, K) < \lambda_0(-K) \leq \{\mu_0, \nu_1\}$$
$$< \lambda_1(-K) < \lambda_1(\gamma, K) < \lambda_1(K) \leq \{\mu_1, \nu_2\}$$
$$\leq \lambda_2(K) < \lambda_2(\gamma, K) < \lambda_2(-K) \leq \{\mu_2, \nu_3\}$$
$$\leq \lambda_3(-K) < \lambda_3(\gamma, K) < \lambda_3(K) \leq \{\mu_3, \nu_4\} \leq \cdots \quad (4.8.5)$$

- Furthermore, for $0 < \alpha < \beta < \pi$ we have

$$\lambda_0(\beta, K) < \lambda_0(\alpha, K) < \lambda_1(\alpha, K) < \lambda_1(\beta, K) < \lambda_2(\beta, K) < \lambda_2(\alpha, K)$$
$$< \lambda_3(\alpha, K) < \lambda_3(\beta, K) < \cdots$$

- If neither of the above cases holds for K then one of them must hold for $-K$. The notation $\{\mu_n, \nu_m\}$ is used to indicate either ν_n or ν_m but no comparison is made between μ_n and ν_m.

PROOF. For special classes of the coupling matrix K these inequalities were established in Bailey, Everitt and Zettl [41], Weidmann [600], Eastham [153]. The general result is proven in Eastham, Kong, Wu and Zettl [145]. □

8. AN ELEMENTARY EXISTENCE PROOF FOR COUPLED BOUNDARY CONDITIONS

These inequalities have a number of interesting consequences.

REMARK 4.8.1. Assume $p > 0$. In Sections 4.5 and 4.6 the Prüfer transformation is used to establish the existence and to characterize all the eigenvalues for separated self-adjoint boundary conditions. The *existence and a characterization of all the eigenvalues for coupled self-adjoint boundary conditions*

$$Y(b) = e^{i\gamma} K Y(a), \quad -\pi < \gamma \leq \pi, \quad K \in SL_2(\mathbb{R}),$$

can be established as follows: Starting with the eigenvalues μ_n and ν_n, $n \in \mathbb{N}_0$, of the separated BC (4.8.1), (4.8.2) the proof of Theorem 4.8.1 in [**145**] (although this is not explicitly pointed out there) actually shows that there is one and only one eigenvalue of (4.8.3) in the interval $(-\infty, \mu_0]$ and it is $\lambda_0(\gamma, K)$; there is exactly one eigenvalue in the interval $[\mu_0, \mu_1]$ and it is $\lambda_1(\gamma, K)$; there is exactly one eigenvalue of (4.8.3) in the interval $[\mu_n, \mu_{n+1}]$ and it is $\lambda_{n+1}(\gamma, K)$, for $n \in \mathbb{N}_0$. Such an algorithm is used by SLEIGN2 to compute the eigenvalues of (4.8.3).

Further consequences of these inequalities are given below.

THEOREM 4.8.2. *Let* $\omega = (a, b, A, B, 1/p, q, w) \in \Omega_{s-a}$ *with* $p > 0$ *and let* $J = [a, b]$, $-\infty < a < b < \infty$. *Fix* a, b, p, q, w. *Then*

$$\lambda_n(A, B) \leq \lambda_n^D, \quad n \in \mathbb{N}_0. \tag{4.8.6}$$

Recall that λ_n^D *denotes the* $n-th$ *Dirichlet eigenvalue. Thus the n-th Dirichlet eigenvalue maximizes all n-th self-adjoint eigenvalues. Equality can hold in (4.8.6) for non-Dirichlet eigenvalues.*

PROOF. We first prove the inequality for separated boundary conditions. Suppose for some separated boundary condition and some $n \in \mathbb{N}_0$ we have $\lambda_n^D < \lambda_n$. Let $u(\cdot, \lambda_n^D)$, $u(\cdot, \lambda_n)$ be eigenfunctions of λ_n^D and λ_n, respectively. Since $u(\cdot, \lambda_n^D)$ has $n+2$ zeros in the closed interval $[a, b]$ (n in the interior and one at each endpoint) it follows from the Sturm Comparison Theorem that $u(\cdot, \lambda_n)$ has $n+1$ zeros in the open interval. This contradicts Theorem 3.6.2. The general case follows from inequalities (4.8.4) and (4.8.5) applied to either K or $-K$. □

Equality can occur in (4.8.6). The next result characterizes all such cases of equality for $n = 0$.

PROPOSITION 4.8.1. *Let the hypotheses and notation of Theorem 4.8.2 hold; let* $\Phi(t, \lambda) = (\phi_{ij}(t, \lambda))$ *be the principal fundamental matrix of the system representation of equation (4.7.1). Then*

$$\lambda_0(A, B) = \lambda_0^D$$

if and only if the boundary condition (is the Dirichlet condition) or the boundary condition matrices A, B *are given by*

$$A = \begin{bmatrix} \phi_{11}(b, \lambda_0^D) & 0 \\ \phi_{21}(b, \lambda_0^D) & \phi_{22}(b, \lambda_0^D) \end{bmatrix}, B = \begin{bmatrix} -1 & 0 \\ -d & -1 \end{bmatrix}, \text{ with } d \leq 0.$$

In canonical form, these conditions are given by the coupling matrix K *where*

$$K = \begin{bmatrix} \phi_{11}(b, \lambda_0^D) & 0 \\ d\,\phi_{11}(b, \lambda_0^D) + \phi_{21}(b, \lambda_0^D) & \phi_{22}(b, \lambda_0^D) \end{bmatrix}, d \leq 0.$$

PROOF. See Corollary 4.5 in Haertzen, Kong, WU and Zettl [**285**]. □

REMARK 4.8.2. We make a number of observations about Proposition 4.8.1.

(1) $\lambda_0(\alpha,\beta) = \lambda_0^D$ if and only if $\alpha = 0$ and $\beta = \pi$. In other words for no *separated* boundary condition other than the Dirichlet condition does equality hold in (4.8.6) when $n = 0$.

(2) For no **complex coupled** boundary condition does equality hold in (4.8.6) when $n = 0$.

(3) For $n = 0$ equality holds in 4.8.6 for some coupled real boundary conditions. All these are characterized in Proposition 4.8.1 and all of them lie on the jump set $\mathbb{J}$. (Recall that this is the set of boundary conditions on which all eigenvalues λ_n have jump discontinuities as functions of the boundary conditions.)

(4) **Friedrichs Extension.** Among all self-adjoint realizations of the Sturm-Liouville equation (4.7.1) with $p > 0$, $w > 0$ there is a special one ("eine ausgezeichnete") often singled out in Applied Mathematics and Mathematical Physics which is called the "Friedrichs extension" in honor of K. O. Friedrichs who constructed it without direct reference to any boundary condition. One of its basic properties is that it preserves the lower bound of the minimal operator associated with equation (4.7.1) in the Hilbert space $L^2(J,w)$; however it is not characterized by this property i.e. there may be other self-adjoint extensions of the minimal operator which preserve its lower bound. This lower bound is λ_0^D. Thus Proposition 4.8.1 characterizes all self-adjoint realizations which preserve the lower bound of the minimal operator and shows that this happens only for the Dirichlet condition and for certain real coupled boundary conditions which lie on the jump set $\mathbb{J}$.

(5) For regular self-adjoint Sturm-Liouville problems with positive leading coefficient and positive weight function there are always an infinite number of extensions of the minimal operator which preserve its lower bound. Except for the Friedrichs extension which is determined by the Dirichlet boundary condition, every such extension is determined by real coupled boundary conditions.

THEOREM 4.8.3. *Let the hypotheses and notation of Theorem 4.8.2 hold. Then*

$$\lambda_n^D \leq \lambda_{n+2}(A,B), \quad n \in \mathbb{N}_0. \tag{4.8.7}$$

PROOF. See Section 4 of [**384**]. □

As the boundary condition matrices A, B vary what values does λ_n assume? The next result answers this question.

THEOREM 4.8.4. *Let $\omega = (a,b,A,B,1/p,q,w) \in \Omega_{s-a}$ with $p > 0$ and $-\infty < a < b < \infty$. Fix $a,b,1/p,q,w$.*

(1) *The range of $\lambda_0(A,B)$ is $(-\infty, \lambda_0^D]$.*
(2) *The range of $\lambda_1(A,B)$ is $(-\infty, \lambda_0^D]$.*
(3) *The range of $\lambda_n(A,B)$ is $(\lambda_{n-2}^D, \lambda_n^D]$ for $n \geq 2$.*

Moreover, these three statements still hold when A, B are restricted to be real.

PROOF. See [**384**]. □

The next theorem is contained in Theorem 4.3.1 but we state it here for the sake of completeness.

THEOREM 4.8.5. *Let* $\omega = (a, b, A, B, 1/p, q, w) \in \Omega_{s-a}$ *with* $B = -I$, $A = K \in SL_2(\mathbb{R})$, *assume* $p > 0$ *and* $-\infty < a < b < \infty$.

- *Let* u_n *be a real eigenfunction of* $\lambda_n(K)$. *Then the number of zeros of* u_n *in the interval* $[a, b)$ *is 0 or 1 if* $n = 0$, *and* $n - 1$, *or* n *or* $n + 1$ *if* $n \geq 1$.
- *Let* u_n *be a (necessarily non-real) eigenfunction of* $\lambda_n(e^{i\gamma} K)$, $0 < \gamma < \pi$. *Then*
  (1) u_n *has no zero in* $[a, b]$;
  (2) *the number of zeros of* $\operatorname{Re} u_n$ *in the interval* $[a, b)$ *is 0 or 1 if* $n = 0$, *and* $n - 1$, *or* n *or* $n + 1$ *if* $n \geq 1$.
  (3) *the number of zeros of* $\operatorname{Im} u_n$ *in the interval* $[a, b)$ *is 0 or 1 if* $n = 0$, *and* $n - 1$, *or* n *or* $n + 1$ *if* $n \geq 1$.

PROOF. See [**384**]. □

THEOREM 4.8.6. *Let* $\omega = (a, b, A, B, 1/p, q, w) \in \Omega_{s-a}$ *with* $p > 0$ *and* $-\infty < a < b < \infty$. *Let (4.8.3) hold with* $\gamma = 0$ *and let* v_n, ν_n *be the eigenvalues of the separated boundary conditions (4.8.1), (4.8.2). Then an eigenvalue* $\lambda_n(K)$ *is geometrically double if and only if there exist* $k, m \in \mathbb{N}_0$ *such that*

$$\lambda_n(K) = \mu_k = \nu_m.$$

PROOF. See Theorem 4.3 in [**384**]. □

REMARK 4.8.3. In Theorem 4.10.1 below the equivalence between the algebraic and geometric multiplicity of self-adjoint SLP eigenvalues will be established. Thus the word "geometrically" can be omitted in Theorem 4.8.7.

THEOREM 4.8.7. *Let* $\omega = (a, b, A, B, 1/p, q, w) \in \Omega_{rc}$ *with* $p \geq 0$ *and* $-\infty < a < b < \infty$. *Given eigenvalues* $\lambda_n(K)$ *and* $\lambda_{n+1}(K)$ *of* K ($B = -I$, $A = K$), *distinct or not, there exist eigenvalues* v_k, ν_m *of the separated boundary conditions (4.8.1), (4.8.2) such that*

$$\lambda_n(K) \leq \{\mu_k, \nu_m\} \leq \lambda_{n+1}(K).$$

PROOF. See Corollary 4.2 in [**384**]. □

9. Monotonicity of Eigenvalues

THEOREM 4.9.1. *Let* $\omega = (a, b, A, B, 1/p, q, w) \in \Omega_{s-a}$, *suppose* $p > 0$, *and fix* a, b, A, B, *with* $-\infty < a < b < \infty$. *Assume that* $\lambda_n(\omega)$ *is a simple eigenvalue of* ω.

(1) *Fix* p, w. *Suppose* $Q \in L^1((a, b), \mathbb{R})$ *and assume that* $Q \geq q$ *a.e. on* $[a, b]$.
    Then for any $n \in \mathbb{N}_0$, $\lambda_n(Q) \geq \lambda_n(q)$. *If* $Q > q$ *on a subset of* $[a, b]$ *having positive Lebesgue measure, then* $\lambda_n(Q) > \lambda_n(q)$, $n \in \mathbb{N}_0$.
(2) *Fix* q, w. *Suppose* $1/P \in L^1((a, b), \mathbb{R})$ *and* $0 < P \leq p$ *a.e. on* $[a, b]$.
    Then for any $n \in \mathbb{N}_0$, $\lambda_n(1/P) \geq \lambda_n(1/p)$; *if* $1/P < 1/p$ *on a subset of* $[a, b]$ *having positive Lebesgue measure, then* $\lambda_n(1/P) < \lambda_n(1/p)$.
(3) *Fix* p, q. *Suppose* $W \in L^1((a, b), \mathbb{R})$ *and* $W \geq w > 0$ *a.e. on* $[a, b]$.
    Let $n \in \mathbb{N}_0$. *Then* $\lambda_n(W) \geq \lambda_n(w)$ *if* $\lambda_n(W) < 0$ *and* $\lambda_n(w) < 0$; *but* $\lambda_n(W) \leq \lambda_n(w)$ *if* $\lambda_n(W) > 0$ *and* $\lambda_n(w) > 0$. *Furthermore, if strict inequality holds in the hypothesis on a set of positive Lebesgue measure, then strict inequality holds in the conclusion.*

PROOF. We give the proof for (1), the proofs of (2) and (3) are similar. Define a function $f : \mathbb{R} \to \mathbb{R}$ by
$$f(t) = \lambda_n(s(t)), \ s(t) = q + t(Q - q), \quad t \in [0, 1].$$
Then $s(t) \in L^1((a, b), \mathbb{R})$ for each $t \in [0, 1]$. From the chain rule in Banach space and the formula for $\lambda'_n(q)$ in Section 4.3 we have
$$f'(t) = \lambda'_n((s(t)) \, s'(t) = \int_a^b |u^2(r, s(t))| \, (Q(r) - q(r)) \, dr \geq 0, \ t \in [0, 1].$$
Hence f is nondecreasing on $[0, 1]$ and $f(1) = \lambda_n(Q) \geq \lambda_n(q) = f(0)$. The strict inequality part of the theorem also follows from this argument. □

10. Multiplicity of Eigenvalues

In this section we show that the algebraic and geometric multiplicities of the eigenvalues of regular self-adjoint SLP with p, w positive are the same. This requires a detailed analysis of the characteristic functions $D(\lambda)$ and $\delta(\lambda)$. Recall that the geometric multiplicity of an eigenvalue is the dimension of its eigenspace i.e. the number of linearly independent eigenfunctions of this eigenvalue. The algebraic multiplicity of an eigenvalue is the order of its zero as a root of the characteristic function $\delta(\lambda)$ constructed in Chapter 3. See the Comments section at the end of this chapter for an alternative definition of algebraic multiplicity used in Functional Analysis.

THEOREM 4.10.1. *Let $\omega = (a, b, A, B, 1/p, q, w) \in \Omega_{s-a}$ and assume that $p > 0$. The algebraic and geometric multiplicities of the eigenvalues of regular self-adjoint Sturm-Liouville problems are the same.*

PROOF. For coupled BC this is given in [**145**]. The separated case is proven in Theorem 4.12 of [**384**]. □

From here on we speak only of the multiplicity of an eigenvalue.

THEOREM 4.10.2. *Let $\omega = (a, b, A, B, 1/p, q, w) \in \Omega_{s-a}$ with $B = -I$ and $A = K \in SL_2(\mathbb{R})$, $J = (a, b)$ and assume that $p > 0$. Fix all components of ω except q and fix $n \in \mathbb{N}_0$. Let*
$$S_1 = \{q \in L^1(J, \mathbb{R}) : \lambda_n(q) \text{ is simple}\};$$
$$S_2 = \{q \in L^1(J, \mathbb{R}) : \lambda_n(q) \text{ is double}\}.$$
Then S_1 is an open set in $L^1(J, \mathbb{R})$ and S_2 is closed and nowhere dense in $L^1(J, \mathbb{R})$. These conclusions hold when q is replaced by either $1/p$ or by w.

PROOF. This follows from Theorem 4.3 of [**394**] and the continuous dependence of λ_n on $1/p$, q and w established in [**384**]. □

11. Green's Function

Here we specialize the Green's function constructed in Section 3.8 to the self-adjoint case. The eigenvalues of regular self-adjoint SLP can be obtained from the spectrum of the integral operator whose kernel is the Green's function. These operators are compact self-adjoint operators in the Hilbert space $L^2(J, w)$. Thus the well developed theory of self-adjoint operators in Hilbert space can be applied to establish the existence of the eigenvalues of regular self-adjoint SLP.

11. GREEN'S FUNCTION

This section is devoted to the study of self-adjoint regular SLP consisting of the equation

$$-(py')' + qy = \lambda w y \text{ on } J = (a,b), \text{ with } -\infty \leq a < b \leq \infty,$$
$$1/p, q, w \in L^1(J, \mathbb{R}), w > 0 \text{ a.e.} \tag{4.11.1}$$

together with boundary conditions

$$AY(a) + BY(b) = 0, A, B \in M_2(\mathbb{C}), \text{rank}(A:B) = 2,$$
$$AEA^* = BEB^*, E = \begin{bmatrix} 0 & -1 \\ 1 & 0 \end{bmatrix}. \tag{4.11.2}$$

Note that we assume $w > 0$ a.e in this section but no sign restriction is made on p. The reason for the restriction to positive weight functions w is so that the space (of equivalence classes of functions f)

$$H = L^2(J, w) = \{f : J \to \mathbb{C}, \int_J |f|^2 w < \infty\},$$

is a Hilbert space.

The next lemma overlaps with Theorem 4.3.1 whose proof was indicated only in general terms. Here we supply some details and discuss the orthogonality of the eigenfunctions.

LEMMA 4.11.1. *All eigenvalues of (4.11.1), (4.11.2) are real and eigenfunctions belonging to different eigenvalues are orthogonal in the space H.*

PROOF. The self-adjoint boundary conditions are either separated or coupled and, if coupled, can be put into the canonical form

$$B = -I, \quad A = e^{i\gamma}K, \quad K \in SL_2(\mathbb{R}).$$

Assume that $\lambda \in \mathbb{C}$ is an eigenvalue for a coupled boundary condition in this canonical form. Multiply the equation of (4.11.1) by $\overline{y}$ and its conjugate by y, subtract to get

$$(\lambda - \overline{\lambda})y\overline{y}w = -\overline{y}(py')' + y(p\overline{y}')' = [y(p\overline{y}') - \overline{y}(py')]' = [y, \overline{y}]'$$

and hence, noting that $K^*EK = E$ we get

$$(\lambda - \overline{\lambda})\int_a^b |y|^2 w = [y, \overline{y}](b) - [y, \overline{y}](a) = (EY(b), Y(b))) - (EY(a), Y(a))$$
$$= (Ee^{i\gamma}KY(a), e^{i\gamma}KY(a)) - (EY(a), Y(a))$$
$$= (K^*EKY(a), Y(a)) - (EY(a), Y(a)) = 0.$$

Since the integral on the left is not zero (here we use the hypothesis that $w > 0$) we conclude that $\lambda - \overline{\lambda} = 0$ i. e. $\lambda = \overline{\lambda}$.

To prove the orthogonality assume λ and μ are distinct eigenvalues with eigenfunctions y and z, respectively. Proceeding as above with $\overline{\lambda}$ replaced by v and $\overline{y}$ by z we get

$$(\lambda - \mu)\int_a^b y\overline{z}w = 0.$$

For the case of separated boundary conditions the proof is similar; for this case we can see that the right-hand side of the equation is zero by a direct substitution. The orthogonality property also follows similarly for this case. □

The next result establishes the existence of eigenvalues for general regular self-adjoint SLP. We first observe that, from Lemma 4.11.1 and the characterization of the eigenvalues as zeros of an entire function in Section 3.2, it follows that the eigenvalues are all isolated points on the real axis. In particular no regular SLP can have all real numbers as eigenvalues, nor all positive rational or irrational numbers.

THEOREM 4.11.1 (The Green's Kernel Operator). *Let (4.11.1), (4.11.2) hold and let Φ denote the primary fundamental matrix of the system formulation of equation 4.11.1). Assume that for some $\mu \in \mathbb{R}$,*

$$\det[A + B\,\Phi(b,a,P,W,\mu)] \neq 0. \tag{4.11.3}$$

Then

(1) *$\lambda = \mu$ is not an eigenvalue of (4.11.1), (4.11.2);*
(2) *the Green's function $K(\cdot,\cdot,\mu) = K_{12}(\cdot,\cdot,\mu,A,B,P,w)$ exists and is hermitian i.e.*

$$K(t,s,\mu) = \overline{K(s,t,\mu)}, \ a \leq s,t \leq b;$$

(3) *the integral operator T defined by $(Tf)(t) = \int_a^b K(t,s,\mu)\,w(s)\,f(s)\,ds,\ t \in J,\ f \in H = L^2(J,w)$ is a compact (completely continuous) self-adjoint operator defined on the space H and mapping H into H;*
(4) *$\lambda \in \mathbb{C}$ is an eigenvalue of the SLP (4.11.1), (4.11.2) if and only if $1/(\lambda-\mu)$ is an eigenvalue of the operator T;*
(5) *the eigenvalue λ of the SLP (4.11.1), (4.11.2) and the eigenvalue $1/(\lambda-\mu)$ of T have the same eigenfunction;*
(6) *the operator T and therefore the boundary value problem (4.11.1), (4.11.2) have a countably infinite number of eigenvalues, these are all real and can be indexed to satisfy (4.3.2) if $p > 0$ on J, and (4.3.1) if p changes sign.*

PROOF. Although the proof given in the classic book by Coddington and Levinson [**123**] uses stronger hypotheses, it can readily be adapted to this case. The hermitian property of the Green's function is a special case of Theorem 3.8.3 and is equivalent to

$$AEA^* = BEB^*,\ E = \begin{bmatrix} 0 & -1 \\ 1 & 0 \end{bmatrix},\ \text{rank}(A:B) = 2. \tag{4.11.4}$$

$\square$

12. Finite Real Spectrum

In this section we specialize the construction of non-self-adjoint problems with finite spectrum, given in Section 3.7 to the self-adjoint case. In this case all eigenvalues are real and much more information is available including eigenvalue inequalities analogous to those given in Section 4.8 for an infinite number of eigenvalues.

Here we construct *self-adjoint* SLP with exactly n eigenvalues for each non-negative integer n (See the example at the end of this section for the case $n = 0$). Clearly this construction cannot allow $r = 1/p$ and w to be positive on any common subinterval. Thus, we choose r and w such that r they are alternatively zero on consecutive subintervals.

We also show that these n eigenvalues can be arbitrarily distributed throughout the real line $\mathbb{R}$: given any k disjoint open sets J_i in $\mathbb{R}$ and any k integers n_i there exists a self-adjoint SLP with exactly n_k eigenvalues in J_i, for $i = 1, 2, ..., k$. This complements the well known classical result that for a self-adjoint regular SLP with positive leading coefficient and weight function, the eigenvalues go to infinity asymptotically as n^2.

Here we consider the equation

$$-(py')' + qy = \lambda w y \text{ on } J = (a,b) \text{ with } -\infty \leq a < b \leq \infty \qquad (4.12.1)$$

where λ is the spectral parameter and the coefficients satisfy the minimal conditions

$$r = 1/p, \; q, \, w \in L^1(J, \mathbb{R}). \qquad (4.12.2)$$

According to Theorems 2.2.1 and 2.2.2 condition (4.12.2) is minimal (in the real-valued i.e. symmetric case) in the sense that it is necessary and sufficient for all initial value problems of equation (4.12.1) to have unique real-valued solutions on J, see [**231**]. Let $r = 1/p$. In this section it is best to think of equation (4.12.1) is terms of its equivalent system

$$y' = r z, \quad z' = (q - \lambda w) y \quad \text{on } J. \qquad (4.12.3)$$

Recall that by a trivial solution of equation (4.12.1) on some interval we mean a solution y which is identically zero and whose quasi-derivative $z = py'$ is also identically zero on this interval.

Consider the two point boundary conditions

$$AY(a) + BY(b) = 0, \; Y = \begin{bmatrix} y \\ py' \end{bmatrix}. \qquad (4.12.4)$$

determined by 2×2 complex matrices A, B satisfying the self-adjointness conditions:

$$AEA^* = BEB^*, \; rank(A : B) = 2. \qquad (4.12.5)$$

For the convenience of the reader we recall the characterization of the eigenvalues as the zeros of the characteristic function constructed in Chapter 3.

LEMMA 4.12.1. *Let (4.12.2) hold, and let $\Phi(t, \lambda) = [\phi_{ij}](t, \lambda)$ denote the fundamental matrix of the system (4.12.3) determined by the initial condition $\Phi(a, \lambda) = I, \lambda \in \mathbb{C}$. Then a complex number λ is an eigenvalue of the SLP (4.12.1), (4.12.4) if and only if*

$$\delta(\lambda) = \det[A + B \, \Phi(b, \lambda)] = 0.$$

Let

$$C = \begin{bmatrix} c_{11} & c_{12} \\ c_{21} & c_{22} \end{bmatrix} := \begin{bmatrix} a_{22}b_{11} - a_{12}b_{21} & a_{11}b_{21} - a_{21}b_{11} \\ a_{22}b_{12} - a_{12}b_{22} & a_{11}b_{22} - a_{21}b_{12} \end{bmatrix} \qquad (4.12.6)$$

Then for $\lambda \in \mathbb{C}$ we have

$$\delta(\lambda) = \det(A) + \det(B) + c_{11}\phi_{11}(b, \lambda) + c_{12}\phi_{12}(b, \lambda) + c_{21}\phi_{21}(b, \lambda) + c_{22}\phi_{22}(b, \lambda). \qquad (4.12.7)$$

PROOF. The characterization of the eigenvalues as zeros of the characteristic equation $\delta(\lambda) = 0$ is contained in Lemma 3.2.2, the formula (4.12.7) follows from a straight-forward computation. □

Recall the two categories of self-adjoint boundary conditions:
Separated BC:

$$\begin{aligned} &A_1 y(a) + A_2 (py')(a) = 0, \quad A_1, A_2 \in \mathbb{R}, \quad (A_1, A_2) \neq (0,0), \\ &B_1 y(b) + B_2 (py')(b) = 0, \quad B_1, B_2 \in \mathbb{R}, \quad (B_1, B_2) \neq (0,0). \end{aligned} \qquad (4.12.8)$$

Coupled BC:

$$Y(b) = e^{i\gamma} K Y(a), \ Y = \begin{bmatrix} y \\ py' \end{bmatrix}, \ -\pi \leq \gamma \leq \pi, \ K \in SL_2(\mathbb{R}), \qquad (4.12.9)$$

$$SL_2(\mathbb{R}) = \{K = (k_{ij}), \ k_{ij} \in \mathbb{R}, \ 1 \leq i, j \leq 2, \ \det(K) = 1\}. \qquad (4.12.10)$$

THEOREM 4.12.1. *Let (4.12.2), (3.7.6), (3.7.7) hold; let $n = 2m+1$ for $m \in \mathbb{N}$. Then the separated BVP (4.12.1), (4.12.8) has exactly $m+1$ real eigenvalues λ_j if $A_2 B_2 \neq 0$, exactly m if $A_2 B_2 = 0$ and $A_2^2 + B_2^2 \neq 0$, and exactly $m-1$ if $A_2 = B_2 = 0$. Furthermore, the eigenfunction associated with each eigenvalue is unique up to constant multiples; and the eigenfunction u_j associated with the j-th eigenvalue λ_j has exactly j zeros in J. (Recall that a "zero" may be an entire subinterval of J.)*

PROOF. The statement about the number of eigenvalues is follows from Corollary 3.7.2, (i). The statement about the numbers of zeros of the eigenfunctions follows from Theorem 4.3.3. $\square$

THEOREM 4.12.2. *Let (4.12.2), (3.7.6), (3.7.7) hold; let $n = 2m+1$ for $m \in \mathbb{N}$. The coupled BVP (4.12.1), (4.12.9), (4.12.10) has exactly $m+1$ eigenvalues if $k_{12} \neq 0$, exactly m if $k_{12} = 0$ and $k_{11} \neq 0$.*

PROOF. This follows from Corollary 3.7.2, (ii), and the observation that if $a_{12} = 0$, then $a_{11} a_{22} = 1$, and hence $a_{22} w_0 + a_{11} w_{2m} \neq 0$. $\square$

The next theorems present analogues of the inequalities established in Section 4.8 for the case when there are only a finite number of eigenvalues. We note that, by Theorems 4.12.3 and 4.12.4, although the numbers of eigenvalues are different for different boundary conditions, the difference is at most two.

For $K \in SL_2(\mathbb{R})$, $K = (k_{ij})$, $\gamma \in (-\pi, \pi]$, we denote by $\lambda_j(e^{i\gamma} K)$ the j-th eigenvalue of the SLP (4.12.1), (4.12.9), (4.12.10). We also let μ_j and ν_j be the j-th eigenvalue of the SLP consisting of (4.12.1), and the separated BC

$$y(a) = 0, \quad k_{22} y(b) - k_{12}(py')(b) = 0 \qquad (4.12.11)$$

and

$$(py')(a) = 0, \quad k_{21} y(b) - k_{11}(py')(b) = 0 \qquad (4.12.12)$$

respectively.

THEOREM 4.12.3. *Let (4.12.2), (3.7.6), (3.7.7) hold and suppose that $K \in SL_2(\mathbb{R})$, $K = (k_{ij})$, and $n = 2m+1$, where m is an odd positive integer. Assume $\gamma \in (-\pi, \pi)$ and $\gamma \neq 0$.*

(1) *If $k_{12} < 0$ and $k_{11} > 0$, then*

$$\nu_0 \leq \lambda_0(K) < \lambda_0(e^{i\gamma} K) < \lambda_0(-K) \leq \{\mu_0, \nu_1\}$$
$$\leq \lambda_1(-K) < \lambda_1(e^{i\gamma} K) < \lambda_1(K) \leq \{\mu_1, \nu_2\} \leq \cdots$$
$$\leq \lambda_{m-1}(K) < \lambda_{m-1}(e^{i\gamma} K) < \lambda_{m-1}(-K) \leq \{\mu_{m-1}, \nu_m\}$$
$$\leq \lambda_m(-K) < \lambda_m(e^{i\gamma} K) < \lambda_m(K).$$

(2) If $k_{12} = 0$ and $k_{11} > 0$, then

$$\nu_0 \leq \lambda_0(K) < \lambda_0(e^{i\gamma}K) < \lambda_0(-K) \leq \{\mu_0, \nu_1\}$$
$$\leq \lambda_1(-K) < \lambda_1(e^{i\gamma}K) < \lambda_1(K) \leq \{\mu_1, \nu_2\} \leq \cdots$$
$$\leq \lambda_{m-1}(K) < \lambda_{m-1}(e^{i\gamma}K) < \lambda_{m-1}(-K) \leq \nu_m.$$

(3) If $k_{12} < 0$ and $k_{11} = 0$, then

$$\lambda_0(K) < \lambda_0(e^{i\gamma}K) < \lambda_0(-K) \leq \{\mu_0, \nu_0\}$$
$$\leq \lambda_1(-K) < \lambda_1(e^{i\gamma}K) < \lambda_1(K) \leq \{\mu_1, \nu_1\} \leq \cdots$$
$$\leq \lambda_{m-1}(K) < \lambda_{m-1}(e^{i\gamma}K) < \lambda_{m-1}(-K) \leq \{\mu_{m-1}, \nu_{m-1}\}$$
$$\leq \lambda_m(-K) < \lambda_m(e^{i\gamma}K) < \lambda_m(K).$$

(4) If $k_{12} < 0$ and $k_{11} < 0$, then

$$\lambda_0(K) < \lambda_0(e^{i\gamma}K) < \lambda_0(-K) \leq \{\mu_0, \nu_0\}$$
$$\leq \lambda_1(-K) < \lambda_1(e^{i\gamma}K) < \lambda_1(K) \leq \{\mu_1, \nu_1\} \leq \cdots$$
$$\leq \lambda_{m-1}(K) < \lambda_{m-1}(e^{i\gamma}K) < \lambda_{m-1}(-K) \leq \{\mu_{m-1}, \nu_{m-1}\}$$
$$\leq \lambda_m(-K) < \lambda_m(e^{i\gamma}K) < \lambda_m(K) \leq \nu_m.$$

(5) If none of these conditions apply to K, then one of them applies to $-K$.

PROOF. Since $r = 1/p$ and w are identically zero on adjacent subintervals by construction, Theorem 4.3.4 does not apply. But if in just one of the subintervals on which either r or w is identically zero we replace either coefficient by $\varepsilon > 0$ then Theorem 4.3.4 applies and guarantees an infinite number of eigenvalues since r and w are both positive on a common subinterval. The idea of the proof is to replace $r = 1/p$ and w by $r+\varepsilon$ and $w+\varepsilon$ for $\varepsilon > 0$, apply Theorem 4.8.1 to 'this ε problem' to get inequalities and then let $\varepsilon \to 0$. We only prove case (1), cases (2)-(4) can be proved in the same way; (5) follows from the other four cases.

From Theorems 4.12.1 and 4.12.2, we see that in case (1), the SLP (4.12.1) with one of the BC's (4.12.9) to (4.12.12) has eigenvalues $\{\lambda_j(e^{i\gamma})\}_{j=0}^m, \{\mu_j\}_{j=0}^{m-1}$, and $\{\nu_j\}_{j=0}^m$. Without loss of generality we may assume that $h = \min\{\lambda_0(e^{i\gamma}), \mu_0, \nu_0\} > 0$. For otherwise, we choose $l > -h$ and consider the equation

$$-(py')' + \tilde{q}y = \tilde{\lambda}wy \qquad (4.12.13)$$

where $\tilde{q} = q + lw$. Let $\tilde{\lambda}_j(e^{i\gamma}) = \lambda_j(e^{i\gamma}) + l$, $j = 0, \ldots, m$; $\tilde{\mu}_j = \mu_j + l, j = 0, \ldots, m-1$; and $\tilde{\nu}_j = \nu_j + l, j = 0, \ldots, m$. Then $\{\tilde{\lambda}_j(e^{i\gamma})\}_{j=0}^m, \{\tilde{\mu}_j\}_{j=0}^{m-1}$, and $\{\tilde{\nu}_j\}_{j=0}^m$ are the eigenvalues of the SLP's consisting of (4.12.13) and one of the BC (4.12.9) to (4.12.12), and $\min\{\tilde{\lambda}_0(e^{i\gamma}), \tilde{\mu}_0, \tilde{\nu}_0\} > 0$. It is easy to see that the inequalities in (1) hold for $\{\lambda_j(e^{i\gamma})\}_{j=0}^m, \{\mu_j\}_{j=0}^{m-1}$, and $\{\nu_j\}_{j=0}^m$ if and only if they hold for $\{\tilde{\lambda}_j(e^{i\gamma})\}_{j=0}^m, \{\tilde{\mu}_j\}_{j=0}^{m-1}$, and $\{\tilde{\nu}_j\}_{j=0}^m$.

To establish the inequalities of case (1), we define for $\epsilon > 0$

$$r_\epsilon(t) = r(t) + \epsilon, \quad w_\epsilon(t) = w(t) + \epsilon \text{ for } t \in J.$$

Now consider the SLP's consisting of (4.12.1) and the same boundary conditions but with p replaced by p_ϵ and w replaced by w_ϵ. Then each of these problems has an infinite number of eigenvalues by Theorem 4.3.4. Denote these by

$\{\lambda_j(e^{i\gamma}K)(\epsilon)\}_{j=0}^\infty, \{\mu_j(\epsilon)\}_{j=0}^\infty$, and $\{\nu_j(\epsilon)\}_{j=0}^\infty\}$. Thus for each $\epsilon > 0$, the inequalities established Theorem 4.8.1 hold for these eigenvalues. Note from Theorem 4.2, (5) and (7) of [**394**] that as $\epsilon \to 0+$ all of $\lambda_j(e^{i\gamma}K)(\epsilon)$, $\mu_j(\epsilon)$ and $\nu_j(\epsilon)$ are increasing and hence bounded away from $-\infty$, then from the continuity of the eigenvalues λ_n as functions of the coefficients $1/p$, q, w it follows that as $\epsilon \to 0+$,

$$\lambda_j(e^{i\gamma}K)(\epsilon) \to \lambda_j(e^{i\gamma}K), \; j=0,\ldots,m,$$

$$\mu_j(\epsilon) \to \mu_j, \; j = 0, \ldots, m-1; \; \nu_j(\epsilon) \to \nu_j, \; j=0,\ldots,m,$$

and all others approach $+\infty$. Hence the same "interlacing" relations hold as in Theorem 4.8.1 except that "<" may become "≤" in the limit process. To see that this does not happen consider the characteristic equation for the coupled BC (4.12.9), (4.12.10). This is equivalent with the equation

$$k_{22}\phi_{11}(b,\lambda) - k_{21}\phi_{12}(b,\lambda) - k_{12}\phi_{21}(b,\lambda) + k_{11}\phi_{22}(b,\lambda) = 2\cos\gamma,$$

which implies that the same strict inequalities must hold. This completes the proof of (1). □

THEOREM 4.12.4. *Let (4.12.2), (3.7.6), (3.7.7) hold and suppose that* $K \in SL_2(\mathbb{R})$, $K = (k_{ij})$, *and* $n = 2m+1$, *where* m *is an even positive integer. Assume* $\gamma \in (-\pi, \pi)$ *and* $\gamma \neq 0$.

(1) *If* $k_{12} < 0$ *and* $k_{11} > 0$, *then*

$$\nu_0 \leq \lambda_0(K) < \lambda_0(e^{i\gamma}K) < \lambda_0(-K) \leq \{\mu_0, \nu_1\}$$
$$\leq \lambda_1(-K) < \lambda_1(e^{i\gamma}K) < \lambda_1(K) \leq \{\mu_1, \nu_2\} \leq \cdots$$
$$\leq \lambda_{n-1}(-K) < \lambda_{n-1}(e^{i\gamma}K) < \lambda_{n-1}(K) \leq \{\mu_{n-1}, \nu_n\}$$
$$\leq \lambda_n(K) < \lambda_n(e^{i\gamma}K) < \lambda_n(-K).$$

(2) *If* $k_{12} = 0$ *and* $k_{11} > 0$, *then*

$$\nu_0 \leq \lambda_0(K) < \lambda_0(e^{i\gamma}K) < \lambda_0(-K) \leq \{\mu_0, \nu_1\}$$
$$\leq \lambda_1(-K) < \lambda_1(e^{i\gamma}K) < \lambda_1(K) \leq \{\mu_1, \nu_2\} \leq \cdots$$
$$\leq \lambda_{n-1}(-K) < \lambda_{n-1}(e^{i\gamma}K) < \lambda_{n-1}(K) \leq \nu_n.$$

(3) *If* $k_{12} < 0$ *and* $k_{11} = 0$, *then*

$$\lambda_0(K) < \lambda_0(e^{i\gamma}K) < \lambda_0(-K) \leq \{\mu_0, \nu_0\}$$
$$\leq \lambda_1(-K) < \lambda_1(e^{i\gamma}K) < \lambda_1(K) \leq \{\mu_1, \nu_1\} \leq \cdots$$
$$\leq \lambda_{n-1}(-K) < \lambda_{n-1}(e^{i\gamma}K) < \lambda_{n-1}(K) \leq \{\mu_{n-1}, \nu_{n-1}\}$$
$$\leq \lambda_n(K) < \lambda_n(e^{i\gamma}K) < \lambda_n(-K).$$

(4) *If* $k_{12} < 0$ *and* $k_{11} < 0$, *then*

$$\lambda_0(K) < \lambda_0(e^{i\gamma}K) < \lambda_0(-K) \leq \{\mu_0, \nu_0\}$$
$$\leq \lambda_1(-K) < \lambda_1(e^{i\gamma}K) < \lambda_1(K) \leq \{\mu_1, \nu_1\} \leq \cdots$$
$$\leq \lambda_{n-1}(-K) < \lambda_{n-1}(e^{i\gamma}K) < \lambda_{n-1}(K) \leq \{\mu_{n-1}, \nu_{n-1}\}$$
$$\leq \lambda_n(K) < \lambda_n(e^{i\gamma}K) < \lambda_n(-K) \leq \nu_n.$$

(5) *If none of these conditions apply to K, then one of them applies to $-K$.*

PROOF. The proof of this theorem is similar to that of Theorem 12.4.3 and hence is omitted. □

COROLLARY 4.12.1. *Let (4.12.2), (3.7.6). (3.7.7) hold and suppose that $K \in SL_2(\mathbb{R})$, $K = (k_{ij})$, and $n = 2m+1$, where m is an even positive integer. If $k_{12} < 0$ or $k_{12} = 0$ and $k_{11} > 0$, then $\lambda_0(K)$ is simple; if $k_{12} > 0$ or $k_{12} = 0$ and $k_{11} < 0$, then $\lambda_0(-K)$ is simple.*

PROOF. This follows from Theorems 12.4.3 and 12.4.4. □

COROLLARY 4.12.2. *Let (4.12.2), (3.7.6). (3.7.7) hold and suppose that $K \in SL_2(\mathbb{R})$, $K = (k_{ij})$, and $n = 2m + 1$, where m is an even positive integer. Let $\gamma \in (-\pi, \pi)$, $\gamma \neq 0$..*

(a) Let u_j be a real eigenfunction for $\lambda_j(K)$. Then the number of zeros of u_j on $[a, b)$ is 0 or 1 if $j = 0$, and $j - 1$ or j or $j + 1$ if $j \geq 1$.

(b) Let u_j be an eigenfunction for $\lambda_j(e^{i\gamma}K)$. Then the number of zeros of $\Re u_j$ on $[a, b)$ is 0 or 1 if $j = 0$, and $j - 1$ or j or $j + 1$ if $j \geq 1$. The same conclusion holds for $\Im u_j$. Moreover, u_j is never zero on $[a, b]$. Recall that a "zero" may be an entire subinterval of J.

PROOF. With the above inequalities in Theorems 4.12.3 and 4.12.4, the proof is similar to that of Theorem 4.8 in [**384**] and hence omitted. □

The next theorem characterizes the "jump-discontinuities" of the eigenvalues for the case when there are only a finite number of them. In the above construction yielding only a finite number of eigenvalues the coefficients r and w (and q) are alternately identically zero on adjacent subintervals. If just one of these zeros is replaced by $\epsilon > 0$ then by Corollary 4.3.1 there are an infinite number of eigenvalues, say $\lambda_n(\varepsilon)$. What happens as $\varepsilon \to 0$? Here we use the notation $\{\lambda_j(K)\}, \{\mu_j(K)\}$, and $\{\nu_j(K)\}$ for the eigenvalues defined above equation (4.12.11) and study the dependence of these eigenvalues on K even though each of the latter two depends only on one column of K.

THEOREM 4.12.5. *Let (4.12.2), (3.7.6), (3.7.7) hold and suppose that $K \in SL_2(\mathbb{R})$, $K = (k_{ij})$, and $n = 2m + 1$, for $m \in \mathbb{N}$. Let $\gamma \in (-\pi, \pi]$. Let $\bar{K}, \tilde{K} \in SL_2(\mathbb{R})$ such that $\bar{k}_{11} = 0$ and $\tilde{k}_{12} = 0$. Let $\mathbb{K} = \{K \in SL_2(\mathbb{R}) : k_{11}k_{12} \neq 0\}$. By $\mathbb{K} \ni K \to \bar{K}+$ we mean that $k_{11} \to 0+$ and $k_{ij} \to \bar{k}_{ij}$ for $(i, j) \neq (1, 1)$. Similar meanings are given for $\mathbb{K} \ni K \to \bar{K}-$, $\mathbb{K} \ni K \to \tilde{K}+$, and $\mathbb{K} \ni K \to \tilde{K}-$. We have*

(1) $\lambda_j(Ke^{i\gamma})$, $j = 0, \ldots, m$, $\mu_j(K)$, $j = 0, \ldots, m-1$, and $\nu_j(K)$, $j = 0, \ldots, m$ *depend continuously on K in $\mathbb{K}$.*

(2) $\lim_{\mathbb{K} \ni K \to \bar{K}+} \nu_0(K) = -\infty$, $\lim_{\mathbb{K} \ni K \to \bar{K}-} \nu_m(K) = \infty$, *and for* $j = 1, \ldots, m$
$$\lim_{\mathbb{K} \ni K \to \bar{K}+} \nu_j(K) = \nu_{j-1}(\bar{K}).$$

(3) $\lim_{\mathbb{K} \ni K \to \tilde{K}+} \lambda_0(e^{i\gamma}K) = -\infty$, $\lim_{\mathbb{K} \ni K \to \tilde{K}-} \lambda_m(e^{i\gamma}K) = \infty$ *and for* $j = 1, \ldots, m$
$$\lim_{\mathbb{K} \ni K \to \tilde{K}+} \lambda_j(e^{i\gamma}K) = \lambda_{j-1}(e^{i\gamma}\tilde{K}),$$

(4) $\lim_{\mathbb{K} \ni K \to \tilde{K}+} \mu_0(K) = -\infty$, $\lim_{\mathbb{K} \ni K \to \tilde{K}-} \mu_{m-1}(K) = \infty$ *and for* $j = 1, \ldots, m - 1$
$$\lim_{\mathbb{K} \ni K \to \tilde{K}+} \mu_j(K) = \mu_{j-1}(\tilde{K}).$$

PROOF. These results can be obtained by comparing the inequalities in Theorems 4.12.3 and 4.12.4. We omit the details. □

REMARK 4.12.1. In all the above results using the hypothesis that in (3.7.7) the partition of the underlying interval $J = (a, b)$:

$$a = a_0 < a_1 < \ldots < a_{n-1} < a_n = b \qquad (4.12.14)$$

requires $n = 2m+1$, $m \in \mathbb{N}$ is not essential. Parallel results can also be established for the case that $n = 2m$, $m \in \mathbb{N}$. For details see [**145**].

13. Comments

(1) These are made separately for each section.
(2) The canonical forms (4.2.3), (4.2.4) and (4.2.5), (4.2.6) for separated self-adjoint problems are well known and can be found in all the books. On the other hand the general canonical form for coupled self-adjoint BC given by (4.2.7), (4.2.8), (4.2.9) are known but seem not to be "well known". These forms can be established from the general characterization of all self-adjoint boundary conditions given in Naimark's classic book [**487**], see Chapter 10 below for details.
(3) For general separated boundary conditions and $p > 0$, $w > 0$ the existence of the eigenvalues and their characterization can be established with the Prüfer transformation and its equation. For the general coupled case there has been only one proof for the existence of the eigenvalues: the operator theory proof. Construct a self-adjoint integral operator whose kernel is the Green's function of the problem, then use the theory of compact (completely continuous) self-adjoint operators in the Hilbert space $L^2(J, w)$ to get the existence of an eigenvalue. Repeating this argument using the restriction of the integral operator to the orthogonal complement of the eigenfunction subspace yields a second eigenvalue etc. Now one shows, using some basic theory of differential equations that this process does not come to a halt in a finite number of steps. Thus one gets infinitely many eigenvalues and also it follows from this construction that all of them are real and are captured by this process. See the book by Coddington and Levinson [**123**] for details. Although their proof is given under more severe hypotheses than we have here it can readily be adapted to fit our hypotheses.

There now is an elementary proof available also for the general coupled regular self-adjoint boundary conditions. It is based on the inequalities in Section 4.8 between the eigenvalues of coupled BC and those for related separated BC. See Remark 4.8.1. Given a coupled self-adjoint boundary condition this proof constructs an interval which contains exactly one eigenvalue; and does this for each eigenvalue. The endpoints of these bracketing intervals are eigenvalues for separated boundary conditions constructed from the coupled condition except, possibly, for the lowest eigenvalue which may be bracketed on the left by $-\infty$. The SLEIGN2 code of the Bailey, Everitt and Zettl (dated 01 December 2000) uses an algorithm for the coupled case based on the inequalities of Section 8.

The results allowing the weight function w to be identically zero on subintervals are due to Atkinson [**21**] and Everitt, Kwong and Zettl [**214**].

The latter also allows w to be identically zero on a subset of the underlying interval J which has positive Lebesgue measure but contains no interval. The former allows $r = 1/p$ to be identically zero on subintervals, i.e. p may be infinite at each point of a subinterval. For recent results on problems with $r = 1/p$ or w identically zero on one or more subintervals see [**583**], [**585**], [**583**].

M. Möller [**474**] showed that the eigenvalues are not bounded below when p changes sign even if there is no subinterval on which p is negative.

(4) Partly this section is a refinement of the results of Section 3.6 to the self-adjoint case. The continuity of λ_n as a function of $1/p, q, w$ is established by Kong, Wu and Zettl in [**384**]. This result is probably not new but this author knows of no reference containing a rigorous proof of the general case with integrable coefficients and both separated and coupled self-adjoint boundary conditions. The characterization of the jump discontinuities of λ_n for both separated and coupled boundary conditions can also be found in [**384**]. Special cases are known but this author knows of no reference containing a proof for general regular self-adjoint SLP. The "jump set" $\mathbb{J}$ given by Definition 4.4.1 and used in Theorem 4.4.2 was discovered by Everitt, Möller and Zettl for separated BC and by Eastham, Kong, Wu and Zettl for coupled BC. Special cases of the separated case, particularly the jump discontinuity at the Dirichlet boundary condition is part of numerical analysis folklore but, again, this author knows of no paper giving a rigorous proof for the general case. The jump at the coupled BC represented by K when $k_{12} = 0$ seems to be new; also the fact that there is no discontinuity of λ_n when $k_{12} \neq 0$.

(5) This section is largely based on Weidmann's book [**600**]. The author thanks an anonymous referee for noting that the original assumption that p_2 and p_1 are positive is not needed in Theorem 4.5.1.

(6) Theorems 4.6.1 and 4.6.2 are standard. Theorems 4.6.3 and 4.6.4 are extensions of results of Dauge and Helffer [**135**], [**136**] due to Kong and Zettl [**392**], [**394**]. Theorem 4.6.5 and the example following it are recent results of Binding and Volkmer [**80**].

Does $\lambda_n(a,b) \to +\infty$ as $b \to a^+$ in general? The answer is yes for $n > 0$ (when $p > 0, w > 0$ on J). For $n = 0$ this limit may be $+\infty, -\infty$, finite or it might not exist.

Can the hypothesis $q/w \in AC_{loc}[a, b')$, $p(b) \geq \delta > 0$, for $b \in (a, b')$ be weakened? eliminated? These questions are studied in a forthcoming paper by Kong, Wu and Zettl [**385**].

(7) Much of this section is based on [**384**].

(8) Theorem 4.8.1 is based on Eastham, Kong, Wu and Zettl [**145**]. Special cases were known from [**41**], [**600**], [**153**]. The characterization of the range λ_n as a function of the boundary conditions is taken from [**384**]. The characterization of all boundary conditions that have the same lowest eigenvalue as the Dirichlet condition given by Proposition 4.8.1 is a special case of a result in Haertzen, Kong, Wu and Zettl [**285**]. From another perspective Proposition 4.8.1 characterizes all self-adjoint extensions of the minimal operator which preserve its lower bound. There are an infinite number of these and all of them are determined by a real coupled

boundary condition except for the Dirichlet condition which determines the Friedrichs extension. (The Dirichlet boundary condition also determines the Friedrichs extension in the higher order regular case, see Niessen and Zettl [**502**].) Given any real coupled boundary condition, Bailey [**35**] constructed a continuum of separated boundary conditions and characterized each coupled eigenvalue λ_n as an extremum of a curve of eigenvalues λ_n as a function of the separated conditions.

The inequalities of Theorem 4.8.1 are for the case when $p > 0$ on J. When p changes sign the eigenvalues are unbounded below and above and inequalities analogues to those of Theorem 4.8.1 are much more difficult to obtain but this has been done in [**115**] by Cao, Kong, Wu and Zettl.

(9) The monotonicity results here are largely due to Kong and Zettl [**394**], [**392**] and to Kong, Wu and Zettl [**384**].

(10) The equivalence of the algebraic and geometric multiplicities is taken from [**145**]; the result of Theorem 4.10.2 on the multiplicity spaces is from [**394**], [**392**].

(11) The results of Section 3.7 are specialized to the self-adjoint case.

(12) This section is based on [**391**].

CHAPTER 5

Regular Left-Definite and Indefinite Problems

Rényi, Alfréd
> If I feel unhappy, I do mathematics to become happy. If I am happy, I do mathematics to keep happy.

P. Turán, "The Work of Alfréd Rényi", Matematikai Lapok 21, 1970, pp. 199-210.

1. Introduction

In Chapter 4 we studied self-adjoint problems with a positive weight function w; we will call these right-definite (RD) problems to distinguish them from the left-definite (LD) problems to be studied in this Chapter. For these LD problems the weight function w is allowed to change sign. There does not seem to be a universally accepted definition of left-definiteness in the literature which encompasses separated and coupled boundary conditions. Here, for a problem with weight function w, we use a definition of left-definite based on a closely related problem with weight function $|w|$. This allows use to use the RD results of Chapter 4 to study LD problems.

It turns out that a problem involving a weight function w which changes sign, is left-definite if and only the lowest eigenvalue of the corresponding right-definite problem with weight function $|w|$ is positive. In this case the spectrum of the LD problem is real. What if the lowest eigenvalue of the corresponding right-definite problem is negative? Then the 'indefinite' problem may have non-real eigenvalues. In Section 5.8 we establish an upper bound for the number of non-real eigenvalues.

In the left-definite case the leading coefficient p must be positive in contrast to the RD case studied in Chapter 4 where p is, in general, allowed to change sign. If p changes sign then the equation (5.1.1) with any self-adjoint boundary condition is not left-definite since the spectrum is not bounded below.

We investigate boundary value problems associated with the regular differential equation

$$-(py')' + qy = \lambda w y \text{ on } J = (a,b), \ -\infty \le a < b \le \infty, \qquad (5.1.1)$$

where p is positive and w is allowed to change sign. Our approach is to "connect" this equation with the corresponding right-definite equation

$$-(py')' + qy = \lambda |w| y \text{ on } J, \qquad (5.1.2)$$

and, more generally, with an associated family of right-definite equations:

$$-(py')' + qy - \lambda w y = \xi |w| y \text{ on } J. \qquad (5.1.3)$$

The coefficients are assumed to satisfy:

$1/p, q, w \in L^1(J, \mathbb{R})$, $p > 0$ on J, $|w| > 0$ on J, and w changes sign on J. (5.1.4)

5. REGULAR LEFT-DEFINITE AND INDEFINITE PROBLEMS

The boundary conditions (BC) used in these problems will always be a subset of the RD self-adjoint conditions studied in Chapter 4, i.e. satisfy the conditions (4.1.3), (4.1.4).

It has been shown (see, for example, [**344**]) that the eigenvalues of certain left-definite problems with *separated* BC can be numbered by the index set

$$\mathbb{Z}^* = \{..., -2, -1, -0, 0, 1, 2, ...\}$$

such that

$$\cdots < \lambda_{-2} < \lambda_{-1} < \lambda_{-0} < 0 < \lambda_0 < \lambda_1 < \lambda_2 < \cdots$$

and for each $n \in \mathbb{Z}^*$ the eigenfunctions (which are unique up to constant multiples) for λ_n have exactly $|n|$ zeros in the open interval (a, b). We discuss general left-definite problems with separated and coupled boundary conditions. These problems have only real eigenvalues, there are an infinite but countable number of them, they are unbounded above and below, and can be indexed to satisfy

$$\cdots \leq \lambda_{-2} \leq \lambda_{-1} \leq \lambda_{-0} < 0 < \lambda_0 \leq \lambda_1 \leq \lambda_2 \leq \cdots$$

with only geometrically double eigenvalues appearing twice.

As a special case of Theorem 4.8.1 we have the well-known [**600**] classical inequalities for the right-definite case:

$$\begin{aligned}\lambda_0^N &\leq \lambda_0^P < \lambda_0^S \leq \{\lambda_0^D, \lambda_1^N\} \leq \lambda_1^S < \lambda_1^P \leq \{\lambda_1^D, \lambda_2^N\} \\ &\leq \lambda_2^P < \lambda_2^S \leq \{\lambda_2^D, \lambda_3^N\} \leq \lambda_3^S < \lambda_3^D \leq \{\lambda_3^D, \lambda_4^N\} \leq \cdots,\end{aligned} \quad (5.1.5)$$

where $\{\lambda_n^N\}$, $\{\lambda_n^P\}$, $\{\lambda_n^S\}$ and $\{\lambda_n^D\}$ denote the Neumann, periodic, semi-periodic and Dirichlet eigenvalues, respectively. Here the notation $\{\lambda_0^D, \lambda_1^N\}$ is used for each of λ_0^D and λ_1^N. (λ_0^D and λ_1^N are not compared with each other.) Analogues of the inequalities in (5.1.5) for the left-definite case when $p = 1$ have been found by Constantin [**128**].

Below we discuss similar inequalities for the general left-definite case for general separated and coupled boundary conditions. These inequalities are comparable to those in Section 4.8 for the right-definite case. We also find upper and lower bounds for each eigenvalue $\lambda_n : n \in \mathbb{Z}^*$ as a function of the boundary conditions. Furthermore our inequalities imply that the asymptotic formulas of [**29**] for the eigenvalues in the separated case also hold in the coupled case.

The continuous and differentiable dependence of λ_n on all parameters of the problem is studied. Analogues of results in Section 4.4 are established. In contrast to the RD case we show that each λ_n is a continuous function of the problem *on the space of left-definite problems*, in particular each λ_n depends continuously on each parameter, i.e., on each of $a, b, 1/p, q, w$ and *on the boundary conditions as long as the problem remains left-definite*. Formulas for the derivatives of λ_n with respect to all parameters are found. We also give some comparison results for λ_n with respect to the coefficients implied by these derivative formulas.

2. Definition and Characterization of Left-Definite Problems

Given a SLP, what does it mean to say it is left-definite? We answer this question here completely for regular SLP with weight functions w that change sign and boundary conditions which may be separated or coupled. For the singular case see Chapter 12.

2. DEFINITION AND CHARACTERIZATION OF LEFT-DEFINITE PROBLEMS

DEFINITION 5.2.1. *Let (5.1.4) hold. The SLP consisting of the equation (5.1.1) and a (separated or a coupled) boundary condition (4.1.3) (4.1.4) is left-definite if the lowest eigenvalue of the corresponding right-definite problem (5.1.2) and the same boundary condition is positive.*

REMARK 5.2.1. By Theorem 5.2.2 below, if a problem is left-definite for some $w \in L^1(J, \mathbb{R})$ then it is left-definite for every $w \in L^1(J, \mathbb{R})$. Thus $|w|$ in (5.1.2) can be replaced by any positive and integrable function on J, see also Proposition 2.6 in [**133**]. In particular, on a bounded interval J, $|w|$ can be replaced by the constant function 1. (Note that the constant function 1 is not integrable on an unbounded interval J). This characterization makes clear the dependence of left-definiteness on the coefficients p, q, the endpoints a, b, and the BC. Note, in particular, that although p is assumed to be positive, there is no sign restriction on q. Also note that the interval J of the regular Sturm-Liouville equation (5.2.1) is allowed to be unbounded.

REMARK 5.2.2. To see that Definition 5.2.1 is a "natural" one, let the matrices A, B satisfy (4.1.4). Let $\mathcal{D}_{\max}$ and $\mathcal{D}(A, B)$ be the linear submanifolds of the Hilbert space $\mathcal{H} = L^2(J, |w|)$ given by

$$\mathcal{D}_{\max} = \left\{ f \in \mathcal{H} : f, pf' \in AC_{loc}(J), \frac{1}{|w|}\left[-(pf')' + qf\right] \in \mathcal{H} \right\},$$
$$\mathcal{D}(A, B) = \{f \in \mathcal{D}_{\max} : AY(a) + BY(b) = 0\}, \ Y = \begin{bmatrix} f \\ pf' \end{bmatrix}. \quad (5.2.1)$$

In (5.2.1) $f(a)$ and $f^{[1]}(a) = (pf')(a)$ are defined by limits:

$$f(a) = \lim_{t \to a^+} y(t), \qquad (pf')(a) = \lim_{t \to a^+} (pf')(t).$$

By Theorem 2.3.1 these limits exist and are finite for any solution y of (5.1.1). Similarly, these limits exist and are finite for any maximal domain function, i.e., for any $f \in \mathcal{D}_{\max}$: let $g = -(pf')' + qf$, then

$$\int_a^b |g| = \int_a^b \frac{|g|}{|w|^{1/2}} \cdot |w|^{1/2} \le \left(\int_a^b \frac{|g|^2}{|w|}\right)^{1/2} \left(\int_a^b |w|\right)^{1/2} < +\infty,$$

i.e., $g \in L^1(J)$; thus, f and pf' have a finite limit at a, since f is a solution of $-(py')' + qy = g$. We have similar statements at the endpoint b. So, $\mathcal{D}(A, B)$ is well-defined.

Define two functionals $\mathcal{R}$ and $\mathcal{L}$ on $\mathcal{D}(A, B)$ as follows:

$$\mathcal{R}f = \int_a^b |f|^2 w, \qquad \mathcal{L}f = \int_a^b [-(pf')'\bar{f} + q|f|^2], \ f \in \mathcal{D}(A, B). \quad (5.2.2)$$

Clearly $\mathcal{R}$ is well defined not only on $\mathcal{D}(A, B)$ but on all of $\mathcal{H}$; the well-definedness of $\mathcal{L}$ can be seen from the Cauchy-Schwarz inequality in $\mathcal{H}$: for any $f \in \mathcal{D}_{\max}$,

$$\int_a^b |-(pf')'\bar{f} + q|f|^2| = \int_a^b \left|\frac{-(pf')' + qf}{|w|^{1/2}} \cdot \bar{f}\right| |w|^{1/2} \le \left\|\frac{-(pf')' + qf}{|w|}\right\| \|f\| < +\infty,$$

where $\|\cdot\|$ denotes the norm of $\mathcal{H}$.

REMARK 5.2.3. Clearly $\mathcal{R}f > 0$ for all $f \not\equiv 0$ in $\mathcal{H}$ or $\mathcal{R}f < 0$ for all $f \not\equiv 0$ in $\mathcal{H}$ if and only if w does not change sign in J. This is the right-definite (RD) case. Similarly, the next result will show that the left-definite (LD) case is equivalent to $\mathcal{L}f > 0$ for all $f \not\equiv 0$ in $\mathcal{D}(A, B)$.

THEOREM 5.2.1. *The Sturm-Liouville problem (5.1.1), (5.1.4) and (4.1.3), (4.1.4) is left-definite if and only if the functional $\mathcal{L}$ is positive definite on $\mathcal{D}(A, B)$, i.e. $\mathcal{L}f > 0$ for all f in $\mathcal{D}(A, B)$, $f \not\equiv 0$.*

PROOF. This follows from the variational characterization of the lowest eigenvalue $\xi_0(|w|)$ of RD problems:

$$\xi_0(|w|) = \inf \frac{\mathcal{L}f}{\int_a^b |f|^2 |w|}, \tag{5.2.3}$$

where the infimum is taken over all $f \not\equiv 0$ in $\mathcal{D}(A, B)$ and is achieved by eigenfunctions of $\xi_0(|w|)$. □

Next we exhibit some subclasses of LD problems.

COROLLARY 5.2.1. *Let (5.1.4) hold. Assume $q \geq 0$ and $\int_a^b q > 0$. Then:*
(i) The Sturm-Liouville problem (5.1.1), with separated boundary conditions (4.2.3), (4.2.4) is left-definite if $\pi/2 \leq \alpha \leq \pi$ and $0 \leq \beta \leq \pi/2$;
(ii) The Sturm-Liouville problem (5.1.1), with coupled boundary conditions (4.2.6), (4.2.7) is left-definite if

$$K = \begin{bmatrix} c & 0 \\ 0 & 1/c \end{bmatrix}, \ c \in \mathbb{R}, \ c \neq 0.$$

PROOF. For both cases, we let $\xi_0(|w|)$ be the least eigenvalue of the corresponding RD problem and y an eigenfunction for ξ_0. Then, by integration by parts we have that

$$\xi_0(|w|) \int_a^b |y|^2 |w| = y^{[1]}(a)\bar{y}(a) - y^{[1]}(b)\bar{y}(b) + \int_a^b [p|y'|^2 + q|y|^2]$$
$$> y^{[1]}(a)\bar{y}(a) - y^{[1]}(b)\bar{y}(b).$$

(i) To prove the case for $\pi/2 < \alpha < \pi$ and $0 < \beta < \pi/2$, we may assume that y is real-valued and note that by (4.2.3), (4.2.4) we have

$$y^{[1]}(a)\bar{y}(a) - y^{[1]}(b)\bar{y}(b) = -|y^{[1]}(a)|^2 \tan\alpha + |y^{[1]}(b)|^2 \tan\beta \geq 0.$$

These two observations imply that $\xi_0(|w|) > 0$. The cases $\alpha = \pi/2$ or $\beta = \pi/2$ can be proven similarly.

(ii) In this case we have that

$$y(b) = ce^{i\theta} y(a) \quad \text{and} \quad y^{[1]}(b) = \frac{1}{c} e^{i\theta} y^{[1]}(a).$$

Hence,

$$y^{[1]}(a)\bar{y}(a) - y^{[1]}(b)\bar{y}(b) = 0.$$

This together with the above observations imply that $\xi_0(|w|) > 0$. □

REMARK 5.2.4. In Corollary 5.2.1 as in much of the literature on LD problems, it is assumed that $q \geq 0$. The next result shows not only that this assumption is not needed in general but also that q can even be unbounded from below. Moreover, for each fixed BC and for any given p and w, there is a potential q yielding a LD problem. Furthermore, Corollary 5.2.2 gives an explicit construction of such $q's$.

COROLLARY 5.2.2. *Let (5.1.4) hold. Denote by $\xi_0 = \xi_0(|w|)$ the least eigenvalue of the right-definite Sturm-Liouville problem consisting of (5.1.2) and self-adjoint BC. Then for any $\epsilon > 0$, the Sturm-Liouville problem consisting of the differential equation*

$$-(py')' + [q - (\xi_0 - \epsilon)|w|]\, y = \lambda w y \text{ on } J \qquad (5.2.4)$$

and the same boundary condition is left-definite.

PROOF. By Theorem 5.2.1 the SLP consisting of (5.2.4) and the same BC is LD if and only if the RD problem consisting of the DE

$$-(py')' + [q - (\xi_0 - \epsilon)|w|]\, y = \xi |w| y \text{ on } J \qquad (5.2.5)$$

and the same BC has only positive eigenvalues. From the definition of $\xi_0(|w|)$ it follows that ϵ is the lowest eigenvalue of the latter. Therefore, the former is LD. $\square$

REMARK 5.2.5. When $w > 0$ a.e. on J, $L^2(J; w)$ is a Hilbert space with the inner product $(f, g) = \int_a^b f \bar{g} w$. This space is widely used to study the spectrum of RD problems. When w changes sign on J, but $w \neq 0$ a.e. on J, $L^2(J; w)$ is a Krein space and the theory of operators in Krein spaces can be applied to study such problems, see [133] and [134]. In the LD case, Hilbert space operator theory can also be applied with a Hilbert space $\mathcal{H}_l$ constructed as follows: since $\mathcal{L}f$ is positive definite on the linear manifold $\mathcal{D}(A, B)$ of $L^2(J; |w|)$, the inner product

$$<f, g> = \int_a^b [-(pf')' + q f]\, \bar{g}$$

induces a norm $\|\cdot\|_l$ on $\mathcal{D}(A, B)$. The Hilbert space $\mathcal{H}_l$ is the completion of $\mathcal{D}(A, B)$ with respect to this norm.

Next we show that left-definiteness does not depend on the weight function w in the DE. In particular, $|w|$ can be replaced by any positive and integrable weight function on J. Hence, when J is bounded, $|w|$ can be replaced by the constant function 1; when J is unbounded $|w|$ can be replaced by functions such as $1/(1 + t^2)$, but not by the constant function 1.

THEOREM 5.2.2. *If the first eigenvalue $\lambda_0 = \lambda_0(w)$ of a right-definite self-adjoint regular Sturm-Liouville problem, (with $p > 0$ on J) is positive for some positive weight function $w \in L^1(J)$, then this is true for every $w \in L^1(J)$, $w > 0$.*

PROOF. (See also [133, 8, Proposition 2.6]) Assume that $\lambda_0(w_0) > 0$ for some $w_0 \in L^1(J, \mathbb{R}_+)$, then 0 is not an eigenvalue of this SLP. Hence by Lemma 3.2.2 the characteristic function of this problem $\delta(\lambda) \neq 0$ for $\lambda = 0$. Since this characteristic function $\delta(\lambda)$ does not depend on $w \in L^1(J, \mathbb{R}_+)$, when $\lambda = 0$ we deduce that $\delta(\lambda) \neq 0$ when $\lambda = 0$ for any $w \in L^1(J, \mathbb{R}_+)$, i.e., 0 is not an eigenvalue of the SLP when the given w_0 is replaced by any other $w \in L^1(J, \mathbb{R}_+)$. By Theorem 4.4.3 $\lambda_0(w)$ is a continuous function of $w \in L^1(J, \mathbb{R}_+)$. Since $L^1(J, \mathbb{R}_+)$ is connected it follows that $\lambda_0(w) > 0$ for any $w \in L^1(J, \mathbb{R}_+)$. $\square$

3. Existence of Eigenvalues

In this section we discuss basic properties of eigenvalues and eigenfunctions including existence and their characterization in terms of 'eigencurves' of an associated family of right-definite problems.

THEOREM 5.3.1. *Assume that the Sturm-Liouville problem consisting of (5.1.1), (5.1.4), (4.1.3), (4.1.4) is left-definite i.e. $\lambda_0(|w|) > 0$. Then all its eigenvalues are real, there exist countably infinitely many positive and negative eigenvalues, they are unbounded from below and from above, and have no finite cluster point.*

(1) *The number $\lambda = 0$ is not an eigenvalue.*
(2) *If the boundary condition is separated, then the eigenvalues can be indexed to satisfy the inequalities*

$$\cdots < \lambda_{-n} < \cdots < \lambda_{-1} < \lambda_{-0} < 0 < \lambda_0 < \lambda_1 < \cdots < \lambda_n < \cdots. \qquad (5.3.1)$$

Moreover, for each $n \in \mathbb{Z}^*$, any eigenfunction for λ_n has exactly $|n|$ zeros in the open interval J.

(3) *If the boundary condition is real and coupled, then the eigenvalues can be indexed to satisfy*

$$\cdots \leq \lambda_{-n} \leq \cdots \leq \lambda_{-1} \leq \lambda_{-0} < 0 < \lambda_0 \leq \lambda_1 \leq \cdots \leq \lambda_n \leq \cdots. \qquad (5.3.2)$$

Each eigenvalue may be geometrically simple or double but there cannot be two consecutive equalities in (5.3.2) since, for any value of λ equation (5.1.1) has exactly two linearly independent solution. Note that λ_n is well defined for each $n \in \mathbb{Z}^*$ but there is some ambiguity in the indexing of the eigenfunctions u_n, $n \in \mathbb{Z}^*$ corresponding to a double eigenvalue since every nontrivial solution of the equation is an eigenfunction in this case. Thus the exact number of zeros of u_n cannot be determined. Nevertheless we can say the following: Given an indexing scheme based on (5.3.2), let u_n be an eigenfunction of λ_n, $n \in \mathbb{Z}^*$, then the number of zeros of u_n in the open interval J is 0 or 1 if $n = \pm 0$, and $n - 1$, n or $n + 1$ if $|n| \geq 1$.

(4) *If the boundary condition is coupled and complex, then the eigenvalues can be indexed to satisfy the inequalities*

$$\cdots < \lambda_{-n} < \cdots < \lambda_{-1} < \lambda_{-0} < 0 < \lambda_0 < \lambda_1 < \cdots < \lambda_n < \cdots. \qquad (5.3.3)$$

Moreover, if u_n is an eigenfunction of $\lambda_n(\gamma, K)$, $0 < \gamma < \pi$ or $-\pi < \gamma < 0$, then the number of zeros of the real part $\operatorname{Re} u_n$ on $[a, b]$ is 0 or 1 if $n = \pm 0$, and $n - 1$, or n or $n + 1$ if $|n| \geq 1$. The same conclusion holds for $\operatorname{Im} u_n$, the imaginary part of u_n. Furthermore, for any $n \in \mathbb{Z}^*$, the complex valued function eigenfunction u_n has no zero in the closed interval $[a, b]$,.

(5) *In all three cases, (2), (3), (4) the eigenvalues satisfy the asymptotic formula:*

$$\frac{\lambda_{\pm n}}{n^2} \to c = \pm \pi^2 \left(\int_a^b \sqrt{\frac{w_\pm(t)}{p(t)}} \right)^{-2} \quad \text{as } n \to +\infty,$$

where w_- and w_+ denote the negative and positive parts of the function w, respectively.

PROOF. See Theorems 3.1, 3.2, 3.3 in [**387**] and the references cited there; the asymptotic formula is established in Atkinson and Mingarelli [**29**]. □

3. EXISTENCE OF EIGENVALUES

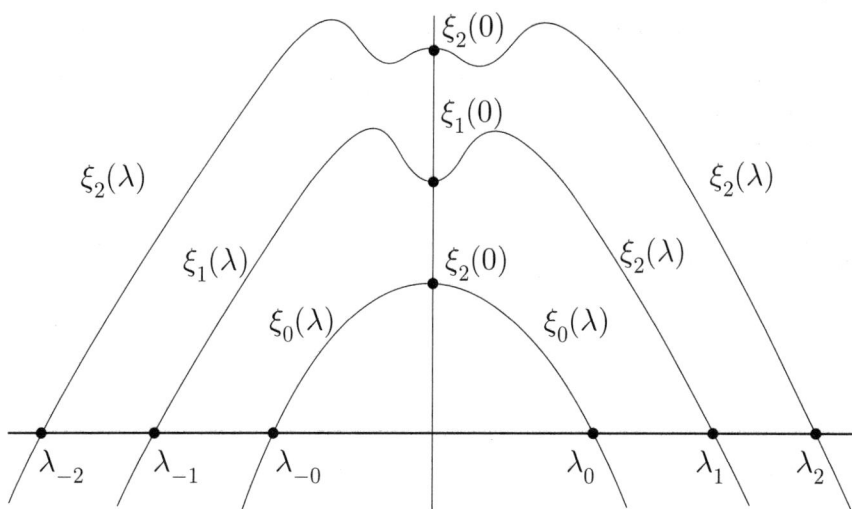

FIGURE 1. The eigenvalues as zeros of the functions $\xi_n(\lambda)$.

The next result characterizes the eigenvalues of left-definite problems as roots of eigencurves of the associated two parameter problem consisting of equation (5.1.3) with the same boundary conditions.

THEOREM 5.3.2. *Assume that the Sturm-Liouville problem consisting of (5.1.1), (5.1.4), (4.1.3), (4.1.4) is left-definite i.e. $\lambda_0(|w|) > 0$. For each $\lambda \in \mathbb{R}$, let $\xi_n(\lambda)$ be the $n-th$ eigenvalue of the right-definite problem consisting of (5.1.3) with the same boundary condition (4.1.3), (4.1.4). Then for each $n \in \mathbb{N}_0$, the function $\xi_n(\lambda)$, $\lambda \in \mathbb{R}$, has exactly one positive root, to be denoted by λ_n, and exactly one negative root, to be given the notation λ_{-n} and each of these roots is an eigenvalue of the problem (5.1.1), (5.1.4), (4.1.3), (4.1.4). Furthermore there are no other eigenvalues.*

PROOF. See Theorems 3.1, 3.2, 3.3 in [**387**]. □

DEFINITION 5.3.1 (λ_n *in the LD Case*). *We use Theorem 5.3.2 to define λ_n uniquely for $n \in \mathbb{Z}^*$ and any left-definite problem. Thus λ_n and λ_{-n} are the unique positive and negative roots of the function $\xi_n(\lambda)$ for any $n \in \mathbb{Z}^*$ including $n = 0$. Recall the definition of $\mathbb{Z}^* = \{... -3, -2, -1, -0, 0, 1, 2, 3...\}$. Note that this defines λ_n uniquely for LD problems. Thus the definition of λ_n for LD problems is based on the definition of ξ_n in the RD case and "connects" the two cases with each other.*

REMARK 5.3.1. Theorem 5.3.2 can be used to construct an algorithm for the numerical computation of the eigenvalues λ_n for any $n \in \mathbb{Z}^*$: Use SLEIGN2 to compute the eigencurve $\xi_n(\lambda)$, then find its positive or negative root with any root finder.

4. Continuous Dependence of Eigenvalues on the Problem

As in the right-definite case to study the continuous and differentiable dependence of eigenvalues on the problem we need to introduce a metric space containing the left-definite problems.

We consider the SLP consisting of the DE

$$-(py')' + qy = \lambda w y \text{ on } J = (a', b'),\ -\infty \le a' < a < b < b' \le \infty, \quad (5.4.1)$$

and the self-adjoint BC

$$A Y(a) + B Y(b) = 0,\ Y = \begin{bmatrix} y \\ py' \end{bmatrix}, \quad (5.4.2)$$

where

$$1/p,\ q,\ w \in L^1(J, \mathbb{R}),\ p > 0 \text{ on } J,\ |w| > 0 \text{ on } J, \quad (5.4.3)$$

while A and B are 2×2 complex-valued matrices such that

$$AEA^* = BEB^*,\ E = \begin{bmatrix} 0 & -1 \\ 1 & 0 \end{bmatrix},\ \text{and } rank(A : B) = 2. \quad (5.4.4)$$

Here C^* is the complex conjugate transpose of the matrix C.

If we abbreviate this SLP as $\omega = (a, b, A, B, 1/p, q, w)$, then the space of the SLP studied here is

$$\Omega = \{(a, b, A, B, 1/p, q, w) : (5.4.1) \text{ to } (5.4.4) \text{ holds}\}. \quad (5.4.5)$$

A natural topology on Ω is the product topology induced from the usual topology on $\mathbb{R}$ and on $L^1(J, \mathbb{R})$ together with any matrix norm topology. More precisely, given $\epsilon > 0$ and $\omega_0 = (a_0, b_0,, A_0, B_0, 1/p_0, q_0, w_0) \in \Omega$, the ϵ-neighborhood of ω_0 is defined to be the set of $\omega = (a, b, 1/p, q, w, A, B) \in \Omega$ satisfying

$$|a-a_0|+|b-b_0|+\int_R \left(|\widetilde{1/p} - \widetilde{1/p_0}| + |\tilde{q} - \tilde{q_0}| + |\tilde{w} - \tilde{w_0}|\right) + \|A - A_0\| + \|B - B_0\| < \epsilon, \quad (5.4.6)$$

where $\|\cdot\|$ is any fixed matrix norm, and $\tilde{f}$ is the extension of $f \in L((a, b), \mathbb{R})$ to $\mathbb{R}$ which equals 0 on $\mathbb{R} \setminus (a, b)$. It is with respect to this topology that we study the dependence of the eigenvalues of an SLP on its parameters.

REMARK 5.4.1. Note that this space Ω is not a vector space but is a metric space in general and a Banach space when the endpoints a, b remain fixed. Also note that it contains the space of right-definite SLP with $p > 0$ and $w > 0$ on J.

For the convenience of the reader we recall from Chapter 4 that the self-adjoint BC are classified into two disjoint classes: separated and coupled. The separated self-adjoint BC have the canonical representation

$$\begin{aligned} \cos\alpha\ y(a) + \sin\alpha\ y^{[1]}(a) &= 0,\ 0 \le \alpha < \pi, \\ \cos\beta\ y(b) + \sin\beta\ y^{[1]}(b) &= 0,\ 0 < \beta \le \pi. \end{aligned} \quad (5.4.7)$$

Below, we may use $(a, b, \alpha, \beta, 1/p, q, w,)$ instead of $(a, b, A, B, 1/p, q, w,)$ to denote the SLP with these separated BC. Each coupled self-adjoint BC has the canonical representation

$$Y(b) = e^{i\gamma} K Y(a), \quad (5.4.8)$$

where
$$-\pi < \gamma \leq \pi, \quad K \in SL(2,\mathbb{R}) \tag{5.4.9}$$
with
$$SL(2,\mathbb{R}) = \left\{ K = \begin{bmatrix} k_{11} & k_{12} \\ k_{21} & k_{22} \end{bmatrix}, \; k_{ij} \in \mathbb{R} : \det K = 1 \right\}.$$

It turns out, as we will see below, that each left-definite problem is an interior point of the space Ω *provided the boundary condition is not on the "jump set" $\mathcal{J}$.* If it is on the jump set, then the problem is not an interior point because an arbitrarily small change in the boundary condition in the direction in which the lowest eigenvalue $\lambda_0(|w|)$ of the associated RD problem "jumps" to $-\infty$ means that these problems are not LD. The "jump set" $\mathcal{J}$ is the set of boundary conditions on which the eigenvalues of the associated right-definite problem have a jump discontinuity as functions of the boundary condition. It is given by Definition 4.4.1 and consists of all separated BC (5.4.7) together with all coupled BC (5.4.8) satisfying the following conditions:

$$\alpha = 0 \text{ or } \beta = \pi; \; k_{12} = 0, \; -\pi < \gamma < \pi. \tag{5.4.10}$$

THEOREM 5.4.1. *Let $\omega = (a, b, , A, B, 1/p, q, w) \in \Omega$. Assume that w changes sign and ω is left-definite, i.e. $\lambda_0(|w|) > 0$. Let $\{\lambda_n : n \in \mathbb{Z}^*\}$ denote the eigenvalues of ω as given in Theorem 5.3.1. Fix $n \in \mathbb{Z}^*$. Then λ_n is a continuous function of ω as ω varies within the subset of LD problems inside Ω. More specifically, we have:*

(1) *ω remains left-definite under any sufficiently small change of the equation (5.1.1), i.e. under any sufficiently small change of $a, b, 1/p, q, w$. And λ_n is a continuous function of each of the variables $a, b, 1/p, q, w$.*

(2) *If the boundary condition is not on the jump set $\mathcal{J}$, then there is a neighborhood $\mathcal{N}$ of ω in Ω such that each problem in $\mathcal{N}$ is also left-definite with weight function which changes sign and $\lambda_n(\rho)$ is a continuous function of $\rho \in \mathcal{N}$.*

(3) *If the boundary condition is on the jump set $\mathcal{J}$, then left-definiteness is preserved only for those small changes of the BC which lie on "the continuity side" of $\mathcal{J}$. (See Theorems 3.3.9, 3.7.3, 3.7.6 and Propositions 3.7.1 and 3.7.2 in [384] for an explanation of the meaning of "the continuity side" of $\mathcal{J}$.) Thus λ_n is continuous at the given jump set boundary condition within the set of LD problems in Ω.*

PROOF. See [387]. □

REMARK 5.4.2. We comment on the contrast between the continuity of λ_n in the left-definite and right-definite cases. In the RD case each λ_n is a continuous function of the coefficients and the endpoints but has a jump discontinuity on the jump set $\mathcal{J}$ of boundary conditions. In the LD case each λ_n is continuous in all parameters of the problem including the boundary conditions as long as *the problem* varies within the space of left-definite problems only. Each LD problem with an indefinite weight function is an interior point of the set of LD problems if the boundary condition is not on $\mathcal{J}$. If the boundary condition is on $\mathcal{J}$, then the problem is not an interior point of the set of LD problems but is on the boundary of this set. However, the problem ceases to be LD as soon as the BC is varied in the direction of the jump discontinuity for RD problems. So each λ_n in the LD case with w changing sign can be embedded in a continuous eigenvalue branch within the space of LD problems but this continuity branch ends abruptly on $\mathcal{J}$. It is interesting to observe that for an LD problem with a boundary condition on the jump set $\mathcal{J}$ each eigenvalue is embedded in the interior of a continuous eigenvalue branch by Theorem 3.5.1 but this branch may not be contained in the set of LD problems and may not be defined by a fixed index n. See Remark 4.1 in [**387**] and for more details, see [**286**] and [**384**].

5. Eigenvalue Inequalities

In this section we discuss inequalities analogous to those established in Section 4.8 for the RD case. In order to compare eigenvalues, for the same or for different boundary conditions, with each other they must be uniquely defined. In the RD case with $p > 0$, the eigenvalues are bounded below and so λ_n has a natural meaning based on the ordering (4.3.7). Since the eigenvalues are unbounded above and below in the LD case the meaning of λ_n is not clear just from the natural ordering (5.3.3) along the real line. For LD problems we use Definition 5.3.1 which is based on Theorem 5.3.2 to determine λ_n uniquely for all $n \in \mathbb{Z}^* = \{..., -2, -1, -0, 0, 1, 2, ...\}$ and note that this definition connects the eigenvalues of an LD problems with those of the corresponding one parameter family of RD problems as given in Theorem 5.3.2. This connection is exploited in proving the inequalities given in this chapter for the LD case.

Just as in the RD case, the inequalities given below show that for any given coupled boundary condition, determined, say, by K and γ, and any one of its eigenvalues, say λ_n, there is an interval containing λ_n and no other eigenvalue of K and γ. The endpoints of this interval are eigenvalues of separated boundary conditions which depend on K. (They are independent of γ.) Thus these inequalities can be used to construct an algorithm for the numerical computation of eigenvalues of coupled boundary conditions for any LD problem, given that the eigenvalues for separated BC are known. The latter can be computed from their Prüfer transformation characterization. Such an algorithm is implemented by SLEIGN2 in the RD case but not for the LD case; see Remark 5.5.1 below.

Next, we establish some inequalities among the eigenvalues of a LD problem with a coupled BC and those for two corresponding separated BC's.

THEOREM 5.5.1. *Let (5.1.4) hold; let $K \in SL(2, \mathbb{R})$. Consider equation (5.1.1) with the following boundary conditions:*

$$y(a) = 0, \qquad k_{22}y(b) - k_{12}y^{[1]}(b) = 0, \tag{5.5.1}$$

$$y^{[1]}(a) = 0, \qquad k_{21}y(b) - k_{11}y^{[1]}(b) = 0, \tag{5.5.2}$$

$$Y(b) = e^{i\gamma}KY(A), \ -\pi < \gamma \leq \pi. \tag{5.5.3}$$

Assume that (5.5.2) is LD. Then (5.5.1) is LD, and (5.5.3) is LD for any γ, $-\pi < \gamma \leq \pi$. Denote the eigenvalues of (5.5.1), (5.5.2), (5.5.3) by $\upsilon_n = \upsilon_n(K)$, $\nu_n = \nu_n(K)$, $\lambda_n(K)$, $n \in \mathbb{Z}^*$, respectively. (These exist by Theorem 5.3.1)

(1) *Suppose that $k_{11} > 0$, $k_{12} \leq 0$. Then each of $\lambda_0(K)$ and $\lambda_{-0}(K)$ is geometrically simple, and for each $\gamma \in (-\pi, \pi)$, $\gamma \neq 0$, we have*

$$\begin{aligned}
\upsilon_0 &\leq \lambda_0(K) < \lambda_0(e^{i\gamma}K) < \lambda_0(-K) \leq \{\mu_0, \nu_1\} \\
&\leq \lambda_1(-K) < \lambda_1(e^{i\gamma}K) < \lambda_1(K) \leq \{\mu_1, \nu_2\} \\
&\leq \lambda_2(K) < \lambda_2(e^{i\gamma}K) < \lambda_2(-K) \leq \{\mu_2, \nu_3\} \\
&\leq \lambda_3(-K) < \lambda_3(e^{i\gamma}K) < \lambda_3(K) \leq \{\mu_3, \nu_4\} \leq \cdots.
\end{aligned} \tag{5.5.4}$$

5. EIGENVALUE INEQUALITIES

and
$$\begin{aligned}
\nu_{-0} &\geq \lambda_{-0}(K) > \lambda_{-0}(e^{i\gamma}K) > \lambda_{-0}(-K) \geq \{\mu_{-0}, \nu_{-1}\} \\
&\geq \lambda_{-1}(-K) > \lambda_{-1}(e^{i\gamma}K) > \lambda_{-1}(K) \geq \{\mu_{-1}, \nu_{-2}\} \\
&\geq \lambda_{-2}(K) > \lambda_{-2}(e^{i\gamma}K) > \lambda_{-2}(-K) \geq \{\mu_{-2}, \nu_{-3}\} \\
&\geq \lambda_{-3}(-K) > \lambda_{-3}(e^{i\gamma}K) > \lambda_{-3}(K) \geq \{\mu_{-3}, \nu_{-4}\} \geq \cdots .
\end{aligned} \quad (5.5.5)$$

(2) *Suppose that $k_{11} \leq 0$, $k_{12} < 0$. Then each of $\lambda_0(K)$ and $\lambda_{-0}(K)$ is geometrically simple, and for each $\gamma \in (-\pi, \pi)$, $\gamma \neq 0$, we have*
$$\begin{aligned}
\lambda_0(K) &< \lambda_0(e^{i\gamma}K) < \lambda_0(-K) \leq \{\mu_0, \nu_0\} \leq \\
\lambda_1(-K) &< \lambda_1(e^{i\gamma}K) < \lambda_1(K) \leq \{\mu_1, \nu_1\} \leq \\
\lambda_2(K) &< \lambda_2(e^{i\gamma}K) < \lambda_2(-K) \leq \{\mu_2, \nu_2\} \leq \\
\lambda_3(-K) &< \lambda_3(e^{i\gamma}K) < \lambda_3(K) \leq \{\mu_3, \nu_3\} \leq \cdots .
\end{aligned} \quad (5.5.6)$$

and
$$\begin{aligned}
\lambda_{-0}(K) &> \lambda_{-0}(e^{i\gamma}K) > \lambda_{-0}(-K) \geq \{\mu_{-0}, \nu_{-0}\} \geq \\
\lambda_{-1}(-K) &> \lambda_{-1}(e^{i\gamma}K) > \lambda_{-1}(K) \geq \{\mu_{-1}, \nu_{-1}\} \geq \\
\lambda_{-2}(K) &> \lambda_{-2}(e^{i\gamma}K) > \lambda_{-2}(-K) \geq \{\mu_{-2}, \nu_{-2}\} \geq \\
\lambda_{-3}(-K) &> \lambda_{-3}(e^{i\gamma}K) > \lambda_{-3}(K) \geq \{\mu_{-3}, \nu_{-3}\} \geq \cdots .
\end{aligned} \quad (5.5.7)$$

(3) *If neither case (1) nor case (2) applies to K, then either case (1) or case (2) applies to $-K$. The notation $\{v, \nu\}$ means both v and ν but in no particular order.*

(4) *In all these cases the eigenvalues satisfy the asymptotic formula:*
$$\lambda_{\pm n} \sim \pm \frac{n^2 \pi^2}{\left(\int_a^b \frac{w_{\pm}(t)}{p(t)} dt\right)^2} \quad \text{as } n \to +\infty. \quad (5.5.8)$$

PROOF. The asymptotic formula (5.5.8) is proven in Atkinson and Mingarelli [29] for separated boundary conditions. The coupled case follows from the separated case and the inequalities in parts (1), (2) and (3).

Parts (1), (2) and (3) are established in Kong, Wu and Zettl [387]; we give the proof of Part (1), Parts (2) and (3) can be proved similarly.

For each $\lambda \in \mathbb{R}$, let $\xi_n(\lambda, e^{i\gamma}K)$, $n \in \mathbb{N}_0$ denote the eigenvalues of the RD problem (5.1.3), (5.5.3); and let $\eta_n(\lambda)$, $\zeta_n(\lambda)$, $n \in \mathbb{N}_0$ denote the eigenvalues of (5.1.3), (5.5.1) and (5.1.3), (5.5.2), respectively.

To prove (1) apply Theorem 4.8.1 to the RD problem consisting of (5.1.3), (5.5.3) to get
$$\begin{aligned}
\zeta_0(\lambda) &\leq \xi_0(\lambda, K) < \xi_0(\lambda, e^{i\gamma}K) < \xi_0(\lambda, -K) \leq \{\eta_0(\lambda), \zeta_1(\lambda)\} \\
&\leq \xi_1(\lambda, -K) < \xi_1(\lambda, e^{i\gamma}K) < \xi_1(\lambda, K) \leq \{\eta_1(\lambda), \zeta_2(\lambda)\} \\
&\leq \xi_2(\lambda, K) < \xi_2(\lambda, e^{i\gamma}K) < \xi_2(\lambda, -K) \leq \{\eta_2(\lambda), \zeta_3(\lambda)\} \\
&\leq \xi_3(\lambda, -K) < \xi_3(\lambda, e^{i\gamma}K) < \xi_3(\lambda, K) \leq \{\eta_4(\lambda), \zeta_4(\lambda)\} \leq \cdots .
\end{aligned} \quad (5.5.9)$$

Note that $\zeta_0(0) > 0$ implies that $\xi_0(0) > 0$ and $\eta_0(0) > 0$ and hence, by Theorem 2.1, both problems (5.1.3), (5.5.1) and (5.1.3), (5.5.2), are LD. By Theorem 5.3.2, for $n \in \mathbb{N}_0$, $\lambda_{\pm n}$, $\mu_{\pm n}$ and $\nu_{\pm n}$ are the positive and negative roots of $\xi_n(\lambda)$, $\eta_n(\lambda)$ and $\zeta_n(\lambda)$, respectively. Since each of $\xi_n(\lambda)$, $\eta_n(\lambda)$ and $\zeta_n(\lambda)$ is continuous in λ and increasing in n, (5.5.9) implies (5.5.4) and (5.5.5). □

REMARK 5.5.1. The eigenvalues of a LD problem with a separated BC can be computed by using their characterization in terms of either the Prüfer transformation or some variant of it. The inequalities of Theorem 5.5.1 can be used to construct an algorithm for computing the eigenvalues of a LD problem with an *arbitrary (self-adjoint) coupled* BC. Note that $\xi_0(0, e^{i\gamma}K) > 0$ implies $\eta_0(0) > 0$. Thus, if the former problem is LD so is the latter. In this case, one computes the μ_n's first, then uses the inequalities of Theorem 5.5.1 with the ν_n's removed to bound each eigenvalue for (5.5.3) between 0 or μ_n and μ_{n+1} for some $n \in \mathbb{N}_0$. The key point here is that there is exactly one eigenvalue for $e^{i\gamma}K$ in such an interval. Now, one applies a root finder to locate this one and only one root of the characteristic function in the interval. Such an algorithm is implemented in the Bailey, Everitt and Zettl code SLEIGN2 for the RD case.

In the RD case by Theorem 4.8.3 we have
$$\lambda_n(A, B) \leq \lambda_n^D, \ n \in \mathbb{N}_0,$$
and by Theorem 4.8.5
$$\lambda_n^D \leq \lambda_{n+2}(A, B), \ n \in \mathbb{N}_0.$$
here λ_n^D denote the Dirichlet eigenvalues and $\lambda_n(A, B)$ are the eigenvalues of an arbitrary self-adjoint, separated or coupled, boundary condition. The next theorem gives analogous results for the left-definite case.

THEOREM 5.5.2 (cf. [13, p. 258]). *Let (5.1.4) hold. Assume that the Sturm-Liouville problem consisting of (5.4.1) and (5.4.4) is left-definite and denote its eigenvalues by λ_n, $n \in \mathbb{Z}^*$. Then, (5.4.1) with Dirichlet boundary conditions is also LD, let λ_n^D, $n \in \mathbb{Z}^*$, denote its eigenvalues. Then we have*

$$\lambda_n \in (0, \lambda_n^D], \ \lambda_{-n} \in [\lambda_{-n}^D, 0) \quad n = 0, 1, \tag{5.5.10}$$
$$\lambda_n \in (\lambda_{n-2}^D, \lambda_n^D], \ \lambda_{-n} \in [\lambda_{-n}^D, \lambda_{-n+2}^D) \quad n = 2, 3, 4, \ldots.$$

PROOF. Let $\{\xi_n^D(\lambda) : n \in \mathbb{N}_0\}$ denote the eigenvalues of (5.1.3) with Dirichlet BC. Since (5.4.1) and (5.4.4) is LD, we have that $\xi_0(0) > 0$. By Theorem 4.8.2
$$\xi_0^D(\lambda) \geq \xi_0(\lambda) \quad \lambda \in \mathbb{R}. \tag{5.5.11}$$
In particular, $\xi_0^D(0) \geq \xi_0(0) > 0$. Thus, equation (5.4.1) with Dirichlet boundary conditions is LD. From (5.5.11) we also get that $\xi_0(\lambda_0^D) \leq \xi_0^D(\lambda_0^D) = 0$. This shows that $\lambda_0 \in (0, \lambda_0^D]$ since λ_0 is the only positive root of $\xi_0(\lambda) = 0$. The rest can be proved similarly. □

6. Differentiability of Eigenvalues

In this section we show that each eigenvalue is a differentiable function of every parameter of the problem and find formulas for the derivatives. These formulas are analogues of the corresponding formulas for the RD case in Chapter 4.

Recall that a map T from a Banach space X into a Banach space Y is differentiable at a point $x \in X$ if there exists a bounded linear map $T' : X \to Y$ satisfying
$$|T(x+h) - T(x) - T'(x)h| = o(h) \text{ as } h \to 0 \text{ in } X. \tag{5.6.1}$$
In the next theorem, the derivative λ_n' is the Frechet derivative in the appropriate Banach spaces.

NOTATION 5.6.1. *If u is an eigenfunction of an eigenvalue λ (determined by the equation and boundary conditions) then $u(t)$ denotes the value of u at t. Some of the derivative formulas in the next theorem involve u or pu' evaluated at endpoints $t = a, b,$ or a_0, b_0.*

THEOREM 5.6.1. *Let Ω be defined by (5.4.5). Suppose that the Sturm-Liouville problem*

$$\omega_0 = (a_0, b_0, 1/p_0, q_0, w_0, A_0, B_0) \in \Omega$$

is left-definite, and its boundary condition is not on the jump set $\mathcal{J}$. For each $n \in \mathbb{N}_0$, and each LD problem $\omega = (a, b, A, B, 1/p, q, w) \in \Omega$ let $u_{\pm n} = u_{\pm n}(\omega)$ denote normalized eigenfunctions for the eigenvalues $\lambda_{\pm n} = \lambda_{\pm n}(\omega)$, i.e., they satisfy

$$\int_a^b |u_{\pm n}|^2 w = \pm 1.$$

Such eigenfunctions exist by Lemma 3.2 of [**387**].

i) Fix all components of ω_0 except a_0 and consider $\lambda_{\pm n}$ as a function of a. Assume that $\lambda_{\pm n}(a_0)$ is geometrically simple or $\lambda_{\pm n}(a)$ is geometrically double for all a in some neighborhood of a_0. Then, $\lambda_{\pm n}(a)$ is differentiable a.e. in some neighborhood $\mathcal{N}_{a_0}$ of a_0 and

$$\lambda'_{\pm n}(a) = \pm \left\{ \frac{1}{p}(a) |pu'_{\pm n}|^2(a) - |u_{\pm n}(a)|^2 [q(a) - \lambda_{\pm n}(a) w(a)] \right\} \text{ a.e. in } \mathcal{N}_{a_0}. \tag{5.6.2}$$

Furthermore, if p, q, w are continuous at a_0 and $p(a_0) \neq 0$, then (5.6.2) holds at for $a = a_0$.

ii) Fix all components of ω_0 except b_0 and consider $\lambda_{\pm n}$ as a function of b. Assume that $\lambda_{\pm n}(b_0)$ is geometrically simple or $\lambda_{\pm n}(b)$ is geometrically double for all b in some neighborhood of b_0. Then, $\lambda_{\pm n}(b)$ is differentiable a.e. in some neighborhood $\mathcal{N}_{b_0}$ of b_0 and

$$\lambda'_{\pm n}(b) = \pm \left\{ -\frac{1}{p}(b) |pu'_{\pm n}|^2(b) + |u_{\pm n}(b)|^2 [q(b) - \lambda_{\pm n}(b) w(b)] \right\} \text{ a.e. in } \mathcal{N}_{b_0}. \tag{5.6.3}$$

Furthermore, if p, q, w are continuous at b_0 and $p(b_0) \neq 0$, then (5.6.3) holds at the point b_0.

iii) Fix all components of ω_0 except $1/p_0$ and consider $\lambda_{\pm n}$ as a function of $1/p$. Assume that $\lambda_{\pm n}(1/p_0)$ is simple. Then, $\lambda_{\pm n}(1/p)$ is simple and continuously differentiable in some neighborhood $\mathcal{N}_{1/p_0}$ of $1/p_0$ in $L^1(\mathbb{R}, \mathbb{R})$ and for any $1/p \in \mathcal{N}_{1/p_0}$,

$$\lambda'_{\pm n}(1/p) h = \pm \left\{ -\int_{a_0}^{b_0} |pu'_{\pm n}|^2 h \right\}, \quad h \in L^1(J, \mathbb{R}). \tag{5.6.4}$$

iv) Fix all components of ω_0 except q_0 and consider $\lambda_{\pm n}$ as a function of q. Assume that $\lambda_{\pm n}(q_0)$ is simple. Then, $\lambda_{\pm n}(q)$ is simple and continuously differentiable in some neighborhood $\mathcal{N}_{q_0}$ of q_0 in $L^1(\mathbb{R}, \mathbb{R})$ and for any $q \in \mathcal{N}_{q_0}$,

$$\lambda'_{\pm n}(q) h = \pm \left\{ \int_{a_0}^{b_0} |u_{\pm n}|^2 h \right\}, \quad h \in L^1(J, \mathbb{R}). \tag{5.6.5}$$

v) Fix all components of ω_0 except w_0 and consider $\lambda_{\pm n}$ as a function of w. Assume that $\lambda_{\pm n}(w_0)$ is simple. Then, $\lambda_{\pm n}(w)$ is simple and continuously differentiable in some neighborhood $\mathcal{N}_{w_0}$ of w_0 in $L^1(\mathbb{R}, \mathbb{R})$ and for any $w \in \mathcal{N}_{w_0}$,

$$\lambda'_{\pm n}(w) h = \pm \left\{ -\lambda_{\pm n}(w) \int_{a_0}^{b_0} |u_{\pm n}|^2 h \right\}, \quad h \in L^1(J, \mathbb{R}). \tag{5.6.6}$$

vi) *Assume that the boundary condition of ω_0 is separable and is represented by the canonical form (5.4.7); in this case, A_0, B_0 in ω_0 are replaced by α_0, β_0. Fix all components of ω_0 except α_0 and consider $\lambda_{\pm n}$ as a function of α, then $\lambda_{\pm n}$ is differentiable and*

$$\lambda'_{\pm n}(\alpha) = \pm \left\{-|u_{\pm n}|^2(a_0) - |pu'_{\pm n}|^2(a_0)\right\}; \ 0 \leq \alpha < \pi. \tag{5.6.7}$$

Fix all components of ω_0 except β_0 and consider $\lambda_{\pm n}$ as a function of β, then $\lambda_{\pm n}$ is differentiable and

$$\lambda'_{\pm n}(\beta) = \pm \left\{|u_{\pm n}|^2(b_0) + |pu'_{\pm n}|^2(b_0)\right\}. \ 0 < \beta \leq \pi. \tag{5.6.8}$$

vii) *Assume that the boundary condition in ω_0 is coupled and is represented in the canonical form (5.4.8); in this case, A_0, B_0 in ω_0 are replaced by γ_0, K_0. Fix all components of ω_0 except γ_0 and consider $\lambda_{\pm n}$ as a function of γ, then $\lambda_{\pm n}$ is a differentiable function of γ and*

$$\lambda'_{\pm n}(\gamma) = \pm[-2\,\text{Im}(u_{\pm n}(b_0)(p\bar{u}'_{\pm n})(b_0))], \tag{5.6.9}$$

where $\text{Im}\,z$ denotes the imaginary part of z.

Fix all components of ω_0 except K_0 and consider $\lambda_{\pm n}$ as a function of $K \in SL(2,\mathbb{R})$. Assume further that $\lambda_{\pm n}(K_0)$ is simple. Then $\lambda_{\pm n}(K)$ is simple and differentiable in some neighborhood $\mathcal{N}_{K_0}$ of K_0 in $SL(2,\mathbb{R})$ and for any $K \in \mathcal{N}_{K_0}$, $H \in \mathbb{R}^{2\times 2}$ with $\text{trace}(H) = 0$ we have

$$\lambda'_{\pm n}(K)KH = \pm \left\{ \begin{array}{cc} p\bar{u}'_\pm(b_0) & -\bar{u}'_\pm(b_0) \end{array} KH \begin{pmatrix} u(b_0) \\ (pu')(b_0) \end{pmatrix} \right\}. \tag{5.6.10}$$

PROOF. The proof of this theorem is similar to the proofs of the corresponding results for the RD case, see Chapter 4 and ([**384**], [**387**], [**383**], [**394**], [**392**]). Here we just mention that for any $K \in SL(2,\mathbb{R}) \subset \mathbb{R}^{2\times 2}$, the tangent space of $SL(2,\mathbb{R})$ at K is $\{KH;\ H \in \mathbb{R}^{2\times 2},\ tr(H) = 0\} \subset \mathbb{R}^{2\times 2}$. □

To see the importance of the assumption in Theorem 5.6.1 that the BC in ω_0 is not on the jump set $\mathcal{J}$, we give the following example.

EXAMPLE 5.6.1. *The SLP consisting of the equation*

$$-y'' = \lambda\,(\text{sgn}\,t)\,y \text{ on } (-1,1)$$

and the Dirichlet BC

$$y(-1) = 0 = y(1)$$

is LD since the least Dirichlet eigenvalue of the Fourier equation

$$-y'' = \lambda\,y \text{ on } (-1,1)$$

positive. Now, in any neighborhood of this LD problem, in the space Ω there is a problem with the same equation and the BC

$$y(-1) - c\,y^{[1]}(-1) = 0, \qquad y(1) = 0$$

for sufficiently small $c > 0$. If $\lambda_0(c)$ denotes the lowest eigenvalue of the Fourier equation (with $w = 1$) then $\lambda_0(c) \to -\infty$ as $c \to 0^+$, by Theorem 4.6.5 (This can also be confirmed by a direct and simple calculation).

In Section 5.5 we discussed the variation of the eigenvalues as the boundary conditions change but the coefficients and endpoints of the interval remain fixed. Next show that these derivative formulas imply the following comparison results on the eigenvalues $\{\lambda_n;\ n \in \mathbb{Z}^*\}$. Here the boundary condition is fixed and the coefficients vary. These kinds of comparison results are sometimes called monotonicity results. In the next theorem **D** denotes the Dirichlet BC.

THEOREM 5.6.2. *Let Ω be defined by (5.4.5).*
(i) Assume that $\omega_k = (a_k, b_k, \mathbf{D}, 1/p, q, w) \in \Omega$ for $k = 0, 1$ satisfy

$$a_0 \leq a_1, \qquad b_0 \geq b_1. \tag{5.6.11}$$

If ω_1 is left-definite, then so is ω_0, and

$$\lambda_n(\omega_0) \leq \lambda_n(\omega_1), \qquad \lambda_{-n}(\omega_0) \geq \lambda_{-n}(\omega_1),\ n \in \mathbb{N}_0. \tag{5.6.12}$$

(ii) Assume that $\omega_k = (a_k, b_k, A, B, 1/p, q, w) \in \Omega$ for $k = 0, 1$ satisfy

$$1/p_0 \geq 1/p_1, \qquad q_0 \leq q_1,\ \text{a.e. on}\ J = (a, b) \tag{5.6.13}$$

If ω_0 is left-definite, then so is ω_1 and (5.6.12) holds.
(iii) Assume that $\omega_k = (a_k, b_k, \alpha_k, \beta_k, 1/p, q, w) \in \Omega$ for $k = 0, 1$ satisfy

$$\pi > \alpha_0 \geq \alpha_1 \geq 0, \qquad 0 < \beta_0 \leq \beta_1 \leq \pi. \tag{5.6.14}$$

If ω_0 is left-definite, then so is ω_1 and (5.6.12) holds.
(iv) Let $\omega_k = (a, b,\,, A, B, 1/p, q, w_k) \in \Omega$ for $k = 0, 1$. Assume that both weight functions w_k change sign on J. If one of ω_0 and ω_1 is left-definite, then so is the other one. Moreover, in this case, if we also have that

$$w_0 \geq w_1 \tag{5.6.15}$$

a.e. on J, then

$$\lambda_n(\omega_0) \leq \lambda_n(\omega_1) \tag{5.6.16}$$

for any $n \in \mathbb{Z}^*$.

PROOF. The proofs are similar to those of the corresponding results in the RD case, see Chapter 4, particularly Theorem 4.9.1 and also see Theorem 4.1 in [**394**]. □

7. T-Left-Definite Problems

In this section we observe that there is a class of problems which are not LD but can be transformed into LD problems by a translation.

DEFINITION 5.7.1. *Let $T \in \mathbb{R}$, and define a functional $\mathcal{L}_T$ on the linear subspace $\mathcal{D}(A, B)$ of $L^2(J, |w|)$ given in (5.2.1) as follows:*

$$\mathcal{L}_T f = \int_a^b [-(pf')'\bar{f} + (q - Tw)|f|^2]. \tag{5.7.1}$$

Then, the SLP consisting of (5.1.1), (5.1.4) and (4.1.3), (4.1.4) is said to be T-left-definite (T-LD) if $\mathcal{L}_T f > 0$ for all f in $\mathcal{D}(A, B)$, $f \not\equiv 0$.

REMARK 5.7.1. From this definition it is easy to see the following:
(i) The SLP consisting (5.1.1), (5.1.4) and (4.1.3) is LD if and only if it is T-LD for $T = 0$.

(ii) The SLP consisting of (5.1.1), (5.1.4) and (4.1.3) is T-LD for some $T \in \mathbb{R}$ if and only if the problem consisting of the equation

$$-(py')' + (q - Tw)y = \lambda wy \quad \text{on } (a,b) \tag{5.7.2}$$

and the BC $AY(a) + BY(b) = 0$ is LD.

(iii) Every T-LD problem with a weight function that changes sign has countably infinitely many eigenvalues, they are all real, are unbounded above and below, and can be indexed to satisfy

$$\cdots \leq \lambda_{-2} \leq \lambda_{-1} \leq \lambda_{-0} < T < \lambda_0 \leq \lambda_1 \leq \lambda_2 \leq \cdots \tag{5.7.3}$$

with only geometrically double eigenvalues appearing twice.

(iv) Using equation (5.7.2) in place of (5.1.1) the LD results of Sections 5.1 through 5.6 hold for the T-LD problems if LD is replaced by T-LD and the number 0 by T, as in (5.7.3).

The next theorem gives two simple characterizations of T-LD problems.

THEOREM 5.7.1. *Let (5.1.1), (5.1.4) and (4.1.3), (4.1.4) hold. Then the following statements are equivalent:*
(i) the problem is T-left-definite for some $T \in \mathbb{R}$;
(ii) the corresponding function $\xi_0(\lambda)$ has a positive value on $\mathbb{R}$;
(iii) the corresponding function $\xi_0(\lambda)$ has two distinct zeros in $\mathbb{R}$.

PROOF. The proof is straightforward and hence omitted. □

COROLLARY 5.7.1. *If the Sturm-Liouville problem (5.1.1), (5.1.4) and (4.1.3), (4.1.4) has two eigenvalues λ_* and λ^* such that each of them has an eigenfunction without any zero on (a,b), then the problem is T-left-definite for any $T \in (\lambda_*, \lambda^*)$.*

PROOF. The proof is straightforward and hence omitted. □

8. Indefinite Problems and Complex Eigenvalues

In Section 5.1 we pointed out that an SLP with real coefficients and self-adjoint boundary conditions may have nonreal eigenvalues when the weight function w changes sign. In this this section we show that when $p > 0$ the number of nonreal eigenvalues is finite and establish an upper bound for this number. *Throughout this section we assume that (5.1.4), the equations (5.1.1), (5.1.2), (we don't need (5.1.3) in this section) and the self-adjoint boundary conditions (4.1.3), (4.1.4) hold. The boundary conditions (4.1.3), (4.1.4), the interval $J = (a,b)$, and the coefficients p, q are fixed and $p > 0$.* The main result of this section gives an upper bound for the number of nonreal eigenvalues of the indefinite problem *(5.1.1), (4.1.3), (4.1.4)* which we will refer to as the 'w problem' in terms of the number of nonpositive eigenvalues of the corresponding right-definite problem *(5.1.2), (4.1.3), (4.1.4)* to be called the '$|w|$ problem'. Note that the boundary conditions, the underlying interval as well as p and q are the same for these two problems. Let S denote an arbitrary self-adjoint realization of the $|w|$ problem in the Hilbert space $L^2(J, |w|)$ according to Chapter 4 and let $D(S)$ denote its domain.

Recall that the sesquilinear form $[\cdot, \cdot]$ is defined by

$$[f, g] = f(p\overline{g}') - \overline{g}(pf'), \quad f, g \in D_{\max} = D_{\max}(p, q, |w|).$$

The next lemma highlights the role of self-adjointness in the boundary conditions.

8. INDEFINITE PROBLEMS AND COMPLEX EIGENVALUES

LEMMA 5.8.1. *Let (5.1.4) hold, let $D(S)$ be a self-adjoint domain of (5.1.2). If $y, z \in D(S)$, then*

$$[y, z]_a^b = [y, z](b) - [y, z](a) = 0. \tag{5.8.1}$$

PROOF. (i) Separated boundary conditions. Represent these in canonical form (4.2.3), (4.2.4). If $\beta \neq \pi/2$, then $y(b) = \tan(\beta)(py')(b)$, $z(b) = \tan(\beta)(pz')(b)$ and consequently

$$[y, z](b) = y(b)(p\bar{z}')(b) - \bar{z}(b)(py')(b)$$
$$= \tan(\beta)(py')(b)(p\bar{z}')(b) - \tan(\beta)(p\bar{z}')(b)(py')(b) = 0,$$

and similarly $[y, z](a) = 0$. If $\beta = \pi/2$, then (4.2.4) reduces to $(py')(b) = 0$, $(pz')(b) = 0$ and (5.8.1) follows. Similarly for $\alpha = \pi/2$. Thus (5.8.1) holds for all separated boundary conditions.

(ii) Coupled boundary conditions. Represent these in the canonical form (4.2.5) with $\gamma \in (-\pi, \pi]$.

Note that

$$[y, z] = (EY, Z), \quad E = \begin{bmatrix} 0 & -1 \\ 1 & 0 \end{bmatrix}, \quad Y = \begin{bmatrix} y \\ py' \end{bmatrix}, \quad Z = \begin{bmatrix} z \\ pz' \end{bmatrix},$$

and where $(\cdot, \cdot)$ denotes the inner product in $\mathbb{C}^2$. Observe that the self-adjointness conditions (4.1.3), (4.1.4) applied to the canonical form (4.2.5) imply that $K^*EK = E$. Thus using (4.2.5) we get

$$[y, z]_a^b = [y, z](b) - [y, z](a) = (EY(b), Z(b)) - (EY(a), Z(a))$$
$$= (Ee^{i\gamma}KY(a), e^{i\gamma}KZ(a)) - (EY(a), Z(a))$$
$$= (e^{i\gamma}e^{-i\gamma}K^*EKY(a), Z(a)) - (EY(a), Z(a))$$
$$= (EY(a), Z(a)) - (EY(a), Z(a)) = 0.$$

This completes the proof. □

THEOREM 5.8.1. *Let (5.1.4), (4.1.3), (4.1.4) hold. If (λ, y) and (μ, z), $\lambda, \mu \in \mathbb{C}$ are eigenpairs of the w problem, i.e of (5.1.1), (4.1.3), (4.1.4), then*

$$(\lambda - \bar{\mu}) \int_a^b y\bar{z}w = 0. \tag{5.8.2}$$

PROOF. Let $My = -(py')' + qy$. Then, recalling that p, q, w are real-valued, we have

$$\bar{z}\lambda wy - y\bar{\mu}w\bar{z} = \bar{z}My - y\overline{Mz} = \bar{z}\{-(py') + qy\} - y\{-(p\bar{z}) + q\bar{z})\}$$
$$= \{yp\bar{z}' - \bar{z}py'\}' = [y, z]'. \tag{5.8.3}$$

Now (5.8.2) follows from (5.8.3) and (5.8.1). □

COROLLARY 5.8.1. *Let the hypothesis and notation of Theorem 5.8.1 hold. If $\lambda \neq \bar{\mu}$, then*

$$\int_a^b y\bar{z}w = 0. \tag{5.8.4}$$

In particular (5.8.4) holds when λ and μ are either both in the upper half-plane $\mathbb{C}^+$ or both in the lower half-plane $\mathbb{C}^-$.

PROOF. This follows directly from (5.8.2). □

The next result gives an upper bound for the number of nonreal eigenvalues of the indefinite w problem in terms of the number of nonpositive eigenvalues of the corresponding right-definite $|w|$ problem.

THEOREM 5.8.2. *Let (5.1.4), (4.1.3), (4.1.4) hold. Suppose the $|w|$ problem (5.1.2) (4.1.3) (4.1.4), has exactly k nonpositive eigenvalues for some integer $k \in \mathbb{N}_0$. Then the w problem (5.1.1) (4.1.3) (4.1.4) has at most $2k$ nonreal eigenvalues.*

PROOF. If the $|w|$ problem has no nonpositive eigenvalue, then the w problem is left-definite with all eigenvalues real by Lemma 5.3.1. This is the case $k = 0$. Note that if λ is an eigenvalue of the w problem then so is its conjugate $\bar{\lambda}$. Thus the nonreal eigenvalues occur in conjugate pairs.

Suppose $k > 0$ and the w problem has $k+1$ eigenvalues $\lambda_0, \lambda_1, ..., \lambda_k$ in the upper half-plane $\mathbb{C}^+ = \{\lambda \in \mathbb{C} : \text{Im}(\lambda) > 0\}$. Let u_j be an eigenfunction of λ_j, $j = 0, ..., k$. Denote the eigenpairs of the $|w|$ problem by (δ_j, z_j), $j = 0, 1, 2, ...$.

Choose a nontrivial $k + 1$ tuple c_j, $j = 0, ..., k$ such that

$$\sum_{j=0}^{k} c_j (u_j, z_m) = 0, \ m = 0, ..., k-1. \tag{5.8.5}$$

Such a choice of c_j is possible since (5.8.5) is a linear system of k equations in $k + 1$ unknowns. Let

$$f = \sum_{j=0}^{k} c_j u_j.$$

Then f has the following properties:
(1) $f \in D(S)$, i.e. f satisfies the self-adjoint boundary conditions that determine S.
(2) $(f, z_m) = 0$, $m = 0, ..., k-1$.
(3) $<f, f> = \int_a^b \bar{f} M f = 0$.

Property (1) is clear since each u_j satisfies the boundary conditions and these are homogeneous. Equation (5.8.5) is equivalent to (2).

To establish (3) note that $<f, g> = \int_a^b \bar{g} M f$ is a sesquilinear form on $D(S)$. By Corollary 5.8.1 we have $\int_a^b |u_j|^2 w = 0$, $j = 0, ..., k$. Since all the eigenvalues λ_j, $j = 0, ..., k$ are in the upper half-plane it follows from Corollary 5.8.1 that $\int_a^b \bar{u}_r u_m w = 0$, for $r \neq m$. Hence from the sesquilinearity of the form $< \cdot, \cdot >$ it follows that $<f, f> = \int_a^b \bar{f} M f = 0$, and property (3) is established.

To complete the proof we use the well-known variational characterization of the eigenvalue δ_k:

$$\delta_k = \inf \frac{\int_a^b \bar{g} M g}{\int_a^b |g|^2 |w|} \tag{5.8.6}$$

where the infimum is taken over all $g \in D(S)$ satisfying $(g, u_j) = 0$, $j = 0, ..., k-1$. Taking $g = f$ yields $\delta_k \leq 0$ since $\int_a^b \bar{f} M f = 0$. This gives $k + 1$ nonpositive eigenvalues of the $|w|$ problem and completes the proof. □

PROBLEM 1. *In general, the upper bound given by Theorem 5.8.2 is far from sharp. Determining the exact number of nonreal eigenvalues of the w problem in terms of the parameters of the problem: J, p, q, w, A, B is a difficult and interesting open problem in Sturm-Liouville theory. Although, as we know from Chapter 3,* *each eigenvalue can be embedded in a continuous eigenvalue branch, without self-adjointness to force the eigenvalues to stay on the real line, a small perturbation may cause any given eigenvalue to move into the upper or lower complex plane.*

9. Comments

In 1918 Richardson [**546**] showed that even the regular SLP consisting of (5.1.1) with a weight function w that changes sign and the Dirichlet BC can have non-real eigenvalues. However, there is an important class of SLP associated with equation (5.1.1) when w changes sign, the so-called "left-definite" problems, whose eigenvalues are all real. Such a class is studied here. There is an extensive literature on 'left-definite' problems; see, for example, Atkinson and Jabon [**27**], Atkinson and Mingarelli [**29**], Bennewitz and Everitt [**65**], Binding and Browne [**69**], Binding and Huang [**78**], Binding and Ye [**82**], [**82**], Binding and Volkmer [**80**], Curgus and Langer [**133**], Daho and Langer [**134**], Fleige [**249**], Haupt [**321**], Hilbert [**324**], Ince [**344**], Kamke [**356**], Mingarelli [**472**], Richardson [**546**] and the references cited in these papers. Nevertheless, compared to the right-definite case, much less is known for left-definite problems.

Our approach is elementary and based on the basic theory of linear ordinary differential equations and the results of Chapter 4, many of which are recent. This approach is motivated by [**69**], [**80**], [**79**], [**128**], [**472**], [**344**]. The two-parameter equation (5.1.3) is a key feature in this approach. With this equation, the eigenvalues of left-definite problems are connected to those of right-definite problems by "eigenvalue curves" depending on the parameter λ.

These comments are made separately for each section.

(1) Many authors, besides those already referenced in this Chapter, have worked on left-definite problems. We mention a few: Allegretto and Mingarelli [**12**], [**11**], Atkinson and Jabon [**28**], Bennewitz and Everitt [**65**], Daho and Langer [**134**], Everitt [**197**], Everitt [**211**], Everitt and Littlejohn [**217**], Everitt, Littlejohn and Wellman [**218**], Haertzen, Kong, Wu and Zettl [**286**], Kaper, Kwong and Zettl [**365**], , Hilbert [**324**], Littlejohn and Wellman [**449**], Littlejohn and Zettl [**448**], Mingarelli [**472**], Niessen [**497**], Niessen and Schneider [**501**], [**500**], [**499**], Pleijel [**519**], [**517**], Von Hoff [**588**], [**587**], [**586**]. For recent papers using the Bennewitz, Everitt, Niessen, Pleijel, Schneider, et al approach, which differs from the method used here and in Chapter 12 below, the reader is referred to the papers of Von Hoff. This approach also yields expansion theorems in contrast to our approach.

Littlejohn and Wellman developed an abstract theory based on self-adjoint symmetric operators in Hilbert space which are *bounded below* and call this a 'left-definite' theory. This theory is not applicable to Sturm-Liouville problems with a weight function that changes sign and therefore is not comparable with the two other approaches mentioned.

(2) The characterization of LD problems in terms of the lowest eigenvalue of the corresponding RD problem with weight function $|w|$ is a direct consequence of the variational characterization of this lowest eigenvalue and is well known. However few authors have studied the case of coupled boundary conditions.

(3) The existence of eigenvalues for separated boundary conditions has been established by many authors; there is a major recent contribution by Binding and Volkmer [**80**]. The primary method used for these problems is based on either the Prüfer transformation directly or some variant of it. This works well for separated boundary conditions but not for coupled ones. That is probably the main reason why so little has been written about coupled BC in the LD case. One advantage of the "eigenvalue curves" approach taken here is that it works equally well for separated and coupled boundary conditions. On the other hand, in contrast to the Niessen, Pleijel, Schneider et al method (see the recent papers by Von Hoff), it does not give expansion results, at least not directly.

(4) In contrast to the RD case the eigenvalues of LD problems are continuous functions of all parameters of the problem, including the boundary conditions. *But this is true only if we remain in the "space of LD problems".* If all other parameters are fixed and the boundary conditions are allowed to vary then if the BC is not on the jump set $\mathbb{J}$ there is a neighborhood of this problem so that every other problem within this neighborhood is LD and the continuous eigenvalue branches remain *locally* in the LD space. If the boundary condition x is on the jump set $\mathbb{J}$ then as the boundary conditions approach x from the "continuity side" of $\mathbb{J}$ the problems remain in the LD space; however as the boundary conditions pass through x into the discontinuous 'side' of $\mathbb{J}$ the problems cease to be LD. So any continuous eigenvalue branch having x as an interior point does not lie inside the LD space. Here the continuous and discontinuous "sides" of $\mathbb{J}$ are those from the RD theory, they are characterized in Section 3 of [**384**]. See also [**286**].

(5) One application of the eigencurve method is in establishing these general LD inequalities from the corresponding ones for the RD case which were obtained by Eastham, Kong, Wu and Zettl in [**145**].

(6) The results of this section are another example of the "transfer" of results from the RD to the LD case via the eigenvalue curve method.

(7) The reach of the results of this chapter can be widened by observing that they can be applied to problems which are not left-definite but can be transformed to LD problems with a transformation which merely translates the eigenvalues along the real axis: $\lambda_n \to \lambda_n + T$ for some real number T and all n. The results can also be extended to semi-left-definite and T-semi-left-definite problems. The former have $\lambda_0(|w|) = 0$.

All the eigenvalues are real for LD problems and the extensions of LD problems mentioned here. Given a regular SLP with real-valued coefficients (no sign conditions on p and w) and self-adjoint boundary conditions - in the RD sense - some open problems are:

(a) Characterize the class of such problems for which all eigenvalues are real.

(b) If there are nonreal eigenvalues, determine the number of nonreal eigenvalues in terms of the parameters of the problem.

(c) Study the continuous and differentiable dependence of the real and nonreal eigenvalues on all parameters of the problem. This is important for theoretical and numerical reasons. When eigenvalues are

computed numerically it is important to know whether the eigenvalues of "nearby" approximations to the problem are "close" to those of the actual problem.
  (d) Find "inclusion" and exclusion regions for the nonreal eigenvalues in the complex plane, i.e. regions which contain one or more eigenvalues and regions which contain no eigenvalues. Describe these regions in terms of the parameters of the problem.
  (e) Find "brackets" for each real eigenvalue; that is, locate each real eigenvalue inside an interval which contains no other eigenvalue. Given such a bracket a root finder can be used with the transcendental characterization $\delta(\lambda) = 0$ to approximate the eigenvalue inside the given bracket. The next question then is: Which eigenvalue is it? Since the number of eigenvalues is countable (possibly finite) - except in some degenerate cases when there are no eigenvalues or they fill up the entire complex plane - how should they be counted? This is an important question not only for comparing eigenvalues of the same problem with each other but for comparing eigenvalues of different, perhaps, similar problems.
  (f) Greenberg and Marletta have a code for computing real and nonreal eigenvalues of SLP. These authors also have a code for computing eigenvalues of higher order problems and for Hamiltonian systems.
(8) For separated boundary conditions all the results are classical going back to the early 20th century, see Mingarelli [**472**]. These classical results have been studied more recently using operator theory in Krein space, see e.g. Curgus and Langer [**133**]. Although in principle this approach works for separated and coupled boundary conditions, the latter have rarely been explicitly studied in the literature with or without Krein space theory. For an exception see paper of Kong, Möller, Wu and Zettl [**382**]. In the Krein space approach in [**133**] the assumption $p > 0$ on J is made. Aside from information obtainable more or less directly from the characteristic function $\delta(\lambda)$ constructed in Chapter 3 very little is known about problems for which both p and w change sign.

REMARK 5.9.1. We summarize the approach used in Chapter 5 to study the existence and properties of eigenvalues and eigenfunctions of regular left-definite problems. Let (5.1.4) hold.

(1) For each fixed $\lambda \in \mathbb{R}$, the SLP consisting of equation (5.1.3) and self-adjoint boundary conditions (4.1.3), is self-adjoint RD and hence has a countably infinite number of eigenvalues $\{\xi_n(\lambda) : n \in \mathbb{N}_0\}$, which are all real, bounded from below and unbounded from above and can be indexed to satisfy

$$\xi_0(\lambda) \leq \xi_1(\lambda) \leq \xi_2(\lambda) \leq \cdots$$

with only the double eigenvalues appearing twice, and

$$\xi_n(\lambda) \to +\infty \text{ as } n \to +\infty.$$

This problem will be called the two-parameter RD problem corresponding to the SLP consisting of (5.1.1), with self-adjoint BC (4.1.3), (4.1.4).

(2) From the continuous dependence of each eigenvalue on the coefficients, see Chapter 4 and Chapter 12, it follows that, for each $n \in \mathbb{N}_0$, $\xi_n(\lambda)$ is a continuous function of λ in $\mathbb{R}$.
(3) Any given $\lambda^* \in \mathbb{R}$ is an eigenvalue of the SLP consisting of (5.1.1), (4.1.3), (4.1.4) if and only if $\xi_n(\lambda^*) = 0$ for some $n \in \mathbb{N}_0$, i.e., if and only if 0 is an eigenvalue of the problem consisting of equation (5.1.3) with $\lambda = \lambda^*$ and the same boundary condition. Moreover, in this case, the two corresponding eigenspaces are the same.
(4) By definition, the SLP consisting of (5.1.1) with (4.1.3), (4.1.4) is LD if and only if $\xi_0(0) > 0$.
(5) If the SLP consisting of equation (5.1.1) with boundary conditions (4.1.3), (4.1.4) is left-definite, then each of its eigenvalues is a root of the equation $\xi_n(\lambda) = 0$ for some $n \in \mathbb{N}_0$ and each such root is an eigenvalue.

Part 3

Oscillation and Singular Existence Problems

Poincaré, Jules Henri (1854-1912)
> Mathematical discoveries, small or great are never born of spontaneous generation. They always presuppose a soil seeded with preliminary knowledge and well prepared by labour, both conscious and subconscious.

Mordell, L.J. (1888-1972)
> Neither you nor I nor anybody else knows what makes a mathematician tick. It is not a question of cleverness. I know many mathematicians who are far abler than I am, but they have not been so lucky. An illustration may be given by considering two miners. One may be an expert geologist, but he does not find the golden nuggets that the ignorant miner does.

In H. Eves Mathematical Circles Adieu, Boston: Prindle, Weber and Schmidt, 1977.

In Part 3 we introduce some of the major topics for the singular SL equation: the oscillation (O) / non-oscillation (NO) and limit-point (LP) / limit-circle (LC) dichotomies. Both of these have a long and illustrious history. Oscillation theory goes back at least to the fundamental paper of Sturm [**569**]; the LP/LC dichotomy to the seminal paper of Weyl [**607**]. We also discuss 'singular initial conditions', this is a topic of recent origin.

CHAPTER 6

Oscillation

Mencken, H.L. (1880-1956)
> It is now quite lawful for a Catholic woman to avoid pregnancy by a resort to mathematics, though she is still forbidden to resort to physics and chemistry.

Notebooks, "Minority Report".

1. Introduction

In this chapter we study the oscillatory properties of the equation

$$My = (py')' + qy = 0 \quad \text{on} \quad J = (a,b), \ -\infty \leq a < b \leq \infty, \tag{6.1.1}$$

where the coefficients are assumed to satisfy

$$1/p, \ q \in L_{loc}(J, \mathbb{R}), \quad p > 0 \text{ a.e. on } J. \tag{6.1.2}$$

REMARK 6.1.1. Note that there is no minus sign in front of the leading coefficient p of the equation (6.1.1). Although this contrasts with the notation in spectral theory, where there is always a minus sign in front of p, and which has been used in Chapters 1 through 5, we follow the usual convention in oscillation theory for the convenience of the reader.

REMARK 6.1.2. Oscillation and nonoscillation (or disconjugacy) results can be extended from one interval to another by a change of the independent variable, see Ahlbrand, Hinton and Lewis [**5**].

2. Principal and Non-Principal Solutions

Recall that, according to Theorem 2.6.1 no nontrivial solution of (6.1.1) can have an accumulation point of zeros in the interior of J nor at a regular endpoint of J. The zeros, if any, of any nontrivial solution of (6.1.1) inside the interval J are isolated. Thus only an endpoint of J can be an accumulation point of zeros of a nontrivial solution of (6.1.1) and this can happen only at a singular endpoint.

DEFINITION 6.2.1 (Principal Solution). Let u, v be real solutions of (6.1.1). Then

- u is called a principal solution at a if
  (1) $u(t) \neq 0$ for $t \in (a, d]$ and some $d \in J$,
  (2) every solution y of (6.1.1) which is not a multiple of u satisfies $u(t)/y(t) \to 0$ as $t \to a^+$.
    Note that $y(t) \neq 0$ for t in some right neighborhood of a by the Sturm Separation Theorem.
- v is called a non-principal solution at a if
  (1) $v(t) \neq 0$ for $t \in (a, d]$ and some $d \in J$,

(2) *v is not a principal solution at a.*

Principal and non-principal solutions at b are defined similarly.

To simplify things we state some definitions and assertions only for one endpoint. Similar definitions and assertions for the other endpoint also hold and are freely used.

LEMMA 6.2.1. *If (6.1.1) has a principal solution u at a, then every non-zero real multiple of u is also a principal solution and no other solution is a principal solution at a.*

PROOF. This follows directly from the definition. □

REMARK 6.2.1. By Lemma 6.2.1 the principal solution u at an endpoint, if it exists, is unique up to real constant multiplicative factors. Non-principal solutions are never unique, since if v is a non-principal solution and u is a principal solution, both at the same endpoint, then $v + cu$ is also a non-principal solution for any $c \in \mathbb{R}$. Simple examples show that the same solution may be principal at one endpoint and non-principal at the other.

Next we restate the definition of oscillation (O) and nonoscillation (NO), Definition 2.6.1, for the convenience of the reader since these concepts are fundamental in this chapter.

DEFINITION 6.2.2. *Let (6.1.1) and (6.1.2) hold. The endpoint a is called oscillatory (O) if there is a nontrivial solution y which has a zero in the interval (a, c) for every c in J. Similarly b is oscillatory if there is nontrivial solution which has a zero in the interval (c, b) for every c in J. By Theorem 2.6.2, the Sturm Separation Theorem, if one nontrivial solution has this property at an endpoint then all nontrivial solutions have this property. Hence oscillation is a property of the equation (6.1.1) and does not depend on any particular solution. An endpoint is nonoscillatory (NO) if it is not oscillatory.*

REMARK 6.2.2. Clearly principal and non-principal solutions do not exist at an oscillatory endpoint.

REMARK 6.2.3. If the equation in (6.1.1) is regular at a then for any solution y, y and py' can be continuously extended to a and principal solutions u exist and satisfy the initial conditions : $u(a) = 0$, $(pu')(a) \neq 0$. Any non-principal solution v at a satisfies $v(a) \neq 0$.

THEOREM 6.2.1. *The equation (6.1.1) is non-oscillatory at a if and only if there exists a principal solution at a.*

PROOF. See p. 547 in Niessen and Zettl [**503**]. □

The next result gives a characterization of principal and non-principal solutions. This will be used below in "regularizing" singular LCNO endpoints.

THEOREM 6.2.2. *Assume that (6.1.1) is non-oscillatory at a. Let u, v be real solutions satisfying $u(t) \neq 0$, $v(t) \neq 0$ for $t \in (a, d]$ and some $d \in J$. Then*

(1) *u is a principal solution at a if and only if*

$$\int_a^d \frac{1}{p\,u^2} = \infty; \qquad (6.2.2)$$

(2) v is a non-principal solution at a if and only if

$$\int_a^d \frac{1}{p\,v^2} < \infty; \qquad (6.2.3)$$

(3) if u is a principal solution and v is a non-principal solution at a, then there exists $c \in \mathbb{R}$, $c \neq 0$, such that

$$u(t) = v(t) \int_a^t \frac{c}{p\,v^2}, \ a < t \leq d; \qquad (6.2.4)$$

(4) if u is a principal solution and v is a non-principal solution at a, then

$$|u(t)\,v(x)| < |u(x)\,v(t)|, \ for\ a < t < x \leq d. \qquad (6.2.5)$$

PROOF. See p. 548 in [**503**]. □

3. Oscillation Criteria

For equation (6.1.1), with coefficients satisfying (6.1.2), the zeros of every nontrivial solution are isolated by Theorem 2.6.1. From this it follows that only an endpoint of J can be an accumulation point of zeros for any nontrivial solution. By Theorem 2.6.2, the Sturm Separation Theorem, if the zeros of one nontrivial solution accumulate at an endpoint then the zeros of *every* nontrivial solution accumulate at that endpoint. This leads to the oscillatory (O) and nonoscillatory (NO) classification of each endpoint of equation (6.1.1). For what coefficients p, q satisfying (6.1.2) is (6.1.1) O at a? at b? Let $O(p, q, b)$ denote the set of equations which are O at b and $NO(p, q, b)$ those which are NO at b. Then the equations (6.1.1) are classified into these two mutually disjoint classes. A similar classification is made at a. By Theorem 2.6.1 all regular endpoints are in the NO class. Many sufficient conditions are known for each of these classes, there are also some necessary and sufficient conditions known but there are no necessary and sufficient conditions known *which can be checked for all equations. Improving known sufficient conditions and looking for checkable necessary and sufficient ones is today, 170 years after the appearance of the seminal paper of Sturm, still a very active field of research.* In this section we give a very brief introduction to this fascinating area.

We mention some of the classical criteria: Let (6.1.1) and (6.1.2) hold with J a half-line $J = (a, \infty)$ for some a, $1 < a < \infty$ and $p = 1$. For O or NO at ∞ we have the following results. From Theorem 2.6.3, the Sturm Comparison Theorem, using the constant coefficient equation

$$y'' + \delta y = 0$$

for comparison, we find that

(1) $q(t) \geq \delta > 0 \Rightarrow O$.
(2) $q(t) \leq 0 \Rightarrow NO$.

The 'gap' between 0 and δ leaves a lot of 'room' and can be narrowed by using the Euler equation

$$y'' + \frac{1}{4t^2} y = 0$$

for the comparison equation as Kneser did in 1893 [**375**]:

(1) $q(t) \geq \frac{1+\delta}{4t^2} \Rightarrow O$.

(2) $q(t) \leq \frac{1}{4t^2} \Rightarrow NO$.

The Kneser criteria can be refined by using the following extension of the Euler equation for comparison:

$$y'' + \left(\frac{1}{4t^2} + \frac{c}{4(t\ln t)^2}\right)y = 0$$

(3) $c > 1 \Rightarrow O$.
(4) $c \leq 1 \Rightarrow NO$.

This process can be continued to get a sequence of stronger results. The next comparison equation is

$$y'' + \left(\frac{1}{4t^2} + \frac{1}{4(t\ln t)^2}\right)y + \frac{c}{4(t\ln(\ln t))^2}y = 0$$

(5) $c > 1 \Rightarrow O$.
(6) $c \leq 1 \Rightarrow NO$.

And so on. It might appear that this leaves 'no room' between these ever narrower criteria. In fact there is still a lot of room 'in between' as we will see below when we discuss 'interval' criteria for oscillation.

With both coefficients present one of the best known criterion is the integral test which we now give.

THEOREM 6.3.1. *Let (6.1.1), (6.1.2) hold.*

- *If*

$$\int_c^b \frac{1}{p} = \infty \quad \text{and} \quad \int_c^b q = \infty \qquad (6.3.1)$$

for some $c \in J$, then the equation (6.1.1) is oscillatory at b.

- *If*

$$\int_a^c \frac{1}{p} = \infty \quad \text{and} \quad \int_a^c q = \infty \qquad (6.3.2)$$

for some $c \in J$, then the equation (6.1.1) is oscillatory at a.

PROOF. We prove the result for the endpoint b only since the proof for a is similar. Assume nonoscillation at b and let u, v be positive principal and nonprincipal solutions, respectively, on $[d, b)$ for some $d \in J$. Noting that $q = -(pv')'/v$ on $[d, b)$ and integrating by parts we get

$$\int_d^t \frac{(pv')'}{v} = \frac{(pv')}{v}(t) - \frac{(pv')}{v}(d) - \int_d^t pv'\frac{(-v')}{v^2} = \frac{(pv')}{v}(t) - \frac{(pv')}{v}(d)$$
$$+ \int_d^t \frac{1}{p}\frac{(pv')^2}{v^2} = -\int_d^t q \to -\infty \quad as \quad t \to b.$$

Hence $(pv')(t)/v(t) \to -\infty$ as $t \to b$ and pv' is negative near b. Therefore v is decreasing in a left neighborhood of b and hence has a limit at b, say,

$$v(t) \to L \quad as \quad t \to b, \quad 0 \leq L < \infty.$$

From this it follows that, for some $h \in [d, b)$, and some $\varepsilon > 0$ we have

$$v^2(t) < L^2 + \varepsilon, \quad \frac{1}{v^2(t)} > \frac{1}{L^2 + \varepsilon}, \quad d \leq h \leq t < b.$$

This implies that
$$\int_h^t \frac{1}{pv^2} \geq \frac{1}{L^2 + \varepsilon} \int_h^t \frac{1}{p}, \quad d \leq h \leq t < b.$$
This and the hypothesis imply that
$$\int_h^b \frac{1}{pv^2} = \infty,$$
contradicting (6.2.3) since v is a nonprincipal solution at b. □

In [**418**], [**413**] Kwong and Zettl introduced a method, they called 'the telescoping principle', which can be used to extend most of the previously known oscillation criteria. The well known criteria including all those mentioned above are either pointwise conditions or integral conditions involving integrals over the entire domain. The idea is very simple: oscillation conditions should restrict the coefficients only on a sequence of intervals converging to the endpoint with no substantive restrictions on the rest of the domain. For instance : given any subinterval (c, d) of J when $p \leq 1$ and $q \geq h$ on (c, d) then for a constant h large enough (depending on the size of the interval (c, d)) every solution must have a zero in the interval (c, d) regardless of the behavior of the coefficients outside (c, d) by the Comparison Theorem. The proofs of the previously known pointwise or integral criteria did not seem to be adjustable to accommodate such conditions. The basic idea of the proofs in [**418**], [**413**], see also [**368**] is to remove the complementary intervals and show that the resulting equation on the 'shorter interval' (which may still be a half line or the whole line) is oscillatory and then show that this implies that the original equation is oscillatory. (The same basic idea is used to get interval LP conditions, see Chapter 7.) We illustrate this idea by extending conditions (6.3.1).

THEOREM 6.3.2. *Let (6.1.2) hold. Suppose that $J_n = (a_n, b_n)$, $n \in \mathbb{N}$, is a sequence of disjoint subintervals of J such that $a < a_1$ and $\{a_n : n \in \mathbb{N}\}$ is an increasing sequence converging to b. Let $I = \cup_{n=1}^\infty J_n$. If*
$$\int_I \frac{1}{p} = \infty \quad \text{and} \quad \int_I q = \infty, \tag{6.3.3}$$
then the equation (6.1.1) is O at b.

PROOF. Let K be the interval whose points are the points of I; in other words, K is constructed from J by removing the subintervals of J which are not in I. Thus in K we have: $b_1 = a_2$, $b_2 = a_3$, $b_3 = a_4$, etc. Consider the equation
$$(pz')' + qz = 0 \text{ on } K. \tag{6.3.4}$$

This equation is well defined since p and q are defined on K and $1/p, q \in L_{loc}(K, \mathbb{R})$. By Theorem 6.3.1 the equation (6.3.4) is O at b. To show that (6.1.1) is O at b let z be a nontrivial solution of (6.3.4) which has a zero in the interval (a_n, b_n) and in an interval (a_m, b_m) with $m > n$ and define a solution y of (6.1.1) by the initial conditions: $y(a_n) = z(a_n)$, $(py')(a_n) = (pz')(a_n)$. Then y and z are the same solution on the interval $[a_n, b_n)$ and hence y has a zero in this interval. Now define a solution y_1 of (6.1.1) by the initial conditions: $y_1(a_m) = z(a_m)$, $(py_1')(a_m) = (pz')(a_m)$. Then $y_1 = z$ on the interval $[a_m, b_m)$. Since $y = z$ on $[a_n, b_n)$ and $y_1 = z$

on $[a_m, b_m)$ then $y = y_1$ on J. Therefore y has an infinite number of zeros in J since z has an infinite number of zeros in K. □

This idea of removing a finite or infinite number of subintervals from the domain J can be used to get a general 'interval' type comparison theorem: Suppose that $\{(c_i, d_i) : i \in \mathbb{N}\}$ is a sequence of disjoint intervals whose left endpoints converge to b. Construct a new interval (a, B) by shrinking each of these intervals to its left endpoint. Then apply any known sufficient condition for oscillation to (a, B) and prove that oscillation of the new equation at B implies oscillation of the original equation at b.

THEOREM 6.3.3 (Shrinking Domain Comparison Theorem). *Let (6.1.2) hold and assume that p, q are piecewise continuous on J. Let S be the union of an infinite number of disjoint subintervals of J whose left endpoints converge to b,*

$$S = \cup_{i=1}^{\infty} (c_i, d_i).$$

Suppose that

$$\int_{c_i}^{d_i} q \geq 0. \tag{6.3.5}$$

Let m denote Lebesgue measure and define

$$B = m((a,b) \backslash S \tag{6.3.6}$$

and define

$$P(s) = p(t), \ Q(s) = q(t), \ if \ s = s(t) = m((a,b) \backslash S \cap (a,b)). \tag{6.3.7}$$

If the equation

$$(Pz')' + Qz = 0 \ on \ (a, B) \tag{6.3.8}$$

is oscillatory at B, then (6.1.1) is oscillatory at b.

PROOF. See Corollary1, p. 21 of Kwong and Zettl [**418**]. □

As an example of the strong oscillation criteria one can get from this idea of shrinking the domain (called 'the telescoping principle' in [**418**]) we state another result from [**418**]. This theorem is for potentials q which oscillate and are not integrable at b. These tend to be the most challenging cases.

THEOREM 6.3.4. *Let $J = (1, \infty)$, $p = 1$ on J, $q \in L_{loc}([1,\infty), \mathbb{R})$ and assume that q is piecewise continuous on J. Set $q_+(t) = \max\{q(t), 0\}$, $q_- = \max\{-q(t), 0\}$, $t \in J$. Assume that for each $\varepsilon > 0$ there is a $\delta > 0$ such for any subinterval I of J of length δ*

$$\text{either} \int_I q_+ < \varepsilon \ or \ \int_I q_- < \varepsilon. \tag{6.3.9}$$

If

$$= \infty \leq \liminf_{t \to \infty} \int_1^t q < \limsup_{t \to \infty} \int_1^t q \leq \infty, \tag{6.3.10}$$

then equation (6.1.1) is oscillatory ay ∞.

PROOF. This is Corollary 5, p. 28, in [**418**]. □

REMARK 6.3.1. Can the technical condition (6.3.9) in Theorem 6.3.4 be eliminated?

REMARK 6.3.2. The reason for the piecewise continuity hypothesis in Theorems 6.3.3 and 6.3.4 is that the proof in [**418**] is based on the Ricatti equation and differential inequality results which use this hypothesis. The assumption (6.3.5) is also used in connection with the Ricatti equations proof.

REMARK 6.3.3. The class of known oscillatory equations can be greatly enlarged by reversing the construction in the Shrinking Intervals Theorem: Instead of shrinking intervals to a point, one can expand points into intervals. Start with any known oscillatory equation (6.1.1), choose a sequence of points $\{b_i : i \in \mathbb{N}\}$ converging to b. Cut the plane at each vertical line $t = b_i$ and pull the two half planes apart forming a gap of arbitrary length. In the gap place any piecewise continuous positive function p and any piecewise continuous potential function q whose integral over the gap is not negative. (In particular q can be identically zero in the gap.) Do this for any finite or infinite number of points b_i. The equation so constructed is oscillatory at its right endpoint. (This endpoint depends on b and the lengths of the gaps.) If $b = \infty$ then the right endpoint for the new oscillatory equation is also ∞. When $b = \infty$ the shrinking interval construction can be used to create a 'shorter' half-line. In the reverse construction we create a 'longer' half line.

EXAMPLE 6.3.1. *It follows readily from Theorem 6.3.4 that the equation*
$$y'' + qy = 0$$
is oscillatory at ∞ when $q(t) = k\sin(t)$ or $q(t) = k\cos(t)$ for any $k \in \mathbb{R}$, $k \neq 0$. More generally for $q(t) = t^c k \sin(t)$ or $q(t) = t^c k \cos(t)$, with $-1 < c \in \mathbb{R}$, $k \neq 0$.

EXAMPLE 6.3.2. *(See problem 35 in Chapter 13, the Examples Chapter, below.) Although we have defined oscillation and nonoscillation only for equations in symmetric form the same definitions apply to equations in 'standard' form:*
$$y'' + ry' + sy = 0. \tag{6.3.11}$$

Recently M.K. Kwong and J.S.W. Wong (in a private communication) have shown that equation (6.3.11) is NO at ∞ for
$$r(t) = \sin t \text{ and } s(t) = \cos(t),$$
but is O when
$$r(t) = \cos(t) \text{ and } s(t) = \sin(t).$$
When put in symmetric form (6.1.1) the first of these equations has
$$p(t) = e^{-\cos(t)}, \ q(t) = \cos(t) e^{-\cos(t)}$$
and the second has
$$p(t) = e^{\sin(t)}, \ q(t) = \sin(t) e^{\sin(t)}.$$

See the problems in Chapter 13 for more examples of equations with oscillatory and nonoscillatory endpoints.

4. Oscillatory Characterizations

In this section we give two characterizations for oscillation; each one is obtained by using related nonlinear equations as a tool.

LEMMA 6.4.1. *Let (6.1.2) hold. Assume y and z are real-valued solutions of (6.1.1). Let $W = ypz' - zpy'$ be the Wronskian of y, z. Then $W(t) = c \in \mathbb{R}$ for all $t \in J$. Let*

$$r = (y^2 + z^2)^{1/2}. \tag{6.4.1}$$

Then

$$(pr')' + qr = \frac{c^2}{pr^3} \text{ on } J. \tag{6.4.2}$$

PROOF. We have $rr' = yy' + zz'$,

$$r'pr' + r(pr')' = py'^2 + pz'^2 + y(py')' + z(pz')' = p(y'^2 + z'^2) - qr^2. \tag{6.4.3}$$

Rearranging the terms in (6.4.3) we get

$$(pr')' + qr = \frac{p}{r}[-r'^2 + y'^2 + z'^2] \tag{6.4.4}$$

The Wronskian is the constant c, hence we have

$$\frac{c^2}{p^2} = (y^2 z'^2 - 2yz'zy' + z^2 y'^2)$$
$$= (y^2 z'^2 - 2yz'zy' + z^2 y'^2 + y^2 y'^2 - y^2 y'^2 + z^2 z'^2 - z^2 z'^2)$$
$$= [(y^2 + z^2)(y'^2 + z'^2) - (y^2 y'^2 + 2yy'zz' + z^2 z'^2)]$$
$$= [r^2(y'^2 + z'^2) - r^2 r'^2] = r^2[y'^2 + z'^2 - r'^2]$$
$$= r^2 \frac{r}{p}[(pr')' + qr]. \tag{6.4.5}$$

Now (6.4.2) follows from (6.4.5). □

LEMMA 6.4.2. *Let (6.1.2) hold, let $c \in J$. Assume r is a positive solution of (6.4.2) on J. Define y, z by*

$$y(t) = r(t)\sin\left(\int_c^t \frac{1}{pr^2}\right), \quad z(t) = r(t)\cos\left(\int_c^t \frac{1}{pr^2}\right). \tag{6.4.6}$$

Then y and z are linearly independent solutions of (6.1.1).

PROOF. This follows from a straightforward computation. □

LEMMA 6.4.3. *Let (6.1.2) hold, let $c \in J$. Assume r is a positive solution of (6.4.2) on J. Then for any $A, \alpha \in \mathbb{R}$*

$$y(t) = Ar(t)\sin\left(\alpha + \int_c^t \frac{1}{pr^2}\right), \quad t \in J. \tag{6.4.7}$$

is a solution of (6.1.1). Conversely, given any real valued solution y of (6.1.1) there exist $A, \alpha \in \mathbb{R}$ such that (6.4.7) holds.

PROOF. This follows from Lemma 6.4.2 and standard trig. identities. □

LEMMA 6.4.4. *Suppose u is a function satisfying the following conditions:*
(1) $u > 0$ on J;
(2) $u \in AC_{loc}(J)$;
(3) $pu' \in AC_{loc}(J)$.

If y is a solution of (6.1.1), then $z = y/u$ is a solution of

$$(pu^2 z')' + (uMu)z = 0 \text{ on } J. \tag{6.4.8}$$

PROOF. This follows from a direct computation. □

THEOREM 6.4.1. *Let (6.1.2) hold, let $c \in J$. Then:*

- *Equation (6.1.1) is oscillatory at b if and only if there exists a function u with the following properties:*
  (1) $u \in AC_{loc}(J)$,
  (2) $pu' \in AC_{loc}(J)$,
  (3) $u > 0$ on J,
  (4) $\int_c^b \frac{1}{pu^2} = +\infty$,
  (5) $\int_c^b uMu = \infty$.
- *Equation (6.1.1) is oscillatory at a if and only if there exists a function u with the following properties:*
  (1) $u \in AC_{loc}(J)$,
  (2) $pu' \in AC_{loc}(J)$,
  (3) $u > 0$ on J,
  (4) $\int_a^c \frac{1}{pu^2} = +\infty$,
  (5) $\int_a^c uMu = \infty$.

PROOF. We establish the case for the endpoint b first. Assume (1)-(5) hold and y is a solution of (6.1.1). Let $z = y/u$, then (6.4.8) holds. From conditions (1) and (2) it follows that Mu is defined. Theorem 6.3.1 and conditions (4), (5) imply that (6.4.8) is oscillatory. Now (3) implies that y is oscillatory.

Now suppose that (6.1.1) is oscillatory at b. Let y_1, y_2 be linearly independent real-valued solutions of (6.1.1) and set

$$r = (y_1^2 + y_2^2)^{1/2}.$$

Then (6.4.2) holds and (1), (2), (3) hold with $u = r$. Choose $d \in J$ and define y by

$$y(t) = r(t) \sin\left(\int_d^t \frac{1}{pr^2}\right), \ t \in J.$$

Then y is a solution of (6.1.1) by Lemma 6.4.2. Hence y is oscillatory and consequently, since $u > 0$ on J, (4) holds. From (6.4.2) we get

$$uMu = \frac{c^2}{pu^2}$$

and (5) follows since $c \neq 0$ by the linear independence of y_1, y_2.

The proof for the endpoint a is similar. □

The next characterization uses the well-known Riccati equation and a comparison theorem for Ricatti equations from [368].

THEOREM 6.4.2. *Let (6.1.2) hold, let $c \in J$. Then (6.1.1) has a positive solution on $[c, b)$ if and only if there exist $q_1, q_2 \in L_{oc}([c, b), \mathbb{R})$ and $r \in \mathbb{R}$ satisfying the following two conditions:*

(1) $-q = q_1 + q_2$ on $[c, b)$.
(2) $(r + \int_c^t q_2)^2 \leq p q_1$ on $[c, b)$.

PROOF. Suppose u is a positive solution of (6.1.1) on $[c, b)$ and let $z = pu'/u$. Then z satisfies the Ricatti equation:

$$z' + \frac{z^2}{p} = -q$$

and the conditions are satisfied with $q_1 = z^2/p$ and $q_2 = z'$. For the proof of the sufficiency see Proposition 4.8 p. 35 in [**368**]. (Note the different sign convention used in [**368**]: $-q$ in place of q.) □

REMARK 6.4.1. Instead of trying to get necessary and sufficient conditions for oscillation, *which can be checked for every equation, one can try a completely different approach. In equation* (6.1.1) *for a given q construct all p such that the equation is oscillatory.* A step in this direction is taken in Zettl [**624**].

REMARK 6.4.2. Of course characterizing oscillation is equivalent to characterizing nonoscillation. Disconjugacy is a strong form of nonoscillation. A second order linear ordinary differential equation is disconjugate on an interval J if it has a positive solution on J. In [**627**], [**623**] Zettl gives a construction of all disconjugate equations of the form

$$y^{(n)} + q_{n-1}y^{(n-1)} + \cdots + q_1 y' + q_0 y = 0, \ n \in \mathbb{N}.$$

5. Comments

(1) With some misgivings we are following the well established practice in oscillation theory of not putting a minus sign "-" in front of the leading coefficient p of equation (6.1.1). This contrasts with the equally well established practice in spectral theory, which we followed in Parts I and II.

(2) This treatment of principal and non-principal solutions is based on Niessen and Zettl [**503**]. According to Hartman [**319**] these terms were coined by Leighton in [**436**]. The principal solution is the "small" solution but apparently Leighton and presumably Hartman, since he followed Leighton's lead, felt that using a term such as small might lead to confusion since the same solution may be small at one endpoint but not the other.

(3) Theorem 6.3.1 is just one of many well known oscillation criteria. In almost all books and papers it is given only for the case when the left endpoint a is finite and regular and the other infinite, i.e. $b = \infty$. The proof given here seems (to this writer) to come closest of all the proofs we are aware of to being from the "book" of Erdos. What other ode proofs are "from the book"? For more oscillation and nonoscillation criteria the reader is referred to the following papers and books and the references cited there. This topic is still an active research area and this list, although extensive, is not in any sense comprehensive: Ahlbrandt, Hinton and Lewis [**5**], Allegretto and Mingarelli [**12**], Barrett [**49**], [**50**], Bender, Stefan and Savage [**62**], Beesack [**57**], Binding and Browne [**74**], Birkhoff [**85**], Binding and Volkmer [**80**], Bochner [**89**], Byers, Harris and Kwong [**113**], Cole [**125**], [**126**], Coles and Willett [**127**], Erbe, Kong and Ruan [**166**], Everitt, Kwong and Zettl [**214**], Fite [**248**], Gesztesy, Simon and Teschl [**271**], Gesztesy and Unal [**272**], Harris [**306**], Hartman [**314**], [**317**], Haupt [**321**], [**325**], Hinton [**330**], Hurwitz [**343**],

Kamenev [**354**], Kamke [**355**], Kneser [**375**], Kong and Zettl [**395**], Kreith [**405**], Kurss [**408**], Kwong and Zettl [**418**], [**413**], Leighton [**435**], [**434**], Leighton and Nehari [**437**], Levin [**439**], [**441**], [**440**], [**438**], Markus and Moore [**459**], Machi and Wong [**457**], Meng and Mingarelli [**470**], Moore [**479**], Morse [**481**], Nehari [**489**], [**490**], Olech, Opial and Wazewski [**504**], Opial [**507**], [**506**], Picone [**516**], Richardson [**545**], [**546**], Sturm [**569**], [**570**], Weidmann [**594**], Willett, [**609**], [**610**], Wintner [**611**], [**613**], [**612**], Wong [**617**], [**615**], [**614**], [**616**], Zlamal [**648**], Zettl [**627**], [**623**], Ashbaugh, Brown and Hinton [**16**], Boruvka [**96**], Birkhoff and Rota [**84**], Bochner [**90**], Coddington and Levinson [**123**], Coppel [**131**], Dunford and Schwartz [**141**], Elias [**165**], Halvorsen and Mingarelli [**289**], Hartman [**319**], Haupt [**322**], Hille [**326**], Hinton and Schaefer [**334**], Ince [**344**], Kaper, Kwong and Zettl [**366**], Kreith [**406**], Mueller-Pfeiffer [**484**], Reid [**541**], [**542**], Swanson [**571**], Weidmann [**600**], Glazman [**277**], Gesztesy and Unal [**272**],

(4) Although we have not seen Theorem 6.4.1 stated in this form, it is based on a reorganization of some material in Coppell's monograph on Disconjugacy [**131**] and of some things in the Kauffman-Read-Zettl monograph [**368**]. (The latter contains an error - a formula claimed holds as stated only if the linearly independent solutions considered there are normalized to have Wronskian 1.)

CHAPTER 7

The Limit-Point, Limit-Circle Dichotomy

Weyl, Hermann (1885-1955)
>Our federal income tax law defines the tax y to be paid in terms of the income x; it does so in a clumsy enough way by pasting several linear functions together, each valid in another interval or bracket of income. An archeologist who, five thousand years from now, shall unearth some of our income tax returns together with relics of engineering works and mathematical books, will probably date them a couple of centuries earlier, certainly before Galileo and Vieta.

The Mathematical Way of Thinking, an address given at the Bicentennial Conference at the University of Pennsylvania, 1940.

1. Introduction

Regular boundary conditions of the type discussed in Part I do not make sense at a singular endpoint because, in general, the solutions and their quasi-derivatives are not defined there. What should take their place? The answer depends on the nature of the singularity: limit-point (LP) or limit-circle (LC). This terminology was introduced by Hermann Weyl in 1910 [**607**] in connection with a geometric construction of concentric circles in the plane each one contained in the preceding one. In the limit these circles converge to a circle - the limit-circle (LC) case - or to a point - the limit-point (LP) case. For the latter case no boundary condition is required or allowed, but the LC case requires boundary conditions to determine self-adjoint extensions. This paper of Weyl is one of the most widely quoted papers in analysis and can be credited with starting the theory of *singular* SLP. Its extension to the fourth and higher order cases had to wait until about 1950 when Glazman published results from his dissertation under M.G. Krein; see the classic book by Naimark [**487**] for a detailed exposition of this work. This landmark paper of Glazman corrected some fundamental errors in earlier works and provided a Hilbert space framework for the study of second and higher order boundary value problems. Although this approach does not make use of the Weyl circles the LP, LC terminology has survived and is still universally used today.

2. System Regularization of the Scalar Equation

In this section we set the stage for the LC/LP definitions.
Consider the equation

$$-(py')' + qy = \lambda w y \text{ on } J, \ \lambda \in \mathbb{C}, \qquad (7.2.1)$$

with
$$J = (a,b),\ -\infty \leq a < b \leq \infty,\ 1/p, q, w \in L_{loc}(J, \mathbb{C}), \quad (7.2.2)$$
and consider the first order system representation of (7.2.1):
$$Y' = (P - \lambda W)Y, \quad P = \begin{bmatrix} 0 & 1/p \\ q & 0 \end{bmatrix}, \quad W = \begin{bmatrix} 0 & 0 \\ w & 0 \end{bmatrix}, \quad Y = \begin{bmatrix} y \\ py' \end{bmatrix}. \quad (7.2.3)$$

THEOREM 7.2.1 (System Regularization of Scalar Equations). *Let (7.2.1) and (7.2.2) hold. Assume that for some $\lambda \in \mathbb{C}$, say $\lambda = \lambda_0$, all solutions $y(\cdot, \lambda_0)$ of equation (7.2.1) satisfy*
$$\int_a^b |y(\cdot, \lambda_0)|^2 |w| < \infty. \quad (7.2.4)$$
Let $\Phi = (\phi_{ij}) = \Phi(\cdot, \cdot, \lambda, P, W)$ be the primary fundamental matrix of (7.2.3). For some $u \in J$ and all $\lambda \in \mathbb{C}$ define
$$Z(\cdot, \lambda) = \Phi^{-1}(\cdot, u, \lambda_0) Y(\cdot, \lambda), \quad (7.2.5)$$
and let
$$G = \begin{bmatrix} -\phi_{11}(\cdot, \lambda_0)\phi_{12}(\cdot, \lambda_0)w & -\phi_{12}^2(\cdot, \lambda_0)w \\ \phi_{11}^2(\cdot, \lambda_0)w & \phi_{11}(\cdot, \lambda_0)\phi_{12}(\cdot, \lambda_0)w \end{bmatrix}. \quad (7.2.6)$$
Then $G \in L(J, \mathbb{C})$ and for all $\lambda \in \mathbb{C}$ we have
$$Z'(\cdot, \lambda) = (\lambda_0 - \lambda)\, G\, Z(\cdot, \lambda) \quad \text{on} \quad J. \quad (7.2.7)$$

PROOF. By hypothesis $\phi_{11} = \phi_{11}(\cdot, u, \lambda_0)$ and $\phi_{12} = \phi_{12}(\cdot, u, \lambda_0)$ are in $L^2(J, |w|)$. This, and the Cauchy-Schwarz inequality imply that
$$\left(\int_J |\phi_{11}||\phi_{12}||w|\right)^2 = \left(\int_J |\phi_{11}||\phi_{12}||w|^{1/2}|w|^{1/2}\right)^2 \leq \int_J |\phi_{11}|^2 |w| \int_J |\phi_{12}|^2 |w| < \infty,$$
and we have shown that $g_{11} \in L^1(J, \mathbb{C})$. Similarly for g_{12}, g_{21}, g_{22} and thus $G \in L(J, \mathbb{C})$. From (7.2.5) and letting $Z = Z(\cdot, \lambda)$ we get
$$Z' = (\Phi^{-1})'Y + \Phi Y' = -\Phi^{-1}\Phi'\Phi^{-1}Y + \Phi^{-1}Y'$$
$$= \Phi^{-1}[\lambda_0 W - P + P - \lambda W]\Phi Z$$
$$= (\lambda_0 - \lambda)\Phi^{-1} W \Phi Z = (\lambda_0 - \lambda)\, G\, Z \quad \text{on} \quad J.$$
□

Next we give two applications of Theorem 7.2.1, one showing the invariance of L^2 solutions for $\lambda \in \mathbb{C}$, the other showing that oscillation is invariant for $\lambda \in \mathbb{R}$ at an LC endpoint.

THEOREM 7.2.2. *Let the hypotheses and notation of Theorem 7.2.1 hold. If all solutions of (7.2.1) are in $L^2(J, |w|)$ for some $\lambda \in \mathbb{C}$, then this holds for every $\lambda \in \mathbb{C}$. If the quasi-derivatives of all solutions of (7.2.1) are in $L^2(J, |w|)$ for some $\lambda \in \mathbb{C}$, then this holds for every $\lambda \in \mathbb{C}$.*

PROOF. Since u is fixed we suppress it in the notation and let $\Phi(t, \lambda_0) = \Phi(t, u, \lambda_0)$. From $Z = \Phi^{-1} Y$ we get
$$y(t, \lambda) = \phi_{11}(t, \lambda_0) z_1(t, \lambda) + \phi_{12}(t, \lambda_0) z_2(t, \lambda), \quad t \in J;$$
$$y^{[1]}(t, \lambda) = \phi_{21}(t, \lambda_0) z_1(t, \lambda) + \phi_{22}(t, \lambda_0) z_2(t, \lambda), \quad t \in J.$$

By Theorem 1.5.2 we have $z_j(a,\lambda) = \lim_{t\to a} z_j(t,\lambda)$ exists and is finite for $j = 1, 2$. Hence z_j is bounded (almost constant) in a right neighborhood of a. Similarly z_j is bounded (and nearly constant) in a left neighborhood of b. Hence $y(\cdot,\lambda)$ is in $L^2(J, |w|)$ since $\phi_{11}(\cdot,\lambda_0)$ and $\phi_{12}(\cdot,\lambda_0)$ are. Similarly $y^{[1]}(\cdot,\lambda)$ is in $L^2(J, |w|)$ since $\phi_{21}(\cdot,\lambda_0)$ and $\phi_{22}(\cdot,\lambda_0)$ are. □

REMARK 7.2.1. We remark that the λ invariance given by Theorem 7.2.2 is established under hypothesis (7.2.2); thus the LC/LP dichotomy in $L^2(J, |w|)$ holds not only for real valued $r = 1/p, q, w$ which may change sign and or be identically zero on subintervals but also $r = 1/p, q, w$ may be complex valued.

3. Endpoint Classifications: R, LC, LP, O, NO, LCNO, LCO

Definitions of regular (R), limit-circle (LC), limit-point (LP), oscillatory (O), nonoscillatory (NO), limit-circle-non-oscillatory (LCNO) and limit-circle-oscillatory (LCO) endpoints are given here. These are all standard and well known.

Consider the equation

$$-(py')' + qy = \lambda\, w\, y, \ \lambda \in \mathbb{C}, \text{ on } J, \qquad (7.3.1)$$

with

$$J = (a, b), \ -\infty \le a < b \le \infty, \ 1/p, q, w \in L_{loc}(J, \mathbb{C}). \qquad (7.3.2)$$

DEFINITION 7.3.1. *The (finite or infinite) endpoint a*

(1) *is regular if, in addition to (7.3.2),*

$$1/p, q, w \in L((a, d), \mathbb{C})$$

holds for some (and hence any) $d \in J$;
(2) *is limit-circle if all solutions of the equation (7.3.1) are in $L^2((a,d), |w|)$ for some (and hence any) $d \in (a, b)$;*
(3) *is LP if it is not LC;*
(4) *is O if $1/p, q, w, \lambda$ are all real-valued and there is a nontrivial real-valued solution with an infinite number of zeros in any right neighborhood of a;*
(5) *is NO if $1/p, q, w, \lambda$ are all real-valued and it is not O;*
(6) *is LCO if it is both LC and O;*
(7) *and is LCNO if it is both LC and NO.*

Similar definitions are made at b. An endpoint is called singular if it is not regular.

REMARK 7.3.1. According to Definition 7.3.1 a regular endpoint is automatically in the LC class. This follows from Theorem 2.3.1 which shows that all solutions have finite limits at regular endpoints and are therefore bounded in a neighborhood. Nevertheless, we will use such phrases as regular or LC below for emphasis since regular and LC endpoints are often placed in different categories in the literature.

Recall that it is shown in Section 7.2 that the LC, LP, classifications are independent of $\lambda \in \mathbb{C}$. At an LP endpoint the O classification, in general, depends on λ as can readily be seen from the Fourier equation: $-y'' = \lambda y$. The next theorem shows that, for an endpoint in the LC case and $p > 0$, the LCO and LCNO classifications are independent of $\lambda \in \mathbb{R}$.

THEOREM 7.3.1. *Let the hypotheses and notation of Theorem 7.2.1 hold. Assume p, q, w are real-valued and $p > 0$. Suppose the endpoint a is LCO for some real $\lambda = \lambda_0$. Then the equation (7.2.1) is LCO for every $\lambda \in \mathbb{R}$.*

PROOF. Note that the Sturm Comparison Theorem only implies LCO for $\lambda > \lambda_0$.

Case 1: $z_j(a) > 0$, $j = 1, 2$. We show that between any three zeros of the solution $\phi_{11}(t, \lambda_0)$ there must be a zero of the solution $y(t, \lambda) = \phi_{11}(t, v) z_1(t, \lambda) + \phi_{12}(t, v) z_2(t, \lambda)$. Sketch the graphs of $\phi_{11}(\cdot, \lambda_0)$ and $\phi_{12}(\cdot, \lambda_0)$ keeping in mind the Sturm Separation Theorem. (This is where the hypotheses that p, q, w are real-valued and $p > 0$ are used.) There are two disjoint subintervals of J such that $y(\cdot, \lambda)$ is positive on one and negative on the other. Hence $y(\cdot, \lambda)$ must have a zero between these intervals. The other cases have similar proofs. □

4. LP and LC Conditions

This section contains an introduction to the LC/LP classification theory of the symmetric equation

$$My \equiv -(py')' + qy = \lambda w y, \quad \lambda \in \mathbb{C}, \text{ on } J, \tag{7.4.1}$$

under the conditions

$$J = (a, b), \ -\infty \leq a < b \leq \infty, \ 1/p, q, w \in L_{loc}(J, \mathbb{R}), \ p > 0, \ w > 0, \text{ a.e. on } J. \tag{7.4.2}$$

Recall that the LC/LP classification at each endpoint is independent of λ and depends solely on the behavior of the coefficients $1/p, q, w$ near the endpoint. The literature on this topic is vast and most of it is for the case $J = [0, \infty)$ and $w = 1$. We restrict ourselves to a few criteria which have relatively simple proofs, state a number of further criteria with references to the literature for proofs, make comments and give still further references in the Comments section at the end of this chapter.

We start with a basic lemma for the equation

$$My \equiv -(py')' + qy = 0, \text{ on } J = (a, b), \ -\infty \leq a < b \leq \infty. \tag{7.4.3}$$

LEMMA 7.4.1. *Let (7.4.3) hold and assume that $1/p, q \in L_{loc}(J, \mathbb{R})$, $p > 0$ a.e. on J.*

(1) *If, for some $c \in J$, $q \geq 0$ on (c, b), then the equation (7.4.3) has a positive increasing solution on (c, b).*
(2) *If, for some $c \in J$, $q \geq 0$ on (a, c), then the equation (7.4.3) has a negative increasing solution on (a, c).*
(3) *if $q \geq 0$ on (a, b) and u is the real valued solution determined by the initial conditions: $u(c) = 0$, $(pu')(c) = 1$ for $c \in J$, then u is increasing in (a, b), positive in (c, b), and negative in (a, c).*

PROOF. We prove part (1) only since the proofs of the other parts are similar. Determine a real valued solution u with the initial conditions

$$u(c) = 0, \ (pu')(c) = 1. \tag{7.4.4}$$

We claim that u is such a solution. From (7.4.4) it follows that u is positive in a right neighborhood of c. By Theorem 2.6.1 the zeros of u (if there are any) in the interval $[c, b)$ are isolated; thus if u is not positive on $[c, b)$ it has a smallest zero, say d, $c < d < b$. By the proof of Theorem 2.6.1, there is a g in (a,c) such that $(pu')(g) = 0$. This and the hypothesis imply that

$$(pu')(t) = 1 + \int_c^t qu \geq 1, \ c \leq t \leq g. \tag{7.4.5}$$

This contradicts $(pu')(g) = 0$ and we may conclude that $u > 0$ on $[c, g]$. Now noting that (7.4.5) holds for all $t \in [c, b)$, we conclude that $(pu') > 0$ on $[c, b)$. Let $c \leq h < k < b$, then

$$u(k) - u(h) = \int_h^k u' = \int_h^k (pu')(\frac{1}{p}) > 0,$$

and the proof is complete. $\square$

THEOREM 7.4.1. *Assume q and w satisfy (7.4.2) and, in addition, suppose there exists a point $c \in J$ and a point $k \in \mathbb{R}$ such that*

$$q \geq k\,w \ on \ (c, b), \ and \ w \notin L^1(c, b). \tag{7.4.6}$$

Then equation (7.4.1) is LP at b for every p satisfying (7.4.2).

PROOF. Assume that (7.4.1) is LC at b. Then all solutions are in $L^2((c,b), w)$ for any λ, in particular for $\lambda = k$. By Lemma 7.4.1 there exists a solution u of (7.4.1) with $\lambda = k$, which is positive and increasing on (c, b). Let $d \in (c, b)$. Then

$$\int_d^b u^2 w \geq u^2(d) \int_d^b w.$$

This contradiction proves that b is LP. $\square$

THEOREM 7.4.2. *Assume q and w satisfy (7.4.2) and, in addition, suppose there exists a point $c \in J$ such that*

$$\frac{q^2}{w} \in L^1(c, b), \ w \notin L^1(c, b). \tag{7.4.7}$$

Then equation (7.4.1) is LP at b for every p satisfying (7.4.2).

PROOF. Assume (7.4.1) is LC at b and let u, v be real valued linearly independent solutions of (7.4.1) for $\lambda = 0$, normalized to satisfy the Wronskian identity

$$u(pv') - v(pu') = 1 \ on \ J. \tag{7.4.8}$$

Integrating $(pu')' = qu$ we get

$$(pu')(t) = (pu')(c) + \int_c^t qu, \ t \in [c, b). \tag{7.4.9}$$

From the Schwarz inequality

$$\left(\int_c^t qu\right)^2 \leq \left(\int_c^t \frac{q^2}{w}\right) \int_c^t u^2 w, \tag{7.4.10}$$

the hypothesis, and (7.4.9) we conclude that (pu') is bounded on (c, b). Similarly, (pv') is bounded on (c, b). This and (7.4.8) imply that, for some positive constant K, we have

$$\sqrt{w} \leq K\,(|u| + |v|)\sqrt{w}. \tag{7.4.11}$$

Squaring (7.4.11) we conclude that, for some constant C,

$$w \leq C(u^2 + v^2)w.$$

This leads to a contradiction since $u, v \in L^2((c,b), w)$ and $w \notin L^1(c,b)$. $\square$

REMARK 7.4.1. It is interesting to note that no condition on p, other than (7.4.2) is needed in Theorems 7.4.1 and 7.4.2, in particular no growth restriction is required. This shows that for a given q and w it is not possible, in general, to find a p such that (7.4.1) is LC. On the other hand it is shown in Section 2, Remark 3.7 of [**368**] and also in ([**624**]) (under weaker conditions) that for a given p there exist q, w such that the LC case holds. This is one indication of the different roles that, p and q, w play in the determination of the LC/LP classification.

Instead of starting with all three coefficients p, q, w and then finding criteria for LP or LC to hold one could use an alternative approach: Start with one or two given coefficient and try to characterize the other(s) for the LC case to hold. The first part of this remark suggests that for this strategy one would want to start with p rather than q or w.

COROLLARY 7.4.1. *Let (7.4.1), (7.4.2) hold. Assume that $h \in L_{loc}(J, \mathbb{R})$ satisfies, for some $c \in J$,*

$$\frac{|h|}{w} \in L^\infty(c,b).$$

Then (7.4.1) is LP at b if and only if the equation

$$-(py')' + (q+h)y = \lambda w y$$

is LP at b.

PROOF. This result is proved in Theorem1 p192 of Naimark [**487**] for the case $w = 1$. The proof given there readily extends to w satisfying the given hypothesis. The general result for arbitrary $w > 0$ also follows from an abstract operator theory result, see Theorem 6, page 33 of [**487**]: If S is a symmetric operator in a Hilbert space H and B is a bounded symmetric operator defined on all of H, then S and $S + B$ have the same deficiency indices. $\square$

COROLLARY 7.4.2. *Let (7.4.1), (7.4.2) hold. Assume there exists a $c \in J$ such that $w \notin L^1(c,b)$ and*

$$\frac{q}{\sqrt{w}} \in L^r(c,b), \ 2 \leq r \leq \infty. \tag{7.4.12}$$

Then (7.4.1) is LP at b for every p satisfying (7.4.2).

PROOF. This follows from Theorems 7.4.2 and 7.4.3 and the following observation: Let $f = q/\sqrt{w}$. Then $f = g + h$ with $g \in L^2(c,b)$ and $h \in L^\infty(c,b)$: define

$$h(t) = \begin{cases} f(t), & \text{if } |f(t)| \leq 1, \text{ for } t \in (c,b), \\ 1, & \text{if } f(t) > 1, \text{ for } t \in (c,b), \\ -1, & \text{if } f(t) < -1, \text{ for } t \in (c,b); \end{cases}$$

and let $g = f - h$ on (c, b). Then $g \in L^2(c,b)$. $\square$

The strategy used in the proofs of Theorem 7.4.1 and 7.4.2 can be used to get a more general result by introducing a parameter function B, then different choices for B lead to different results.

THEOREM 7.4.3. *Let (7.4.1), (7.4.2) hold. Assume that for some $c \in J$, $w \notin L^1(c,b)$, and for some positive function $B \in AC_{loc}[c,b)$ the following three conditions are satisfied:*

(1)
$$\int_c^b \left(\frac{w}{pB}\right)^{1/2} = \infty, \tag{7.4.13}$$

(2) *there is a* $k \in \mathbb{R}$ *such that*
$$q \geq -k\, B\, w, \text{ on } [c,b), \tag{7.4.14}$$

(3) *there is a* $K \in \mathbb{R}$ *such that*
$$\left|\frac{\sqrt{p}B'}{B^{3/2}\sqrt{w}}\right| \leq K < \infty, \text{ on } [c,b). \tag{7.4.15}$$

Then b *is LP.*

PROOF. Suppose that b is LC. For $\lambda = 0$ choose real valued linearly independent solutions u, v satisfying the Wronskian identity (7.4.8). Multiplying this identity by $(\sqrt{w})/\sqrt{pB}$ we get
$$\left[\left(\frac{p}{B}\right)^{1/2} v'\right](u\sqrt{w}) - \left[\left(\frac{p}{B}\right)^{1/2} u'\right](v\sqrt{w}) = \frac{\sqrt{w}}{\sqrt{pB}} \text{ on } [c.b). \tag{7.4.16}$$

From the equation (7.4.1) with $\lambda = 0$ we get $-(pu')'u/B = -qu^2/B$. In the rest of the proof K_1, K_2, will denote constants. Integrating and using (7.4.14),
$$-\int_c^t \frac{q}{B} u^2 \leq k \int_c^t u^2 w < K_1, \text{ for } t \in [c,b).$$

Hence
$$-\int_c^t \frac{q}{B} u^2 = \left(\frac{-pu'u}{B}\right)(t) + \left(\frac{-pu'u}{B}\right)(c)$$
$$+ \int_c^t \frac{pu'^2}{B} - \int_c^t \frac{pu'uB'}{B^2} < K_1 < \infty, \; t \in [c,b).$$

Let
$$H(t) = \int_c^t \frac{pu'^2}{B}, \text{ for } t \in [c,b).$$

We want to show that $H(t)$ has a finite limit at b. From (7.4.15) and the Schwarz inequality, we get
$$\left|\int_c^t \frac{pu'uB'}{B^2}\right| \leq \int_c^t \left|\frac{pu'uB'}{B^2}\right| \leq \int_c^t \left|\frac{\sqrt{p}B'}{B^{3/2}\sqrt{w}}\right|\left|\frac{\sqrt{p}u'u\sqrt{w}}{B^{1/2}}\right|$$
$$\leq K H^{1/2}(t) \left(\int_c^t u^2 w\right)^{1/2} < K_2 H^{1/2}(t), \; t \in [c,b).$$

Therefore, for some constants K_3, K_4 we have
$$\left(\frac{-pu'u}{B}\right)(t) + H(t) - K_3 H^{1/2}(t) < K_4. \tag{7.4.17}$$

If $H(t) \to \infty$ as $t \to b$, then from (7.4.17) it follows that u and (pu') have the same sign on some interval (d,b), $c < d < b$. But this implies that u^2 is strictly increasing in (d,b) and hence
$$\int_d^t u^2 w \geq u^2(d) \int_d^t w.$$

This contradiction completes the proof. $\square$

THEOREM 7.4.4. *Let (7.4.1), (7.4.2) hold. Assume that b is NO when $\lambda = 0$ and suppose that for some $c \in J$, $\sqrt{w}/\sqrt{p} \notin L^1(c,b)$. Then b is LP.*

PROOF. Let v be a nonprincipal solution at b and assume that b is LC. Choose d in J so that $v(t) \neq 0$ in $[d,b)$. From the Schwarz inequality we have

$$\left(\int_d^b \frac{\sqrt{w}}{\sqrt{p}}\right)^2 \leq \int_d^b \frac{1}{pv^2} \int_d^b v^2 w.$$

The second integral on the right is finite by the LC assumption, the first by Theorem 6.2.2. This contradicts the hypothesis and completes the proof. □

The next example is a classic in the LP/LC theory.

EXAMPLE 7.4.1. *If $q \in L_{loc}(J, \mathbb{R})$, $J = (1, \infty)$, and $q(t) \geq -kt^2$, then*

$$y'' + qy = \lambda y \text{ on } J,$$

is LP at ∞.

PROOF. Here $w = 1 = p$. Take $B(t) = t^2$, $t \in J$ and note that all three conditions of Theorem 7.4.3 hold; also $w \notin L(J)$. □

Example 7.4.1 is interesting for a number of different reasons. The next remarks highlight a few of these reasons.

REMARK 7.4.2. (1) $q(t) = -t^2$ for $t \in J$ is in the LP case at ∞.
(2) $q(t) = -t^{2+\delta}$ for $t \in J$ is in the LC case at ∞ for every $\delta > 0$.
(3) $q(t) = -[t\ln(t)]^2$ for $t \in J$ is in the LP case at ∞.
(4) $q(t) = -[t\ln(\ln(t))]^2$ for $t \in J$ is in the LP case at ∞, etc.

REMARK 7.4.3. Notwithstanding points (3) and (4) of Remark 7.4.2 one might expect that there is a growth restriction on q from below for the LP case to hold at ∞. It is remarkable that there is no such growth restriction: In the monograph by Kauffman, Read and Zettl [368], see Remark 4.15, p .37, it is shown that, given any continuous function f on $J = (1, \infty)$, there exists a continuous function $q < f$ on J such that $y'' + qy = \lambda y$ is LP at ∞.

EXAMPLE 7.4.2. *Consider*

$$-(t^\alpha y')' + kt^\beta y = \lambda y \text{ on } J = (1, \infty), \; k, \alpha, \beta \in \mathbb{R}. \qquad (7.4.18)$$

Equation (7.4.18) is LP at ∞ if
(1) $k \geq 0$ for any α, β.
(2) $k < 0$ and $\beta \leq 0$ for any α.
(3) $0 < \beta \leq 2 - \alpha$ and $0 \leq \alpha < 2$, for any k.
(4) $2 < \alpha$ and $0 < \beta < \alpha - 2$, for any k.
(5) $k < 0$, $\alpha \geq 2$, $\beta = 2 - \alpha$ and $k \geq (3 - 2\alpha)/4$.

Equation (7.4.18) is LC at ∞ if
(1) $k < 0$, $\alpha > 2$, and $\beta > 2 - \alpha$.
(2) $k < 0$, $\alpha < 2$, and $\beta < 2 - \alpha$.
(3) $k < 0$, $\alpha \geq 2$, $\beta = 2 - \alpha$, and $k < (3 - 2\alpha)/4$.

PROOF. Note that this gives a complete LP/LC classification of (7.4.18). On the critical line $b = 2 - \alpha$, $\alpha > 2$, the LP/LC classification depends on the constant k and the critical value of k is $(3 - 2\alpha)/4$. □

REMARK 7.4.4. The LP conditions given by Theorem 7.4.3 are much too strong. 'Natural' LP conditions at an endpoint should restrict the coefficients only on a sequence of intervals converging to that endpoint, with no 'essential' restriction on the complements of these intervals. If there is a solution u big enough on all these intervals then the LP case may hold: If E is a subset of J such that $E \cap (c, b)$ has positive Lebesgue measure for every $c \in J$ and $u > \delta > 0$ on E, then

$$\int_c^b u^2 w \geq \delta \int_c^t w,$$

and the LP case holds when $w \notin L^1(c,b)$. Theorem 2.5 in [**368**] is a far reaching extension of Theorem 7.4.3 which covers 'interval' type : LP criteria for $J = (a, \infty)$, $-\infty < a < \infty$ and $w = 1$. Chapter 2 of [**368**] has many other results and interesting examples. Another result worth noting is: Given the LC case (with $J = (a, \infty)$, $-\infty < a < \infty$ and $w = 1$) and any positive number ε it is possible to change the coefficients only on a set with Lebesgue measure less than ε such that the changed equation is LP.

REMARK 7.4.5. In Example 7.4.1
(1) $q(t)$ can 'oscillate' back and forth across the $-t^2$ curve and still be in the LP case.
(2) $q(t) = -t^2$ for $t \in E \subset J$ is in the LP case at ∞, for certain subsets E of J, e.g. for E a countable union of intervals whose left endpoints converge to ∞; with some restrictions on E and no 'essential' restrictions on the complement of E in J.

For (5) see Eastham and Zettl [**157**] and [**368**]; for proofs of the other statements see Chapter 2 of the monograph [**368**].

To illustrate LP criteria we mention some more results from [**368**] .

EXAMPLE 7.4.3. *In Example 7.4.1 the LP case at ∞ holds for*
(1)
$$q(t) = -t^2 + te^t \cos(e^t),$$
(2)
$$q(t) = -t^3 \sin^4(t).$$

EXAMPLE 7.4.4. *In (7.4.1) the LP case at ∞ holds for $J = (1, \infty)$, $w = 1$, $p(t) = t^\alpha$, $q(t) = t^\beta \sin(t^\gamma)$ when $0 \leq \alpha \leq 2$, $\beta \geq 0$, $\gamma \geq 0$.*

5. Comments

(1) These are made separately for each section.
(2) The "system regularization" of LC endpoints based on the fundamental matrix U is not new. It has been used by Fulton and by Fulton and Krall [**259**].

In connection with boundary value problems the transformation $Z = U^{-1}Y$ has been used by Fulton and by Fulton and Krall [**259**].

The proof of the invariance of the O classification with respect to all real λ at an LC endpoint given in Theorem 7.2.3 is not the standard one and may be new.

This "oscillation theory proof" is simpler than the usual spectral theory proof.

(3) The definitions of R, LC, LP, O, NO, LCNO, LCO are standard except, as pointed out in Part I and above, we classify an infinite endpoint as regular if $1/p, q, w$ are integrable in a neighborhood of this point. This contrasts with the usual practice in the literature. For a definition of oscillation for difference equations see [**395**].

The standard proof of the invariance of the $L^2(J, w)$ solutions with respect to λ is based on the variation of parameter formula [**123**].

The invariance of the LCO case with respect to real λ follows from the spectral theory of ordinary differential operators; see [**600**]. In general for the symmetric case i.e. p, q, w real and $1/p, q, w \in L_{loc}(a, d), a < d < b$, $p \geq 0$, $w > 0$ there exists a σ_0, $-\infty \leq \sigma_0 \leq \infty$, such that the equation (7.3.1) is O at a for $\lambda > \sigma_0$ and is NO for $\lambda < \sigma_0$. Examples show that for $\lambda = \sigma_0$ the equation can be O or NO. This "oscillation number" σ_0 is also the starting point of the essential (continuous) spectrum of every self-adjoint realization of the equation. The case $\sigma_0 = -\infty$ is interpreted as meaning that the essential spectrum is not bounded below; the case $\sigma_0 = \infty$ means that the essential spectrum is empty. The latter holds for all SLP for which each endpoint is either R or LC since in this case the spectrum is discrete. The essential spectrum is the same for all self-adjoint realizations of the equation (7.3.1). For a proof of this statement as well as further information the reader is referred to [**600**]. As we have seen in Part II for regular problems the discrete spectrum depends heavily on the boundary conditions; this is true in the singular case as well, except when both endpoints are LP in which case there are no boundary conditions.

(4) This section contains a very brief introduction to LP criteria for the general equation with all three coefficients p, q, w present and an arbitrary interval with finite or infinite endpoints. The case $J = (a, \infty)$ with a a finite regular endpoint, and with $w = 1$ of Theorem 7.4.2 is known as Levinson's theorem. The extension given here uses basically the same proof, the key observation is that the condition $w \notin L^1$ is needed; for the half line case this is automatic, since $w = 1$ is not integrable but $w = 1$ is in L^1 on a bounded interval. Evans and Zettl [**179**] showed that, for smooth coefficients, the hypothesis of Levinson's theorem is sufficient to guarantee that all powers of the second order expression are in the (higher order version of the) LP case.

There is hardly any literature on the LP/LC dichotomy when all three coefficients are present. Most of the literature just considers the half line case $J = (a, \infty)$ with a a finite regular endpoint, and with $w = 1$. A few authors have studied the case when both p and q are present and $p > 0$ case but most just set $p = 1$. There seems to be no literature on LP/LC criteria when p changes sign. Although a problem with $p > 0$ or $w > 0$ can be transformed into one with $p = 1 = w$ it may not be clear how the conditions transform. A prime illustration of this situation is the well known Molchanov criterion for the discreteness of the spectrum when $p = 1$ and $w = 1$. What is the corresponding criterion when all three coefficients p, q, w are present?

There is a vast literature on LC/LP criteria, we now cite a few papers and books. The interested reader may want to consult the references in these cited items: Ahlbrandt, Hinton and Lewis [**5**], Atkinson, Eastham and McLeod [**22**], Behncke and Focke [**58**], Bennewitz [**63**], Bradley [**99**], Eastham [**148**], [**147**], [**151**], Eastham and Thompson [**156**], Eastham and Zettl [**157**], Evans, Kwong and Zettl [**173**], Evans [**170**], [**169**], Everitt [**192**], [**191**], [**194**], [**187**], [**198**], Everitt, Giertz and McLeod [**203**], Everitt, Giertz and Weidmann [**204**], Everitt, Knowles and Read [**209**], Evans and Zettl [**177**], [**179**], [**178**], Harris [**302**], [**318**], Hinton [**328**], Kalf [**353**], Knowles [**377**], [**378**], [**376**], Kurss [**408**], Kwong [**409**], [**411**], [**410**], Kwong and Zettl [**424**], Levinson [**442**], Niessen [**498**], [**497**], Niessen and Zettl [**503**], Ong [**505**], Patula and Waltman [**509**], Patula and Wong [**510**], Pleijel [**519**], [**517**], [**518**], Read [**537**], [**536**], [**534**], [**535**], Walter [**590**], Weidmann [**595**], Wong and Zettl [**618**], Zettl [**637**], [**644**], [**636**], [**639**], [**624**], [**640**], Boruvka [**96**], Coddington and Levinson [**123**], Dunford and Schwatz [**141**], Eastham [**152**], Hartman [**319**], Hille [**326**], Hinton and Schaefer [**334**], Jorgens [**350**], Kaper, Kwong and Zettl [**366**], Muller-Pfeiffer [**484**], Naimark [**487**], Weidmann [**600**], [**288**].

CHAPTER 8

Singular Initial Value Problems

All the effects of Nature are only the mathematical consequences of a small number of immutable laws. - Pierre-Simon Laplace

1. Introduction

Initial conditions specified at a regular point do not make sense at a singular point. If the singularity is of limit-circle type then there is an analogue for such conditions and an existence-uniqueness theorem can be established which reduces to the well known theorem for regular points.

2. Scalar Regularization with Regularizing Functions

Here we construct a pair of functions u, v which we call "regularizing" functions since they can be used to "regularize" singular equations with LCNO endpoints. Unlike the system regularization of Section 6.2 this regularization maps a scalar equation into a scalar equation.

Consider the equation

$$My = -(py')' + qy = \lambda w y \text{ on } J, \ \lambda \in \mathbb{R}, \qquad (8.2.1)$$

with

$$J = (a,b), \ -\infty \leq a < b \leq \infty, \ 1/p, q, w \in L_{loc}(J, \mathbb{R}), \ p > 0, \ w > 0, \quad \text{a.e.} \qquad (8.2.2)$$

THEOREM 8.2.1. *Let (8.2.1), (8.2.2) hold. Assume each endpoint is either regular or LCNO. Let $D_{\max}$ be the maximal domain i.e.*

$$D_{\max} = \{y \in L^2(J, w) : y, y^{[1]} \in AC_{loc}(J), \ w^{-1}[-(py')' + qy] \in L^2(J, w)\}.$$

Then there exist functions $u, v \in D_{\max}$ satisfying the following conditions:

(1) *They are real valued.*
(2) *For some real $\lambda = \lambda_a$, u is a principal solution at a and v is a nonprincipal solution at a.*
(3) *For some real $\lambda = \lambda_b$, u is a principal solution at b and v is a nonprincipal solution at b.*
(4) *These functions u, v need not be solutions through the interior of (a, b), and, in case $\lambda_a = \lambda_b$, they need not be the same solution near a and near b.*
(5) $[u, v](a) = \lim_{t \to a+}[u, v](t) = 1,$
(6) $[u, v](b) = \lim_{t \to b-}[u, v](t) = 1,$
(7) $v > 0$ *on* $J = (a, b).$

PROOF. See Section 6.2 for the definition of principal and non-principal solution; and see Lemma 7 in Niessen and Zettl [503] for a proof of Theorem 8.2.1. □

REMARK 8.2.1. In Theorem 8.2.1

$$[u,v](t) = u(t)(p\overline{v}')(t) - \overline{v}(t)(pu')(t), \ t \in J; \tag{8.2.3}$$

denotes the classical Lagrange sesquilinear form. Since y and (py') are continuous on J each individual term is well defined at each interior point t of the underlying interval J. But these individual terms $u, (pu'), \overline{v}, (p\overline{v}')$ are not, in general, defined at an endpoint of J; they may "blow up" there or be wildly oscillatory. Nevertheless, the Lagrange form $[\cdot,\cdot]$ has a finite limit at each endpoint; thus the "blow ups" or "wild oscillations" of the individual terms must cancel each other.

REMARK 8.2.2. The hypothesis that an endpoint is LCNO is not needed in order for the Lagrange form $[\cdot,\cdot]$ to have a finite limit at that endpoint. Given any $y, z \in D_{\max}$, $[y, z](a)$ exists as a finite limit at any endpoint a, regular, LCNO, LCO or LP, see Lemma 10.2.3 below.

DEFINITION 8.2.1. *Functions u and v satisfying properties (1) through (7) of Theorem 8.2.1 are called "regularizing functions" of the equation (8.2.1) on (a,b). The reason for this terminology will become clear in the next section and in Section 9.6 where we show that with the help of such functions u, v, particularly v, one can construct a regular equation which is "equivalent" in a natural sense to the singular equation (8.2.1). This equation is called singular if at least one endpoint of the underlying interval J is singular; the equation is said to be regular if both endpoints are regular.*

3. Factorization of Solutions near an LCNO Endpoint

A singular equation with an LCNO endpoint can be "regularized" using the function v from a regularizing pair u, v of functions as defined in the preceding section.

THEOREM 8.3.1. *Let (8.2.1), (8.2.2) hold. Assume that the left endpoint a is LCNO and let u, v be a pair of regularizing functions at a i.e. on (a,d) for some $d \in (a,b)$. Define*

$$P = v^2 p, \quad W = v^2 w, \quad Q = w v M v \text{ on } J,$$

and consider the equation

$$-(Pz')' + Qz = \lambda W z \text{ on } J. \tag{8.3.1}$$

Then we have

$$1/P, \ Q, \ W \in L(a,d), \ a < d < b,$$

i.e. the equation (8.3.1) is regular at a and:

(1) *If y is a solution of (8.2.1) on (a,d), then $z = y/v$ is a solution of (8.3.1) on (a,d). Conversely, if z is a solution of (8.3.1) on (a,d) then $y = vz$ is a solution of (8.2.1) on (a,d).*
(2) *The limits*

$$z(a) = \lim_{t \to a^+} z(t); \ (Pz')(a) = \lim_{t \to a^+} (Pz')(t)$$

exist and are finite. Thus the solution z and its quasi-derivative (Pz') can be continuously extended to the (finite or infinite) endpoint a.
(3) *Note that v is independent of $\lambda \in \mathbb{R}$ but does depend on $1/p, q, w$ and on the endpoint a i.e. on some neighborhood (a,d) for $d \in (a,b)$.*

(4) *The one-to-one mappings*

$$y(t,\lambda) = v(t)\, z(t,\lambda) \text{ and } (py')(t,\lambda) = v(t)\,(Pz')(t,\lambda)$$

can be given more explicitly, using the notation from Chapter 2, by

$$y(t,c,h,k,1/p,q,w,\lambda) = v(t)z(t,c,h/v(c),kv(c) - h(pv')(c),1/P,Q,W,\lambda)$$

$$(py')(t,c,h,k,1/p,q,w,\lambda)$$
$$=(pv')(t)(Pz')(t,c,h/v(c),\, kv(c) - h(pv')(c),1/P,Q,W,\lambda)$$

for $c \in J$, and $h,k \in \mathbb{C}$.

PROOF. See Niessen and Zettl [**503**]. Although the explicit formulas for the 1-1 map $y \to vz$ are not given in this paper they can easily be obtained from there. □

REMARK 8.3.1. Note that, in general, y and v do not exist at the point a. Thus we have

$$z(a) = \frac{y}{v}(a);\ (Pz')(a) = (vpy' - ypv')(a) = [v,y](a)$$

but neither the numerator y nor the denominator v, nor the individual terms in $(Pz')(a)$, can be evaluated separately at a. Of course there is an entirely analogous theorem and remark for the endpoint b.

If each endpoint is either R or LCNO then Theorem 8.3.1 holds on the entire interval J.

THEOREM 8.3.2. *Let the notation of Theorem 8.3.1 hold. If each endpoint of the interval J is either R or LCNO, then there exist a pair of regularizing functions u,v defined on all of J such that the conclusions of Theorem 8.3.1 hold on the whole interval J.*

PROOF. See Niessen and Zettl [**503**]. □

The next example illustrates these results.

EXAMPLE 8.3.1. *For the classical Legendre equation*

$$-((1-t^2)\,y')' = \lambda\,y \text{ on } (-1,1),$$

the principal solution for both endpoints can be taken to be

$$u(t) = 1,\ -1 < t < 1,$$

and for a positive nonprincipal solution at both endpoints one can take

$$v(t) = \begin{cases} -(1/2)\ln((1-t)/(1+t)) & 1/2 \le t < 1, \\ (1/2)\ln((1-t)/(1+t)) & -1 < t \le -1/2. \end{cases}$$

Note that $v > 0$ near $+1$ and near -1 and so, by Theorem 8.3.2, v can be extended over the whole interval so that v is in the maximal domain on the entire interval J and $v > 0$ on J. Every solution y can be factored as follows:

$$y(t,\lambda) = v(t)\, z(t,\lambda),\ -1 < t < 1,\ \lambda \in R$$

where z is continuous on the closed interval $[-1,1]$. This shows, in particular, that the asymptotic behavior of solutions of the Legendre equation satisfying a fixed initial condition is independent of $\lambda \in \mathbb{C}$.

4. Limit-Circle "Initial Value Problems"

In this section we define singular analogues of the regular initial conditions: $y(a) = h$, $(py')(a) = k$ when a is an LC endpoint. These are defined in terms of the Lagrange sesquilinear form $[\cdot,\cdot]$, see (8.2.3) and Remark 8.2.1. The proof of Theorem 8.4.1 is based on the system regularization discussed in Section 7.2 and is similar to the proof of Theorem 7.2.1.

THEOREM 8.4.1. *Let (8.2.1), (8.2.2) hold. Assume the left endpoint a is R or LC. Suppose u, v are real valued linearly independent solutions on some interval $(a, d]$ for some fixed real λ_0. Given any $\lambda \in \mathbb{R}$ and any $h, k \in \mathbb{R}$ the singular initial value problem consisting of the equation*

$$-(py')' + qy = \lambda w y \text{ on } J$$

and the singular "initial condition"

$$[y, u](a) = h, \ [y, v](a) = k$$

has a unique real solution y on J. Similarly at b.

PROOF. Since u, v are linearly independent solutions near a we have $[u, v](a) \neq 0$ and we can assume that $[u, v](a) = 1$. Let

$$U = \begin{bmatrix} u & v \\ pu' & pv' \end{bmatrix}, \ Z = U^{-1}Y, \ Y' = (P - \lambda W)Y, \ U' = (P - \lambda_0 W)U,$$

$$P = \begin{pmatrix} 0 & 1/p \\ q & 0 \end{pmatrix}, \ W = \begin{bmatrix} 0 & 0 \\ w & 0 \end{bmatrix}, \ Y = \begin{bmatrix} y \\ py' \end{bmatrix}$$

Note that U is a fundamental matrix solution for a fixed λ_0 but Y is a vector solution for an arbitrary $\lambda \in \mathbb{R}$. A direct computation reveals that

$$Z' = (\lambda_0 - \lambda)(U^{-1}WU)Z = (\lambda_0 - \lambda) G Z \text{ on } (a, d]$$

where

$$G = U^{-1}WU = \begin{bmatrix} -uvw & -v^2 w \\ u^2 w & uvw \end{bmatrix} \in L((a, d), \mathbb{R}).$$

Note that $G \in L((a, d), \mathbb{R})$ follows from the Cauchy-Schwarz inequality coupled with the assumption that a is in the LC case. Hence by Theorem 1.5.3 all initial value problems

$$Z' = (\lambda_0 - \lambda) G Z \text{ on } (a, d], \ Z(a) = C,$$

have a unique solution. From $Y = UZ$, $Z = \begin{bmatrix} z_1 \\ z_2 \end{bmatrix}$ we get

$$u z_1 + v z_2 = y, \ (pu') z_1 + (pv') z_2 = py' \text{ both on } (a, d].$$

Using Cramer's rule we get

$$z_1(t) = \begin{vmatrix} y(t) & v(t) \\ (py')(t) & (pv')(t) \end{vmatrix} = [y, v](t), \ a < t < d,$$

$$z_2(t) = \begin{vmatrix} u(t) & y(t) \\ (pu')(t) & (py')(t) \end{vmatrix} = -[y, u](t), \ a < t < d.$$

By letting $t \to a$ we get $z_1(a) = k$, $z_2(a) = -h$. Since this holds for arbitrary h, k the proof is complete. □

5. Comments

(1) These are given separately for each section.
(2) Theorem 8.2.1 is taken from Niessen and Zettl [**503**].
(3) Theorems 8.3.1 and 8.3.2 are also adapted from [**503**] although phrased in terms of factoring rather than regularizing.

Example 8.3.1 illustrates Theorem 8.3.2 by showing that the classical Legendre equation, which is singular at both endpoints, is "equivalent" to a regular equation on the same interval. The leading coefficient P, the potential Q and the weight function W of the equivalent regular equation are not bounded on $(-1, 1)$ but, nevertheless, satisfy the regularity conditions $1/P, Q, W \in L(-1, 1)$. The Legendre equation and its regularization are equivalent in the sense that the transformation

$$y(t, \lambda) = v(t)\, z(t, \lambda)$$

($\lambda_0 = 0$ in this case) maps solutions y of the Legendre equation into solutions z of the regular equation in a 1-1 onto manner. All solutions z of the regular equation are continuous on the compact interval $[-1, 1]$. So the singular behavior is contained in the transformation function v. Since v is independent of λ this shows that the singular behavior of the Legendre equation is independent of $\lambda \in \mathbb{C}$. The invariance of the LC classification with respect to $\lambda \in \mathbb{C}$ and the invariance of the LCNO classifications with respect to $\lambda \in \mathbb{R}$ are merely specific instances of this general invariance property.

There is a 1-1 onto correspondence between the regular self-adjoint boundary conditions of the regularized equation and of the singular Legendre equation. This correspondence leaves the spectrum invariant but not the eigenfunctions; these are related by

$$y_n = v\, z_n,\ n \in \mathbb{N}_0.$$

Of course, these remarks apply to all other equations where each endpoint is either regular or LCNO. See Chapter 10 for details and Chapter 14 for more examples.

Part 4

Singular Boundary Value Problems

Titchmarsh, E. C.
> Perhaps the most surprising thing about mathematics is that it is so surprising. The rules which we make up at the beginning seem ordinary and inevitable, but it is impossible to foresee their consequences. These have only been found out by long study, extending over many centuries. Much of our knowledge is due to a comparatively few great mathematicians such as Newton, Euler, Gauss, or Riemann; few careers can have been more satisfying than theirs. They have contributed something to human thought even more lasting than great literature, since it is independent of language.

In N. Rose Mathematical Maxims and Minims, Raleigh, NC: Rome Press Inc., 1988.

CHAPTER 9

Two-Point Singular Boundary Value Problems

Kronecker, Leopold (1823-1891)
   God made the integers, all else is the work of man.
      (Die Ganzen Zahlen hat Gott gemacht, alles andere ist Menschenwerk.)
   Jahresberichte der Deutschen Mathematiker Vereinigung.

1. Introduction

At a singular endpoint solutions y and their quasi-derivatives (py') do not exist, in general. Thus regular two point boundary conditions, such as those studied in Part I e.g. $AY(a) + BY(b) = 0$ do not make sense in the singular case. What can take their place? We will see below that the Lagrange sesquilinear form $[\cdot, \cdot]$ plays an important role in answering this question, especially for LC endpoints. This form involves "combinations" of maximal domain functions which include solutions and their quasiderivatives. As we will see below these combinations remain finite even when the individual members within the form do not.

2. The Minimal and Maximal Domains and Lagrange Form

Consider the differential expression M defined by

$$My = -(py')' + qy, \quad \text{for} \quad 1/p, q \in L_{loc}(J, \mathbb{C}), \quad J = (a,b), \quad -\infty \le a < b \le \infty. \tag{9.2.1}$$

The expression My is defined a.e. for functions y such that y, and py' are in $AC_{loc}(J)$; we refer to this as the expression domain of M.

In this section, although we allow p and q to be complex valued we will assume that the weight function w is real and positive. This is so that we can use operator theory in the Hilbert space $H = L^2(J, w)$.

The maximal domain $D_{\max} = D_{\max}(M, w, J)$ of M on J with weight function w satisfying

$$w \in L_{loc}(J, \mathbb{R}), \quad w > 0, \tag{9.2.2}$$

is defined by

$$D_{\max} = \{y : J \to \mathbb{C} : y, \, py' \in AC_{loc}(J), \, y, \, w^{-1} My \in L^2(J, w)\}. \tag{9.2.3}$$

The preminimal domain $D_0 = D_0(M, w, J)$ is defined by

$$D_0 = \{y \in D_{\max} : y \text{ has compact support in } J\}. \tag{9.2.4}$$

The maximal and preminimal operators $S_{\max}$ and $S'_{\min}$, respectively, are defined by

$$S_{\max} f = w^{-1} M f, \quad f \in D_{\max}; \quad S'_{\min} f = w^{-1} M f, \quad f \in D_0. \tag{9.2.5}$$

The preminimal operator $S'_{\min}$ is closable and its closure, denoted by $S_{\min}$, is called the minimal operator. Its domain is denoted by $D_{\min}$.

The adjoint expression of M is given by
$$M^+ z = -(\bar{p}z')' + \bar{q}z \quad \text{on} \quad J. \tag{9.2.6}$$

For y in the expression domain of M and z in the expression domain of M^+ the Lagrange sesquilinear form $[\cdot,\cdot]$ is given by
$$[y,z] = yp\bar{z}' - \bar{z}py'. \tag{9.2.7}$$

LEMMA 9.2.1 (The Lagrange Identity). *For any y in the expression domain of M and z in the expression domain of M^+ we have*
$$\bar{z}My - y\overline{M^+z} = [y,z]'. \tag{9.2.8}$$

PROOF. This can be verified by a direct computation. □

LEMMA 9.2.2 (Green's Formula). *For any y in the expression domain of M, z in the expression domain of M^+ and $\alpha, \beta \in J$, $\alpha < \beta$ we have*
$$\int_\alpha^\beta \{\bar{z}My - y\overline{M^+z}\} = [y,z](\beta) - [y,z](\alpha). \tag{9.2.9}$$

PROOF. This is obtained by integrating (9.2.8). □

LEMMA 9.2.3. *For any y in the expression domain of M, z in the expression domain of M^+ we have*
$$\int_a^b \{\bar{z}My - y\overline{M^+z}\} = [y,z](b) - [y,z](a). \tag{9.2.10}$$

PROOF. This follows from (9.2.9) by taking limits as $\alpha \to a$ and $\beta \to b$. That these limits exist and are finite can be seen from definition of the expression domains of M and M^+, see (9.2.3). □

We can now give the important adjoint relationships between the minimal and maximal operators of M and M^+.

THEOREM 9.2.1. *Let (9.2.1) to (9.2.6) hold. Then the domains of the minimal and maximal operators of M and M^+ are all dense in the Hilbert space $H = L^2(J,w)$ and we have*
  (1) $S^*_{\min}(M) = S_{\max}(M^+)$;
  (2) $S^*_{\max}(M^+) = S_{\min}(M)$;
  (3) $S^*_{\min}(M^+) = S_{\max}(M)$;
  (4) $S^*_{\max}(M) = S_{\min}(M^+)$.

PROOF. Theorem 9.2.1 follows from the Lagrange identity and Green's formula of Lemmas 9.2.1 and 9.2.2. For details see the proof in Naimark's classic book [**487**]. Although the latter is given only for the symmetric case with unit weight function it can readily be adapted to yield a proof of Theorem 9.2.1. See also Zettl [**633**], Everitt and Zettl [**239**]. □

Since $S \subset T$ implies that $T^* \subset S^*$ we have the following relationships for operators S between the minimal and maximal operators:
$$S_{\min}(M) \subset S(M) \subset S_{\max}(M);$$
$$S^*_{\max}(M) = S_{\min}(M^+) \subset S^*(M) \subset S^*_{\min}(M) = S_{\max}(M^+).$$

3. Transcendental Characterization of Eigenvalues

Consider

$$M y = -(py')' + qy = \lambda w y \quad \text{on} \quad J = (a,b),$$
$$-\infty \leq a < b \leq \infty, \quad 1/p, q, w \in L_{loc}(J, \mathbb{C}). \quad (9.3.1)$$

Note that we are not assuming that w is positive nor that it is real-valued in this section.

For singular boundary conditions, which are defined at non LP endpoints only, we consider

$$A \begin{bmatrix} W(y,v)(a) \\ -W(y,u)(a) \end{bmatrix} + B \begin{bmatrix} W(y,v)(b) \\ -W(y,u)(b) \end{bmatrix} = 0, \quad A, B \in M_2(\mathbb{C}), \quad (9.3.2)$$

where u, v are solutions of (9.3.1) for some $\lambda = \lambda_0 \in \mathbb{C}$ satisfying

$$W(u,v)(t) = \det U(t) = 1, \ U(t) = \begin{bmatrix} u & v \\ pu' & pv' \end{bmatrix}(t), \quad t \in J. \quad (9.3.3)$$

A complex number λ is an eigenvalue of the singular two-point boundary value problem (9.3.1) to (9.3.3) if the equation (9.3.1) has a nontrivial solution y satisfying (9.3.2). If each endpoint is regular or LC (Recall that we defined the LC case in Section 7.3 for complex valued coefficients.) then, as in the regular case, the eigenvalues of this singular problem can be characterized as the zeros of an entire function. This characterization is given in the next theorem. This theorem uses the system regularization established in Theorem 7.2.1 to construct a characteristic function for singular boundary value problems.

REMARK 9.3.1. Note that (9.3.3) is merely a normalization since the Wronskian of two solutions of equation (9.3.8) for the *same* value of λ is constant on the entire interval J and this constant is nonzero for linearly independent solutions.

THEOREM 9.3.1 (Singular Non-Self-Adjoint Boundary Conditions). *Let u, v be linearly independent solutions of equation (9.3.1) for some $\lambda = \lambda_0 \in \mathbb{C}$. Let (9.3.1) to (9.3.3) hold and assume each endpoint is either regular or LC.*
Define

$$G = \begin{bmatrix} -uvw & -v^2 w \\ u^2 w & uvw \end{bmatrix}. \quad (9.3.4)$$

Then

(1) $G \in L^1(J, \mathbb{C})$ *and* $\Phi(b, a, \lambda - \lambda_0, G)$ *exists and is an entire function of λ where* $\Phi(\cdot, \cdot, \lambda - \lambda_0, G)$ *is the primary fundamental matrix of the system*

$$Z'(\cdot, \lambda) = (\lambda_0 - \lambda) G Z(\cdot, \lambda) \quad \text{on} \quad J. \quad (9.3.5)$$

(2) $\lambda \in \mathbb{C}$ *is an eigenvalue of the singular SLP (9.3.1) to (9.3.3) if and only if*

$$\delta(\lambda) = \det(A + B \ \Phi(b.a, \lambda - \lambda_0, G)) = 0. \quad (9.3.6)$$

(3) $\lambda \in \mathbb{C}$ *is an eigenvalue of geometric multiplicity two if and only if*

$$A + B \ \Phi(b, a, \lambda - \lambda_0, G) = 0. \quad (9.3.7)$$

PROOF. Equation (9.3.1) is equivalent to the first order system

$$Y' = (P - \lambda W)Y, \quad P = \begin{bmatrix} 0 & p^{-1} \\ q & 0 \end{bmatrix}, \quad W = \begin{bmatrix} 0 & 0 \\ w & 0 \end{bmatrix}$$

Let

$$U = \begin{bmatrix} u & v \\ pu' & pv' \end{bmatrix}, \quad Z = U^{-1}Y;$$

then (9.3.5) holds. Part (1) now follows from Sections 1.2 and 1.5.

To prove part (2) consider (9.3.5) with the boundary condition

$$AZ(a) + BZ(b) = 0.$$

This is a *regular* boundary value problem. It has a nontrivial solution Z if and only if the algebraic system

$$[A + B\,\Phi(\cdot\cdot, \lambda - \lambda_0, G)]Z(a) = 0$$

has a nontrivial solution for $Z(a)$.

For the proof of part (3) we observe that two linearly independent solutions of the algebraic system for $Z(a)$ yield two linearly independent solutions $Z(t)$ of the differential system and conversely.

To complete the proof we relate the boundary condition for Z i.e. $AZ(a) + BZ(b) = 0$ to (9.3.2). This can be done with Cramer's rule: Let $Z = \begin{bmatrix} z_1 \\ z_2 \end{bmatrix}$ and solve the system $Y = UZ$ for z_1, z_2 to get

$$z_1 = W(y, v) = y(pv') - v(py'), \quad z_2 = -W(y, u) = u(py') - y(pu') \quad \text{on} \quad J. \quad (9.3.8)$$

Since the system for Z i.e. (9.3.5) is regular z_1 and z_2 can be continuously extended to the endpoints a, b. This shows that the boundary value problem for Z i.e (9.3.5) with $AZ(a) + BZ(b) = 0$ is equivalent with the SLP (9.3.1) to (9.3.3) and the proof is complete. □

REMARK 9.3.2. By Theorem 1.5.2, both $z_1(a)$ and $z_1(b)$ exist and are finite even though the individual terms inside $W(y, u)$ and $W(y, v)$ may not exist at a or b. In particular, if one term (e.g. $y(pv')$) blows up the other ($v(py')$) must blow up at the same rate so that the difference remains finite at a and at b.

The Wronskian of two solutions of equation (9.3.8) for the *same* value of λ is constant on the entire interval J. If u, v are solutions of (9.3.1) for *different* values of λ then their 'Wronskian' is "nearly constant" near an LC endpoint. Specifically we have

COROLLARY 9.3.1. *Let (9.3.1) hold. Assume the endpoint a is either regular or LC. If u is a solution of (9.3.1) for some $\lambda = \lambda_1$ and v is a solution of (9.3.1) for $\lambda = \lambda_2$, then*

$$W(u, v)(a) = \lim_{t \to a^+} W(u, v)(t) = \lim_{t \to a^+} (upv' - vpu')(t)$$

exists and is finite. A similar result holds at b.

PROOF. This follows from the change of dependent variable transformation $Z = U^{-1}Y$ as in the proof of Theorem 9.3.1. □

THEOREM 9.3.2. *Let the hypotheses and notation of Theorem 9.3.1 hold. Then $\delta(\lambda)$ is an entire function of λ and therefore exactly one of the following four cases holds for the boundary value problem (9.3.1) to (9.3.3):*

(1) *Every complex number is an eigenvalue.*
(2) *No complex number is an eigenvalue.*
(3) *There are exactly n eigenvalues for some $n \in \mathbb{N}$.*
(4) *There are an infinite but countable number of eigenvalues.*

PROOF. This follows directly from the well known fact that the zeros of an entire function (if there are any) are isolated in the finite complex plane. □

THEOREM 9.3.3. *Let the hypotheses and notation of Theorem 9.3.1 hold. Assume that*

$$A = e^{i\gamma}K, \quad -\pi \leq \gamma \leq \pi, \quad K = (k_{ij}) \in SL_2(\mathbb{R}), \quad B = -I. \qquad (9.3.9)$$

Define for all $\lambda \in \mathbb{C}$

$$\begin{aligned} D(\lambda) &= k_{11}\phi_{22}(b, a, \lambda - \lambda_0, G) + k_{22}\phi_{11}(b, a, \lambda - \lambda_0, G) \\ &\quad - k_{12}\phi_{21}(b, a, \lambda - \lambda_0, G) - k_{21}\phi_{12}(b, a, \lambda - \lambda_0, G). \end{aligned} \qquad (9.3.10)$$

Then

(1) *the number $\lambda \in \mathbb{C}$ is an eigenvalue of the singular boundary value problem (9.3.1) to (9.3.3) if and only if*

$$D(\lambda) = 2\cos\gamma; \qquad (9.3.11)$$

(2) *the number $\lambda \in \mathbb{C}$ is a geometrically double eigenvalue of the singular boundary value problem (9.3.1), (9.3.3) if and only if*

$$\Phi(b, a, \lambda - \lambda_0, G) = e^{i\gamma}K. \qquad (9.3.12)$$

PROOF. Using (9.3.5) and proceeding as in the proof of Lemma 3.2.6 we get

$$\delta(\lambda) = 1 + e^{2i\gamma} - e^{i\gamma}D(\lambda)$$

and the conclusion for (1) follows. Part (2) follows directly from (9.3.6). □

4. Green's Function

Here we construct the Green's function of *singular* scalar problems directly in terms of a Green's function of a *regular* system. This regular system is constructed with the "system regularization" technique used in Section 7.2 and in Theorem 9.3.1. The scalar Green's function is then obtained from the upper right component of the Green's matrix of the regular system.

THEOREM 9.4.1. *Let the hypotheses and notation of Theorem 9.3.1 hold. Let $K(t, s, \lambda - \lambda_0, G, A, B)$ denote the Green's (matrix) function of the regular (system) boundary value problem*

$$Z' = (\lambda - \lambda_0)\, G\, Z + F, \quad AZ(a) + BZ(b) = 0, \quad A, B \in M_2(\mathbb{C}). \qquad (9.4.1)$$

Then

(1)
$$K(t,s,\lambda,P,W,A,B) = U(t)\,K(t,s,\lambda-\lambda_0,G,A,B)\,U^{-1}(s), \quad t,s \in J, \quad (9.4.2)$$

is the Green's function of the singular system boundary value problem

$$Y' = (P - \lambda W)Y + H, \quad A\begin{bmatrix} W(y,v)(a) \\ -W(y,u)(a) \end{bmatrix} + B\begin{bmatrix} W(y,v)(b) \\ -W(y,u)(b) \end{bmatrix} = 0, \quad (9.4.3)$$

(2) K_{12} is the Green's function of the singular scalar boundary value problem

$$-(py')' + qy = \lambda wy + h, \quad A\begin{bmatrix} W(y,v)(a) \\ -W(y,u)(a) \end{bmatrix} + B\begin{bmatrix} W(y,v)(b) \\ -W(y,u)(b) \end{bmatrix} = 0. \quad (9.4.4)$$

Specifically we have:

(1) For any $\lambda \in \mathbb{C}$ which is not an eigenvalue of (9.4.3) when $H = 0$, the problem (9.4.3) has a unique solution for any H satisfying

$$UH \in L(J,\mathbb{C}) \quad (9.4.5)$$

and this solution is given by

$$Y(t) = \int_a^b K(t,s,\lambda,P,W)H(s)\,ds, \quad t \in J. \quad (9.4.6)$$

(2) For any $\lambda \in \mathbb{C}$ which is not an eigenvalue of (9.4.4) when $h = 0$ the problem (9.4.4) has a unique solution for any h such that $H = \begin{bmatrix} 0 \\ h \end{bmatrix}$ satisfies (9.4.5) and this solution is given by

$$y(t) = \int_a^b K_{12}(t,s,\lambda,P,W)\,h(s)\,ds, \quad t \in J. \quad (9.4.7)$$

PROOF. Using the construction of the regular Green's function in Section 3.7 the proof proceeds as in Theorem 9.3.1. □

Using the transformation employed by Theorem 9.4.1 to transform a singular scalar or system boundary value problem to a regular system of the form (9.4.1) we can use the adjointness conditions developed in Section 3.8 for regular systems and extend them to the singular case for both system and scalar boundary value problems. This is done in the next theorem.

THEOREM 9.4.2. *Let the hypotheses and notation of Theorem 9.3.1 hold.*

Let $K(t,s,\lambda,P,W,A,B)$ be the matrix Green's function of the singular boundary value problem (9.4.3) and let $K(t,s,\overline{\lambda},\overline{P},\overline{W},C,D)$ be the matrix Green's function of the singular problem

$$Z' = (\overline{P} - \overline{\lambda W})Z + H, \quad C\begin{bmatrix} -W(y,u)(a) \\ W(y,v)(a) \end{bmatrix} + D\begin{bmatrix} -W(y,u)(b) \\ W(y,v)(b) \end{bmatrix} = 0. \quad (9.4.8)$$

Then

$$K(t,s,\overline{\lambda},\overline{P},\overline{W},C,D) = V(t)\,K(t,s,\overline{\lambda}-\overline{\lambda_0},\overline{G},C,D)\,V^{-1}(s), \quad t,s \in J, \quad (9.4.9)$$

$$V = \begin{bmatrix} -v & u \\ -(pv') & (pu') \end{bmatrix}; \quad (9.4.10)$$

and the following statements are equivalent for all λ such that λ is not an eigenvalue of (9.4.3) and $\overline{\lambda}$ is not an eigenvalue of (9.4.8):

(1)
$$AEC^* = BED^*, \qquad (9.4.11)$$

(2)
$$K(t,s,\lambda,P,W,A,B) = EK^*(s,t,\overline{\lambda},\overline{P},\overline{W},C,D)E, \qquad (9.4.12)$$

(3)
$$K_{12}(t,s,\lambda,P,W,A,B) = \overline{K}_{12}(s,t,\overline{\lambda},\overline{P},\overline{W},C,D). \qquad (9.4.13)$$

PROOF. This follows from Theorem 3.8.3, Theorem 9.4.1 and the observations that
$$E\overline{U}^{*-1} = \begin{bmatrix} v & -u \\ v^{[1]} & -u^{[1]} \end{bmatrix} = -V; \quad \overline{U}^*E = -V^{-1}.$$

□

5. Comments

The definition of the maximal domain and of the Lagrange sesquilinear form is standard. These play an important role in the theory of boundary value problems. The construction of the characteristic function and the consequent transcendental characterization of the eigenvalues as well as the construction of the Green's function are all based on the system regularization. This regularization reduces the construction of Green's functions for *singular* boundary value problems to that of *regular* systems so that the results of Section 3.8 can be applied. As in the regular case the singular scalar Green's function is then obtained from the upper right-hand corner of the singular Green's matrix.

In Theorem 9.4.2, given that λ is not an eigenvalue of (9.4.3), the assumption that $\overline{\lambda}$ is not an eigenvalue of the adjoint problem (9.4.8) is superfluous. This will be shown in a forthcoming paper by Ridenhour and Zettl [**548**]. In this paper adjoint boundary conditions are found *explicitly* and singular Green's function*s* for higher order problems are constructed.

CHAPTER 10

Singular Self-Adjoint Problems

Hilbert, David (1862-1943)
    Wir müssen wissen.
        Wir werden wissen.
[Engraved on his tombstone in Göttingen.]
    Physics is much too hard for physicists.
C. Reid, 'Hilbert', London: Allen and Unwin, 1970.
    The art of doing mathematics consists in finding that special case which contains all the germs of generality.
In N. Rose Mathematical Maxims and Minims, Raleigh, NC: Rome Press Inc., 1988.
    One can measure the importance of a scientific work by the number of earlier publications rendered superfluous by it.
In H. Eves Mathematical Circles Revisited, Boston: Prindle, Weber and Schmidt, 1971.
    Mathematics knows no races or geographic boundaries; for mathematics, the cultural world is one country.
In H. Eves Mathematical Circles Squared, Boston: Prindle, Weber and Schmidt, 1972.

1. Introduction

In this chapter we discuss singular self-adjoint SLP with a positive weight function w in some detail. The basic underlying theory is based on the theory of self-adjoint operators in the Hilbert space $H = L^2(J, w)$. The case when w changes sign is studied in Chapters 11 and 12 by different methods.

A number of results, definitions etc, in this chapter are special cases of the corresponding results, definitions etc, from Chapter 9. These are given here for the convenience of the reader so that readers interested only in the self-adjoint case need not necessarily be familiar with all of Chapter 9.

2. The Lagrange Form

LaGrange, Joseph-Louis
The reader will find no figures in this work. The methods which I set forth do not require either constructions or geometrical or mechanical reasonings: but only algebraic operations, subject to a regular and uniform rule of procedure. Preface to Mécanique Analytique.
[said about the chemist Lavoisier:]
It took the mob only a moment to remove his head; a century will not suffice to reproduce it.

H. Eves An Introduction to the History of Mathematics, 5th Ed., Saunders.
When we ask advice, we are usually looking for an accomplice.

Consider the differential expression M defined by

$$My = -(py')' + qy, \text{ for } r = 1/p, \, q \in L_{loc}(J, \mathbb{R}), \, J = (a,b), \quad -\infty \leq a < b \leq \infty. \tag{10.2.1}$$

The expression My is defined a.e. for functions y such that y and py' are in $AC_{loc}(J)$; we refer to this as the expression domain of M. The maximal domain $D_{\max} = D_{\max}(M, w, J)$ of M on J with weight function $w \in L_{loc}(J, \mathbb{R}), \, w > 0$ is defined by

$$D_{\max} = \{y : J \to \mathbb{C} : y, \, py' \in AC_{loc}(J), \, y, \, w^{-1}My \in L^2(J, w)\}. \tag{10.2.2}$$

We refer to the expression M as being symmetric because it is equal to its adjoint expression: $M = M^+$ (see Chapter 9) and because the minimal operator it generates (see Section 10.3 below) is a symmetric operator in the Hilbert space H.

For y and z in the expression domain of M the Lagrange sesquilinear form $[\cdot, \cdot]$ is given by

$$[y, z] = yp\bar{z}' - \bar{z}py'. \tag{10.2.3}$$

LEMMA 10.2.1 (The Lagrange Identity). *For any y and z in the expression domain of M we have*

$$\bar{z}My - y\overline{Mz} = [y, z]'. \tag{10.2.4}$$

PROOF. This can be verified by a direct computation. □

LEMMA 10.2.2 (Green's Formula). *For any y, z in the expression domain of M and $\alpha, \beta \in J, \, \alpha < \beta$ we have*

$$\int_\alpha^\beta \{\bar{z}My - y\overline{Mz}\} = [y, z](\beta) - [y, z](\alpha). \tag{10.2.5}$$

PROOF. This is obtained by integrating (10.2.4). □

The Lagrange form also exists at the endpoints of the underlying interval J and the values of this form at the endpoints are of critical importance in the characterization of self-adjoint realizations of the SL equation when one or both endpoints are singular.

LEMMA 10.2.3. *For any y, z in $D_{\max}$ both limits*

$$[y, z](b) = \lim_{t \to b^-} [y, z](t), \quad [y, z](a) = \lim_{t \to a^+} [y, z](t) \tag{10.2.6}$$

exist and are finite.

PROOF. Fix $\alpha \in J$ and let $\beta \to b^-$ in (10.5.2). It follows from the definition of $D_{\max}$ that both integrals on the left of (10.5.2) have a finite limit at b. Hence the first limit in (10.2.6) exists and is finite. Letting $\alpha \to a$ we see that the second limit of (10.2.6) exists and is finite. □

3. The Minimal and Maximal Domains and Self-Adjoint Operators

DEFINITION 10.3.1 (The Maximal and Minimal Operators). *Let the maximal domain $D_{\max}$ and the expression M be defined by (10.2.1), (10.2.2) and let*

$$w \in L_{loc}(J, \mathbb{R}), w > 0.$$

Define

$$S_{\max} f = w^{-1} M f, \text{ for } f \in D_{\max}, \qquad (10.3.1)$$

$S'_{\min} f = M f$, $f \in D_{\max}$, f *has compact support in* J.

Then $S_{\max}$ is called the maximal operator of (M, w) on J, $S'_{\min}$ is called the pre-minimal operator and the minimal operator $S_{\min}$ of (M, w) on J is defined as the closure of $S'_{\min}$. The pre-minimal operator is closable and so $S_{\min}$ is well defined as given in the next lemma.

LEMMA 10.3.1. *The maximal and minimal domains are dense in the Hilbert space $H = L^2(J, w) = \{f : J \to \mathbb{C}, \int_J |f|^2 w < \infty\}$, the pre-minimal operator is closable so that the minimal operator $S_{\min}$ is a closed, symmetric, densely defined operator and the operators $S_{\min}, S_{\max}$ are an adjoint pair in the sense that*

$$S^*_{\min} = S_{\max} \quad \text{and} \quad S^*_{\max} = S_{\min}. \qquad (10.3.2)$$

Hence any self-adjoint extension of $S_{\min}$ is also a self-adjoint restriction of $S_{\max}$ and conversely.

PROOF. See [487], [600]. □

From (10.3.2) it is clear that any self-adjoint extension S of the minimal operator $S_{\min}$ satisfies

$$S_{\min} \subset S = S^* \subset S_{\max}. \qquad (10.3.3)$$

We will see below that any operator S satisfying (10.3.3) can be determined by two-point boundary conditions. These, however, are vacuous at an LP endpoint. To describe these conditions it is convenient to take cases depending on the LP/LC classification of the endpoints. Here LC/LP will mean that the left endpoint a is LC and the right endpoint b is LP, etc.

An operator S satisfying (10.3.3) is called a self-adjoint extension of $S_{\min}$ on J, or a self-adjoint restriction of $S_{\max}$ on J, or simply, a self-adjoint realization of the equation $My = \lambda wy$ on J or a self-adjoint realization of (M, w) on J, or a self-adjoint realization of the SL equation.

Since we also will study some non-self-adjoint operators in Chapters 11 and 12 we now review the definition of the spectrum of closed, densely defined, linear operators in Hilbert space; see the book by Weidmann [599] for more details. Let T be a closed (not necessarily self-adjoint) linear operator with dense domain $D(T)$ on the Hilbert space H. Let I denote the identity operator on H. A number λ in $\mathbb{C}$ is an eigenvalue of T if there exists a $u \in H$, $u \neq 0$, such that $Tu = \lambda u$. In this case $T - \lambda I$ is not $1-1$ and the null space of $T - \lambda I$ is not empty, its dimension is the geometric multiplicity of the eigenvalue λ. Each non zero element of the null space of $T - \lambda I$ is called an eigenfunction of λ.

If $\lambda \in \mathbb{C}$ is not an eigenvalue of T, then

$$R(T, \lambda) = (T - \lambda I)^{-1}$$

is well defined and is a closed linear operator on H but its domain may not be all of H. Let
$$\rho(T) = \{\lambda \in \mathbb{C} : T - \lambda I \text{ is } 1-1 \text{ and onto}\}.$$
The set $\rho(T)$ is called the resolvent set of T and the spectrum of T, $\sigma(T)$, is defined by
$$\sigma(T) = \mathbb{C}\backslash\rho(T).$$
In other words, the spectrum of T consists of all complex numbers which are not in the resolvent set $\rho(T)$. Various parts of the spectrum: absolutely continuous, continuous, discrete, essential, point, singular continuous, are studied. In this monograph we consider only the discrete and essential spectrum. The discrete spectrum consists of all isolated eigenvalues of finite geometric multiplicity and is denoted by $\sigma_d(T)$. The rest of the spectrum is called the essential spectrum and is denoted by $\sigma_e(T)$. Thus
$$\sigma_e(T) = \sigma(T)\backslash\sigma_d(T).$$
For non-self-adjoint operators there are other definitions of the essential spectrum; see the paper by Evans, Lewis and Zettl [**175**] for several different such definitions and a comparison between them; also see the book by Edmunds and Evans [**164**].

4. Operator Theory Characterization and Self-Adjoint Boundary Conditions

The operators S satisfying (10.3.3) differ only by their domains. In this section we describe these domains. This description depends on the classification of the endpoints as regular (R), singular limit-circle (LC) or limit-point (LP). It is remarkable that all self-adjoint domains can be described by 'two-point' boundary conditions with the understanding that these conditions are vacuous in the LP/LP case and depend only on the non LP endpoint in the LC/LP and LP/LC cases.

We discuss all self-adjoint realizations of the Sturm-Liouville equation

$$My = -(py')' + qy = \lambda wy \text{ on } J = (a,b), \ -\infty \leq a < b \leq \infty, \quad (10.4.1)$$

under the conditions

$$\frac{1}{p}, q, w \in L_{loc}(J, \mathbb{R}), \ w > 0 \text{ a.e. on } J, \quad (10.4.2)$$

in the Hilbert space $H = L^2(J, w)$.

REMARK 10.4.1. Note that no sign restriction is placed on p. The reason for the sign restriction on w is so that $H = L^2(J, w)$ is a Hilbert space and the theory of self-adjoint operators in Hilbert space can be applied.

We give a number of characterizations and also canonical forms of the self-adjoint domains. These are given in terms of boundary conditions which can be separated or coupled, regular or singular. Of critical importance to the description of these self-adjoint domains in terms of boundary conditions is the Lagrange sesquilinear form (10.2.3).

Recall the Green's formula from Section 10.2:

$$\int_\alpha^\beta \{\overline{g}Mf - f\overline{Mg}\} = [f,g](\beta) - [f,g](\alpha), \ (f,g \in D_{\max}; \ a \leq \alpha < \beta \leq b). \quad (10.4.3)$$

4. OPERATOR THEORY AND SELF-ADJOINT BOUNDARY CONDITIONS

Our starting point for the description of the self-adjoint domains is the characterization based on the theory of self-adjoint operators in the Hilbert space $H = L^2(J, w)$. For details of this as well as other 'well known' facts used here the reader is referred to the classic book by Naimark [**487**], see also [**645**].

For $\lambda \in \mathbb{C}$, let R_λ denote the range of $S_{\min} - \bar{\lambda}I$, I being the identity operator on H; let $N_\lambda = R_\lambda^\perp$ and let

$$N^+ = N_i, \ N^- = N_{-i}, \ i = \sqrt{-1}, \ d^+ = \dim N^+, \ d^- = \dim N^-. \qquad (10.4.4)$$

The spaces N^+, N^- are called the deficiency spaces of the minimal operator $S_{\min}$ and d^+, d^- are called its deficiency indices. These are related to the equation (10.2.1) as follows:

$$N_\lambda = \{y \in H : S_{\max} y = w^{-1} My = \lambda y, \ \lambda = \pm i\}. \qquad (10.4.5)$$

Thus N^+, N^- consist of the solutions of (10.2.1) for $\lambda = i$ and $\lambda = -i$, respectively, which lie in the space H. Thus d^+, d^- are the number of linearly independent solutions of this equation which lie in H for $\lambda = i$ and $\lambda = -i$, respectively. From the reality of the coefficients p, q, w it follows that $d^+ = d^-$. We call its common value the deficiency index of M on J, denote it by d, and observe that

$$0 \leq d \leq 2.$$

Next we present a few technical lemmas to be used below.

LEMMA 10.4.1 (Patching Lemma). *(a)* $D_{\min} = \{ y \in D_{\max} : [y, z](b) - [y, z](a) = 0, \text{ for all } z \in D_{\max}\}$.
(b) The endpoint $c = a$ *or* $c = b$ *is LP if and only if* $[y, z](c) = 0$ *for all* $y, z \in D_{\max}$.
(c) If $c, d \in J$ *then for any* $\alpha, \beta, \gamma, \delta \in \mathbb{C}$ *there exists a function* $g \in D_{\max}$ *such that*

$$g(c) = \alpha, \ (pg')(c) = \beta, \ g(d) = \gamma, \ (pg')(d) = \delta. \qquad (10.4.6)$$

PROOF. See Naimark [**487**]. □

We will refer to part (c) of Lemma 10.4.1 as the 'Naimark Patching Lemma' or just the 'Patching Lemma' It is quite useful in 'patching together' functions defined on subintervals of J to obtain a function in the maximal domain on J.

The next Lemma plays an important role in this section. We will refer to it as the 'Lagrange bracket' or just the 'bracket' decomposition; see Fulton [**258**], Fulton and Krall [**259**].

LEMMA 10.4.2 (Bracket Decomposition). *Let* $y, z, u, v \in D_{\max}$. *If* $[v, u](c) = 1$ *for some* c, $a \leq c \leq b$, *then*

$$[y, z](c) = [y, v](c)[\bar{z}, \bar{u}](c) - [y, \bar{u}](c)[\bar{z}, v](c). \qquad (10.4.7)$$

PROOF. Note that

$$[y, z] = (\bar{z}, p\bar{z}') \begin{pmatrix} 0 & -1 \\ 1 & 0 \end{pmatrix} \begin{pmatrix} y \\ py' \end{pmatrix}$$

$$= (\bar{z}, p\bar{z}') \begin{pmatrix} (pv') & (p\bar{u}') \\ -v & -\bar{u} \end{pmatrix} \begin{pmatrix} -(p\bar{u}') & \bar{u} \\ (pv') & -v \end{pmatrix} \begin{pmatrix} y \\ py' \end{pmatrix}$$

$$= ([\bar{z}, v], [\bar{z}, \bar{u}]) \begin{pmatrix} -[y, \bar{u}] \\ [y, v] \end{pmatrix} = [y, v][\bar{z}, \bar{u}] - [y, \bar{u}][\bar{z}, v], \qquad (10.4.8)$$

holds at each t, $a < t < b$. So if $c \in J$ the proof is complete. If c is an endpoint, say d, take the limit as $t \to d$ on both sides of (10.4.8). This completes the proof. □

Using Green's formula and basic operator theory in H yields the following characterization of the self-adjoint domains:

THEOREM 10.4.1. *(i) $d = 0$. In this case the minimal operator $S_{\min}$ is itself a self-adjoint operator and has no proper self-adjoint extensions in H. This case occurs if and only if both endpoints are LP.*

(ii) $d = 1$. This case occurs if and only if one endpoint is LP and the other LC or R. If S is a self-adjoint extension of $S_{\min}$ then there exists a function $g \in D(S) \subset D_{\max}$ satisfying the following three conditions :

(1) *g is not in $D_{\min}$;*
(2) *$[g, g](b) - [g, g](a) = 0$;*
(3) *$D(S) = \{f \in D_{\max} : [f, g](b) - [f, g](a) = 0\}$.*

Conversely, given $g \in D_{\max}$ satisfying (1) and (2), $D(S)$ defined by (3) is a self-adjoint domain.

(iii) $d = 2$. This case occurs if and only if neither endpoint is LP. If S is a self-adjoint extension of $S_{\min}$ then there exist $g_1, g_2 \in D(S) \subset D_{\max}$ satisfying the following three conditions:

(1) *no nontrivial linear combination of g_1 and g_2 is in $D_{\min}$;*
(2) *$[g_j, g_k](b) - [g_j, g_k](a) = 0$, $j, k = 1, 2$;*
(3) *$D(S) = \{y \in D_{\max} : [y, g_j](b) - [y, g_j](a) = 0, j = 1, 2.\}$.*

Conversely, given $g_1, g_2 \in D_{\max}$ satisfying (1) and (2), $D(S)$ defined by (3) is a self-adjoint domain.

PROOF. See Theorem 4, pp. 75-76 in Naimark [**487**]. □

We comment on the three cases separately.

REMARK 10.4.2. $d = 0$. In the case when both endpoints are LP there is no boundary condition since $[y, g](b) - [y, g](a) = 0$ for all maximal domain functions; thus (3) is 'automatically' satisfied and therefore imposes no restriction on $D_{\max}$. Hence $D_{\max} = D_{\min}$ by Lemma 10.4.1 (a) and consequently the minimal operator $S_{\min}$ is itself a self-adjoint operator and has no proper self-adjoint extensions.

REMARK 10.4.3. $d = 1$. At an LP endpoint c, $[y, g](c) = 0$ is automatically satisfied; hence condition (3) imposes a restriction only at the non LP endpoint. This restriction is a genuine 'boundary condition' since the Lagrange bracket $[y, g]$ is not zero for all maximal domain functions at the non LP endpoint. Thus (3) is a separated boundary condition at the non LP endpoint, (2) is a condition on this condition and (1) says there is exactly one such linearly independent condition.

Condition (3) depends on $D_{\max}$ and therefore on the coefficients p, q, w of the differential equation (10.4.1). Can this dependence be eliminated? The answer is an unqualified yes if the non LP endpoint is regular and a qualified yes if it is singular. In the latter case (3) has an explicit representation in terms of two linearly independent solutions of (10.4.1) for any particular real value of λ; thus if such solutions are known for *any* real value of λ (including $\lambda = 0$) there is an explicit representation of (3). In most, but not quite all, of the celebrated equations (10.4.1) of Applied Mathematics and Mathematical Physics such explicit solutions can be found. For details see the Examples Chapter 14, or [**43**], [**42**].

4. OPERATOR THEORY AND SELF-ADJOINT BOUNDARY CONDITIONS

REMARK 10.4.4. $d = 2$. Conditions (3) are conditions on the 'boundary' points a, b, (1) says that there are exactly two such linearly independent conditions, and (2) gives conditions on these conditions. When all three conditions are satisfied we speak of 'a self-adjoint boundary condition'. Conditions (3) seem to link or 'couple' the endpoints together. We will see below that (3) may consist of two separate conditions, independent of each other, one at each endpoint, in this case we speak of a separated self-adjoint boundary condition, or (3) consists of coupled conditions. It cannot consist of one separated and one coupled condition. In the coupled case the coupling parameters may be real or complex in a sense made precise below.

As in case $d = 1$, conditions (3) depend on $D_{\max}$ and therefore on the coefficients p, q, w of the differential equation (10.4.1). Can this dependence be eliminated? The answer is an unqualified yes if both endpoints are regular, and a qualified yes if one or both are singular. If one endpoint is regular and the other singular (3) has a representation at the regular endpoint which is completely independent of the equation (10.4.1) and a representation at the singular endpoint in terms of two linearly independent solutions of (10.4.1) for *any* particular real value of λ, including $\lambda = 0$. If both endpoints are singular the representation may require two different pairs of linearly independent solutions, one pair for each endpoint in some neighborhood of that endpoint. In most, but not quite all, of the celebrated equations (10.4.1) of Applied Mathematics and Mathematical Physics such explicit solutions can be found. In general the two pairs of solutions, one pair for each endpoint, are then patched together through the 'middle' of the interval J to get maximal domain functions. For details see the Examples Chapter, Chapter 14, below or [**43**], [**42**].

REMARK 10.4.5. For both cases $d = 1$ and $d = 2$ actually only the asymptotic behavior of the solutions at the singular endpoints is required for the explicit representation of the self-adjoint boundary conditions and this is independent of $\lambda \in \mathbb{R}$.

To make the conditions of Theorem 10.4.1 more explicit a number of technical lemmas are used. The number and form of these explicit conditions depends on the endpoint classification of (10.4.1) and is given in the following subsections.

4.1. Regular Endpoints. In this subsection we clarify how Theorem 10.4.1 gives rise to the self-adjoint boundary conditions of Chapter 4, thus justifying the use of the word 'self-adjoint' in Chapter 4.

REMARK 10.4.6. Recall from Chapter 2 that the significance of an endpoint being regular is that if d is regular then the limits

$$y(d) = \lim_{t \to d} y(t),\ (py')(d) = \lim_{t \to d}(py')(t)$$

both exist and are finite for any solution y of the nonhomogeneous equation

$$-(py')' + qy = f,\ f \in L^1(J).$$

In general, these limits do not exist at a singular endpoint in both the LC and LP cases.

There is a stronger result:

LEMMA 10.4.3. *Assume an endpoint $d = a$ or $d = b$ is regular, then the limits*

$$y(d) = \lim_{t \to d} y(t),\ (py')(d) = \lim_{t \to d}(py')(t) \tag{10.4.9}$$

both exist and are finite for any function $y \in D_{\max}$.

PROOF. Suppose $y \in D_{\max}$. Let $f = w^{-1}My$. Assume a is regular and $I = (a, c)$, $a < c < b$. Then $My = wf$ and $wf \in L^1(I)$ since

$$\left(\int_I w|f|\right)^2 = \left(\int_I \sqrt{w}\sqrt{w}|f|\right)^2 \leq \int_I w \int_I w|f|^2.$$

Since a is a regular point the first integral on the right is finite, and the second is finite because $y \in D_{\max}$. Hence the result follows from Remark 10.4.6. The proof for the endpoint b is similar. □

The next Lemma is closely related to Lemma 10.4.1.

LEMMA 10.4.4. *If a and b are both regular, then for any $\alpha, \beta, \gamma, \delta \in \mathbb{C}$ there exists a function $g \in D_{\max}$ such that*

$$g(a) = \alpha, \ (pg')(a) = \beta, \ g(b) = \gamma, \ (pg')(b) = \delta. \tag{10.4.10}$$

Furthermore, if $f \in D_{\max}$ is in $D_{\min}$ then $f(a) = (pf')(a) = 0 = f(b) = (pf')(b)$.

PROOF. See Naimark [**487**]. □

THEOREM 10.4.2. *Assume each endpoint is regular. Let $A, B \in M_2(\mathbb{C})$, the set of 2×2 matrices over $\mathbb{C}$, satisfying the following conditions:*

$$\mathrm{rank}(A : B) = 2; \tag{10.4.11}$$

$$AEA^* = BEB^*, \ E = \begin{bmatrix} 0 & -1 \\ 1 & 0 \end{bmatrix}. \tag{10.4.12}$$

Define

$$D(S) = \{y \in D_{\max} : AY(a) + BY(b) = 0, \ Y = \begin{pmatrix} y \\ py' \end{pmatrix}\}. \tag{10.4.13}$$

Then $D(S)$ is a self-adjoint domain and all self-adjoint domains are generated this way i.e. given any self-adjoint domain there exist matrices $A, B \in M_2(\mathbb{C})$ satisfying conditions (10.4.11) and (10.4.12) such that $D(S)$ is given (10.4.13).

PROOF. In this case $d = 2$. Let A, B satisfy (10.4.11) and (10.4.12) and let $D(S)$ be defined by (10.4.13). Choose $g_1, g_2 \in D_{\max}$ such that

$$g_1(a) = \bar{a}_{12}, \ (pg_1')(a) = -\bar{a}_{11}, \ g_1(b) = -\bar{b}_{12}, \ (pg_1')(b) = \bar{b}_{11}, \tag{10.4.14}$$

$$g_2(a) = \bar{a}_{22}, \ (pg_2')(a) = -\bar{a}_{21}, \ g_2(b) = -\bar{b}_{22}, \ (pg_2')(b) = \bar{b}_{21}. \tag{10.4.15}$$

Such a choice is possible by (10.4.10). Condition (1) of Theorem 10.4.1 follows from (10.4.11) since $f = cg_1 + dg_2 \in D_{\min}$ implies, by Lemma 10.4.4, that $f(a) = 0 = f(b) = (pf')(a) = (pf')(b)$ and this implies that $cr_1 + dr_2 = 0$ where r_1, r_2 are the rows of the matrix $(A : B)$. Hence $c = 0 = d$ and condition (1) holds. Condition (2) follows from (10.4.12). Now note that (3) is equivalent to (10.4.13) in view of (10.4.14), (10.4.15). Conversely, given a self-adjoint domain $D(S)$ defined by (3) of Theorem 10.4.1 we construct matrices A, B satisfying (10.4.11) and (10.4.12) and reverse the above argument to show that (3) is equivalent to (10.4.13). □

We can now give the canonical forms of the regular self-adjoint boundary conditions stated in Lemma 4.2.1.

4. OPERATOR THEORY AND SELF-ADJOINT BOUNDARY CONDITIONS

THEOREM 10.4.3 (Canonical Form of Self-Adjoint Boundary Conditions). *Assume $A, B \in M_2(\mathbb{C})$ satisfy (10.4.11) and (10.4.12) and $D(S)$ is given by (10.4.13). Then either*

(1) *there exist $A_1, A_2 \in \mathbb{R}$ with $(A_1, A_2) \neq (0,0)$ and $B_1, B_2 \in \mathbb{R}$ with $(B_1, B_2) \neq (0,0)$ such that*

$$D(S) = \{y \in D_{\max} : A_1 y(a) + A_2 (py')(a) = 0 \text{ and } B_1 y(b) + B_2 (py')(b) = 0\}; \quad (10.4.16)$$

or

(2) *there exist $\gamma \in (-\pi, \pi)$ and a*

$$K \in SL(2, \mathbb{R}) = \{K \in M_2(\mathbb{R}) : \det K = 1\} \quad (10.4.17)$$

where $M_2(\mathbb{R})$ is the set of 2×2 matrices over $\mathbb{R}$, such that

$$D(S) = \{y \in D_{\max} : Y(b) = e^{i\gamma} K Y(a), \ Y = \begin{pmatrix} y \\ py' \end{pmatrix}\}. \quad (10.4.18)$$

Conversely, given either (10.4.16) or (10.4.18) there exist $A, B \in M_2(\mathbb{C})$ satisfying (10.4.11) and (10.4.12) such that (10.4.13) holds.

PROOF. Note that $\gamma = \pi$ reduces to $\gamma = 0$ by replacing K by $-K$. Condition (10.4.12) when written out in terms of the components of $A = (a_{ij})$, $B = (b_{ij})$ becomes

$$\begin{aligned}
a_{11}\overline{a_{22}} - a_{12}\overline{a_{21}} &= b_{11}\overline{b_{22}} - b_{12}\overline{b_{21}}, \\
a_{11}\overline{a_{12}} - a_{12}\overline{a_{11}} &= b_{11}\overline{b_{12}} - b_{12}\overline{b_{11}}, \\
a_{21}\overline{a_{22}} - a_{22}\overline{a_{21}} &= b_{21}\overline{b_{22}} - b_{22}\overline{b_{21}}, \\
a_{22}\overline{a_{11}} - a_{21}\overline{a_{12}} &= b_{22}\overline{b_{11}} - b_{21}\overline{b_{12}}.
\end{aligned} \quad (10.4.19)$$

Assume A is singular. Then (10.4.12) implies that B is singular. Hence there is a nonsingular matrix G such that GA has the form

$$GA = \begin{pmatrix} A_1 & A_2 \\ 0 & 0 \end{pmatrix}.$$

Similarly there is a nonsingular matrix F such that

$$FB = \begin{pmatrix} 0 & 0 \\ B_1 & B_2 \end{pmatrix}.$$

Multiplying by GF reduces the condition $AY(a) + BY(b) = 0$ to the one in (10.4.16) but with A_j, B_j possibly complex valued. If $A_1 \neq 0$ multiply $A_1 y(a) + A_2 (py')(a) = 0$ by $\overline{A_1}$ to get an equivalent condition with a new A_1 which is real. Now from (10.4.19) it follows that the new A_2 is also real. If $A_1 = 0$ multiply by $\overline{A_2}$ to get both coefficients A_1, A_2 real.

Assume A is nonsingular. Then from (10.4.12) it follows that B is nonsingular. Hence $AY(a) + BY(b) = 0$ is equivalent with $CY(a) - Y(b) = 0$ where $C = -B^{-1}A$. Note that (10.4.13) holds with A replaced by C and B by $-I$ where I is the identity matrix. In other words, in this case we can assume that $B = -I$. We use the notation $C = A$ and $B = -I$.

Set

$$a_{jr} = e^{i\gamma_{jr}} k_{jr}, \ k_{jr} \in \mathbb{R}, \ -\pi < \gamma_{jr} \leq \pi, \ j, r = 1, 2. \quad (10.4.20)$$

Condition (10.4.12) becomes
$$a_{11}\overline{a_{22}} - a_{12}\overline{a_{21}} = 1,$$
$$a_{11}\overline{a_{12}} - a_{12}\overline{a_{11}} = 0,$$
$$a_{21}\overline{a_{22}} - a_{22}\overline{a_{21}} = 0,$$
$$a_{22}\overline{a_{11}} - a_{21}\overline{a_{12}} = 1. \qquad (10.4.21)$$

From $a_{11}\overline{a_{12}} = e^{i(\gamma_{11}-\gamma_{12})}k_{11}k_{12} = e^{-i(\gamma_{11}-\gamma_{12})}k_{11}k_{12} = a_{12}\overline{a_{11}}$ it follows that $\gamma_{11} = \gamma_{12}$ if $k_{11}k_{12} \neq 0$. Similarly we conclude that $\gamma_{21} = \gamma_{22}$ if $k_{21}k_{22} \neq 0$.

From this and from the first and last equations in (10.4.21) we get
$$e^{i(\gamma_{11}-\gamma_{22})}k_{11}k_{22} - e^{i(\gamma_{12}-\gamma_{21})}k_{21}k_{12} = e^{i(\gamma_{11}-\gamma_{22})}[k_{11}k_{22} - k_{12}k_{21}] = 1,$$
$$e^{i(\gamma_{22}-\gamma_{11})}k_{11}k_{22} - e^{i(\gamma_{21}-\gamma_{12})}k_{21}k_{12} = e^{i(\gamma_{22}-\gamma_{11})}[k_{11}k_{22} - k_{12}k_{21}] = 1, \quad (10.4.22)$$

provided that $k_{11}k_{12} \neq 0$ and $k_{21}k_{22} \neq 0$; but the latter two conditions must hold in view of (10.4.21). Thus we may conclude that
$$\gamma_{11} = \gamma_{22} = \gamma_{21} = \gamma_{12}; \text{ and } \det K = k_{11}k_{22} - k_{12}k_{21} = 1.$$

This completes the proof of Theorem 10.4.3. $\square$

We illustrate one theoretical application of the canonical representation of the coupled boundary conditions by characterizing all *real* self-adjoint extensions of the minimal operator.

DEFINITION 10.4.1. *Suppose A is a symmetric densely defined linear operator in a Hilbert space $H = L^2(J,w)$. A linear operator B is called a real self-adjoint extension of A if B is a self-adjoint extension of A in H with the following properties:*
 (1) $g \in D(B)$ implies $\overline{g} \in D(B)$,
 (2) $B(\overline{g}) = \overline{Bg}$.

COROLLARY 10.4.1. *Assume S is a self-adjoint extension of the minimal operator $S_{\min}$ in H. Then S is a real self-adjoint extension of $S_{\min}$ in H if its domain $D(S)$ is given by either (i) separated boundary conditions (10.4.16) or (ii) coupled boundary conditions (10.4.18) with $\gamma = 0$. (Note that $\gamma = \pi$ reduces to $\gamma = 0$ by replacing K by $-K$.)*

PROOF. This follows directly from the representations (10.4.16) and (10.4.18) and the reality of the coefficients of equation (10.4.1). $\square$

4.2. One Regular and one LP Endpoint. This case is commonly studied in Quantum Mechanics. We start with some technical lemmas.

LEMMA 10.4.5. *(i) If a is regular and b is LP, then a function $y \in D_{\max}$ is in $D_{\min}$ if and only if the following two conditions are satisfied:*
 (1) $y(a) = 0$ and $(py')(a) = 0$;
 (2) $[y,z](b) = 0$ for all $z \in D_{\max}$.

(ii) If b is regular and a is LP, then a function $y \in D_{\max}$ is in $D_{\min}$ if and only if the following two conditions are satisfied:
 (1) $y(b) = 0$ and $(py')(b) = 0$;
 (2) $[y,z](a) = 0$ for all $z \in D_{\max}$.

PROOF. See Naimark [**487**]. $\square$

4. OPERATOR THEORY AND SELF-ADJOINT BOUNDARY CONDITIONS

THEOREM 10.4.4. *Assume a is regular and b is LP. Then $d = 1$. If $A_1, A_2 \in \mathbb{R}$ and $(A_1, A_2) \neq (0,0)$ then*

$$D(S) = \{y \in D_{\max} : A_1 y(a) + A_2 (py')(a) = 0\} \tag{10.4.23}$$

is a self-adjoint domain. Conversely, if $D(S)$ is a self-adjoint domain then there exist $A_1, A_2 \in \mathbb{R}$ satisfying $(A_1, A_2) \neq (0,0)$ such that $D(S)$ is given by (10.4.26). There is a similar result when b is regular and a is LP.

PROOF. The proof of Theorem 10.4.5 given below covers this case since a regular endpoint can be viewed as a special case of an LC endpoint. To specialize the proof of Theorem 10.4.5 to this case choose the real valued solutions u, v by the initial conditions

$$u(a) = 0 = (pv')(a), \; (pu')(a) = 1 = -v(a). \tag{10.4.24}$$

□

4.3. One LC and one LP Endpoint. Since we are dealing with an LC endpoint in this subsection conditions of the form (10.4.23) may not make sense if the LC endpoint is singular. To get appropriate replacements we need an extension of part (c) of Lemma 10.4.1.

LEMMA 10.4.6. *Assume a is LC and b is LP. Let $c \in J$, and fix a $\lambda = r \in \mathbb{R}$. Suppose u, v are nontrivial real valued solution of equation (10.4.1) on (a, c) for $\lambda = r$, normalized to satisfy*

$$[u, v](t) = 1, \; t \in (a, c). \tag{10.4.25}$$

Then for any $\alpha, \beta \in \mathbb{C}$, there exists a $g \in D_{\max}$ such that

$$[g, u](a) = \alpha, \; [g, v](a) = \beta. \tag{10.4.26}$$

Furthermore, g can be taken to be a (possibly complex) linear combination of u, v in (a, c).

Moreover, if a is a regular endpoint then for any $\alpha, \beta \in \mathbb{C}$, there exists a $g \in D_{\max}$ such that

$$g(a) = \alpha, \; (pg')(a) = \beta. \tag{10.4.27}$$

There is a similar result when b is regular and a is LP.

PROOF. Let $g = \beta u - \alpha v$. Then $[g, u](t) = \alpha$ for $t \in (a, c)$ and hence also at a. Similarly $[g, v](a) = \beta$. Now choose $d \in (c, b)$ and use the Patching Lemma to construct a maximal domain function g_m which agrees with g on (a, c). To prove the moreover statement determine u, v by the initial conditions: $u(a) = 0, (pu')(a) = 1$; $v(a) = -1, (pv')(a) = 0$ and evaluate $[g, u](a), [g, v](a)$. □

THEOREM 10.4.5. *Assume a is LC and b is LP. Then $d = 1$. Let $c \in J$, and fix a $\lambda = r \in \mathbb{R}$. Suppose that u, v are nontrivial real valued solution of equation (10.4.1) on (a, c), normalized to satisfy $[u, v](t) = 1, \; t \in (a, c)$. Then*

$$D(S) = \{y \in D_{\max} : A_1 [y, u](a) + A_2 [y, v](a) = 0, \; A_1, A_2 \in \mathbb{R}, \; (A_1, A_2) \neq (0,0)\} \tag{10.4.28}$$

is a self-adjoint domain. Conversely, if $D(S)$ is a self-adjoint domain then there exist $A_2, A_2 \in \mathbb{R}$ with $(A_1, A_2) \neq (0,0)$, such that (10.4.28) holds. There is a similar result when b is LC and a is LP.

PROOF. We prove the converse first. By Lemma 10.4.1, $[f,g](b) = 0$ for all $f,g \in D_{\max}$. Suppose $D(S)$ is a self-adjoint domain. By Theorem 10.4.1 there exists a $g \in D_{\max}$ such that (i), (ii), (iii) of Theorem 10.4.1 for case $d=1$ hold. For $y \in D_{\max}$ we have, using the above lemmas,

$$[y,g](b) - [y,g](a) = -[y,g](a) = -[y,u](a)[\overline{g},v](a) + [y,v](a)[\overline{g},u](a). \quad (10.4.29)$$

Set
$$C_1 = -[\overline{g},v](a), \; C_2 = [\overline{g},u](a). \quad (10.4.30)$$

Then (10.4.28) holds with $A_1 = C_1$, $A_2 = C_2$.

Applying part (ii) of Theorem 10.4.1, Lemma 10.4.1, and the decomposition lemma, we get

$$[g,g](b) - [g,g](a) = -[g,g](a) = -[g,u](a)[\overline{g},v](a) + [g,v](a)[\overline{g},u](a) = 0. \quad (10.4.31)$$

This and (10.4.30) imply that
$$-\overline{C_2}C_1 = -\overline{C_1}C_2. \quad (10.4.32)$$

Set $A_1 = \overline{C_1}C_1$, then A_1 is real and from (10.4.30) it follows that $A_2 = \overline{C_1}C_2$ is real. If $C_1 = 0 = C_2$, then $[y,g](a) = 0$ by (10.4.29) for every $y \in D_{\max}$ but this contradicts that a is LC. Therefore (10.4.28) holds.

Conversely, suppose (10.4.28) holds. To prove that $D(S)$ is a self-adjoint domain we find a maximal domain function g such that the conditions (i), (ii), (iii) of Theorem 10.4.1 hold. Let $g = A_1 u + A_2 v$. Then g is a nontrivial real valued solution on the interval (a,c). Note that g may not be in $D_{\max}$ but choose $d,h \in (a,c)$, $d < h$ and use the Patching Lemma to extend g to a function f in $D_{\max}$ by choosing f so that $f(d) = g(d)$, $(pf')(d) = (pg')(d)$ and $f(t) = 0$ for $t \geq h$. We continue to use the notation g for this extended function f since we are only interested in values of g near a where $g = f$. Using Lemma 10.4.5 (together with the above note) and the definition of g near a we have

$$[g,g](b) - [g,g](a) = -[g,g](a) = 0.$$

To show that g is not in $D_{\min}$, choose $d,h \in (a,c)$, $d < h$ and use the Patching Lemma to extend both u,v to maximal domain functions. We continue to use the notation u,v for these extended functions. Note that $[g,u](a) = -A_2$ and $[g,v](a) = A_1$. Since not both of A_1, A_2 are zero, g in $D_{\min}$ would contradict part (a) of Lemma 10.4.1. $\square$

REMARK 10.4.7. Note that no sign condition is assumed on the leading coefficient p in Theorem 10.4.1. Also note that we assumed only that the left endpoint a is LC, it may be oscillatory or nonoscillatory even when p is positive. Note also that the minimal operator $S_{\min}$ is automatically unbounded above and below if p changes sign [**474**].

Since r is arbitrary in Theorem 10.4.5 and the normalization $[u,v](t) = 1$, $t \in (a,c)$ is satisfied by many pairs u,v there is a wide choice of function u,v in (10.4.28). Moreover, we will see below that u,v need not be solutions but can be maximal domain functions. This leads to the next definition.

DEFINITION 10.4.2. A pair of real-valued functions $\{f,g\}$ is called a (BC) basis at a if $f,g \in D_{\max}$ and satisfy $[g,f](a) = 1$.

For a fixed BC bases u, v at a by varying the constants A_1, A_2 over the real numbers but with at least one of them not zero we obtain all self-adjoint domains from (10.4.28).

But keeping A_1, A_2 fixed and varying the BC bases u, v changes the self-adjoint domain. The next lemma studies this change.

LEMMA 10.4.7. *Assume a is LC and b is LP. Suppose u, v are a BC bases at a, and u_1, v_1 are another BC bases at a. Suppose (10.4.28) holds. Then*

$$D(S) = \{y \in D_{\max} : C_1[y, u_1](a) + C_2[y, v_1](a) = 0,$$
$$C_1 = A_1[u, v_1](a) + A_2[v, v_1], \; C_2 = -A_1[u, u_1](a) - A_2[v, u_1](a)\} \quad (10.4.33)$$

A similar result holds if a is LP and b is LC.

PROOF. By the Decomposition Lemma for $y \in D_{\max}$ we have

$$[y, u](a) = [y, u_1](a)[u, v_1](a) - [y, v_1](a)[u, u_1](a),$$
$$[y, v](a) = [y, u_1](a)[v, v_1](a) - [y, v_1](a)[v, u_1](a).$$

Hence

$$A_1[y, u](a) + A_2[y, v](a) = A_1\{[y, u_1](a)[u, v_1](a) - [y, v_1](a)[u, u_1](a)\}$$
$$+ A_2\{[y, u_1](a)[v, v_1](a) - [y, v_1](a)[v, u_1](a)\}$$
$$= \{A_1[u, v_1](a) + A_2[v, v_1](a)\}[y, u_1](a)$$
$$- \{A_1[u, u_1](a) + A_2[v, u_1](a)\}[y, v_1](a). \quad (10.4.34)$$

By reversing the roles of u, v and u_1, v_1 we see that $(A_1, A_2) \neq (0, 0)$ if and only if $(C_1, C_2) \neq (0, 0)$. This proves the case when a is LC and b is LP; the proof of the case when b is LC and a is LP is entirely similar. □

4.4. LC Endpoints. In this section we give a representation of the self-adjoint domains when neither endpoint is LP. The results are stated for singular LC endpoints but this includes the case when one or both endpoints are regular. The special case when both endpoints are regular is discussed in Subsection 10.4.1; the case when one endpoint is singular LC and the other regular is discussed in Subsection 10.4.3. We will indicate how the general LC/LC results can be specialized to the case when one or both endpoints are regular.

A detailed proof of the canonical representation of the coupled singular boundary conditions is given here since we don't know of a detailed proof in the literature even though these representations have been used, see [**43**], [**645**]. A characterization of all *real* self-adjoint Sturm-Liouville operators is derived from this canonical representation of the boundary conditions.

We continue to study boundary value problems for the equation

$$My = -(py')' + qy = \lambda wy \text{ on } J = (a, b), \; -\infty \leq a < b \leq \infty, \; \lambda \in \mathbb{C}, \quad (10.4.35)$$

with coefficients which are only locally Lebesgue integrable:

$$1/p, q, w \in L_{loc}(J, \mathbb{R}), \; w > 0 \text{ a.e. on } J. \quad (10.4.36)$$

In this case the equation (10.4.35) may be singular at the endpoints a or b. We remind the reader of a few basic facts, these facts are repeated and highlighted here because they play an important role in this subsection.

Except when specifically stated otherwise:

throughout this subsection we assume that each endpoint is LC. (10.4.37)

Recall that

$$D_{\max} = \{f \in H = L^2(J,w) : f, pf' \in AC_{loc}(J),\ w^{-1}Mf \in H\}. \quad (10.4.38)$$

Of critical importance to the description self-adjoint boundary conditions is the Lagrange sesquilinear form given by

$$[f,g] = fp\bar{g}' - \bar{g}pf',\ (f,g \in D_{\max}). \quad (10.4.39)$$

Observe that Green's formula:

$$\int_\alpha^\beta \{\bar{g}Mf - f\overline{Mg}\} = [f,g](\beta) - [f,g](\alpha),\ (f,g \in D_{\max};\ \alpha,\beta \in J), \quad (10.4.40)$$

holds and that it follows from (10.4.40) that the limits

$$\lim_{\beta \to b^-} [f,g](\beta)\ ;\ \lim_{\alpha \to a^+} [f,g](\alpha) \quad (10.4.41)$$

exist and are finite for all $f,g \in D_{\max}$, and in particular, for all solutions y of (10.4.35) for any $\lambda \in \mathbb{C}$.

The representation of the self-adjoint boundary conditions depends on a 'BC basis' at each endpoint. In general these bases are different at the two endpoints. Next we restate and expand Definition 10.4.2.

DEFINITION 10.4.3. *A pair of real-valued functions $\{f,g\}$ is called a (BC) basis at a if $f,g \in D_{\max}$ and satisfy $[g,f](a) = 1$. Similarly a pair of real-valued functions $\{h,k\}$ is called a BC basis at b if h,k are in $D_{\max}$ and satisfy $[k,h](b) = 1$. A pair of real-valued functions $\{f,g\}$, $f,g \in D_{\max}$ is called a (BC) basis on $J = (a,b)$ if it is a BC basis at a and at b.*

We remark that BC bases exist: just take real-valued linearly independent solutions of (10.4.35) for any particular real value of λ and normalize their Wronskian to be 1 to get a basis on J. Or, more generally, take real-valued linearly independent solutions of (10.4.35) for some real $\lambda = \lambda_a$ on some interval (a,c), $c \in J$ and normalize their Wronskian to be 1; then take real-valued linearly independent solutions of (10.4.35) for some real $\lambda = \lambda_b$ on some interval (d,b), $d \in J$ and normalize their Wronskian to be 1, then patch these together using the Patching Lemma 10.4.1 to obtain maximal domain functions. But note that, while this construction provides a plethora of BC bases in terms of solutions, such bases, in general, need not be solutions near the endpoints. For example f,g might be constructed from the first term of the asymptotic expansion of solutions when solutions are not known in closed form for any λ, see Examples 4 and 8 of Chapter 14 for an illustration.

Let $\{f,g\}$ be a BC basis at a and $\{h,k\}$ a BC basis at b. For matrices $A, B \in M_2(\mathbb{C})$ we consider the two point boundary condition

$$A \begin{pmatrix} [y,f](a) \\ [y,g](a) \end{pmatrix} + B \begin{pmatrix} [y,h](b) \\ [y,k](b) \end{pmatrix} = \begin{pmatrix} 0 \\ 0 \end{pmatrix}. \quad (10.4.42)$$

Note that the boundary conditions (10.4.42) are well defined since the Lagrange form exists for any maximal domain functions, see (10.4.41). Hence for any complex number λ it makes sense to ask if (10.4.35) has a nontrivial solution y satisfying the boundary conditions (10.4.42), e.g. if λ is an eigenvalue of (10.4.35), (10.4.42).

Since there are many BC bases, it is natural to ask how the boundary conditions change when the BC basis is changed. The next result is the 'change of BC bases theorem', it describes how the BC matrices A, B change when the basis is changed.

THEOREM 10.4.6 (Change of BC bases theorem). *Let $\{f, g\}$ be a BC basis at a and $\{h, k\}$ a BC basis at b. Assume that $\{f_1, g_1\}$ and $\{h_1, k_1\}$ are other BC bases at a and b, respectively. Let*

$$A_1 = AC,\ C = \begin{pmatrix} -[f, g_1](a) & [f, f_1](a) \\ -[g, g_1](a) & [g, f_1](a) \end{pmatrix},$$

$$B_1 = BD,\ D = \begin{pmatrix} -[h, k_1](b) & [h, h_1](b) \\ -[k, k_1](b) & [k, h_1](b) \end{pmatrix}. \tag{10.4.43}$$

Then (10.4.42) is equivalent to

$$A_1 \begin{pmatrix} [y, f_1](a) \\ [y, g_1](a) \end{pmatrix} + B_1 \begin{pmatrix} [y, h_1](b) \\ [y, k_1](b) \end{pmatrix} = \begin{pmatrix} 0 \\ 0 \end{pmatrix}. \tag{10.4.44}$$

PROOF. This follows from a direct computation using Lemma 10.4.2. See also Theorem 3.3 in [384] □

Since both endpoints may be singular we need an extension of Lemma 10.4.6. An important special case of this extension has been established by Krall and Zettl, see Lemma 2 in [400]. We state this extension as

PROPOSITION 10.4.1. *Let (10.4.35), (10.4.36) hold and assume each endpoint is LC. Let $\{u, v\}$ be a BC basis on J. Given any $\alpha, \beta, \gamma, \delta \in \mathbb{C}$ there exists a $g \in D_{\max}$ such that*

$$[g, u](a) = \alpha,\ [g, v](a) = \beta,\ [g, u](b) = \gamma,\ [g, v](b) = \delta.$$

PROOF. First we establish the case when $\gamma = 0 = \delta$ and u, v are solutions of (10.4.35) on some interval (a, c), $c \in J$ for some $\lambda \in \mathbb{R}$. Let $g_1 = \beta u - \alpha v$ on $(a, c]$. Then

$$[g_1, u] = \beta[u, u] - \alpha[v, u] = \alpha \text{ on } (a, c] \text{ and hence also at } t = a;$$
$$[g_1, v] = \beta[u, v] - \alpha[v, v] = \beta \text{ on } (a, c] \text{ and hence also at } t = a.$$

Choose d such that $c < d < b$. Now extend g_1 to the whole interval J such that the extended g_1 and (pg_1') are absolutely continuous on $[c, d]$, g_1 is in $D_{\max}(J)$ and $g_1 = 0$, $(pg_1') = 0$ both on $[d, b)$. Then the conclusion holds with $g = g_1$ and $\gamma = 0 = \delta$. Similarly we construct a g_2 such that the conclusion holds for $g = g_2$ and $\alpha = 0 = \beta$. Setting $g = g_1 + g_2$ we obtain the conclusion for the case when u, v are solutions in a neighborhood of each endpoint. The general case then follows from the Bracket Decomposition Lemma 10.4.2. □

An important special case of the next result was established by Krall and Zettl in [400] and its proof uses the same basic strategy as the proof of Theorem 2 in [400].

PROPOSITION 10.4.2. *Let (10.4.35), (10.4.36) hold. Let $\{u, v\}$ be a BC basis on J. Let the matrices $A, B \in M_2(\mathbb{C})$ satisfy (10.4.11), (10.4.12). Then $D(S) = \{y \in D_{\max} : (10.4.42) \text{ holds}\}$ is a self-adjoint domain and all self-adjoint domains are generated this way.*

PROOF. Although the basic strategy of the proof is similar to that of the regular case given by Theorem 10.4.2 it is more technical and uses Proposition 10.4.1. We outline the proof here. Since, in general, the endpoints are singular, equations (10.4.14) and (10.4.15) are replaced by the assignments:

$$\overline{a_{11}} = -[g_1, v](a),\ \overline{a_{12}} = [g_1, u](a),\ \overline{b_{11}} = [g_1, v](a),\ \overline{b_{12}} = -[g_1, u](a);$$
$$\overline{a_{21}} = -[g_2, v](a),\ \overline{a_{22}} = [g_2, u](a),\ \overline{b_{21}} = [g_2, v](a),\ \overline{b_{22}} = -[g_2, u](a).$$

These assignments can be made by Proposition 10.4.1 with $g_1, g_2 \in D_{\max}$. With these assignments made we proceed as in the proof of the regular case given by Theorem 10.4.2 and show that condition (3) of case (iii) of Theorem 10.4.1 for $d = 2$ is equivalent with the boundary conditions (10.4.42). Condition (1) is equivalent with the linear independence of the two equations in (10.4.42) and condition (2) is equivalent with (10.4.12). □

The next result is the basic existence theorem for the eigenvalues.

PROPOSITION 10.4.3. *Let (10.4.35), (10.4.36) hold. Let $\{f, g\}$ be a BC basis at a and $\{h, k\}$ a BC basis at b. Let the matrices $A, B \in M_2(\mathbb{C})$ satisfy (10.4.11), (10.4.12). Then the spectrum of the SLP (10.4.35), (10.4.42) is real, discrete, and consists of an infinite but countable number of eigenvalues.*

PROOF. By Proposition 10.4.2 the problem, represented by the operator S, is self-adjoint and hence its spectrum is real. It is well known [**600**], [**487**] that, since neither endpoint is LP, the spectrum is discrete and consists of an infinite but countable number of eigenvalues. □

Next we show that the SLP (10.4.35), (10.4.42) can be represented as a boundary value problem for a regular system.

The next theorem is 'almost' a special case of Theorem 9.3.1. The word almost is used here because the boundary conditions (9.3.2) are given in terms of 'boundary condition' functions which are solutions over the entire interval J, whereas the boundary conditions (10.4.42) are given in terms of BC bases which may be solutions only near the endpoints. (And they may be different pairs of solutions near the two endpoints for the same λ or for different values of λ.) Both proofs are based on Lemma 10.4.2, the Bracket Decomposition Lemma, and the system regularization discussed in Section 7.2.

THEOREM 10.4.7. *Let (10.4.36), (10.4.37) hold, let $r \in \mathbb{R}$ and let u, v be real-valued linearly independent solutions of (10.4.35) with $\lambda = r$, normalized to make their Wronskian $[v, u] = 1$. Let*

$$U = \begin{bmatrix} v & u \\ pv' & pu' \end{bmatrix},\ G = U^{-1}WU = \begin{bmatrix} -v\,u\,w & -u^2\,w \\ v^2\,w & v\,u\,w \end{bmatrix}. \tag{10.4.45}$$

For $\lambda \in \mathbb{C}$, consider the first order system

$$Z' = (r - \lambda)GZ \quad \text{on } J. \tag{10.4.46}$$

Then

  (1) *The system (10.4.46) is regular and consequently $Z(a, \lambda)$ and $Z(b, \lambda)$ exist as finite limits.*

(2) For each $\lambda \in \mathbb{C}$, $Z(t,\lambda)$ is a (vector or matrix) solution of (10.4.46) if and only if
$$Y(t,\lambda) = U(t)Z(t,\lambda), \ a < t < b, \tag{10.4.47}$$
is a (vector or matrix) solution of equation (7.2.3).

(3) Let (10.4.47) hold with $Y = \begin{pmatrix} y \\ py' \end{pmatrix}$, $Z = \begin{pmatrix} z_1 \\ z_2 \end{pmatrix}$. Then y is a solution of (10.4.35) which satisfies the (singular) boundary condition (10.4.42) if and only if Z satisfies (10.4.46) and the regular BC.
$$A_r Z(a) + B_r Z(b) = 0, \tag{10.4.48}$$
where
$$A_r = -AC(a), \ C(a) = \begin{pmatrix} [f,v](a) & [f,u](a) \\ [g,v](a) & [g,u](a) \end{pmatrix},$$
$$B_r = -BD(b), \ D(b) = \begin{pmatrix} [h,v](b) & [h,u](b) \\ [k,v](b) & [k,u](b) \end{pmatrix}. \tag{10.4.49}$$

PROOF. A direct computation establishes (1). The Schwartz inequality and the hypothesis (10.4.47) imply that each component of G is in $L^1(J)$, proving (2). Hence the BC (10.4.48) is well-defined. To prove part (3), let $Z = \begin{pmatrix} z_1 \\ z_2 \end{pmatrix}$ be a vector solution of (10.4.46), apply Cramer's rule to (10.4.47) to get
$$z_1(t,\lambda) = [y,u](t,\lambda), \ z_2(t,\lambda) = -[y,v](t), a \le t \le b. \tag{10.4.50}$$
The equivalence of (10.4.42) with (10.4.48) then follows from the bracket decomposition Lemma 10.4.2 and (10.4.49). □

For a fixed BC basis $\{f,g\}$ at a the Lagrange brackets $[y,f](a,\lambda)$, $[y,g](a,\lambda)$ exist as finite limits for any solution y of (10.4.35) for any λ by (10.4.41). Can these brackets assume arbitrary values and is their dependence on λ analytic?

LEMMA 10.4.8. Let (10.4.35), (10.4.36) hold. Let $\{f,g\}$ be BC basis at a. Let $c,d \in \mathbb{C}$. For any $\lambda \in \mathbb{C}$ there exists a unique solution $y = y(\cdot, \lambda)$ of (10.4.35) such that
$$[y,f](a,\lambda) = c \text{ and } [y,g](a,\lambda) = d. \tag{10.4.51}$$
Furthermore the brackets $[y,f](t,\lambda)$ and $[y,g](t,\lambda)$ exist and are entire functions of λ for any fixed t, $a \le t \le b$. There is a similar result for the endpoint b.

PROOF. From the bracket decomposition Lemma 10.4.2 we get, for any $y \in D_{\max}$, and, in particular, for any solution y of (10.4.35) for any $\lambda \in \mathbb{C}$
$$\begin{pmatrix} [y,f](a,\lambda) \\ [y,g](a,\lambda) \end{pmatrix} = \begin{pmatrix} [f,v](a) & [f,u](a) \\ [g,v](a) & [g,u](a) \end{pmatrix} \begin{pmatrix} [y,u](a,\lambda) \\ [y,v](a,\lambda) \end{pmatrix} = C(a) \begin{pmatrix} [y,u](a,\lambda) \\ [y,v](a,\lambda) \end{pmatrix} \tag{10.4.52}$$
and $\det C(a) = 1$. Note that f, g, u, v and hence $C(a)$ do not depend on λ. From Theorem 10.4.7 and the theory of regular systems, it follows that for any $c, d \in \mathbb{C}$, $z_1(a,\lambda) = c$, $z_2(a,\lambda) = d$ determines a unique solution $Z = \begin{pmatrix} z_1 \\ z_2 \end{pmatrix}$ of (10.4.46) on (a,b) for any $\lambda \in \mathbb{C}$ and $z_1(b,\lambda)$, $z_2(b,\lambda)$ are entire functions of λ. Hence from (10.4.50) we may conclude that for any $c, d \in \mathbb{C}$, the 'singular initial condition'

$[y, u](a, \lambda) = c$, $[y, v](a, \lambda) = d$ determines a unique solution $Y = \begin{pmatrix} y \\ py' \end{pmatrix}$ on (a, b) for any $\lambda \in \mathbb{C}$ and $[y, u](b, \lambda)$, $[y, v](b, \lambda)$ are entire functions of λ. It follows from (10.4.52) that the same result holds for $[y, f](a, \lambda) = c$, $[y, g](a, \lambda)$. There is a similar argument for 'initial conditions' at the endpoint b. This concludes the proof. □

REMARK 10.4.8. The unique solution $y(\cdot, \lambda)$ of (10.4.35) satisfying (10.4.52) is defined on the open interval (a, b) but not, in general, at the endpoints a, b. Thus $[y, f](a, \lambda)$ may be viewed as a substitute for $y(a, \lambda)$; similarly $[y, g](a, \lambda)$ may be viewed as a replacement for $(py')(a, \lambda)$. Note that we are using the notation $[y, f](a, \lambda)$ for $[y(\cdot, \lambda), f](a)$, etc. If a is regular, then f, g can be chosen so that $[y, f](a, \lambda) = y(a, \lambda)$ and $[y, g](a, \lambda) = (py')(a, \lambda)$. Similar remarks apply at the endpoint b. With such a choice the results of this subsection can be specialized to the case when one endpoint is regular and the other singular or when both are regular.

THEOREM 10.4.8. *Let the hypotheses and notation of Proposition 10.4.1 hold. Let $\psi_1 = \psi_1(\cdot, \lambda)$, $\psi_2 = \psi_2(\cdot, \lambda)$ be the unique solutions of (10.4.35) satisfying, for each $\lambda \in \mathbb{C}$,*

$$[\psi_1, f](a, \lambda) = 1 \text{ and } [\psi_1, g](a, \lambda) = 0; \ [\psi_2, f](a, \lambda) = 0 \text{ and } [\psi_2, g](a, \lambda) = 1. \quad (10.4.53)$$

Define for all $\lambda \in \mathbb{C}$,

$$\delta(\lambda) = \det\left(A + B \begin{pmatrix} [\psi_1, h] & [\psi_2, h] \\ [\psi_1, k] & [\psi_2, k] \end{pmatrix}\right)(b, \lambda). \quad (10.4.54)$$

Then δ is an entire function, and λ is an eigenvalue of the SLP (10.4.35), (10.4.42) if and only if $\delta(\lambda) = 0$.

PROOF. The existence and uniqueness of solutions ψ_1, ψ_2 determined by the singular 'initial conditions' (10.4.53) and the entire dependence of δ on λ follows from Lemma 10.4.8, see (10.4.51). Let $y = c\psi_1 + d\psi_2$ and consider

$$A \begin{pmatrix} [y, f](a, \lambda) \\ [y, g](a, \lambda) \end{pmatrix} + B \begin{pmatrix} [y, h](b, \lambda) \\ [y, k](b, \lambda) \end{pmatrix}$$
$$= A \begin{pmatrix} [c\psi_1 + d\psi_2, f](a, \lambda) \\ [c\psi_1 + d\psi_2, g](a, \lambda) \end{pmatrix} + B \begin{pmatrix} [c\psi_1 + d\psi_2, h](b, \lambda) \\ [c\psi_1 + d\psi_2, k](b, \lambda) \end{pmatrix}$$
$$= \left(A \begin{pmatrix} [\psi_1, f](a, \lambda), [\psi_2, f](a, \lambda) \\ [\psi_1, g](a, \lambda), [\psi_2, g](a, \lambda) \end{pmatrix} + B \begin{pmatrix} [\psi_1, f](b, \lambda), [\psi_2, f](b, \lambda) \\ [\psi_1, g](b, \lambda), [\psi_2, g](b, \lambda) \end{pmatrix}\right) \begin{pmatrix} c \\ d \end{pmatrix}$$
$$= \left(A + B \begin{pmatrix} [\psi_1, f](b, \lambda), [\psi_2, f](b, \lambda) \\ [\psi_1, g](b, \lambda), [\psi_2, g](b, \lambda) \end{pmatrix}\right) \begin{pmatrix} c \\ d \end{pmatrix} = \begin{pmatrix} 0 \\ 0 \end{pmatrix}.$$

This algebraic system has a nontrivial solution for c, d if and only if $\delta(\lambda) = 0$. □

DEFINITION 10.4.4. *The function $\delta(\lambda)$ given by (10.4.54) is a characteristic function of the singular SLP (10.4.35), (10.4.42). The algebraic multiplicity of an eigenvalue λ is the order of it as a root of the characteristic equation $\delta(\lambda) = 0$. We write $\delta(\lambda) = \delta(\lambda, A, B, f, g, h, k)$ to indicate the dependence of δ on the BC matrices A, B and on the BC basis $\{f, g\}$ at a and $\{h, k\}$ at b. Thus we refer to $\delta(\lambda)$ defined by (10.4.54) as a singular characteristic function.*

4. OPERATOR THEORY AND SELF-ADJOINT BOUNDARY CONDITIONS

REMARK 10.4.9. It is shown in Section 10.11 that the algebraic and geometric multiplicities of any eigenvalue are the same for any regular or singular self-adjoint SLP with regular or LC endpoints. From this it follows that the algebraic multiplicity does not depend on the representation of the self-adjoint domain in terms of (A, B, f, g, h, k). However note that when the boundary bases $\{f, g\}, \{h, k\}$ are changed, the BC matrices A, B must be changed accordingly.

Next we give a representation of the characteristic function $\delta(\lambda)$ of the singular problem (10.4.35), (10.4.42) in terms of a characteristic function of the regular system (10.4.46), (10.4.48). See identity (10.4.57) below.

REMARK 10.4.10. Note that the system (10.4.46) does not reduce to a scalar Sturm-Liouville equation (10.4.35) because the coefficient matrix G does not have the same form as P in (7.2.3). In particular g_{11} and g_{22} are not the zero function. Nevertheless, it follows from Theorem 10.4.7 that regular boundary value problems for the system (10.4.46) are equivalent to singular problems for the scalar Sturm-Liouville equation (10.4.35). Equivalent in the sense that they have the same eigenvalues and their eigenfunctions are related as shown by Theorem 10.4.7.

DEFINITION 10.4.5. Let the hypotheses and notation of Theorem 10.4.8 hold. For each $\lambda \in \mathbb{C}$, and every s, $a \leq s \leq b$, let $\Phi_r(t, s, \lambda)$ be the fundamental matrix solution of (10.4.46) determined by the initial condition

$$\Phi_r(s, s, \lambda) = I. \tag{10.4.55}$$

Thus, $\Phi_r(t, s, \lambda)$ is defined for all t, $a \leq t, s \leq b$. For any $A, B \in M_2(\mathbb{C})$, let A_r, B_r be given by (10.4.49) and define

$$\delta_r(\lambda) = \delta_r(\lambda, A_r, B_r) = \det[A_r + B_r \, \Phi_r(b, a, \lambda)], \; \lambda \in \mathbb{C}. \tag{10.4.56}$$

This function $\delta_r(\lambda)$ is called a characteristic function of the system boundary value problem (10.4.46), (10.4.48); we write $\delta_r(\lambda, A_r, B_r)$ to indicate the dependence of δ on A, B and r.

REMARK 10.4.11. The function $\delta_r(\lambda)$ given by (10.4.56) is not to be confused with $\delta(\lambda)$ given by (10.4.54). The relationships between these functions will be established below.

LEMMA 10.4.9. Let the hypotheses and notation of Theorem 10.4.8 hold and let $\delta_r(\lambda)$ be defined by (10.4.56). Then

(1) $\delta_r(\lambda)$ is an entire function of λ.
(2) λ is an eigenvalue of the regular boundary value problem (10.4.46), (10.4.48) if and only if $\delta_r(\lambda) = 0$.

PROOF. Part (1) follows from the well-known theory of regular boundary value problems and (2) follows from a direct computation. □

THEOREM 10.4.9. Let the notation and hypotheses of Theorems 10.4.8 and 10.4.7 hold; let matrices $A, B \in M_2(\mathbb{C})$ satisfy (10.4.11), (10.4.12), and let A_r, B_r given by (10.4.49). Let $\delta(\lambda) = \delta(\lambda, A, B)$ be given by (10.4.54) and let $\delta_r(\lambda) = \delta_r(\lambda, A_r, B_r)$ be given by (10.4.56). Then A_r, B_r satisfy (10.4.11), (10.4.12) and

$$\delta(\lambda, A, B) = -\delta_r(\lambda, A_r, B_r). \tag{10.4.57}$$

PROOF.

$$\delta(\lambda) = \det\left(A + B\begin{pmatrix} [\psi_1,h] & [\psi_2,h] \\ [\psi_1,k] & [\psi_2,k] \end{pmatrix}(b,\lambda)\right)$$

$$= \det\left(A\begin{pmatrix} [\psi_1,f] & [\psi_2,f] \\ [\psi_1,g] & [\psi_2,g] \end{pmatrix}(a,\lambda) + B\begin{pmatrix} [\psi_1,h] & [\psi_2,h] \\ [\psi_1,k] & [\psi_2,k] \end{pmatrix}(b,\lambda)\right)$$

$$= \det(A\begin{pmatrix} [f,v] & [f,u] \\ [g,v] & [g,u] \end{pmatrix}(a)\begin{pmatrix} [\psi_1,u] & [\psi_2,u] \\ [\psi_1,v] & [\psi_2,v] \end{pmatrix}(a,\lambda)$$
$$+ B\begin{pmatrix} [h,v] & [h,u] \\ [k,v] & [k,u] \end{pmatrix}(b)\begin{pmatrix} [\psi_1,u] & [\psi_2,u] \\ [\psi_1,v] & [\psi_2,v] \end{pmatrix}(b,\lambda)$$

$$= \det\begin{pmatrix} A\,C(a)\begin{pmatrix} [\psi_1,u] & [\psi_2,u] \\ [\psi_1,v] & [\psi_2,v] \end{pmatrix}(a,\lambda) \\ +B\,C(b)\begin{pmatrix} [\psi_1,u] & [\psi_2,u] \\ [\psi_1,v] & [\psi_2,v] \end{pmatrix}(b,\lambda) \end{pmatrix}. \qquad (10.4.58)$$

Let

$$D = \begin{pmatrix} [\psi_1,u] & [\psi_2,u] \\ [\psi_1,v] & [\psi_2,v] \end{pmatrix}(a,\lambda)$$

and let Z be the solution of (10.4.46) determined by the initial condition $Z(a,\lambda) = D$ and note that $Z(t,\lambda) = \Phi(t,a,\lambda)\,D$, $a \le t \le b$. Hence we get from (10.4.58) and (10.4.52)

$$\delta(\lambda, A, B) = -\det(A_r\Phi_r(a,a,\lambda)\,D + B_r\Phi_r(b,a,\lambda)\,D)$$
$$= -\det(A_r\,\Phi_r(a,a,\lambda) + B_r\,\Phi_r(b,a,\lambda))$$
$$= -\delta_r(\lambda, A_r, B_r).$$

In the penultimate step we used $\det D = 1$ which follows from Lemma 10.4.2.

It remains to show that A_r, B_r satisfy (10.4.11), (10.4.12). From Lemma 10.4.2 it follows that $\det C(a) = 1 = \det C(b)$ and that

$$C(a)EC^*(a) = E = C(b)EC^*(b).$$

Hence the self-adjointness properties (10.4.11), (10.4.12) are preserved:

$$rank(A|B) = rank(A_r|B_r) \text{ and } A_r E A_r^* = B_r E B_r^*.$$

This completes the proof of Theorem 10.4.9. □

REMARK 10.4.12. We comment on the remarkable identity (10.4.57). Note that the left hand side is independent of r and of the fundamental matrix U associated with r. But the matrices A_r, B_r depend on U and hence on r as we have indicated with the notation. Thus as r, and possibly U, are changed the identity (10.4.57) holds provided the matrices A_r, B_r are chosen according to (10.4.11), (10.4.12). The self-adjointness conditions (10.4.11), (10.4.12) are preserved for A_r, B_r. This identity plays a critical role in the proof of the equivalence between the geometric and algebraic multiplicities of the eigenvalues of singular problems. See Section 10.11 below.

4.5. Canonical Boundary Conditions. In this subsection we give a canonical form of the coupled LC boundary conditions; this is then used to define an alternate version of the characteristic function. This alternate version is a special case of the version used in Theorem 9.3.3 (see also [41]) and parallels the version used in Floquet theory in the regular case.

4. OPERATOR THEORY AND SELF-ADJOINT BOUNDARY CONDITIONS

Just as in the regular case the singular self-adjoint boundary conditions (10.4.42) fall into two disjoint classes: the separated conditions and the coupled ones. And there is a canonical representation for each of these classes analogous to the regular case. The separated conditions

$$A_1[y,f](a) + A_2[y,g](a) = 0, \ A_1, A_2 \in \mathbb{R}, \ (A_1, A_2) \neq (0,0);$$
$$B_1[y,h](b) + B_2[y,k](b) = 0, \ B_1, B_2 \in \mathbb{R}, \ (B_1, B_2) \neq (0,0). \tag{10.4.59}$$

have the canonical representation

$$\cos(\alpha)[y,f](a) - \sin(\alpha)[y,g](a) = 0, \ 0 \leq \alpha < \pi;$$
$$\cos(\beta)[y,h](b) - \sin(\beta)[y,k](b) = 0, \ 0 < \beta \leq \pi. \tag{10.4.60}$$

And the canonical representation of the coupled BC is given by:

$$\begin{pmatrix} [y,h](b) \\ [y,k](b) \end{pmatrix} = e^{i\gamma} K \begin{pmatrix} [y,f](a) \\ [y,g](a) \end{pmatrix}, \ -\pi < \gamma \leq \pi, \ K = (k_{ij}), \ k_{ij} \in \mathbb{R}, \det K = 1. \tag{10.4.61}$$

As in the regular case, it follows directly from these representations of the BC that S is a *real* self-adjoint extension of the minimal operator $S_{\min}$ if and only if it is determined by either the separated BC or the coupled BC with $\gamma = 0$. (Note that $\gamma = \pi$ corresponds to replacing K by $-K$.)

How does the representation of the coupled BC change when the BC basis f, g at a changes?

LEMMA 10.4.10. *Let $\{f, g\}$ be a BC basis at a and $\{h, k\}$ a BC basis at b. If f_1, g_1 is another boundary condition basis at a, then (10.4.61) is equivalent with*

$$\begin{pmatrix} [y,h](b) \\ [y,k](b) \end{pmatrix} = e^{i\gamma} K \begin{pmatrix} -[f,g_1](a) & [f,f_1](a) \\ -[g,g_1](a) & [g,f_1](a) \end{pmatrix} \begin{pmatrix} [y,f_1](a) \\ [y,g_1](a) \end{pmatrix}. \tag{10.4.62}$$

There is a similar result when the BC basis at b changes.

PROOF. This follows directly from Theorem 10.4.6. Let

$$C = \begin{pmatrix} -[f,g_1](a) & [f,f_1](a) \\ -[g,g_1](a) & [g,f_1](a) \end{pmatrix} \tag{10.4.63}$$

and note that Lemma 10.4.2 and the normalization of f, g imply $\det C = 1$ so that for $K_1 = KC$ we have $\det K_1 = 1$ consistent with (10.4.61). □

THEOREM 10.4.10. *Let the notation and the hypotheses of Theorem 10.4.7 hold. Assume that $B = -I$ and $A = e^{i\gamma} K$, $-\pi < \gamma \leq \pi$, $K \in SL(2, \mathbb{R})$. Let $K = (k_{ij})$ and for each $\lambda \in \mathbb{C}$, let $\Phi(t, s, \lambda)$ be the fundamental matrix solution of (10.4.46) determined by the initial condition $\Phi(s, s, \lambda) = I$, $a \leq s \leq b$.*
Define

$$D(\lambda, K) = k_{11}\phi_{22}(b,a,\lambda) + k_{22}\phi_{11}(b,a,\lambda) - k_{12}\phi_{21}(b,a,\lambda) - k_{21}\phi_{12}(b,a,\lambda). \tag{10.4.64}$$

Then λ is an eigenvalue of (10.4.35) with boundary condition

$$\begin{pmatrix} [y,u](b) \\ [y,v](b) \end{pmatrix} = e^{i\gamma} K \begin{pmatrix} [y,u](a) \\ [y,v](a) \end{pmatrix}, \tag{10.4.65}$$

if and only if

$$D(\lambda, K) = 2\cos(\gamma). \tag{10.4.66}$$

PROOF. Note that $\det \Phi(b, a, \lambda) = 1$ by Abel's Theorem since $trace(G) = 0$. Expanding the determinant in the definition of $\Delta(\lambda)$ and letting $\Phi = (\phi_{ij})$ we get

$$\begin{aligned}
\delta(\lambda) &= \det[e^{i\gamma} K - \Phi(b, a, \lambda)] \\
&= \begin{vmatrix} e^{i\gamma} k_{11} - \phi_{11} & e^{i\gamma} k_{12} - \phi_{12} \\ e^{i\gamma} k_{21} - \phi_{21} & e^{i\gamma} k_{22} - \phi_{22} \end{vmatrix} (b, a, \lambda) \\
&= [(e^{i\gamma} k_{11} - \phi_{11})(e^{i\gamma} k_{22} - \phi_{22}) - (e^{i\gamma} k_{12} - \phi_{12})(e^{i\gamma} k_{21} - \phi_{21})](b, a, \lambda) \\
&= [e^{2i\gamma}[k_{11} k_{22} - k_{12} k_{21}] - e^{i\gamma}[k_{11} \phi_{22} + k_{22} \phi_{11} - k_{12} \phi_{21} - k_{21} \phi_{12}] \\
&\quad + \phi_{11} \phi_{22} - \phi_{12} \phi_{21}](b, a, \lambda) \\
&= e^{2i\gamma} - e^{i\gamma} D(\lambda, K) + 1.
\end{aligned}$$

Now dividing by $e^{i\gamma}$ yields

$$e^{-i\gamma} \delta(\lambda) = e^{i\gamma} + e^{-i\gamma} - D(\lambda, K).$$

Hence $\delta(\lambda) = 0$ if and only if $D(\lambda, K) = e^{i\gamma} + e^{-i\gamma} = 2\cos(\gamma)$. □

REMARK 10.4.13. The characterization of the eigenvalues for coupled BC given by (10.4.66) was proved in Bailey, Everitt and Zettl [**41**] for the case when $p > 0$ and used by Everitt and Nasri-Roudsari [**229**] as the definition of algebraic multiplicity. Theorem 10.4.10 shows that our definition of algebraic multiplicity is equivalent with the one given in [**41**] and used in [**229**].

Now that we know all the self-adjoint realizations S of (M, w) on J we summarize some of the basic general properties their spectrum $\sigma(S)$. Let $\sigma_d(S)$ denote the discrete spectrum i.e. the set of isolated eigenvalues, if any, of S. Set $\sigma_e = \sigma \setminus \sigma_d$, then σ_e is called the essential spectrum; in some of the literature this is also referred to as the continuous part of the spectrum. Either one, but not both, of σ_d and σ_e may be empty. There may be eigenvalues in σ_e, these are said to be embedded in the essential spectrum.

PROPOSITION 10.4.4. *Let (10.4.1), (10.4.2) hold. Let S be a self-adjoint realization of (M, w) on J. (Here each endpoint may be regular, LP or LC.) Then:*
  (1) *The spectrum of S is a closed subset of the reals $\mathbb{R}$ which is not bounded; it may be bounded from above or below but not both.*
  (2) *$\sigma_e(S) = \sigma_e(S_{\min})$, and so the essential spectrum of all self-adjoint extensions is the same. In particular this means that the essential spectrum depends only on p, q, w not on the boundary condition. The discrete spectrum depends on the boundary condition.*
  (3) *The essential spectrum is also invariant under an $L^1(J, \mathbb{R})$ perturbation of q (See Hinton and Shaw [**336**]).*
  (4) *The operator $S_{\min}$ is bounded below if and only if $\sigma(S)$ is bounded below for each self-adjoint realization S; (But not uniformly for all S).*
  (5) *Each one of the sets $\sigma(S), \sigma_e(S)$ is a closed subset of $\mathbb{R}$. Either one of the two sets $\sigma_d(S), \sigma_e(S)$, but not both, may be empty.*

PROOF. This is well known; see [**600**], [**277**], [**487**]. □

REMARK 10.4.14. We see from this Proposition that the essential spectrum does not depend on the boundary conditions; thus $\sigma_e = \sigma_e(J, p, q, w)$. The eigenvalues do depend on the boundary condition. When $p > 0$ on J, there may be no eigenvalues to the left of the essential spectrum, a finite number of them, or an infinite number of them. In the case of an infinite number of eigenvalues to the left of $\sigma_0 = \inf \sigma_e$ these can have no accumulation point other than, possibly, σ_0 or $-\infty$. There also may be eigenvalues embedded in the continuous spectrum as well as in gaps of the continuous spectrum. The essential spectrum may have no gaps, a finite number of them, or an infinite number of them. A remarkable result of Hartmann [**318**] states that any closed set of real numbers which is not bounded above is the spectrum of a Sturm-Liouville operator. See also Halvorsen [**287**]. In particular the spectrum of an SLP can be a Cantor-like set which is not bounded above. For specific examples of SLP which illustrate some of these features (but not the Cantor-like behavior of the spectrum) the reader is referred to the examples discussed in Chapter 14, the examples chapter, below; many of these examples are part of the SLEIGN2 'package' which can be downloaded from **http://www.math.niu.edu./~zettl/SL2/**.

5. The Friedrichs Extension

If there is one proper self-adjoint extension of the minimal operator then there are an infinite number of such extensions. When the minimal operator is bounded below, Friedrichs singled out one of these which he called "ausgezeichnet" and which has come to be known as "the Friedrichs extension". He constructed a self-adjoint extension of $S_{\min}$ which has the same lower bound as $S_{\min}$. (However, having the same lower bound does not characterize the Friedrichs extension since there are, in general, other self-adjoint extensions which also have the same lower bound. See Proposition 4.8.1 for a characterization of *all* self-adjoint extensions for regular Sturm-Liouville problems.) To get the Friedrichs extension you must use the Friedrichs construction or some equivalent version of it. This construction works for any symmetric, densely defined, and bounded below operator in a Hilbert space. When it is applied to $S_{\min}$ it makes no explicit use of boundary conditions. Thus the question arises: What boundary conditions determine the Friedrichs extension of the minimal operator $S_{\min}$?

DEFINITION 10.5.1 (The Friedrichs Extension). *Suppose S is a densely defined symmetric (but not necessarily closed) operator in a Hilbert space H which is bounded below. Let $D(S_F)$ denote the set of all $y \in D(S^*)$ for which there exists a sequence $\{y_n \in D(S) : n \in N\}$ such that*

(1) $y_n \to y$ in H as $n \to \infty$,
(2) $(S(y_n - y_m), y_n - y_m) \to 0$ as $n, m \to \infty$.

Define the operator S_F by

$$S_F y = S^* y, \text{ for } y \in D(S_F).$$

Then S_F is called the Friedrichs extension of S. According to the well known result of Friedrichs [**255**], *S_F is a self-adjoint operator with the same lower bound as S.*

Friedrichs himself [**256**] addressed the question of which boundary condition determines the Friedrichs extension for regular SLP? For regular SLP the answer in general is the Dirichlet condition; see [**502**], [**352**], [**51**], [**554**]. We take up the singular case next.

THEOREM 10.5.1. *Let (10.4.1), (10.4.2) hold and assume that $p > 0$ on J. If the minimal operator $S_{\min}$ is bounded below with lower bound c, then equation (10.4.1), (10.4.2) is NO at a and at b for any $\lambda < c$. Conversely, if the equation*

(10.4.1), (10.4.2) is NO at a for some real λ_a and at b for some real λ_b, then the minimal operator $S_{\min}$ is bounded below.

PROOF. See Niessen and Zettl [**503**], Corollary 2.1 and Theorem 4.2. □

THEOREM 10.5.2. *Let (10.4.1), (10.4.2) hold and assume that $p > 0$ on J. Suppose (M, w) is in the LC case at a. If $My = \lambda wy$ is NO at a for some real λ, then $My = \lambda w y$ is NO at a for every real λ.*

PROOF. See Theorem 4.1 in Niessen and Zettl [**503**]. For quite a different proof see Section 6.2. □

THEOREM 10.5.3. *Let (10.4.1), (10.4.2) hold and assume that $p > 0$ on J. Let the maximal domain $D_{\max}$, the Lagrange form $[\cdot, \cdot]$ and the expression M be defined as above. Assume that each endpoint is either regular or LCNO. Then the minimal operator is bounded below and thus has a Friedrichs extension S_F. Let u_a be a principal solution at a for some real λ_a and let u_b be a principal solution at b for some real λ_b. Then the domain $D(S_F)$ of S_F is given by*

$$D(S_F) = \{y \in D_{\max} : [y, u_a](a) = 0 = [y, u_b](b)\}. \tag{10.5.1}$$

Note that the Friedrichs extension is independent of the particular principal solution chosen for λ_a and is independent of $\lambda_a \in \mathbb{R}$; similarly at b.

PROOF. See Niessen and Zettl [**503**], Theorems 4.2 and 4.3. □

REMARK 10.5.1. At a regular endpoint, say a, $[y, u_a](a) = 0$ reduces to $y(a) = 0$. Thus the BC (10.5.1) can be viewed as the singular analogue of the regular Dirichlet boundary condition. It is interesting to observe, however, that in the case of the classical Legendre equation on the interval $(-1, 1)$ the Friedrichs condition (10.5.2) reduce to

$$(py')(-1) = 0 = (py')(1).$$

Thus, for this example, the Friedrichs condition looks like the regular Neumann boundary condition even though it is the analogue of the regular Dirichlet condition. See the Legendre example in the Examples Chapter for more details.

It is also shown in [**503**] that the condition (10.5.1) is equivalent to

$$\lim_{t \to a^+} \frac{y(t)}{v_a(t)} = 0 = \lim_{t \to b^-} \frac{y(t)}{v_b(t)}, \tag{10.5.2}$$

where v_a, v_b are arbitrary non-principal solutions at a and b, respectively, for arbitrary real λ_a, λ_b.

6. Nonoscillatory Endpoints

We now study the case when each endpoint is either regular or LCNO. The properties of the eigenvalues and eigenfunctions in this case are similar to the regular case. In fact Niessen and Zettl [**503**] have shown that, given any SLP with endpoints which are either regular or LCNO (and with positive leading coefficient p and positive weight function w) there exists a regular SLP which has exactly the same spectrum as this singular problem and furthermore the eigenfunctions of the given singular problem $\{y_n : n \in N_0\}$ are related to the eigenfunctions $\{z_n : n \in N_0\}$ of the corresponding regular problem by the equation

$$y_n(t) = v(t)\, z_n(t),\ t \in (a, b),\ n \in N_0, \tag{10.6.1}$$

for some function v in the maximal domain of the singular problem which satisfies $v(t) > 0$ for $t \in (a, b)$. Since each z_n is a solution of a regular problem on (a, b) the eigenfunction z_n and its quasi-derivative $z_n^{[1]}$ can be continuously extended to the endpoints by Theorem 1.5.2. Hence the singular behavior at each endpoint is contained in v. In particular, this shows that at each endpoint the singular (e.g. asymptotic) behavior of all eigenfunctions (in fact of solutions for all real λ) is the same. Of course y_n and z_n have exactly the same zeros on (a, b).

THEOREM 10.6.1. *Let (10.4.1), (10.4.2) hold and assume that $p > 0$ on J. Assume that each endpoint is either R or LCNO. Let S be a self-adjoint realization of (M, w) on J.*

(1) *Then the spectrum of S is discrete and bounded below. It consists of a countably infinite sequence $\{\lambda_n : n \in \mathbb{N}_0\}$ of only real eigenvalues tending to $+\infty$ which can be ordered to satisfy*

$$-\infty < \lambda_0 \leq \lambda_1 \leq \lambda_2 \leq \lambda_3 \leq \ldots \to \infty. \tag{10.6.2}$$

Here the eigenvalues are counted according to their multiplicity. Each eigenvalue can have multiplicity one, in which case it is called simple, or two, in which case it is called double. Therefore, in (10.6.2), equality cannot hold for more than two consecutive terms.

(2) *Let y_n be a real eigenfunction of λ_n. Then y_0 has either one or no zero in (a, b), y_n has either $n - 1, n$, or $n + 1$ zeros in (a, b) for $n > 0$. If $[y_n, u_a](a) = 0$, where u_a is a principal solution at a, for some $\lambda_a \in \mathbb{R}$ then y_n has at most n zeros in (a, b).*

(3) *Let $N(\lambda)$ denote the number of eigenvalues of S in the interval $(-\infty, \lambda]$. Then*

$$\frac{N(\lambda)}{\sqrt{\lambda}} \to \frac{1}{\pi} \int_a^b \sqrt{\frac{w(t)}{p(t)}} dt < \infty, \text{ as } \lambda \to \infty, \tag{10.6.3}$$

and

(4)
$$\frac{\lambda_n}{n^2 \pi^2} \to \frac{1}{\left(\int_a^b \sqrt{\frac{w(t)}{p(t)}} dt\right)^2}, \text{ as } n \to \infty. \tag{10.6.4}$$

The finiteness of the integral in (10.6.4) and in (10.6.3) is a consequence of the assumption that each endpoint is R or LCNO (and $p > 0$, $w > 0$).

PROOF. See Theorem 5.2 in [**503**]. □

THEOREM 10.6.2. *Let (10.4.1), (10.4.2) hold. Assume that each endpoint is either R or LCNO. Let u, v be a pair of regularizing functions as in Theorem 8.2.1 and Definition 8.2.1, normalized to satisfy : $[u, v](a) = 1 = [u, v](b)$. Let S with spectrum $\sigma(S)$ be the self-adjoint realization determined by the normalized separated boundary conditions*

$$\cos(\alpha) [y, u](a) + \sin(\alpha) [y, v](a) = 0, \ 0 \leq \alpha < \pi, \tag{10.6.5}$$

$$\cos(\beta) [y, u](b) + \sin(\beta) [y, v](b) = 0, \ 0 < \beta \leq \pi. \tag{10.6.6}$$

Then

(1) *the spectrum of S is discrete: $\sigma(S) = \{\lambda_n : n \in N_0\}$ and each eigenvalue λ_n is simple. These eigenvalues are bounded below and can be ordered to satisfy*

$$-\infty < \lambda_0 < \lambda_1 < \lambda_2 < \lambda_3 < \ldots \to \infty; \qquad (10.6.7)$$

(2) *if y_n is an eigenfunction corresponding to λ_n, then y_n has exactly n zeros in the open interval (a,b) for any $n \in N_0$;*
(3) *$\lambda_n(\alpha, \beta)$ is a continuous function of α, β, $n \in N_0$;*
(4) *$\lambda_n(\alpha, \beta)$ is strictly decreasing in α for each fixed β and strictly increasing in β for each fixed α, $n \in N_0$;*
(5) *with the understanding that terms involving $\lambda_{-1}, \lambda_{-2}$ are not present we have the following inequalities for $\alpha \in [0, \pi)$, $\beta \in (0, \pi]$, and $n \in N_0$*

$$\lambda_{n-2}(0, \pi) < \left\{ \begin{array}{c} \lambda_{n-1}(0, \beta) \\ \lambda_{n-1}(\alpha, \pi) \end{array} \right\} < \lambda_n(\alpha, \beta) \leq \left\{ \begin{array}{c} \lambda_n(0, \beta) \\ \lambda_n(\alpha, \pi) \end{array} \right\} \leq \lambda_n(0, \pi). \quad (10.6.8)$$

PROOF. The regular case is known, see [**600**]. The singular case then follows from the regular case and the transformation employed in [**503**] to "regularize" singular LCNO endpoints. □

THEOREM 10.6.3 (Canonical Coupled Boundary Conditions). *Let (10.4.1), (10.4.2) hold. Assume that each endpoint is either R or LCNO. Let θ, ψ be a pair of regularizing functions as constructed in Theorem 7.2.1 and defined in Definition 7.2.1 with θ being a principal solution at both endpoints and ψ a positive nonprincipal solution at both endpoints. (Recall form Section 7.2 that neither ψ nor θ need be solutions on the whole interval J). Then the canonical form of all coupled self-adjoint boundary conditions is*

$$Y(b) = e^{i\gamma} K Y(a), \quad -\pi < \gamma \leq \pi, \qquad (10.6.9)$$

where

$$Y = \begin{bmatrix} [y, \theta] \\ [y, \psi] \end{bmatrix}, \quad \theta, \psi \in D_{\max}, \quad \theta, \psi \text{ real}, \quad K = (k_{ij}) \in SL_2(\mathbb{R}),$$
$$[\theta, \psi](a) = 1 = [\theta, \psi](b); \quad Y(a) = \lim_{t \to a^+} Y(t), \quad Y(b) = \lim_{t \to b^-} Y(t). \quad (10.6.10)$$

Fix p, q, w, a, b and let $\lambda_n(\gamma, K)$, $n \in N_0$ denote the eigenvalues for BC 10.6.9; when $\gamma = 0$ this notation is abbreviated to $\lambda_n(K)$. Let μ_n and ν_n $n \in N_0$ denote the eigenvalues for the separated boundary conditions

$$[y, \theta](a) = 0, \quad k_{22}[y, \theta](b) - k_{12}[y, \theta](b) = 0; \qquad (10.6.11)$$

$$[y, \psi](a) = 0, \quad k_{21}[y, \psi](b) - k_{11}[y, \psi](b) = 0, \qquad (10.6.12)$$

respectively.

- *Suppose that*

$$k_{12} < 0 \quad \text{and} \quad k_{11} \leq 0. \qquad (10.6.13)$$

Then

(1) *$\lambda_0(K)$ is simple;*
(2) *$\lambda_0(K) < \lambda_0(-K)$;*

(3) *the following inequalities hold for $-\pi < \gamma < 0$ and $0 < \gamma < \pi$:*

$$-\infty < \lambda_0(K) < \lambda_0(\gamma, K) < \lambda_0(-K) \leq \{\mu_0, \nu_0\}$$
$$\leq \lambda_1(-K) < \lambda_1(\gamma, K) < \lambda_1(K) \leq \{\mu_1, \nu_1\}$$
$$\leq \lambda_2(K) < \lambda_2(\gamma, K) < \lambda_2(-K) \leq \{\mu_2, \nu_2\}$$
$$\leq \lambda_3(-K) < \lambda_3(\gamma, K) < \lambda_3(-K) \leq \{\mu_3, \nu_3\} \leq \ldots. \qquad (10.6.14)$$

Furthermore, for $0 < \alpha < \beta < \pi$ we have

$$\lambda_0(\alpha, K) < \lambda_0(\beta, K) < \lambda_1(\beta, K) < \lambda_1(\alpha, K) < \lambda_2(\alpha, K) < \lambda_2(\beta, K)$$
$$< \lambda_3(\beta, K) < \lambda_3(\alpha, K) < \ldots. \qquad (10.6.15)$$

- *Suppose that*
$$k_{12} \leq 0 \quad \text{and} \quad k_{11} > 0. \qquad (10.6.16)$$

Then

(1) $\lambda_0(K)$ *is simple;*
(2) $\lambda_0(K) < \lambda_0(-K)$;
(3) *the following inequalities hold for $-\pi < \gamma < 0$ and $0 < \gamma < \pi$:*

$$-\infty < \nu_0 < \lambda_0(K) < \lambda_0(\gamma, K) < \lambda_0(-K) \leq \{\mu_0, \nu_1\}$$
$$\leq \lambda_1(-K) < \lambda_1(\gamma, K) < \lambda_1(K) \leq \{\mu_1, \nu_2\}$$
$$\leq \lambda_2(K) < \lambda_2(\gamma, K) < \lambda_2(-K) \leq \{\mu_2, \nu_3\}$$
$$\leq \lambda_3(-K) < \lambda_3(\gamma, K) < \lambda_3(-K) \leq \{\mu_3, \nu_4\} \leq \ldots. \qquad (10.6.17)$$

Furthermore, for $0 < \alpha < \beta < \pi$ we have

$$\lambda_0(\alpha, K) < \lambda_0(\beta, K) < \lambda_1(\beta, K) < \lambda_1(\alpha, K) < \lambda_2(\alpha, K) < \lambda_2(\beta, K)$$
$$< \lambda_3(\beta, K) < \lambda_3(\alpha, K) < \ldots. \qquad (10.6.18)$$

PROOF. In Niessen and Zettl [**503**] it is shown for any singular problem with each endpoint either R or LCNO there exists a regular problem which has exactly the same eigenvalues. This result then follows from the regular case; see Section 4.8. The regular case is established by Eastham, Kong, Wu and Zettl in [**145**]. See also Bailey, Everitt and Zettl [**41**], Weidmann [**600**]. □

The next result characterizes the eigenvalues of singular SLP consisting of the canonical form (10.6.9) of the coupled self-adjoint boundary conditions *but for the general equation with real or complex valued coefficients*. Recall that we defined the LC case for complex w in Section 7.3, see also Chapter 9. The next theorem is related to Theorem 9.3.3.

THEOREM 10.6.4. *Consider the SLP consisting of the equation*

$$-(py')' + qy = \lambda wy \text{ on } J = (a, b), \ -\infty \leq a < b \leq \infty, \qquad (10.6.19)$$

with
$$1/p, q, w \in L_{loc}(J, \mathbb{C}). \qquad (10.6.20)$$

Assume each endpoint is either R or LC and the boundary conditions are given by (10.6.9), (10.6.10). For each $\lambda \in \mathbb{C}$ determine unique solutions $u = u(\cdot, \lambda)$, $v = v(\cdot, \lambda)$ by the "singular initial conditions"

$$[u, \theta](a, \lambda) = 0, \ [u, \varphi](a, \lambda) = 1, \ [v, \theta](a, \lambda) = 1, \ [v, \varphi](a, \lambda) = 0. \qquad (10.6.21)$$

Such solutions $u(\cdot, \lambda), v(\cdot, \lambda)$ *exist by Theorem 8.4.1. Let* $K \in SL_2(\mathbb{R})$. *Choose* γ, $-\pi < \gamma \leq \pi$, *and for* $\lambda \in \mathbb{C}$ *define*

$$D(K, \lambda) = k_{11}[u(\cdot, \lambda), \psi](b) + k_{22}[v(\cdot, \lambda), \theta](b)$$
$$- k_{12}[v(\cdot, \lambda), \psi](b) - k_{21}[u(\cdot, \lambda), \theta](b). \quad (10.6.22)$$

Then $\lambda \in \mathbb{C}$ *is an eigenvalue of the BVP (10.6.9), (10.6.10), (10.6.19), (10.6.20) if and only if*

$$D(K, \lambda) = 2\cos(\gamma). \quad (10.6.23)$$

PROOF. See Theorem 3.1 in [**41**]. Although Theorem 10.6.4 is stated there only for the case when p, q, w are real valued and $w > 0$ the proof given there holds with no significant changes when p, q, w are complex valued. □

COROLLARY 10.6.1. *For any* $n \in N_0$ *we have* $\lambda_n(-\gamma, K) = \lambda_n(\gamma, K)$.

PROOF. This follows directly from the characterization (10.6.22) since $\cos(\gamma) = \cos(-\gamma)$. □

THEOREM 10.6.5. *Let (10.4.1), (10.4.2) hold and assume that each endpoint is either R or LCNO. Suppose*

$$v, pv' \in AC_{loc}(J), \; v > 0 \text{ on } J; \quad (10.6.24)$$

let

$$P = v^2 p, \quad W = v^2 w, \quad Q = vMv, \quad (10.6.25)$$

and consider the equation

$$Nz = -(Pz')' + Qz = \lambda W z \quad \text{on} \quad J. \quad (10.6.26)$$

Let $T = T_v$ *be the transformation defined by*

$$Tf = f/v$$

for complex-valued functions f *defined on* J. *Note that* T *is 1-1 and* $T^{-1} = T_{1/v}$. *Then*

(1) y *is a solution of (10.4.1) if and only if* $z = y/v$ *is a solution (for the same* λ*) of (10.6.26).*

(2)
$$P \geq 0, \quad W > 0, \quad P^{-1}, Q, W \in L_{loc}(J, \mathbb{R}). \quad (10.6.27)$$

(3) *For any subinterval* I *of* J,

$$Q, \; W \in L^1(I, \mathbb{R}), \quad \text{if} \quad v \in L^2(I, w). \quad (10.6.28)$$

(4)
$$P^{-1} \in L^1(I, \mathbb{R}), \quad (10.6.29)$$

for $I = (a, c)$, $a < c < b$, *if* v *is a non-principal solution at* a *for some* $\lambda_a \in \mathbb{R}$; *and for* $I = (c, b)$, $a < c < b$, *if* v *is a non-principal solution at* b *for some* $\lambda_b \in \mathbb{R}$.

(5)
$$N = TMT^{-1}. \quad (10.6.30)$$

(6) The equation
$$S(N) = T\, S(M)\, T^{-1} \qquad (10.6.31)$$
defines a 1-1 and onto correspondence between the symmetric extensions $S(N)$ of the minimal operator $S_{\min}(N, W, J)$ of (N, W) on J in the Hilbert space $L^2(J, W)$ and the symmetric extensions of the minimal operator $S_{\min}(M, w, J)$ of (M, w) on J in the Hilbert space $L^2(J, w)$.

(7) The equation (10.6.31) also defines a 1-1 and onto correspondence between the self-adjoint extensions $S(N)$ of the minimal operator $S_{\min}(N, W, J)$ of (N, W) on J in the Hilbert space $L^2(J, W)$ and the self-adjoint extensions of the minimal operator $S_{\min}(M, w, J)$ of (M, w) on J in the Hilbert space $L^2(J, w)$.

(8)
$$S_{\min}(N, W, J) = T\, S_{\min}(M, w, J)\, T^{-1}; \quad S_{\max}(N, W, J) = T\, S_{\max}(M, w, J)\, T^{-1}. \qquad (10.6.32)$$

(9) For any f, g in the domain of $S_{\max}(M, w, J)$, Tf, Tg are in the domain of $S_{\max}(N, W, J)$ and
$$(S_{\max}(N, W, J)Tf, Tg)_W = (S_{\max}(M, w, J)f, g)_w. \qquad (10.6.33)$$

Here $(\cdot, \cdot)_W$, $(\cdot, \cdot)_w$ denote the usual inner products in the Hilbert spaces $L^2(J, W)$, and $L^2(J, w)$, respectively.

(10) For any f, g in the domain of $S_{\max}(M, w, J)$, Tf, Tg are in the domain of $S_{\max}(N, W, J)$ and
$$[Tf, Tg]_N = [f, g]_M. \qquad (10.6.34)$$

Here $[\cdot, \cdot]_N$ and $[\cdot, \cdot]_M$ denote the Lagrange forms of N and M, respectively.

(11) Let u, v be a pair of regularizing functions of (M, w) on J, normalized to satisfy
$$[u, v](a) = -1 = [u, v](b). \qquad (10.6.35)$$
Let $S(M, w, A, B)$ denote the self-adjoint operator realization of (M, w, J) determined by the matrices A, B satisfying the self-adjointness conditions (10.4.5). Let $S(N, W, A, B)$ be the operator realization of (N, W, J) determined by the same matrices A, B. Then the spectrum of both operators is discrete and identical i.e.,
$$\sigma(S(M, w, A, B)) = \sigma(S(N, W, A, B)$$
and the eigenfunctions y_n of $S(M, w, A, B)$ are related to the eigenfunctions z_n of $S(N, W, A, B)$ by
$$y_n(t) = v(t)\, z_n(t), \quad t \in J, \quad n \in \mathbb{N}_0. \qquad (10.6.36)$$

PROOF. Theorem 10.6.5 combines a number of results from Niessen and Zettl in [**503**], see in particular Lemmas 3.1, 3.2 and pages 559-566, especially (5.14) on page 565. Although it is more general than some of the results in [**503**] the proofs are essentially the same. □

REMARK 10.6.1. A function v which is a non-principal solution at a for some real λ_a and at b for some real λ_b and satisfies (10.6.24) can be constructed as follows: Choose a non-principal solution at a for some real λ_a which is positive in some interval (a, c), $a < c < b$; also choose a non-principal solution at b which is positive on some interval (d, b), $a < d < b$. These two pieces can be patched together to construct a function v in the maximal domain by using the Naimark Lemma. But this classical result of Naimark does not guarantee the v has no zero in the interior of the interval J. For this an extension of the Naimark Lemma is needed. Such an extension is given by Lemmas 2.3 and 3.6 of Niessen and Zettl in [**503**]. Since principal and non-principal solutions u, v are linearly independent we have $[u, v](d) \neq 0$ for $d = a$ and $d = b$; thus the normalization (10.6.35) can be achieved by multiplication by a non-zero constant. Possibly different constants for the two endpoints, then the two pieces near the endpoints are patched together to construct a maximal domain function by Lemma 10.4.1, the Patching Lemma.

REMARK 10.6.2. Using a function v constructed as in Theorem 8.2.1 the equation (10.6.26) is regular on J. It can be called a "regularization" of the singular equation (10.4.1), (10.4.2) with one or both endpoints in the LCNO class. The solutions of these two equations are related to each other by the equation

$$y(t, \lambda) = v(t) z(t, \lambda), \quad t \in J, \quad \lambda \in \mathbb{R}. \tag{10.6.37}$$

Note that v does not depend on λ. Since z is a solution of a regular equation on $J = (a, b)$ $z(t, \lambda)$ exists and is finite for $t = a$ and $t = b$; thus the singularity of y is contained in v and is independent of λ.

REMARK 10.6.3. From the perspective of inverse spectral theory, part (11) of Theorem 10.6.5 implies that the spectrum alone cannot distinguish between regular and singular problems. Also the eigenfunctions of a singular problem may have 'regular' behavior at the endpoints; see the Legendre problem in Chapter 14. This is a singular problem on $J = (-1, 1)$ whose eigenfunctions are polynomials.

7. Oscillatory Endpoints

The oscillatory behavior at one endpoint of the symmetric equation with $p > 0$ and $w > 0$ has a dramatic effect on the character of the spectrum. In the LP case the O classification depends on λ. In this case there exists a σ_0, $-\infty \leq \sigma_0 \leq \infty$, such that the equation is O for all $\lambda > \sigma_0$ and NO for all $\lambda < \sigma_0$. Here $\sigma_0 = -\infty$ means that the equation is O for all real λ; $\sigma_0 = +\infty$ means equation is NO for all real λ. If neither endpoint is LP then the spectrum is discrete and the O classification at one or both endpoints implies that the eigenvalues are unbounded below as well as above. Any approximation of a singular problem with an LCO endpoint with a regular problem must take into account the fact that the spectrum of the singular problem is not bounded below whereas the spectrum of any approximating regular problem no matter how "close" to the singular one, is bounded below.

The FORTRAN code SLEIGN2 deals with this phenomenon by using a sequence of regular problems to approximate a given singular one and then "jumping" the eigenvalue indexes as the endpoint of the regular approximating problems move closer to the singular endpoint.

Consider

$$My = -(py')' + qy = \lambda wy \quad \text{on} \quad J, \tag{10.7.1}$$

$$J = (a, b), \quad -\infty \leq a < b \leq \infty, \quad p^{-1}, q, w \in L(J, \mathbb{R}), \quad p \geq 0, \quad w > 0. \tag{10.7.2}$$

THEOREM 10.7.1. *Let (10.7.1), (10.7.2) hold and assume each endpoint is LC. Then:*

(1) The spectrum of every self-adjoint realization of (10.7.1) is discrete.
(2) The equation (10.7.1) is oscillatory at a (b) for some real λ if and only if it is oscillatory at a (b) for every real λ.
(3) The minimal operator $S_{\min}(M, w, J)$ is bounded below if and only if both endpoints are non-oscillatory.
(4) The spectrum of each self-adjoint realization of (10.7.1) is bounded below if and only if each endpoint is non-oscillatory.
(5) If the minimal operator $S_{\min}(M, w, J)$ is bounded below with lower bound α, then the equation (10.7.1) is non-oscillatory at a and at b for every $\lambda < \alpha$. Furthermore, for $\lambda < \alpha$ every nontrivial solution of (10.7.1) has at most one zero in the open interval (a, b).

PROOF. These results are well known; see Weidmann [600], Niessen and Zettl (Theorem 2.3, Corollary 2.1 and Theorem 4.2). □

According to Theorem 10.7.1 if both endpoints are LC and at least one of them is O, then the spectrum of every self-adjoint realization S of (10.7.1) is discrete i.e. consists entirely of eigenvalues:

$$\sigma(S) = \sigma_d(S) = \{\lambda_n : n \in \mathbb{Z}\}, \quad \lambda_n \to \infty \quad \text{as} \quad n \to \infty; \quad \lambda_n \to -\infty \quad \text{as} \quad n \to -\infty.$$

In contrast to the LCNO case there is no asymptotic result such as (4.3.7). For a given problem the rate at which the eigenvalues go to $+\infty$ may be quite different than the rate at which they go to $-\infty$. Also examples show that rates at which the eigenvalues approach $+\infty$ may be dramatically different for different problems; similarly for $-\infty$. For many problems the eigenvalues go to $-\infty$ at a rate which is at least exponential. Computing a large number of negative eigenvalues for these problems is a challenge for SLEIGN2; see [39] for an illustration of this phenomenon.

8. Behavior of Eigenvalues near a Singular Boundary

For this section we change the notation for the interval J from

$$J = (a, b)$$

to

$$J = (a', b').$$

The reason for this change of notation is that we wish to consider "approximations" of a singular SLP on an interval (a', b') by a sequence of regular SLP on truncated intervals (a_r, b_r) where

$$-\infty \leq a' < a_r < b_r < b' \leq \infty \qquad (10.8.1)$$

and the sequence $\{a_r : r \in N\}$ converges decreasingly to a' and the sequence $\{b_r : r \in N\}$ converges increasingly to b'. By S and S_r we denote self-adjoint realizations of (M, w) on the intervals (a', b') and (a_r, b_r), respectively. Thus S and S_r are self-adjoint operators in the Hilbert spaces $H = L^2((a', b'), w)$ and $H_r = L^2((a_r, b_r), w)$, respectively. Similarly the spectrum and eigenvalues are denoted by $\sigma(S), \sigma(S_r), \lambda_n(S), \lambda_n(S_r)$, $n \in N_0$, (or $n \in \mathbb{Z}$, in case $\sigma(S)$ is not bounded below), $r \in N$. Below, the "inherited" operators S_r^i are defined by "inherited" boundary conditions. These play a special role in the approximation of the singular spectrum with eigenvalues of regular problems.

THEOREM 10.8.1. *Let (10.4.1), (10.4.2) hold with $J = (a', b')$ and let (10.8.1) hold. Let S, S_r be arbitrary self-adjoint realizations of (M, w) on the intervals (a', b') and (a_r, b_r), respectively, for $r \in N$.*
For all endpoint classifications: *If $\sigma(S)$ is not bounded below, then*

$$\lambda_n(S_r) \to -\infty, \text{ as } r \to \infty, \text{ for each } n \in N_0. \tag{10.8.2}$$

PROOF. See Everitt and Zettl [**225**]. □

DEFINITION 10.8.1 (Inherited Boundary Conditions and Operators). *The inherited boundary conditions are defined as follows:*

- *Near an LP endpoint the inherited BC is the Dirichlet condition.*
- *Assume a' is LC and b' is LP. Then there is no singular boundary condition at b' and all singular self-adjoint BC at a' have the form*

$$A_1 [y, u](a') + A_2 [y, v](a') = 0, \; A_1, A_2 \in \mathbb{R}, \; (A_1, A_2) \ne (0, 0), \tag{10.8.3}$$

  where $u, v \in D_{\max}$ are real valued and satisfy $[u, v](a') = 1$.

  The inherited BC on (a_r, b_r) are obtained by replacing a' by a_r in (10.8.3) and by using the Dirichlet conditions $y(b_r) = 0$ at b_r. Note that, although u, v and their quasi-derivatives may not be defined at a' they are well defined at any point $a' < a_r < b_r < b'$ and the Lagrange sesquilinear forms $[y, u](a')$ and $[y, v](a')$ are well defined as finite limits by Green's formula (see Section 9.2).

- *Assume b' is LC and a' is LP. Then there is no singular boundary condition at a' and all singular self-adjoint BC at b' have the form*

$$B_1 [y, u](b') + B_2 [y, v](b') = 0, \; B_1, B_2 \in \mathbb{R}, \; (B_1, B_2) \ne (0, 0), \tag{10.8.4}$$

  where $u, v \in D_{\max}$ are real valued and satisfy $[u, v](b') = 1$.

  The inherited BC on (a_r, b_r) are obtained by replacing b' by b_r in these conditions and by using the Dirichlet conditions $y(a_r) = 0$ at a_r. Note that, although u, v and their quasi-derivatives may not be defined at b' they are well defined at any point b_r, $a' < a_r < b_r < b'$ (and (10.8.4) is well defined by Green's formula).

- *Each endpoint is either regular or LC. In this case we have self-adjoint realizations determined by both separated and coupled BC. Let the BC on (a', b') be determined by (10.4.5) but with our changed notation a' for a and b' for b. To obtain the inherited BC just replace a, b in (10.4.5) by (a_r, b_r). Note that although the same matrices A, B occur in the BC on (a', b') and on (a_r, b_r) when these inherited boundary conditions are written in the usual form for regular BC as in Chapter 4 the coefficient matrices, say $A = A(a_r), B = B(b_r)$ depend on values of u, v and their quasi-derivatives at a_r, b_r. In particular as the endpoint a_r or b_r is changed, e.g. in the code SLEIGN2, the inherited boundary conditions change accordingly.*

- *The inherited operators and their spectral quantities are identified with the superscript i: S_r^i, $\lambda_n^i(a_r, b_r)$.*

DEFINITION 10.8.2 (The Starting Point of the Essential Spectrum). *For any operator S let*

$$\sigma_0 = \inf \sigma_e, \; -\infty \le \sigma_0 \le \infty, \tag{10.8.5}$$

where σ_e denotes the essential spectrum. Here $\sigma_0 = -\infty$ is interpreted to mean that the essential spectrum is not bounded below and $\sigma_0 = \infty$ means that the essential spectrum is empty i.e. the spectrum consists entirely of isolated eigenvalues. In the latter case we say that the spectrum is discrete.

REMARK 10.8.1. The starting point of the essential spectrum i.e. σ_0 is also the "oscillation point" of the equation, see [**277**]. For $\lambda > \sigma_0$ the equation is oscillatory and for $\lambda < \sigma_0$ the equation is nonoscillatory. For $\sigma_0 = -\infty$ this is taken to mean that the equation is oscillatory for all real λ; if $\sigma_0 = +\infty$ this is interpreted to mean that the equation is nonoscillatory for all real λ. For $\lambda = \sigma_0$, $-\infty < \sigma_0 < \infty$, the equation may be O or NO.

THEOREM 10.8.2. *Let (10.4.1), (10.4.2) hold with $J = (a', b')$ and let (10.8.1) hold. Let S be a self-adjoint realization of (M, w) on (a', b') and let S_r^i be the inherited operator on (a_r, b_r). Assume that $\sigma(S)$ is bounded below and discrete. Let $\sigma(S) = \{\lambda_n(S) : n \in N_0\}$, $\sigma(S_r^i) = \{\lambda_n(S_r^i) : n \in N_0\}$. Then*

$$\lambda_n(S_r^i) \to \lambda_n(S), \text{ as } r \to \infty, \text{ for each } n \in N_0. \tag{10.8.6}$$

PROOF. See Everitt, Marletta and Zettl [**226**]. □

REMARK 10.8.2. We comment on the contrast between (10.8.6) and (10.8.2). This markedly different behavior of the eigenvalues of regular problems shows the enormous influence that the spectrum of a singular problem has on the regular problems which are "close" to the singular one; in this case by virtue of the fact that the endpoints of the regular problem are close to the endpoints of the singular one. To understand the behavior of the eigenvalues of *regular* problems one needs a perspective which includes the *singular* case. This is even more interesting when viewed in the light of the asymptotic formula (10.6.4) according to which the eigenvalues on each fixed interval (a_r, b_r) go to $+\infty$ asymptotically as n^2.

THEOREM 10.8.3. *Let (10.4.1), (10.4.2) hold with $J = (a', b')$ and let (10.8.1) hold. Let S be a self-adjoint realization of (M, w) on (a', b') and let S_r^i be the inherited operator on (a_r, b_r). Assume each endpoint is either R or LC and at least one endpoint is O. Then*

$$\sigma(S) = \{\lambda_n : n \in \mathbb{Z}\} \text{ and } \lambda_n(S_r^i) \to -\infty, \text{ as } r \to \infty, n \in N_0. \tag{10.8.7}$$

Nevertheless we have: Given any $\lambda_k \in \sigma(S)$ there exists an increasing (index) sequence of positive integers $n(r, k)$, depending on r and on λ_k such that

$$\lambda_{n(r,k)}(S_r^i) \to \lambda_k \text{ as } r \to \infty. \tag{10.8.8}$$

PROOF. This is contained in Theorem 4.1 of [**38**]; see also Remark 1 (ii) on pages 15-16. □

In this case, i.e. when the eigenvalues are unbounded below as well as above, we follow the SLEIGN2 convention that λ_0 denotes the smallest nonnegative eigenvalue. This makes the indexing scheme unique.

THEOREM 10.8.4. *Let (10.4.1), (10.4.2) hold with $J = (a', b')$ and let (10.8.1) hold. Let S be a self-adjoint realization of (M, w) on (a', b') and let S_r^i be the inherited operator on (a_r, b_r). Assume that*

$$-\infty < \sigma_0 < \infty. \tag{10.8.9}$$

Then at least one endpoint is LP.

(1) If S has no eigenvalue less than σ_0, then
$$\lambda_n(S_r^i) \to \sigma_0, \quad r \to \infty, \ n \in \mathbb{N}_0. \tag{10.8.10}$$

(2) If S has exactly one eigenvalue below σ_0, say $\lambda_0 < \sigma_0$, then
$$\lambda_0(S_r^i) \to \lambda_0, \text{ and } \lambda_n(S_r^i) \to \sigma_0, \text{ as } r \to \infty, \ n \geq 1. \tag{10.8.11}$$

(3) If S has exactly two eigenvalues, say λ_0 and λ_1, below σ_0, then
$$\lambda_0(S_r^i) \to \lambda_0, \ \lambda_1(S_r^i) \to \lambda_1, \ r \to \infty; \ \lambda_n(S_r^i) \to \sigma_0, \ r \to \infty, \ n \geq 2. \tag{10.8.12}$$

(4) etc.

(5) If S has an infinite number of eigenvalues $\{\lambda_n : n \in \mathbb{N}_0\}$ to the left of σ_0, then
$$\lambda_n(S_r^i) \to \lambda_n, \ r \to \infty, \ n \in \mathbb{N}_0. \tag{10.8.13}$$

PROOF. See Everitt, Marletta and Zettl [225]. □

REMARK 10.8.3. Theorem 10.8.4 can be used to detect the number of eigenvalues to the left of the essential spectrum. Using one of the numerical codes such as SLEIGN2 or the Fulton and Pruess code SLEDGE [261] together with the "exact spectral convergence" result in Bailey, Everitt, Weidmann and Zettl [38] one can ascertain which of the possibilities (10.8.9) to (10.8.13) holds.

REMARK 10.8.4. The so called Coffee-Evans equation has attracted a good deal of attention in the literature because it has the interesting feature that, despite the asymptotic behavior (10.6.4), the lower eigenvalues occur in clusters of three which are close together. It is clear from Theorem 9.8.4 how to construct examples of regular problems with clusters of three million or three billion eigenvalues as close together as you please. This also shows that it easy to construct examples of *regular* SLP which will defeat any numerical code for the computation of eigenvalues. See Zettl [626] for some illustrations of this. (This paper can be downloaded from the author's web page following the instructions for downloading SLEIGN2 given at the end of Section 10.4 above.)

REMARK 10.8.5. Using Theorem 10.8.4 together with interval analysis methods developed at the University of Karlsruhe, particularly Lohner's AWA ode solver, along with SLEIGN2 - a "standard numerical analysis" code, Brown, McCormack and Zettl [107], [106] have developed a new method for (rigorously) proving the existence of eigenvalues below the essential spectrum.

9. Approximating a Singular Problem with Regular Problems

Can the spectrum of singular problems be approximated with eigenvalues of regular problems? If so, how? These questions are studied in this section. We continue to use the notation of Section 10.8; in particular, for the inherited boundary conditions and inherited operators.

THEOREM 10.9.1. *Let (10.4.1), (10.4.2) hold with $J = (a', b')$ and let (10.8.1) hold. Let S be a self-adjoint realization of (M, w) on (a', b') and let S_r^i be the inherited operator on (a_r, b_r). The sequence of inherited operators $\{S_r^i : r \in \mathbb{N}\}$ is spectral included for S i.e. given any $\lambda \in \sigma(S)$ there exists an $n(r, \lambda) \in \mathbb{N}_0$ for each $r \in \mathbb{N}$ such that*
$$\lambda_{n(r,\lambda)}(S_r^i) \to \lambda, \text{ as } r \to \infty. \tag{10.9.1}$$

PROOF. This is contained in [**38**]. □

THEOREM 10.9.2. *Let (10.4.1), (10.4.2) hold with $J = (a', b')$ and let (10.8.1) hold. Let S be a self-adjoint realization of (M, w) on (a', b') and let S_r^i be the inherited operator on (a_r, b_r). Assume that $\sigma(S)$ is bounded below. Then the sequence of inherited operators $\{S_r^i : r \in \mathbb{N}\}$ is spectral exact for S below $\sigma_0(S)$, i.e. it is spectral included and if the convergence (10.9.1) holds for some $\lambda < \sigma_0$, then $\lambda \in \sigma(S)$.*

PROOF. This is contained in [**38**]. □

By Theorem 10.9.1 any point of the spectrum of a singular problem can, in principle, be approximated arbitrarily closely by eigenvalues from the inherited sequence of regular problems. In practice this isn't feasible since there are an infinite and generally uncountable number of points in the spectrum of the singular problem; so finding an index sequence for each one is a hopeless task.

Nevertheless Theorems 10.9.1 and 10.9.2 together can be used to approximate the spectrum of many singular SLP quite effectively. This is done by approximating the discrete spectrum below the essential spectrum and, not the individual points of the essential spectrum, but the spectral bands and gaps. It is remarkable that for so many singular problems the first few spectral bands and gaps - or the absence of gaps - can be detected and approximated from the distribution of a few thousand eigenvalues of the inherited problems which can be computed with SLEIGN2 or SLEDGE. For illustrations of this scheme see Zettl [**638**]. This paper can be downloaded as part of the SLEIGN2 package of files from:

www.math.niu.edu/~zettl/SL2

10. Green's Function

In view of the special interest in the self-adjoint case we give the explicit necessary and sufficient conditions on the boundary conditions for the Green's function to be hermitian, which is equivalent to self-adjointness, for problems with no LP endpoint.

Consider the SLP consisting of the equation

$$My = -(py')' + qy = \lambda wy \quad \text{on} \quad J, \tag{10.10.1}$$

$$J = (a, b), \quad -\infty \leq a < b \leq \infty, \quad 1/p, q, w \in L_{loc}(J, \mathbb{R}), \quad w > 0, \tag{10.10.2}$$

together with boundary conditions

$$AY(a) + BY(b) = 0, \; Y = \begin{bmatrix} [y, u] \\ [y, v] \end{bmatrix}, \; A, B \in M_2(\mathbb{C}), \; \text{rank}(A : B) = 2. \tag{10.10.3}$$

THEOREM 10.10.1. *Let (10.10.1), (10.10.2) hold, let $\lambda \in \mathbb{R}$. Assume that neither endpoint is LP. Let u, v be real-valued maximal domain functions satisfying*

$$[u, v](a) = 1 = [u, v](b). \tag{10.10.4}$$

Let $K(t, s, \lambda) = K(t, s, \lambda, P, W, A, B)$ be the matrix Green's function of the boundary value problem (10.10.1), (10.10.2), (10.10.3) as constructed in Section

8.4. Then $K(t,s,\lambda)$ exists for every $\lambda \in \mathbb{C}$ which is not an eigenvalue. Define an integral operator T by

$$(Tf)(t) = \int_a^b K_{12}(t,s,\lambda)f(s)\,ds, \quad t \in J, \quad f \in H = L^2(J,w). \tag{10.10.5}$$

Then T is a compact operator from H into H and for each λ which is not an eigenvalue, the following four statements are equivalent:

(1)
$$AEA^* = BEB^*,\ E = \begin{bmatrix} 0 & -1 \\ 1 & 0 \end{bmatrix}; \tag{10.10.6}$$

(2)
$$K(t,s,\lambda) = E\,K(s,t,\lambda)\,E, \quad t,s \in J; \tag{10.10.7}$$

(3)
$$K_{12}(t,s,\lambda) = \overline{K}_{12}(s,t,\lambda), \quad t,s \in J. \tag{10.10.8}$$

(4) T is a self-adjoint operator from H into H; i.e. $T = T^*$.

PROOF. The equivalence of the first three properties is a special case of Theorem 9.4.2, the equivalence of (4) with (3) is well known. $\square$

11. Multiplicity of Eigenvalues

In this section we study the relationships between the algebraic and geometric multiplicities of eigenvalues of SLP when each endpoint is either regular or LC. We use the notation and results established in Section 10.4, particularly Subsection 10.4.4.

THEOREM 10.11.1. *Let the hypotheses and notation of Theorem 10.4.7 hold. Assume that ρ is an eigenvalue of the regular system (10.4.46), (10.4.48) on (a,b) with geometric multiplicity 2. Then for $a < c < d < b$ and c sufficiently close to a, d sufficiently close to b there exist matrices $A(c)$, $B(d)$ satisfying the self-adjointness conditions (10.4.11), (10.4.12) such that the regular scalar SLP consisting of the equation (10.4.35) on (c,d) with the boundary condition*

$$A(c)Y(c) + B(d)Y(d) = 0 \text{ on } (c,d) \tag{10.11.1}$$

has ρ as an eigenvalue of geometric and algebraic multiplicity two.

PROOF. By Theorem 4.1 of [**388**] the boundary condition (10.4.48) is equivalent to

$$A = \Phi(b,a,\lambda),\ B = -I. \tag{10.11.2}$$

Let

$$A(c) = \Phi(b,a,\rho)\,\Phi^{-1}(c,a,\rho),\ B(d) = -I \tag{10.11.3}$$

and note that

$$\Phi(t,c,\rho) = \Phi(t,a,\rho)\,\Phi^{-1}(c,a,\rho),\ c \leq t \leq d \tag{10.11.4}$$

is the fundamental matrix of (10.4.46) determined by the initial condition $\Phi(c,c,\rho) = I$. Hence the characteristic function of (10.4.35), (10.11.2), is given by

$$s(\lambda,(c,d),A(c),B(d)) = \det[A(c) - \Phi(d,c,\lambda)]. \tag{10.11.5}$$

In particular, $s(\rho,(c,d),A(c),B(d)) = 0$. From the continuity of Φ it follows that as $c \to a$, $d \to b$ we have

$$s(\lambda,(c,d),A(c),B(d)) \to \delta(\lambda,(a,b),A,B). \tag{10.11.6}$$

Note that on the interval (c,d) the problem (10.4.35), (10.11.2) is equivalent to a regular self-adjoint SLP and hence by Theorem 2.1 of [**380**] the algebraic multiplicity of the eigenvalue ρ of this regular problem on the interval (c,d) is 2. From this and (10.11.6) it follows that $\Delta'(\rho,(a,b),A,B) = 0$. Hence the algebraic multiplicity of ρ is at least two. □

THEOREM 10.11.2. *Let (10.4.36), (10.4.37) hold, let $\{f,g\}$ be a boundary condition bases at a and $\{h,k\}$ a boundary condition bases at b. Then the algebraic multiplicity of each eigenvalue of the boundary value problem (10.4.35), (10.4.42) is the same as its geometric multiplicity.*

PROOF. Assume that ρ is an eigenvalue of (10.4.35), (10.4.42) with algebraic multiplicity two. By the "Continuity Principle", see [**285**], all nearby problems have two eigenvalues, counting multiplicity. In particular, for $a < c < d < b$ and c sufficiently close to a, d sufficiently close to b the "inherited problem" consisting of (10.4.35) with the boundary condition

$$AZ(c) + BZ(d) = 0 \text{ on } (c,d), \tag{10.11.7}$$

has two eigenvalues, counting multiplicity, say $\rho_1(c,d), \rho_2(c,d)$ such that

$$\rho_j(c,d) \to \rho,\ j=1,2;\ as\ c \to a,\ d \to b. \tag{10.11.8}$$

On (c,d) each problem is equivalent to a self-adjoint and regular SLP. Let $y_1 = y_1(c,d)$, $y_2 = y_2(c,d)$ be eigenfunctions with eigenvalues $\rho_1(c,d), \rho_2(c,d)$ satisfying (10.11.8). These eigenvalues may or may not be distinct. If they are distinct then their eigenfunctions are orthogonal; if they are not distinct their eigenfunctions can be chosen to be orthogonal. Thus, in either case, we have, or can arrange to have,

$$\int_c^d y_1 \overline{y_2}\, w = 0. \tag{10.11.9}$$

We normalize these eigenfunctions by choosing a fixed h, $a < c < h < d < b$, letting $Y_j = \begin{bmatrix} y \\ py' \end{bmatrix}$, $Z_j = U^{-1} Y_j$ and requiring that

$$\|Z_j(h,(c,d))\|_2 = 1,\ j=1,2. \tag{10.11.10}$$

Note that his normalization is with respect to the Euclidean $2 - norm$, and y_j, Y_j, Z_j depend on the interval (c,d) but we sometimes omit this interval in the notation for simplicity. There exist sequences c_n, d_n and vectors K_j such that

$$c_n \to a,\ d_n \to b,\ Z_j(h,(c_n,d_n)) \to K_j,\ and\ \|K_j\|_2 = 1,\ j=1,2. \tag{10.11.11}$$

Determine solutions Z_j of (10.4.46) by the initial condition

$$Z_j^*(h) = K_j,\ j=1,2. \tag{10.11.12}$$

Each Z_j can be extended to $[a,b]$ as a solution of (10.4.46) and as a consequence of the continuous dependence of solutions of regular systems on initial conditions, it follows that

$$Z_j(c_n,d_n) \to Z_j^*\ on\ [a,b],\ j=1,2. \tag{10.11.13}$$

and this convergence is uniform. Here we use the notation $[a,b]$ even when a or b may be infinite and at a finite or infinite endpoint the solutions are defined as a limit. For a proof of the uniform convergence of (10.11.13) on bounded or unbounded intervals; see [**645**], Theorem 2.12. It follows that $Z_j^*, j=1,2$ satisfies the boundary condition (10.4.48). It remains to show that Z_1^*, Z_2^* are linearly independent.

Let y_1, y_2 be the eigenfunctions satisfying (10.11.9),

$$Y_j = \begin{bmatrix} y_j \\ py_j' \end{bmatrix}, \ Y_j = UZ_j, \ j=1,2, \ Z_1 = \begin{bmatrix} z_{11} \\ z_{21} \end{bmatrix}, \ Z_2 = \begin{bmatrix} z_{12} \\ z_{22} \end{bmatrix}. \qquad (10.11.14)$$

From (10.11.9), (10.11.14) we have

$$\int_{c_n}^{d_n} [v\, z_{11} + u\, z_{21}][v\, z_{12} + u\, z_{22}]w = 0. \qquad (10.11.15)$$

Form (10.11.15), the uniform convergence of (10.11.13) on $[a,b]$ and the fact that $u, v \in L^2(J, w)$ we conclude that

$$\int_a^b [v\, z_{11}^* + u\, z_{21}^*][v\, z_{12}^* + u\, z_{22}^*]w = 0. \qquad (10.11.16)$$

Therefore Z_1^*, Z_2^* are linearly independent and the proof is complete for the system boundary value problem (10.4.46), (10.4.48) on (a,b). The result for the scalar problem (10.4.35), (10.4.42) follows because it is equivalent to this system and Z_1^*, Z_2^* are linearly independent if and only if y_1, y_2 are linearly independent. □

12. Summary of Spectral Properties

In this section we expand on Proposition 10.4.4 and Remark 10.4.14 to summarize spectral properties of self-adjoint regular and singular Sturm-Liouville problems with positive weight function. We consider self-adjoint realizations of the equation

$$My = -(py')' + qy = \lambda wy \quad \text{on} \quad J, \qquad (10.12.1)$$

with the following conditions on the coefficients

$$J = (a,b), \quad -\infty \leq a < b \leq \infty, \quad 1/p, q, w \in L_{loc}(J, \mathbb{R}), \quad w > 0, \qquad (10.12.2)$$

together with self-adjoint boundary conditions. These boundary conditions are characterized in Section 10.4, see also Chapter 4 for the regular case. Recall that the self-adjoint realizations of (10.12.1) are operators S in the Hilbert space $H = L^2(J, w)$ satisfying

$$S_{\min} \subset S = S^* \subset S_{\max}.$$

An eigenvalue is simple if it has exactly one linearly independent eigenfunction, otherwise it is double. The geometric multiplicity of an eigenvalue is the number of its linearly independent eigenfunctions; the algebraic multiplicity of an eigenvalue is the order of its root as a zero of the characteristic function defined in Section 10.4. (See the Comments for Section 10.11 at the end of this chapter for a different definition of multiplicity.)

THEOREM 10.12.1. *Let (10.12.1), (10.12.2) hold. (The spectrum of every self-adjoint operator in a Hilbert space is real.)*

(1) *If p changes sign on J, then the spectrum of every self-adjoint realization of (10.12.1) is unbounded above and below. (This holds even if there is no subinterval of J on which p is negative.)*
(2) *If neither endpoint is LP then the spectrum of every self-adjoint realization of (10.12.1) is discrete. It may be bounded below or above, but not both.*

(3) If $p > 0$ on J and each endpoint is either regular or LCNO, then the spectrum of every self-adjoint realization of (10.12.1) is discrete and bounded below. Let S be such a self-adjoint realization and denote its spectrum by $\sigma = \sigma(S)$. Then $\sigma = \{\lambda_n : n \in \mathbb{N}_0\}$ can be ordered and indexed to satisfy

$$-\infty < \lambda_0 \leq \lambda_1 \leq \lambda_2 \leq \lambda_3 \leq ... \qquad (10.12.3)$$

and $\lambda_n \to \infty$, as $n \to \infty$. Equality cannot hold for two consecutive terms since (10.12.1) has exactly two linearly independent solutions for each λ. The algebraic multiplicity of each eigenvalue is the same as its geometric multiplicity.

(4) If $p > 0$ on J, each endpoint is either regular or LCNO and the boundary conditions are separated, then all eigenvalues are simple and strict inequality holds throughout (10.12.3). Furthermore, if u_n is an eigenfunction of λ_n, then u_n has exactly n zeros in the open interval J for any $n \in \mathbb{N}_0$.

(5) If $p > 0$ on J, each endpoint is either regular or LCNO and the boundary conditions are nonreal coupled, then all eigenvalues are simple and strict inequality holds throughout (10.12.3). If u_n is an eigenfunction of λ_n, then u_n is complex valued and has no zero in the closed interval $[a, b]$; the number of zeros of $\text{Re}(u_n)$ on the half open interval $[a, b)$ is 0 or 1 if $n = 0$, and $n - 1$ or n or $n + 1$ if $n \geq 1$; the number of zeros of $\text{Im}(u_n)$ on the half open interval $[a, b)$ is 0 or 1 if $n = 0$, and $n - 1$ or n or $n + 1$ if $n \geq 1$.

(6) If $p > 0$ on J, each endpoint is either regular or LCNO and the boundary conditions are real coupled, then each eigenvalue may be simple or double. (Note that the eigenvalues are uniquely determined by (10.12.3) but there is some ambiguity in the meaning of u_n if λ_n is a double eigenvalue. Since linearly independent solutions of (10.12.1) may not have the same number of zeros in J, the exact number of zeros of u_n cannot be determined.) If u_n is a real valued eigenfunction of λ_n, then the number of zeros of u_n on the open interval J is 0 or 1 if $n = 0$, and $n - 1$ or n or $n + 1$ if $n \geq 1$.

(7) If $p > 0$ on J and (at least) one endpoint is LCO, then the spectrum of every self-adjoint realization of (10.12.1) is unbounded above and below. If λ is an eigenvalue and u is an eigenfunction of λ, then u has an infinite number of zeros in J.

(8) If $p > 0$ on J and (at least) one endpoint is LP, then $\sigma_e(S) = \sigma_e(S_{\min})$ for every self-adjoint realization S of (10.12.1). In particular, the essential spectrum does not depend on the boundary conditions and therefore depends only on the coefficients p, q, w (more accurately on $1/p, q, w$). The discrete spectrum $\sigma_d(S)$ depends on the boundary conditions. Either one of $\sigma_e(S)$, $\sigma_d(S)$, but not both, may be empty. Let $\sigma_0 = \inf \sigma_e(S_{\min})$. There are three possibilities for σ_0:

(i) $\sigma_0 = -\infty$. In this case σ_e may be the whole line or it may consist of disjoint closed intervals separated by 'gaps'. The number of gaps may be finite or infinite. If there is an eigenvalue, then its eigenfunctions have an infinite number of zeros in J.

(ii) $\sigma_0 = +\infty$. In this case the spectrum of every self-adjoint realization of (10.12.1) is discrete and unbounded above. Either the spectrum of every self-adjoint realization is bounded below (but there is no uniform lower bound for all self-adjoint realizations) or the spectrum of no self-adjoint realization is bounded below. If the spectrum is unbounded below, then every eigenfunction has an infinite number of zeros in J. If S is a self-adjoint realization and its spectrum $\sigma(S)$ is bounded below then $\sigma(S) = \{\lambda_n : n \in \mathbb{N}_0\}$, the eigenvalues λ_n are all simple, can be ordered to satisfy:

$$-\infty < \lambda_0 < \lambda_1 < \lambda_2 < ... \qquad (10.12.4)$$

and $\lambda_n \to \infty$ as $n \to \infty$. In this case, if u_n is an eigenfunction of λ_n then u_n has exactly n zeros in J.

(iii) $-\infty < \sigma_0 < \infty$. In this case σ_0 is also the oscillation number of the equation (10.12.1): For $\lambda < \sigma_0$ each nontrivial solution of (10.12.1) has either no zero or a finite number of zeros in J. If $\lambda > \sigma_0$, then every solution of (10.12.1) has an infinite number of zeros in J. If $\lambda = \sigma_0$, then both cases occur: (i) every solution has an infinite number of zeros or (ii) no nontrivial solution has an infinite number of zeros. There may be no eigenvalues below σ_0 or a finite number of such eigenvalues or an infinite number. All eigenvalues below σ_0 are simple. If there are an infinite number of eigenvalues below σ_0, they must accumulate at σ_0. If λ_n are the, finite or infinite number of, eigenvalues below σ_0, ordered as in (10.12.4), and u_n is an eigenfunction of λ_n, then u_n has exactly n zeros in J.

(9) Suppose $J = (-\infty, \infty)$ and $p > 0$ on J. Assume that each of p, q, w is $r-$periodic ($0 < r < \infty$ and r is the fundamental period). Then there is a unique self-adjoint realization S of (10.12.1), $S = S_{\min} = S_{\max}$ and

$$\sigma_e(S) = \cup_{n=0}^{\infty} J_n,$$

$$J_0 = [\lambda_0^P, \lambda_0^S], \ J_1 = [\lambda_1^S, \lambda_1^P], \ J_2 = [\lambda_2^P, \lambda_2^S],$$
$$J_3 = [\lambda_3^S, \lambda_3^P], \ J_4 = [\lambda_4^P, \lambda_4^S], \ ...$$

Here λ_n^P, λ_n^S are the periodic and semi-periodic eigenvalues of (10.12.1) on the fundamental interval $J = [0, r]$, respectively. In particular $\sigma_0 = \inf \sigma_e(S_{\min}) = \lambda_0^P$. The gaps consist of the open intervals: $(\lambda_0^S, \lambda_1^S)$, $(\lambda_1^P, \lambda_2^P)$, $(\lambda_2^S, \lambda_3^S)$, $(\lambda_3^P, \lambda_4^P)$, ... If λ_0^S is a double eigenvalue, then the 'first gap is missing'; if λ_1^P is a double eigenvalue, then the 'second gap is missing'; etc. (By Theorem 10.6.4 λ_0^P is simple.) (The open interval $(-\infty, \lambda_0^P)$ is also considered a gap by some authors.) If all gaps are missing, $\sigma_e(S) = [\lambda_0^P, \infty)$. There may be no gaps, a finite number of them or an infinite number of them. The closed intervals J_n are called the spectral bands of S. The spectrum of S, $\sigma(S)$, may or may not contain one or more eigenvalues. There may be eigenvalues below σ_0, in one or more spectral bands, and in one or more gaps. Each gap may contain a finite or infinite number of eigenvalues of S.

(10) Suppose $J = (a, \infty)$, $-\infty < a < \infty$ and $p > 0$ on J. Assume that each of p, q, w is $r-$periodic with fundamental interval $[a, a+r]$. Then the endpoint a is regular and ∞ is LP. In this case there are an infinite number of self-adjoint realization of (10.12.1), each determined by a boundary condition at a:

$$\cos(\alpha) y(a) + \sin(\alpha)(py')(a) = 0, \ \alpha \in [0, \pi).$$

If S is any one of these, then the results of (9) above hold for S with λ_n^P, λ_n^S denoting the periodic and semi-periodic eigenvalues of (10.12.1) on the fundamental interval $J = [a, a + r]$, respectively.

PROOF. See the comments for Section 10.12 below in Section 10.13. □

13. Comments

(1) In Chapter 10 we give a detailed, but not comprehensive, account of the theory of singular self-adjoint Sturm-Liouville problems with a positive weight function when neither endpoint is in the LP case. See Chapter 16 for a few comments and references to problems with an LP endpoint.
(2) This is routine.
(3) This is standard; see [**600**], [**487**].
(4) The structure of singular limit circle boundary conditions is well known, see [**600**], [**487**], [**7**], [**400**] but it is not easy to find a clear and comprehensive treatment of them in the literature. We hope that we have made positive contribution in this area. The conditions (10.4.11), (10.4.12), (10.4.13) characterize *all* self-adjoint boundary conditions for the case when each endpoint is either R or LC (either LCO or LCNO). These conditions fall into two classes: separated and coupled. Theorem 10.4.3 gives the canonical form of these conditions for both the separated and coupled cases. When studying the dependence of eigenvalues on the problem it is convenient to have canonical representations.

For a different representation of singular self-adjoint BC see the treatment of Dunford and Schwartz [**141**] using functionals.

(5) The characterization of the Friedrichs extension given here is based on Niessen and Zettl [**503**]. This in turn is based on Rellich [**544**]. Also see Kaper, Kwong and Zettl [**360**], Kalf [**352**], Rosenberg [**554**]. The characterization of the boundary conditions which determine the Friedrichs extension depends on the principal (small) solution at each endpoint.

In the higher order case the situation is much more complicated. For instance in the fourth order case the Friedrichs boundary condition depends on a pair of solutions which is 'small' relative to other pairs. Thus one is led to work with systems. See the paper by Marletta and Zettl [**465**] for a characterization of the Friedrichs extension in terms of boundary conditions for singular higher order problems. Prior to the appearance of Marletta and Zettl [**465**] few results on the Friedrichs extension for higher order equations were known. Two exceptions are Baxley [**51**] who studied the Friedrichs extension of powers of some very special Sturm-Liouville operators and Zettl [**638**] who also studied a very special class of problems. An illustrative example is, with $n > 1$,

$$(-1)^n y^{(2n)} \pm \frac{c}{t} y = \lambda y \text{ on } 0 < t < 1,\ c \in \mathbb{R},\ c \neq 0.$$

For the regular case very general results are known, see Niessen and Zettl [**502**], Möller and Zettl [**477**].

(6) Möller [**474**] has shown that, for regular as well as singular SLP, if p changes sign on the underlying interval then the spectrum is not bounded below. This holds even if there is no subinterval of the underlying interval

on which p is negative. The corresponding result for very general higher order problems was established by Möller and Zettl in [**476**].

The inequalities of Theorem 10.6.3 are derived from the corresponding inequalities of Chapter 4 for the regular case and the transformation from the singular LCNO to the regular case employed by Niessen and Zettl in [**503**].

For earlier work see Jörgens [**349**], Rellich [**544**]. These were extended to the singular case by Niessen and Zettl in [**502**], then further extended to more general K by Bailey, Everitt and Zettl [**41**], and finally the general case was established by Eastham, Kong, Wu and Zettl in [**145**].

In [**41**] there is also the characterization of the eigenvalues given by (10.6.23) of Theorem 10.6.4 for the case when the expression M is symmetric and $w > 0$. It was observed by Zettl in [**645**] that this characterization holds for complex-valued p, q, w but with self-adjoint coupled boundary conditions.

For an extension of the asymptotic formula (10.6.4) to some problems with 'smooth' coefficients, see Harris [**305**], Hartman [**319**], and the references therein.

(7) This brief section is based on Niessen and Zettl [**502**].

(8) This section is based on Everitt, Marletta and Zettl [**225**] which in turn is based, to a considerable extent, on Bailey, Everitt, Weidmann and Zettl [**38**] and on Eastham, Kong, Wu and Zettl in [**145**].

(9) This section is also based on [**38**]. For a contribution to partial differential equations of the concepts of spectral inclusion and spectral exactness, see Brown and Marletta [**103**].

(10) This section is merely a specialization of some of the results from Section 8.4 to the self-adjoint case, together with a summary of some well known results.

(11) This section is based on Kong, Wu and Zettl [**388**].

(12) Part (1) is proved by M. Möller in [**474**]. Part (2) follows from the abstract theory of self-adjoint operators in Hilbert space: The spectrum is discrete if the resolvent operator is compact. See [**487**], [**599**], [**141**]. In (3), for regular problems the first sentence is proved in the books by Naimark [**487**], Weidmann [**600**], see also Möller and Zettl [**476**]. The LCNO case then follows from the transformation employed by Niessen and Zettl in [**503**] which transforms LCNO problems to regular problems while leaving the spectrum invariant. The equivalence of the algebraic and geometric multiplicity is established in Section 10.11. For LCNO endpoints this also follows from the regular case established in Section 4.10 since it is easy to see that the multiplicity is left invariant by the above mentioned transformation. (In Section 10.11 the LCO case is also established, this case does not follow from Section 4.10.) In parts (4), (5) and (6), as in part (3), the singular LCNO case follows from the regular case established in Chapter 4. For a proof of (7) see [**502**]. In (8) the paragraph preceding the three cases is well known [**487**], [**599**], [**141**]. The three cases are also well known [**599**]. Determining (i) the number of eigenvalues below σ_0, (ii) the number in gaps in the essential spectrum, and (iii) the presence

or absence of eigenvalues embedded in the essential spectrum are all research problems of current interest. For some results, see Glazman [**277**], Eastham [**153**], Rofe-Beketov [**552**], [**552**], [**88**], Brown, McCormack and Zettl [**107**],[**106**], and the references therein. For (9) and (10) see the well known books by Weidmann [**599**] and Eastham [**153**]. (The latter assumes much more stringent conditions on the coefficients but most of the results in [**153**] extend readily to those of Theorem 10.12.1.) In the general periodic coefficient case, λ_0^P is simple by Theorem 10.6.4. The Meissner equation, see Example 25 in the Examples chapter below, has an infinite number of simple and an infinite number of double periodic eigenvalues; all its semi-periodic eigenvalues are simple. Thus for the Meissner problem infinitely many gaps are present and infinitely many are absent. When are there exactly n gaps for any $n \in \mathbb{N}$? When are the eigenvalues λ_7^P, λ_{13}^P, λ_{16}^P, λ_{23}^P double and all other periodic eigenvalues simple? Similar questions can be asked for any other real coupled boundary condition with coupling matrix $K \in SL(2,\mathbb{R})$, this for regular and singular problems and with the indices $7, 13, 16, 23$ replaced by any finite or infinite sequence of positive integers. Since the essential spectrum is independent of the boundary conditions which of the three cases in (9) and (10) holds depends only on the coefficients.

For what coefficients $1/p, q, w$ is $\sigma_0 = -\infty$? Not much seems to be known about this case other than conditions implying $\sigma_e = (-\infty, \infty)$ [**277**], [**141**], [**319**].

For what coefficients $1/p, q, w$ is $\sigma_0 = +\infty$, in other words under what conditions on p, q, w is the spectrum discrete? This question leads to two subclasses of problems: (i) Find conditions on p, q, w such that the spectrum is discrete and unbounded above and below, (ii) Find conditions on p, q, w such that the spectrum is discrete and bounded below; Hinton and Lewis denote this class of problems by BD. Relatively little is known about problem (i), a lot is known about property BD. For (i) see Example 30 in Chapter 13 below, the Littlewood-McLeod example and Littlewood [**450**], McLeod [**467**], [**468**], [**469**]. For results on property BD see [**277**], [**487**], [**599**], [**277**], [**141**], [**319**], [**484**], Hinton and Lewis [**333**], Lewis [**445**], Evans, Kwong and Zettl [**174**], Kwong and Zettl [**415**], and the references in these books and papers. (This is a very incomplete and not up to date list.) One of the best known BD results is the Molchanov criterion : Let $J = (0, \infty)$, $p = 1 = w$ and assume that q is bounded below on J; then property BD holds if and only if, for some h, $0 < h < \infty$,

$$\int_t^{t+h} q \to \infty, \text{ as } t \to \infty.$$

The Molchanov criterion has been extended to include the coefficients p and q by Müller-Pfeiffer [**484**] and by Kwong and Zettl [**415**]. It is interesting to note that these extensions are far from trivial and the resulting conditions are complicated. This illustrates an interesting point: Many authors prove their results only for the case when $p = 1 = w$ and then claim their result holds for the general case when all three coefficients p, q, w are present because this general equation (when $p > 0$, $w > 0$) can be transformed into an equation with leading coefficient and weight function both equal to 1. But how do the conditions on the potential of the transformed equation transform back to the original equation with all three coefficients p, q, w present? The Molchanov criterion illustrates that the answer to this question may be quite complicated. I believe that the claim "we can assume without loss of generality that $p = 1 = w \cdots = w''$" is often misleading.

For what coefficients $1/p, q, w$ is σ_0 finite? If it is finite, what is it? As discussed in points (9) and (10) for r-periodic coefficients the answer to both questions is $\sigma_0 = \lambda_0^P$. In general not much is known. For some results see Hinton [**329**], Harris [**303**], [**297**], Everitt [**193**], Everitt, Kwong and Zettl [**174**] and the references in these papers.

The list of references given for point (12) is nearly random and not up to date. Much more is known but also there is much more to learn. There are numerous research problems in this area but the interested reader is advised to search the literature for more up to date information before embarking on such a research project.

Hinton and Shaw [**338**], [**336**], [**335**] have shown that the essential spectrum is invariant under L^1 perturbations of the potential q, Markus and Moore [**459**] have studied the almost periodic case. In the general case when the coefficients are not periodic, almost periodic or L^1 perturbations of such coefficients, not much is known about the essential spectrum. Some problems: (i) What is σ_0? Is σ_e a half line? Are there gaps in the essential spectrum? an infinite number? a finite number? How many? Where are they located? Are there eigenvalues in gaps? How many and in which gaps? Are there embedded eigenvalues? How many?

CHAPTER 11

Singular Indefinite Problems

Gardner, Martin

> Biographical history, as taught in our public schools, is still largely a history of boneheads: ridiculous kings and queens, paranoid political leaders, compulsive voyagers, ignorant generals – the flotsam and jetsam of historical currents. The men who radically altered history, the great scientists and mathematicians, are seldom mentioned, if at all.

In G. Simmons Calculus Gems, New York: McGraw Hill, 1992.

> Mathematics is not only real, but it is the only reality. That is that entire universe is made of matter, obviously. And matter is made of particles. It's made of electrons and neutrons and protons. So the entire universe is made out of particles. Now what are the particles made out of? They're not made out of anything. The only thing you can say about the reality of an electron is to cite its mathematical properties. So there's a sense in which matter has completely dissolved and what is left is just a mathematical structure.

Gardner on Gardner: JPBM Communications Award Presentation. Focus-The Newsletter of the Mathematical Association of America v. 14, no. 6, December 1994.

1. Introduction and Results

In this chapter we study the spectrum of Sturm-Liouville problems (SLP) associated with the differential equation

$$-(py')' + qy = \lambda wy \text{ on } J = (a,b), -\infty \leq a < b \leq \infty, \tag{11.1.1}$$

with coefficients satisfying the basic conditions

$$\frac{1}{p}, q, w \in L_{\text{loc}}(J, \mathbb{R}), \ p > 0, \ |w| > 0 \text{ a.e. on } J, \ w \text{ changes sign on } J; \tag{11.1.2}$$

and suitable boundary conditions. The form of these boundary conditions depends on the classification of the endpoints a, b of the interval J as regular, or singular and, when singular, whether limit-point or limit-circle in the space

$$H = L^2(J, |w|) = \{f : J \to \mathbb{C} : \int_J |f|^2 \, |w| < \infty\},$$

with inner product and norm given by

$$(f, g) = \int_J f \bar{g} \, |w|, \quad \|f\|^2 = \int_J |f|^2 \, |w|.$$

For details see Section 7.3 or [**645**].

Our approach is based on the well established theory based on the equation

$$-(py')' + qy = \lambda |w|\, y \text{ on } J. \tag{11.1.3}$$

This is the so-called right-definite case which can be studied using operator theory in the Hilbert space H. Let S be a self-adjoint realization of (11.1.3) in H, i.e.

$$S_{\min} \subseteq S = S^* \subseteq S_{\max}$$

where $S_{\min}$, $S_{\max}$ are the minimal and maximal operators associated with (11.1.1), respectively. Let $D(S)$ denote the domain of such an operator S. We refer to such domains as self-adjoint domains in H. If both endpoints are in the limit-point case in H then $S_{\min} = S = S^* = S_{\max}$ and there is no proper self-adjoint restriction of $S_{\max}$ and hence there are no boundary conditions. In all other cases the operators S are determined by restricting the domain $D(S_{\max})$ of $S_{\max}$ with self-adjoint boundary conditions. The theory of self-adjoint operators in H yields a characterization of all self-adjoint domains. For details, see Chapter 10. Also see [**487**], [**645**].

To "transfer" results from the right-definite theory with weight function $|w|$ to corresponding problems for the indefinite weight function w we use results and methods of operator theory in the Krein space $K = L^2(J, w)$. This is the space of all (equivalence classes) of functions from H but with the indefinite inner product

$$[f, g] = \int_a^b f\, \overline{g}\, w, \ f, g \in H. \tag{11.1.4}$$

Thus the Hilbert space H and the Krein space K consist of the same set of elements but have different inner products. Operators between these spaces are "connected" by means of the so-called fundamental symmetry operator $\mathcal{J}$ defined by

$$(\mathcal{J}f)(t) = f(t)\, sgn(w(t)), \ f \in H. \tag{11.1.5}$$

We are particularly interested in the self-adjoint realizations T of (11.1.1) in the Krein space K. These are given by

$$T = \mathcal{J}S \tag{11.1.6}$$

where S is a self-adjoint realization of (11.1.3). Curgus and Langer [**133**] have shown that equation (11.1.6) determines a 1-1 onto correspondence between the self-adjoint realizations of (11.1.3) in the Hilbert space H and the self-adjoint realizations of (11.1.1) in the Krein space K. Therefore the self-adjoint boundary conditions that determine S, obtained from the right-definite theory in H with weight function $|w|$, are precisely the same boundary conditions that determine the self-adjoint operators T in the Krein space K with 'weight function' w.

Note that the operators T given by (11.1.6) map H into H and thus are operators in this space. But as an operator in H, T is not self-adjoint or even symmetric. However, observe that the spectrum of T considered as an operator in H is the same as the spectrum of T considered as an operator in K; this follows directly from the definition of the spectrum and from the fact that the topology of K is the same as that of H, both being generated by the norm of H. *Thus by the spectrum of the Sturm-Liouville problem consisting of the equation (11.1.1) together with the boundary conditions that determine S and therefore also T we mean the spectrum of T. Note that this reduces to the right-definite case when $w > 0$ on J.*

How is the spectrum of S, $\sigma(S)$, related to the spectrum of T, $\sigma(T)$? This is the question we study in this chapter. There seems to be no simple answer to this question. The spectrum of S is real, the spectrum of T may not be. (The spectrum of a self-adjoint operator in a Krein space is not necessarily real.) In 1918 Richardson [**546**] showed that, even in the regular case, there are such problems with nonreal eigenvalues.

Of particular interest is the existence and number of nonreal eigenvalues. Our first theorem identifies a class of problems which have only real spectrum.

THEOREM 11.1.1. *Let (11.1.2) hold, let S be a self-adjoint realization of (11.1.3) in the Hilbert space $H = L^2(J, |w|)$ and let the operator T in H be defined by (11.1.6), (11.1.5). Assume that $\inf(\sigma(S)) > 0$. Suppose there exist (nondegenerate) subinterval J_+, J_- of J such that w is positive a.e. on J_+ and negative a.e. on J_-. Then $\sigma(T)$ is real and is unbounded above as well as below.*

PROOF. To be given in Section 11.5. □

The unboundedness of the spectrum, above and below, when $w > 0$ and p changes sign, was established by Möller in [**474**].

There is an extensive literature on regular problems satisfying the hypotheses of Theorem 11.1.1; these are called left-definite. For some recent results see Kong, Wu and Zettl [**387**], Binding and Volkmer [**79**], Binding and Browne [**75**], Volkmer [**579**], Schneider and Vonhoff [**586**], Čurgus and Langer [**133**]. More information can be obtained from the references of these papers.

Our second main result gives conditions for the essential spectrum to be real and gives an upper bound for the number of nonreal eigenvalues. Recall that a closed operator in Banach spaces is called a Fredholm operator if its null space is finite dimensional and its range has finite codimension (this implies that its range is closed). The essential spectrum of a closed operator T in a Banach space, denoted by $\sigma_e(T)$, is the set of all $\lambda \in \mathbb{C}$ such that $T - \lambda I$ is not a Fredholm operator, where I denotes the identity operator.

THEOREM 11.1.2. *Let the operators S and T be defined as in Theorem 11.1.1 above. Assume that for some $\varepsilon > 0$, there are exactly m points of $\sigma(S)$, counting multiplicity, to the left of ε, for some $m \in \mathbb{N}_0 = \{0, 1, 2, 3, ...\}$. Then the essential spectrum of T is real and T has at most $2m$ nonreal eigenvalues, counting multiplicity.*

PROOF. This theorem is a special case of the abstract Theorem 11.3.1 below. □

Since the coefficients p, q, w are real, the nonreal eigenvalues occur in conjugate pairs.

We do not know if the essential spectrum of T is always real. However, under an additional condition we can give an affirmative answer.

THEOREM 11.1.3. *Let the operators S and T be defined as in Theorem 11.1.1 above. Assume that J is the union of finitely many intervals $J_1, \ldots, J_n$ such that w does not change sign on J_k for $k = 1, \ldots, n$. Then the essential spectrum of T is real.*

PROOF. To be given in Section 11.5. □

The next result shows that there is a close relationship between the discreteness of the spectra of S and T.

THEOREM 11.1.4. *Let the operators S and T be defined as in Theorem 11.1.1 above. If the essential spectrum of S, $\sigma_e(S)$, is empty, then either $\sigma_e(T)$ is empty or consists of the entire complex plane. In particular, $\sigma_e(S)$ is empty when each endpoint is either regular or limit-circle.*

PROOF. The first part follows directly from Proposition 11.3.3 below. The last sentence follows from this proposition and the well known fact [**487**] that $\sigma_e(S)$ is empty, i.e. the spectrum is discrete, when each endpoint is either regular or limit-circle. □

For regular problems with separated boundary conditions the upper bound of the number of nonreal eigenvalues given by Theorem 11.1.2 is well known, see [**133**], [**472**].

Our proofs are heavily dependent on the Krein space theory approach. A detailed treatment of such an approach has been undertaken by Čurgus and Langer in the fundamental paper [**133**]. This is based on Langer's theory of definitizable operators in Krein spaces and a deep understanding of the spectral theory of self-adjoint operators is required. Čurgus and Langer are mainly interested in the spectral resolution and characterizations of singular critical points. As a by-product, they have some information about non-real eigenvalues.

Since we are particularly interested in the non-real spectrum, we give a direct description of the underlying Krein space theory. Our interest in this particular topic leads to a treatment which only uses basic Krein space theory. Apart from the right Krein space, which is the space $L^2(J,w)$, we also construct "a left Krein space", which reduces to the Hilbert space $D(S^{\frac{1}{2}})$ for strictly positive definite operators S. We show that under certain assumptions, which correspond to the conditions in [**133**] that the quadratic form $[Tf,f]$ has only finitely many negative squares, the left Krein space reduces to a Pontryagin space.

For the convenience of the reader, we present the basic definitions and facts on Krein and Pontryagin spaces. For more indepth results we refer the interested reader to the monographs of Bognar [**91**] and Azizov and Iohvidov [**30**]. Most of the results on Krein spaces established here are known to specialists in this area. However, we feel it is appropriate to present these results in as simple and self-contained a manner as possible to readers mainly interested in differential equations. In this regard we observe that in [**133**] Čurgus and Langer start with a study of minimal operators in Krein space and develop the theory of self-adjoint extensions of these—in parallel with the GKN Hilbert space theory. However, since boundary conditions are self-adjoint for the weight w if and only if they are self-adjoint for the weight $|w|$ nothing is lost by using the 1-1, onto, correspondence given by equation (11.1.6). Thus the reader can go directly from the self-adjoint, regular or singular, separated or coupled, boundary conditions obtained from the Hilbert space theory for $|w|$ - discussed above in considerable detail in Chapters 4 and 10, to the same boundary conditions for w is the Krein space setting of $L^2(J,w)$.

The characterization of $(S^{\frac{1}{2}})$ used below is due to Krein [**403**], [**404**] but since this work is not easily accessible, we give a self contained proof of it here.

This Chapter is organized as follows: Following this introduction in Section 1, Section 2 gives a short introduction to Krein spaces, and Section 3 contains an overview of self-adjoint operators in Krein spaces and their spectra. In Section 4 a left Krein space is constructed which is related to the "left-definite Hilbert space theory". Section 5 contains proofs of Theorems 11.1.1 and Theorem 11.1.3.

2. Krein Spaces

In this section we give basic definitions and results on Krein spaces. For more details we refer the interested reader to [**91**], [**345**], [**30**].

For a linear operator T we use $D(T)$, $R(T)$ and $N(T)$ to denote its domain, range and nullspace, respectively.

A space $(K, [\cdot, \cdot])$ is called a Krein space if K is a vector space and

$$[\cdot, \cdot] : K \times K \to \mathbb{C} \qquad (11.2.1)$$

is a sesquilinear form such that there are linear submanifolds K_+ and K_- with $K = K_+ \dotplus K_-$ (algebraic direct sum) such that $(K_+, [\cdot, \cdot])$ as well as $(K_-, -[\cdot, \cdot])$ are Hilbert spaces and $[K_+, K_-] = \{0\}$, i.e. $[f, g] = 0$ for all $f \in K_+$ and $g \in K_-$. We write $K = K_+ \oplus K_-$ and call this a fundamental decomposition of the Krein space K. Fundamental decompositions are not unique.

Note that $[f, f] \in \mathbb{R}$ for any $f \in K$. Unless $K_+ = \{0\}$ or $K_- = \{0\}$, there are elements $f \in K \setminus \{0\}$ such that $[f, f] = 0$. Such an element f is called neutral.

If K_- is finite dimensional, then the Krein space K is called a Pontryagin space and κ, the dimension of K_-, is called the index of this Pontryagin space. If K_+ is finite dimensional, then we can consider the Krein space $(K, -[\cdot, \cdot])$ instead, and it is therefore no restriction to assume that K_- is finite dimensional if at least one of K_+, K_-, has this property.

The space $H = K$, equipped with the inner product

$$(f, g) = [f_+, g_+] - [f_-, g_-], \qquad (11.2.2)$$

where $f, g \in K$, $f = f_+ + f_-$, $g = g_+ + g_-$, $f_+, g_+ \in K_+$, $f_-, g_- \in K_-$, is a direct sum of Hilbert spaces and therefore a Hilbert space itself, to be called the associated Hilbert space of K (with respect to the decomposition $K = K_+ \dotplus K_-$). The inner product of H is not unique since it depends on the decomposition of K. The topology of the Krein space K is defined to be the topology generated by the norm of this Hilbert space H. It is this topology which determines the continuity of linear operators in K and their resolvent sets and hence their spectra.

The map $\mathcal{J}$ on K given by

$$\mathcal{J}f = f_+ - f_-, \; f = f_+ + f_-, \; f_+ \in K_+, \; f_- \in K_-, \qquad (11.2.3)$$

is linear and continuous and satisfies $\mathcal{J}^2 = I$, $\mathcal{J}^* = \mathcal{J}$, where $\mathcal{J}^*$ is the adjoint of $\mathcal{J}$ in the associated Hilbert space H. This map $\mathcal{J}$ is called a fundamental symmetry; it connects the Krein space inner product $[\cdot, \cdot]$ with the Hilbert space inner product $(\cdot, \cdot)$ by means of the formulas:

$$[f, g] = (\mathcal{J}f, g), \; (f, g) = [\mathcal{J}f, g], \; f, g \in K. \qquad (11.2.4)$$

Conversely, if $\mathcal{J}$ is a bounded self-adjoint linear operator on a Hilbert space H such that $\mathcal{J}^2 = I$, then by the spectral theory of self-adjoint operators in Hilbert space $[f, g] = (\mathcal{J}f, g)$ defines a Krein space structure on H, $\mathcal{J}$ is a fundamental symmetry, and $N(\mathcal{J} - I) \oplus N(\mathcal{J} + I)$ is its fundamental decomposition.

Of particular interest in this paper are the Hilbert and Krein spaces constructed as follows:

Let $J = (a,b)$, $-\infty \leq a < b \leq \infty$, be an interval and $w \in L_{\text{loc}}(J, \mathbb{R})$, i.e. w is a locally Lebesgue integrable real valued function, with $w(t) \neq 0$ for almost all $t \in J$. Then it is well known that the set of (equivalence classes of) measurable functions f on J such that
$$\int_J |w| \, |f|^2 < \infty$$
is a Hilbert space, denoted $L^2(J, |w|)$, with inner product
$$(f,g) = \int_J |w| \, f \, \overline{g} \,. \tag{11.2.5}$$
The linear operator $\mathcal{J}$ defined by
$$(\mathcal{J} f)(t) = sgn(w(t)) \, f(t) = \frac{w(t)}{|w(t)|} f(t), \ t \in J, \ f \in H, \tag{11.2.6}$$
is continuous, self-adjoint and satisfies $\mathcal{J}^2 = I$. Hence, with respect to the inner product
$$[f,g] = \int_J w \, f \, \overline{g} \,, \tag{11.2.7}$$
$L^2(J, |w|)$ becomes a Krein space which we denote by $L^2(J, w)$. Its fundamental symmetry $\mathcal{J}$ is generated by multiplication by $sgn(w)$ and its fundamental decomposition is given by
$$L^2(J, w) = L^2(J_-, |w|) \oplus L^2(J_+, |w|), \tag{11.2.8}$$
where
$$J_\pm = \{t \in J : \pm w(t) > 0\}.$$
Note that $J_\pm$ is not necessarily an interval and need not even contain an interval.

3. Self-Adjoint Operators in Krein Spaces

In this section we discuss some of the basic theory of self-adjoint operators in a Krein space $(K, [\cdot, \cdot])$ with fundamental symmetry $\mathcal{J}$ and its associated Hilbert space $(H = (K, (\cdot, \cdot))$. Let T be a densely defined linear operator from H into H and T^* its Hilbert space adjoint. The Krein space adjoint T^+ of T is defined by
$$[Tf, g] = [f, T^+ g]$$
for $f \in D(T)$ and $g \in D(T^+)$, where $D(T^+)$ is the set of all $g \in K$ such that the map $f \to [Tf, g]$ is continuous. Note that
$$T^+ = \mathcal{J} T^* \mathcal{J}.$$

A densely defined linear operator T in a Krein space is called Hermitian if $T \subset T^+$. As in the Hilbert space case this is equivalent to requiring that $[Tf, f] \in \mathbb{R}$ for all $f \in D(T)$. The operator T is called self-adjoint in the Krein space K if $T = T^+$. Note that T is self-adjoint in K if and only if $\mathcal{J} T$ is self-adjoint in H. Thus if T is self-adjoint in K then it is closed and its resolvent set is given by
$$\rho(T) = \{\lambda \in \mathbb{C} : T - \lambda \text{ is 1-1 and onto}\}.$$
This follows from the definition of the resolvent set and the Closed Graph Theorem.

In contrast to the Hilbert space case, the spectrum of self-adjoint operators in Krein spaces is, in general, not real. But the points in the spectrum occur in complex conjugate pairs:

3. SELF-ADJOINT OPERATORS IN KREIN SPACES

PROPOSITION 11.3.1. *If T is a self-adjoint operator in a Krein space, then its spectrum is symmetric with respect to the real axis.*

PROOF. Since a closed linear operator is 1-1, onto if and only if its Hilbert space adjoint has this property, it is clear that any of $T - \lambda$, $T^* - \bar\lambda$, $JT^*J - \bar\lambda = T^+ - \bar\lambda = T - \bar\lambda$ being 1-1, onto implies that all these operators are 1-1, onto. Thus $\lambda \in \rho(T)$ if and only if $\bar\lambda \in \rho(T)$. Therefore $\rho(T)$ is symmetric with respect to the real axis, and hence so is the spectrum of T, $\sigma(T) = \mathbb{C} \setminus \rho(T)$. □

Let $\mathbb{C}^+ = \{\lambda \in \mathbb{C} : \operatorname{Im} \lambda > 0\}$, $\mathbb{C}^- = \{\lambda \in \mathbb{C} : \operatorname{Im} \lambda < 0\}$.

PROPOSITION 11.3.2. *Let T be a self-adjoint operator in a Krein space $(K, [\cdot, \cdot])$ and let L be a finite dimensional subspace of K which is invariant under the operator T such that all the eigenvalues of the restriction of T to L are either in $\mathbb{C}^+$ or in $\mathbb{C}^-$. Then $[f, f] = 0$ for all $f \in L$. In particular, $[f, f] = 0$ for all eigenvectors f corresponding to nonreal eigenvalues.*

PROOF. We only establish the case for $\mathbb{C}^+$ since the case for $\mathbb{C}^-$ is similar. Let I_L denote the identity map on L and let $Q = T$ restricted to L. For any $r > 0$ such that the eigenvalues of Q all are inside the semi-circle in the upper half-plane of radius r centered at the origin, we have that

$$I_L = -\frac{1}{2\pi i} \int_{-r}^{r} (Q - \alpha I)^{-1} d\alpha + \frac{1}{2\pi} \int_0^\pi \left(I - \frac{e^{-i\varphi}}{r} Q\right)^{-1} d\varphi.$$

Since

$$\frac{1}{2\pi} \int_0^\pi \left(I - \frac{e^{-i\varphi}}{r} Q\right)^{-1} d\varphi \to \frac{1}{2} I_L$$

as $r \to \infty$, it follows that

$$\lim_{r \to \infty} -\frac{1}{2\pi i} \int_{-r}^{r} (Q - \alpha I)^{-1} d\alpha = \frac{1}{2} I_L.$$

Thus for any $f \in L$ we have that

$$[f, f] = \left[\frac{i}{\pi} \int_{-\infty}^{\infty} (Q - \alpha)^{-1} d\alpha\, f, f\right] = \frac{i}{\pi} \int_{-\infty}^{\infty} [(Q - \alpha)^{-1} f, f]\, d\alpha.$$

Since T and therefore Q is symmetric, $[(Q - \alpha)^{-1} f, f] \in \mathbb{R}$ for all real α, the above identity can only hold if both sides are zero. □

THEOREM 11.3.1. *Let T be a self-adjoint operator in a Krein space with fundamental symmetry J, and let $S = JT$. Assume the spectrum of S is finite below some positive number ε, i.e. $\sigma(S) \cap (-\infty, \varepsilon)$ consists of at most a finite number of eigenvalues of finite multiplicity. Let m be the total multiplicity of these eigenvalues. (If there are no eigenvalues below ε, then $m = 0$.) Then the essential spectrum of T (if any) is real, the nonreal part of the spectrum of T (if any) consists of finitely many nonreal eigenvalues, and the root subspaces of these eigenvalues have finite dimensions and their total dimension is at most $2m$. In particular the nonreal eigenvalues have finite multiplicities and their total multiplicity is at most $2m$.*

PROOF. First we show that the essential spectrum of T is real. In a manner similar to the proof of Proposition 11.3.1 we establish that $T - \lambda$ is Fredholm if and only if $T - \bar\lambda$ is Fredholm and thus if and only if $N(T - \lambda)$ and $N(T - \bar\lambda)$ are finite dimensional and $R(T - \lambda)$ and $R(T - \bar\lambda)$ are closed, since the finite codimensionality of the last two spaces immediately follows from, e.g., $R(T^* - \overline{\lambda}) = \overline{R(T^* - \overline{\lambda})} = N(T - \lambda)^\perp$ and $R(T - \overline{\lambda}) = \mathcal{J} R(T^* - \overline{\lambda}) \mathcal{J}$.

Thus $\lambda \in \sigma_e(T)$ if and only if $N(T - \lambda)$ or $N(T - \overline{\lambda})$ is infinite dimensional or $T - \lambda$ or $T - \overline{\lambda}$ not open, and hence there is an infinite dimensional submanifold M of $D(T)$ such that either $\|(T - \lambda)f\| \leq \frac{\varepsilon \operatorname{Im}(\lambda)}{3|\lambda|}$ for all $f \in M$ with $\|f\| = 1$ or $\|(T - \overline{\lambda})f\| \leq \frac{\varepsilon \operatorname{Im}(\lambda)}{3|\lambda|}$ for all $f \in M$ with $\|f\| = 1$. Here and in the following the norm is the one associated with the Hilbert space inner product $(\cdot, \cdot)$.

Assume there is $\lambda \in \sigma_e(T) \setminus \mathbb{R}$. Without loss of generality we may assume the first case. Then
$$\operatorname{Im}[(T - \lambda)f, f] = -\operatorname{Im}\lambda [f, f]$$
and
$$|[(T - \lambda)f, f]| \leq \|\mathcal{J}(T - \lambda)f\| \|f\| = \|(T - \lambda)f\| \|f\|$$
yield
$$|[f, f]| \leq \frac{\varepsilon}{3|\lambda|} \text{ for } f \in M, \|f\| = 1.$$
Then
$$(Sf, f) = [Tf, f] = [(T - \lambda)f, f] + \lambda [f, f]$$
$$\leq \frac{\varepsilon |\operatorname{Im}\lambda|}{3|\lambda|} + \frac{\varepsilon}{3} < \varepsilon$$
for all $f \in M$, with $\|f\| = 1$. But this contradicts the minimax principle since $\sigma(S) \cap (-\infty, \varepsilon)$ consists of at most finitely many eigenvalues of finite multiplicity.

By Proposition 11.3.2 the submanifold L spanned by all eigenvectors belonging to eigenvalues in the upper half-plane $\mathbb{C}^+$ consists of neutral elements. By the polarization formula for inner products, the inner product is identically zero on L. From $TL \subset L$, $[Tf, f] = 0$ for all $f \in L$ and the minimax principle, it follows that $\dim L \leq m$. Since the total multiplicity of an isolated eigenvalue λ of T coincides with the total multiplicity of the eigenvalue $\overline{\lambda}$ of $T^* = T$, it follows that the total multiplicity of all nonreal eigenvalues is at most $2m$.

To show that there is no other non-real spectrum, we observe that if $\lambda \in \sigma(T) \setminus \sigma_e(T)$, then the observations at the beginning of this proof imply that $N(T - \lambda) \neq \{0\}$ or $N(T - \overline{\lambda}) \neq \{0\}$. But since there are only finitely many nonreal eigenvalues, there is $\mu \in \mathbb{C} \setminus \mathbb{R}$ such that $N(T - \mu) = \{0\}$ and $N(T - \overline{\mu}) = \{0\}$. Thus $T - \mu$ and $T - \overline{\mu}$ have index 0, and by the index stability theorem, see [**367**, Theorem Iv.5.22], the index must be 0 in $\mathbb{C}^+$ and $\mathbb{C}^-$, and more generally, on $\mathbb{C} \setminus \sigma_e(T)$. So every point in $\mathbb{C} \setminus \sigma_e(T)$ must be an eigenvalue. $\square$

PROPOSITION 11.3.3. *Let K be a Krein space, T a self-adjoint operator in K, $\mathcal{J}$ a fundamental symmetry on K, and let $S = \mathcal{J}T$. Assume that $\rho(T)$, the resolvent set of T, is not empty. If the essential spectrum of S is empty, then the essential spectrum of T is empty.*

PROOF. Since S is self-adjoint in a Hilbert space, $\mathbb{C} \setminus \mathbb{R} \subset \rho(S)$. Hence we can choose $\lambda \in \rho(T)$ and $\mu \in \rho(S)$. Since $\sigma_e(S) = \emptyset$, the spectral theorem for self-adjoint operators shows that
$$S = \sum_{j=1}^{\infty} \lambda_j P_j$$

with mutually orthogonal finite rank projections P_j and $|\lambda_j| \to \infty$ as $j \to \infty$, and thus

$$(S - \mu)^{-1} = \sum_{j=1}^{\infty} (\lambda_j - \mu)^{-1} P_j$$

is compact. Hence, in view of $(T - \mu \mathcal{J})^{-1} = (S - \mu)^{-1} \mathcal{J}$,

$$(T - \mu \mathcal{J})^{-1} - (T - \lambda)^{-1} = (T - \mu \mathcal{J})^{-1}(T - \lambda - T + \mu \mathcal{J})(T - \lambda)^{-1}$$
$$= (T - \mu \mathcal{J})^{-1}(\mu \mathcal{J} - \lambda)(T - \lambda)^{-1}$$

is compact, and hence also $(T - \lambda)^{-1}$ is compact. This completes the proof since an operator with compact resolvent has no essential spectrum, see e.g. [**367**, Theorem III.6.29]. □

The following example shows that $\rho(T) = \emptyset$ can occur even if $\sigma_e(S) = \emptyset$.

EXAMPLE 11.3.1. Let $Ly = y'$ on $L^2(0,1)$ with domain $D(L) = H^1(0,1) = \{y \in L^2(0,1) : y \in AC_{loc}(0,1), y' \in L^2(0,1)\}$. Then L is a closed operator with dense domain and $D(L^*) = \{y \in H^1(0,1) : y(0) = 0 = y(1)\}$. Let $H = (H^1(0,1))^2$ and let $\mathcal{J} : H \to H$ be given by

$$\mathcal{J} = \begin{pmatrix} 0 & I \\ I & 0 \end{pmatrix},$$

where I is the identity operator on $L^2(0,1)$. Then H becomes a Krein space K with fundamental symmetry $\mathcal{J}$. Define $T : K \to K$ by

$$T = \begin{pmatrix} L & 0 \\ 0 & L^* \end{pmatrix}$$

Since

$$\mathcal{J} T = \begin{pmatrix} 0 & L^* \\ L & 0 \end{pmatrix}$$

is self-adjoint in H, T is self-adjoint in K. For each $\lambda \in \mathbb{C}$, $y(t) = e^{\lambda t}$ is in $D(L)$ and satisfies $Ly = \lambda y$. Thus $\rho(T)$ is empty.

On the other hand, $D(LL^*)$ is the set of all $y \in H^2(0,1) = \{y \in H^1(0,1) : y' \in H^1(0,1)\}$ satisfying $y(0) = 0 = y(1)$; and $LL^* y = y''$. Thus LL^* is a regular, self-adjoint Sturm-Liouville operator with Dirichlet boundary conditions and its spectrum is discrete consisting entirely of simple isolated eigenvalues. By Theorem 1.2 in [**473**], $\sigma(\mathcal{J} T)$ is discrete and $\sigma_e(\mathcal{J} T) \subset \{0\}$. But $\dim N(\mathcal{J} T) = \dim N(T) = \dim N(L) = 1$, and consequently the essential spectrum of $\mathcal{J} T$ is empty.

4. A Construction of Left-Definite Krein Spaces

Given an invertible self-adjoint operator T in a Krein space K we construct a new Krein space associated with T. This new space is called the left-definite Krein space associated with T and is denoted by K_T. We continue to use the notation from the previous sections.

THEOREM 11.4.1. *Let T be an invertible self-adjoint operator in a Krein space with fundamental symmetry $\mathcal{J}$ and let $S = \mathcal{J} T$. Let $U = |S|^{1/2}$. Then there is an inner product $[\cdot, \cdot]_1$ on $D(U)$ such that $K_T = (D(U), [\cdot, \cdot]_1)$ is a Krein space with the following properties:*

(1) $D(T)$ is dense in $D(U)$ with respect to the Hilbert space norm generated by the inner product $[\cdot,\cdot]_1$,
(2) $[Tf,g] = [f,g]_1$ for all f,g in $D(T)$,
(3) $R = T^{-1}|_{D(U)}$ is a continuous self-adjoint operator in the Krein space $\mathcal{K}_T$.

PROOF. It is known, see e.g. [**367**, Theorem VI.6.23], that $D(U)$ is a Hilbert space with respect to the inner product
$$(f,g)_1 = [\mathcal{J}Uf, Ug] = (Uf, Ug). \tag{11.4.1}$$
From the functional calculus for self-adjoint operators in Hilbert space it follows that $\mathcal{J}_1 := \operatorname{sgn}(S)$ is unitary on the Hilbert space $D(U)$ and $\mathcal{J}_1^2 = I$. Thus
$$[f,g]_1 = (\operatorname{sgn}(S)f, g)_1 = (U\operatorname{sgn}(S)f, Ug) \tag{11.4.2}$$
defines a Krein space structure on $D(U)$. Part (1) follows from the fact that $D(T)$ is dense on the Hilbert space $D(U)$, see Theorem VI.6.23 in [**367**, Theorem VI.6.23].

(2) For $f, g \in D(T)$, from (11.4.2) it follows that
$$[f,g]_1 = (U\operatorname{sgn}(U)f, g) = [Tf, g]. \tag{11.4.3}$$

(3) The inclusions
$$T^{-1}D(U) \subset D(T) = D(|S|) \subset D(|S|^{\frac{1}{2}}) \tag{11.4.4}$$
show that R maps $D(U)$ into itself. Also, the continuity of the embedding $\mathcal{K}_T \hookrightarrow \mathcal{K}$ shows that R has a closed graph. Thus R is continuous by the closed graph theorem. Let $f, g \in D(T)$, then $T^{-1}f, T^{-1}g \in D(T)$ and by (2)
$$[Rf,g]_1 = [TT^{-1}f, g] = [f,g] = \overline{[g,f]} = \overline{[Rg,f]_1} = [f, Rg]_1. \tag{11.4.5}$$
Hence R is symmetric on $D(T)$. Since R is continuous on $D(U)$ and $D(T)$ is dense in $D(U)$, R is self-adjoint in $D(U)$. $\square$

In the following $\|\cdot\|$ and $\|\cdot\|_1$ denote the norms associated with the Hilbert space inner products on $\mathcal{K}$ and on $D(U)$, respectively.

THEOREM 11.4.2. *Let the notation and hypotheses of Theorem 11.4.1 hold. Then, for each $\lambda \in \mathbb{C} \setminus \{0\}$, $\lambda \in \sigma(T)$ if and only if $\frac{1}{\lambda} \in \sigma(R)$.*

PROOF. Let $\lambda \neq 0$ be an eigenvalue of T. Then there is a $y \in D(T)$, $y \neq 0$, such that $Ty = \lambda y$. Then $y, Ty \in D(T) \subseteq D(R)$, and $y = \lambda Ry$. Thus $\frac{1}{\lambda}$ is an eigenvalue of T.

Conversely, let $\lambda \neq 0$ be an eigenvalue of R. Then there is $y \in D(U)$, $y \neq 0$, such that $Ry = \lambda y$. Then $y, Ry \in R(R) \subseteq D(T)$ and $y = \lambda Ty$. Thus $\frac{1}{\lambda}$ is an eigenvalue of T.

Now, let $\lambda \in \rho(T) \setminus \{0\}$. By what we have already shown, $R - \frac{1}{\lambda}$ is injective. Let $f \in D(U)$. Since $T - \lambda$ is surjective, there is a $g \in D(T) \subseteq D(U)$ such that $(T - \lambda)g = -\lambda Tf$, which implies that
$$(R - \frac{1}{\lambda})g = (T^{-1} - \frac{1}{\lambda})g = f. \tag{11.4.6}$$
This shows that $R - \frac{1}{\lambda}$ is also surjective, and hence $1/\lambda \in \rho(R)$.

Finally, let $\lambda \in \rho(R) \setminus \{0\}$. From the first part of the proof it follows that $T - 1/\lambda$ is injective. Let $g \in \mathcal{K}$ and put $h = -\lambda T^{-1}g \in D(T)$. Since $R - \lambda$ is surjective, there is an $f \in D(R)$ such that $(R - \lambda)f = h$. Since $f = (Rf - h)/\lambda \in D(T)$, we have that

$$Tf = \frac{1}{\lambda}f - \frac{1}{\lambda}Th = \frac{1}{\lambda}f + g. \tag{11.4.7}$$

This shows that $T - \frac{1}{\lambda}$ is also surjective and hence $\frac{1}{\lambda} \in \rho(T)$. □

The importance of Theorem 11.4.2 is due to the fact that $K_T = (D(U), [\cdot, \cdot]_1)$ may be a Pontryagin space or a Hilbert space.

COROLLARY 11.4.1. *Under the assumptions and with the notation of Theorem 11.4.1 suppose that $\sigma(S) \cap (-\infty, 0)$ consists of at most finitely many negative eigenvalues of finite multiplicity. Let $m \geq 0$ be the total multiplicity of these negative eigenvalues. Then K_T is a Pontryagin space with Pontryagin index m, and $\sigma(T) \setminus \mathbb{R}$ consists of at most finitely many eigenvalues of total multiplicity not exceeding $2m$.*

PROOF. Let $(E_t)_{t \in \mathbb{R}}$ be the spectral family of S. Then

$$(I - E_0)D(U) \oplus E_0 D(U) \tag{11.4.8}$$

is a fundamental decomposition of the Krein space K_T, and $E_0 D(U) = R(E_0)$ has dimension m. Since $0 \notin \sigma(S)$, there is $\varepsilon > 0$ such that $\sigma(S) \cap (-\infty, \varepsilon) = \sigma(S) \cap (-\infty, 0)$ consists of finitely many eigenvalues of finite multiplicity. By Theorem 11.3.1, $\sigma(T) \setminus \mathbb{R}$ consists of at most finitely many eigenvalues with total multiplicity not exceeding $2m$. A routine modification of the proof of Theorem 11.4.2 yields that the multiplicities of the eigenvalues $\frac{1}{\lambda} \in \sigma(S)$ and $\lambda \in \sigma(T)$ coincide. □

REMARK 11.4.1. It would be desirable to have an estimate of the magnitude of the non-real eigenvalues of T, or at least of their imaginary parts, in terms of the negative eigenvalues of S. The following example shows that, in general, such an estimate does not exist.

EXAMPLE 11.4.1. For $c > 0$, $d > 0$, let

$$T = \begin{pmatrix} c & c+d \\ -(c+d) & -c \end{pmatrix}, \quad \mathcal{J} = \begin{pmatrix} 1 & 0 \\ 0 & -1 \end{pmatrix}$$

on $\mathbb{C}^2$. Then $S = \mathcal{J}T$ is self-adjoint with eigenvalues $2c + d$, and $-d$, whereas the eigenvalues of T are $\pm i\sqrt{2cd + d^2}$. Note that for fixed d and c varying in $(0, +\infty)$, the negative eigenvalue of S is bounded but the imaginary parts of the nonreal eigenvalues of T are not bounded.

5. Proof of Theorems 11.1.1 and 11.1.3

We now proceed with the proof of Theorem 11.1.1 using the notation from previous sections.

PROOF OF THEOREM 11.1.1. In view of Corollary 11.4.1, $\sigma(T)$ is real and the space K_T from Theorem 11.4.1 is a Hilbert space. In terms of the operator R from Theorems 11.4.1 and 11.4.2 we have to show that $\sigma(R) \cap (-\varepsilon, 0) \neq \emptyset$ and $\sigma(R) \cap (0, \varepsilon) \neq \emptyset$ for all $\varepsilon > 0$. Assume there is some $\varepsilon > 0$ such that $\sigma(R) \cap (-\varepsilon, 0) = \emptyset$ or $\sigma(R) \cap (0, \varepsilon) = \emptyset$. This means that for $\lambda = -\frac{2}{\varepsilon}$ or $\lambda = \frac{2}{\varepsilon}$ the estimate

$$\left\| \left(R - \frac{1}{\lambda}\right)f \right\|_1 \geq \frac{1}{|\lambda|} \|f\|_1$$

holds for all $f \in D(R)$. This is equivalent to

$$[(R - \frac{1}{\lambda})f, (R - \frac{1}{\lambda}f)]_1 \geq \frac{1}{\lambda^2}[f, f]_1.$$

An application of part (2) of Theorem 11.4.1 shows that

$$[(\lambda - T)f, (\lambda R - I)f] \geq [Tf, f]$$

for $f \in D(T)$. This leads to

$$\lambda^2[f, Rf] - 2\lambda[f, f] \geq 0$$

for all $f \in D(T)$. Since R is a restriction of T^{-1} and T^{-1} is invertible in $L^2(J, w)$, it follows by continuity that

$$\lambda^2[f, T^{-1}f] - 2\lambda[f, f] \geq 0$$

for all $f \in L^2(J, w)$. Putting $g = T^{-1}f$, it follows that

$$\lambda^2[Tg, g] - 2\lambda[Tg, Tg] \geq 0 \tag{11.5.1}$$

for all $g \in D(T)$.

Now let T_+ be the minimal operator associated with (11.1.1) on $L^2(J_+, w)$. Note that $L^2(J_+, w)$ is a Hilbert space and that T_+ is unbounded above. Hence there is $y \in D(T_+) \subset D(T)$ such that $\|Ty\| = \|T_+y\| > \lambda\|y\|$ for $\lambda = \frac{2}{\varepsilon}$. Then

$$\lambda^2[Ty, y] - 2\lambda[Ty, Ty] \leq \lambda^2\|Ty\|\,\|y\| - 2\lambda\|Ty\|^2$$
$$= \|Ty\|(\lambda^2\|y\| - 2\lambda\|Ty\|)$$
$$< -\lambda^2\|Ty\|\,\|y\| < 0,$$

which contradicts (11.5.1) for $\lambda = \frac{2}{\varepsilon}$.

Let T_- be the minimal operator associated with (11.1.1) on $L^2(J_-, -w)$. Note that $L^2(J_+, -w)$ is a Hilbert space and that T_- is unbounded below. Hence there is $y \in D(T_-) \subset D(T)$ such that $\|Ty\| = \|T_-y\| > -\lambda\|y\|$ for $\lambda = -\frac{2}{\varepsilon}$. Then

$$\lambda^2[Ty, y] - 2\lambda[Ty, Ty] \leq \lambda^2\|Ty\|\,\|y\| + 2\lambda\|Ty\|^2$$
$$= \|Ty\|(\lambda^2\|y\| + 2\lambda\|Ty\|)$$
$$< -\lambda^2\|Ty\|\,\|y\| < 0,$$

which contradicts (11.5.1) for $\lambda = -\frac{2}{\varepsilon}$.

Proof of Theorem 11.1.3. Let T_k, $k = 1, \ldots, n$ be the minimal operators associated with the restriction of (11.1.1) to J_k. Then the direct sum $T_1 \oplus \cdots \oplus T_n$ on $L^2(J_1, w) \oplus \cdots \oplus L^2(J_n, w)$ can be identified with an operator $\widetilde{T} \subset T$. Since each of the T_j is a minimal operator in the Hilbert space $L^2(J_k, |w|)$ (we may replace λ with $-\lambda$, if necessary, $\sigma_e(T_j) \subset \mathbb{R}$. That is, $T_k - \lambda$ is a Fredholm operator for all $\lambda \in \mathbb{C} \setminus \mathbb{R}$. Thus also $(T_1 - \lambda) \oplus \cdots \oplus (T_n - \lambda)$ is Fredholm, and so is $\widetilde{T} - \lambda$. Since $\widetilde{T} \subset T$, $T - \lambda$ has a finite codimensional and thus closed range for $\lambda \in \mathbb{C} \setminus \mathbb{R}$, see [**279**, IV.1.3]. This together with $N(T - \lambda) = R(T - \bar{\lambda})^\perp$ shows that $T - \lambda$ is a Fredholm operator for $\lambda \in \mathbb{C} \setminus \mathbb{R}$. $\square$

6. Comments

Although operator theory in the Krein space $K = L^2(J, w)$, when w changes sign, has been used [133] to study Sturm-Liouville problems, new concrete results about specific classes of Sturm-Liouville problems are hard to find. Although only basic Krein space theory is used here a number of new results are obtained which give new information about large and well identified classes of SLP.

A proof of Theorem 11.1.3 can also be based on the multiple interval theory developed by Everitt and Zettl [243], [238] to be discussed in Chapter 13: Consider the minimal operator on each subinterval and use the abstract Hilbert space result which says that the essential spectrum of the direct sum of a finite number of closed symmetric operators is the union of the essential spectra of these operators.

CHAPTER 12

Singular Left-Definite Problems

Einstein, Albert (1879-1955)
  Everything should be made as simple as possible, but not simpler.
Reader's Digest, October 1977.
  Imagination is more important than knowledge.
  On Science.
  The most beautiful thing we can experience is the mysterious. It is the source of all true art and science.
  What I Believe.
  Since the mathematicians have invaded the theory of relativity, I do not understand it myself anymore.
In A. Sommerfelt "To Albert Einstein's Seventieth Birthday" in Paul A. Schilpp (ed.) Albert Einstein, Philosopher-Scientist, Evanston, 1949.
  Do not worry about your difficulties in mathematics, I assure you that mine are greater.
  It is nothing short of a miracle that modern methods of instruction have not yet entirely strangled the holy curiousity of inquiry.
In H. Eves Return to Mathematical Circles, Boston: Prindle, Weber and Schmidt, 1988.
  The search for truth is more precious than its possession.
The American Mathematical Monthly, vol. 100, no. 3.

1. Introduction

In this chapter we study singular left-definite (LD) Sturm-Liouville problems for the equation

$$M y := -(py')' + qy = \lambda w y \text{ on } J = (a,b), \ -\infty \leq a < b \leq \infty, \quad (12.1.1)$$

where the weight function w changes sign. The existence of eigenvalues is established by elementary means, i.e. without using Hilbert space or Krein space operator theory.

Our approach is based on the method of "spectral-curves" generated by the equation

$$-(py')' + (q - \lambda w) y = \xi |w| y \text{ on } J \quad (12.1.2)$$

and a self-adjoint domain, where ξ and λ are spectral parameters. For each $\lambda \in \mathbb{R}$ and each $n \in \mathbb{N}_0$, $\xi_n(\lambda)$ is defined by applying the well known max-min principle from the right-definite theory to equation (12.1.2) together with a self-adjoint boundary condition (or self-adjoint domain); see Lemma 12.2.3 below. These spectral curves $\xi_n(\lambda)$, $n \in \mathbb{N}_0$, yield information about the spectrum of LD problems associated with (12.1.1) and the same domain, i.e. the same boundary condition. In particular we will see how these curves determine and characterize eigenvalues of left definite problems. For regular problems these spectral-curves reduce to the eigencurves studied in Chapter 5.

Since the special case $\lambda = 0$ in (12.1.2) plays a critical role, so we highlight the equation

$$My = -(py')' + qy = \xi \, |w| \, y \text{ on } J. \tag{12.1.3}$$

We call (12.1.3) the RD equation associated with (12.1.1) and (12.1.2) the one parameter family of RD equations associated with (12.1.1).

Recall from Section 11.1, in particular from (11.1.6) and (11.1.5) that the equation

$$T = \mathcal{J}S \tag{12.1.4}$$

determines a 1-1 onto correspondence between the self-adjoint realizations S of (12.1.3) in the Hilbert space $H = L^2(J, |w|)$ and the self-adjoint realizations T of (12.1.1) in the Krein space $K = L^2(J, w)$. The spectrum of a such a self-adjoint operator T may not be real. Such an operator T maps H into H and therefore can be viewed as an operator in H. It follows from the definition of spectrum and the fact that the topology of H is the same as that of K that the spectrum of T is the same regardless of whether T is considered as an operator in H or in K. Of course, considered as an operator in H, T is not self-adjoint or even symmetric when w changes sign. Thus from this point of view, the main results of this chapter can be viewed as determining the spectrum of a class of non-self-adjoint operators in the Hilbert space H.

In this chapter we identify and study a class of operators T which have real spectrum; this is the so called class of left-definite operators. These are the operators T whose corresponding operator S has its spectrum bounded below by a positive constant, see Lemma 12.3.3 below. The operator S is a self-adjoint realization of (12.1.3) in the Hilbert space H and the operator T is a self-adjoint realization of (12.1.1) in the Krein space K *but both are determined by the same boundary condition, say D. Just as the spectrum of S, $\sigma(S)$, is the spectrum of* (12.1.3) with the boundary condition D, so the spectrum of T, $\sigma(T)$, is the spectrum of the boundary value problem consisting of equation (12.1.1) with the boundary condition D.

The relationship between the spectra of S and T seems to be rather complicated. But we will see in the next few sections that much can be learned about the spectrum of a left-definite operator T from the one parameter family of right-definite operators $S(\lambda)$ associated with equation (12.1.2).

1.1. A Two Parameter LC/LP Dichotomy. Although we are primarily interested in Sturm-Liouville Problems with real valued coefficients we allow the coefficients to take complex values in this subsection. The reason for this is to establish the celebrated LC/LP dichotomy of Weyl for the general two-parameter equation (12.1.2) with complex coefficients satisfying the minimal conditions:

$$1/p, \, q, \, w \in L_{loc}(J; \mathbb{C}). \tag{12.1.5}$$

1. INTRODUCTION

LEMMA 12.1.1. *Let (12.1.5) hold. Let $c \in J$. Assume that all solutions of (12.1.2) are in $L^2((a,c), |w|)$ for some pair $\{\lambda, \xi\} = \{\lambda^*, \xi^*\} \in \mathbb{C}^2$. Then this is true for all pairs $\{\lambda, \xi\} \in \mathbb{C}^2$. Similarly for the endpoint b.*

PROOF. The basic idea of the proof is to apply the system regularization method discussed in Section 7.2 to the two parameter equation (12.1.2). By Theorem 7.2.2 if all solutions of (12.1.2) are in $L^2((a,c), |w|)$, for some pair $\{\lambda, \xi\} = \{\lambda^*, \xi^*\} \in \mathbb{C}^2$, then this is true for all pairs $\{\lambda, \xi\} = \{\lambda^*, \xi\} \in \mathbb{C}^2$. Therefore, it is sufficient to show that if all solutions are in $L^2((a,c), |w|)$ for some $\{\lambda^*, \xi^*\} \in \mathbb{C}^2$, then this is true for all $\{\lambda, \xi^*\}$ with $\lambda \in \mathbb{C}$.

To prove this, let u_1, u_2 be linearly independent solutions of (12.1.2) with $\{\lambda, \xi\} = \{\lambda^*, \xi^*\}$ such that $\det \begin{pmatrix} u_1 & u_2 \\ pu_1' & pu_2' \end{pmatrix}(t) \equiv 1$, for all $t \in J$. Then by hypothesis we have $u_1, u_2 \in L^2((a,c), |w|)$. Let

$$Y = \begin{pmatrix} y \\ py' \end{pmatrix}, \quad P = \begin{pmatrix} 0 & 1/p \\ q - \xi^* |w| & 0 \end{pmatrix}, \quad W = \begin{pmatrix} 0 & 0 \\ w & 0 \end{pmatrix}.$$

The system formulation of equation (12.1.2) with $\xi = \xi^*$ is given by

$$Y' = (P - \lambda W)Y. \tag{12.1.6}$$

Note that $U = \begin{pmatrix} u_1 & u_2 \\ pu_1' & pu_2' \end{pmatrix}$ is a fundamental matrix of (12.1.2) for $\lambda = \lambda^*$ with $\det U(t) \equiv 1$, for all $t \in J$. For any vector or matrix solution $Y(t, \lambda)$ of (12.1.6) we define

$$X(t, \lambda) = U^{-1}(t) Y(t, \lambda), \quad t \in J. \tag{12.1.7}$$

Then a routine computation shows that $X(t, \lambda)$ is a vector or matrix solution of the equation

$$X' = (\lambda^* - \lambda) G X \quad \text{on } J, \tag{12.1.8}$$

where

$$G = U^{-1} W U = \begin{pmatrix} -u_1 u_2 w & -u_2^2 w \\ u_1^2 w & u_1 u_2 w \end{pmatrix}.$$

It follows from the hypotheses and the Schwarz inequality that each component of G is in $L(a, c)$. Hence (12.1.8) is regular at a. This implies that every solution $X(t)$ has a finite limit at a and hence is bounded on (a, c). Note from (12.1.7) that Y is a solution of (12.1.6) if and only if X is a solution of equation (12.1.8). Hence for every scalar solution y of (12.1.2) there exists a vector solution $X = \begin{pmatrix} x_1 \\ x_2 \end{pmatrix}$ of (12.1.8) such that

$$y = u_1 x_1 + u_2 x_2.$$

Hence $y \in L^2((a,c), |w|)$. The proof for the endpoint b is similar. □

COROLLARY 12.1.1. *If all solutions of (12.1.2) are in $L^2(J, |w|)$ for some pair $\{\lambda, \xi\} = \{\lambda^*, \xi^*\} \in \mathbb{C}^2$, then this is true for all pairs $\{\lambda, \xi\} \in \mathbb{C}^2$.*

PROOF. This is an immediate consequence of Lemma 12.1.1. □

REMARK 12.1.1. Note that only the minimal conditions (12.2.1) are imposed on the coefficients of equation (12.1.2) in Lemma 12.2.1. So each of $1/p, q, w$ can be complex-valued and, if real-valued, it may change sign. Also, each of $1/p, q, w$ can be identically zero on one or more subintervals of J.

2. An Associated One Parameter Family of Right Definite Operators

In the rest of this chapter we assume that the coefficients of (12.1.1) are real valued and satisfy

$$1/p, q, w \in L_{loc}(J; \mathbb{R}), \quad p > 0, |w| > 0 \text{ a.e. on } J, \quad w \text{ changes sign on } J. \quad (12.2.1)$$

Recall that "w changes sign on J" means that w assumes positive and negative values on subsets of J with positive Lebesgue measure. Such a w is also referred to as "indefinite".

Throughout this chapter we let $H = L^2(J, |w|)$ be the Hilbert space with the inner product $(f, g) = \int_a^b f \bar{g} |w|$, and $AC_{loc}(J, \mathbb{C})$ the set of complex valued functions which are absolutely continuous on each compact subinterval of J.

Consider the maximal and minimal domains and maximal and minimal operators associated with equation (12.1.2) given by

$$D_{\max}(\lambda) = \{f \in H : f, pf' \in AC_{loc}(J; \mathbb{C}), (Mf - \lambda w f)/|w| \in H\},$$
$$S_{\max}(\lambda)f = (Mf - \lambda w f)/|w| \text{ for all } f \in D_{\max}(\lambda),$$

and

$$S_{\min}(\lambda) = S_{\max}^*(\lambda), \quad D_{\min}(\lambda) = \text{domain of } S_{\min}(\lambda), \quad (12.2.2)$$

where $S_{\max}^*(\lambda)$ denotes the adjoint of the operator $S_{\max}(\lambda)$ in H.

Recall that by a self-adjoint realization of (12.1.2) in H we mean an operator $S(\lambda)$ satisfying

$$S_{\min}(\lambda) \subset S(\lambda) = S^*(\lambda) \subset S_{\max}(\lambda).$$

Such a self-adjoint realization is determined completely by its domain and can be viewed either as a restriction of the maximal operator $S_{\max}(\lambda)$ or an extension of the minimal operator $S_{\min}(\lambda)$. We will refer to the domains of self-adjoint realizations as self-adjoint domains.

The special case $\lambda = 0$ plays an important role; for this case we abbreviate the notation to: $S_{\max} := S_{\max}(0)$, $S_{\min} := S_{\min}(0)$, and S for arbitrary self-adjoint realizations $S(0)$.

LEMMA 12.2.1. *Let (12.2.1) hold. For each $\lambda \in \mathbb{R}$ we have*

$$D_{\max}(\lambda) = D_{\max}(0) \text{ and } D_{\min}(\lambda) = D_{\min}(0).$$

Hence $D_{\max}(\lambda)$ and $D_{\min}(\lambda)$ can be written as $D_{\max}$ and $D_{\min}$, respectively.

If S is a self-adjoint realization of (12.1.3) in H with domain D, then for each $\lambda \in \mathbb{R}$ the operator $S(\lambda) : D \to H$ defined by

$$S(\lambda) f = \frac{Mf - \lambda w f}{|w|}, \quad f \in D, \quad (12.2.3)$$

is a self-adjoint realization of (12.1.2), and all self-adjoint realizations of (12.1.2) are obtained this way, i.e., they are determined by the domain of a self-adjoint realization S of (12.1.3).

Furthermore, if $S(\lambda)$ is bounded below for some $\lambda \in \mathbb{R}$, then it is bounded below for every $\lambda \in \mathbb{R}$.

PROOF. From its definition we see that the maximal domain $D_{\max}(\lambda)$ is independent of λ. Hence the minimal domain $D_{\min}(\lambda)$ is also independent of λ.

Suppose S is a self-adjoint realizations of (12.1.2) in H with domain D. Define an operator $\Phi(\lambda) : H \to H$ by $\Phi(\lambda) f = \lambda(sgn w) f$. Then $\Phi(\lambda)$ is self-adjoint on H

2. AN ASSOCIATED ONE PARAMETER FAMILY OF RIGHT DEFINITE OPERATORS

for any $\lambda \in \mathbb{R}$. We observe that $S(\lambda) = S - \Phi(\lambda)$. Hence $S(\lambda)$ is self-adjoint on D. Conversely, if $S(\lambda)$ is self-adjoint in H with domain D, then S is self-adjoint with domain D.

To prove the furthermore statement assume that for some $\lambda = \lambda^*$, there exists $c \in \mathbb{R}$ such that $(S(\lambda^*)f, f) \geq c(f, f)$ for all $f \in D$. Then for any $\lambda \in \mathbb{R}$ and $f \in D$

$$(S(\lambda)f, f) = (\frac{1}{|w|}(Mf - \lambda^* wf), f) - ((\lambda - \lambda^*)\frac{w}{|w|}f, f) \geq (c - |\lambda - \lambda^*|)(f, f),$$

i.e., $S(\lambda)$ is bounded below by $c - |\lambda - \lambda^*|$. □

REMARK 12.2.1. It follows from Lemma 12.2.1 that the self-adjoint domains are invariant with respect to $\lambda \in \mathbb{R}$ and thus, in particular, are given by those when $\lambda = 0$. Note however that, although their domains are independent of λ, the self-adjoint operators $S(\lambda)$ depend on λ. This dependence is critical in our approach to the study of left-definite problems using an associated family of right-definite problems.

For the convenience of the reader we now state the well-known min-max principle for RD problems, which plays an important role in this chapter. We denote by $G_n(V)$ the set of n-dimensional subspaces of a vector space V.

LEMMA 12.2.2. *Suppose $S : D \to H$ is a self-adjoint operator and is bounded below. Let $\sigma_e(S)$ be the essential spectrum of S. For each $n \in \mathbb{N}_0$, define*

$$s_n(S) = \inf\{\sup\{(Sf, f) : f \in F \cap U\} : F \in G_{n+1}(D)\},$$

where U is the unit sphere in H. Then for any fixed $n \in \mathbb{N}_0$, either

$s_n(S) = \inf \sigma_e(S)$, *in this case, $s_n(S) = s_{n+1}(S) = s_{n+2}(S) = \cdots$, there are at most $n + 2$ eigenvalues of S, counting multiplicity, less than or equal to $\inf \sigma_e(S)$, and these eigenvalues are among $s_0(S)$, $s_1(S)$, ..., $s_{n+1}(S)$; or*

$s_n(S) < \inf \sigma_e(S)$, *in this case, there are at least $n + 1$ eigenvalues $s_0(S)$, $s_1(S)$, ..., $s_n(S)$ of S, counting multiplicity, strictly less than $\inf \sigma_e(S)$ (or ∞ if $\sigma_e(S) = \emptyset$), and $s_n(S)$ is the $(n + 1)$-th eigenvalue of S, counting multiplicity.*

PROOF. See [**141**], p. 1543. □

Although the characterization of the self-adjoint domains in terms of boundary conditions is discussed in Chapter 10 (also see [**645**], [**386**]), we summarize it here to make more explicit the singular boundary conditions for equation (12.1.1) to which our results in this chapter apply.

The number of BC's needed and allowed to determine self-adjoint domains depends on the LC/LP classification of the endpoints. Recall that the Lagrange sesquilinear form is given by

$$[f, g] = f(p\overline{g}') - \overline{g}(pf'), \ f, g \in D_{\max}.$$

PROPOSITION 12.2.1. *The boundary conditions determining self-adjoint domains for (12.1.3) are as follows:*

1. Suppose both endpoints are LP. In this case, no BC's are needed or allowed and $S_{\min}$ is self-adjoint and has no proper self-adjoint extension in H.

2. Suppose a is LC and b is LP. In this case, there exist real-valued $u, v \in D_{\max}$ such that $[u, v](a) \neq 0$. For $\alpha \in [0, \pi)$, define
$$D = \{y \in D_{\max} : \cos\alpha \, [y, u](a) - \sin\alpha \, [y, v](a) = 0\}.$$
Then D is a self-adjoint domain and all self-adjoint domains are obtained this way.

3. Suppose a is LP and b is LC. In this case, there exist real-valued $u, v \in D_{\max}$ such that $[u, v](b) \neq 0$. For $\beta \in (0, \pi]$, define
$$D = \{y \in D_{\max} : \cos\beta \, [y, u](b) - \sin\beta \, [y, v](b) = 0\}.$$
Then D is a self-adjoint domain and all self-adjoint domains are obtained this way.

4. Suppose both endpoints are LC. In this case, there exist real-valued $u, v \in D_{\max}$ such that $[u, v](a) = [u, v](b) = -1$. Define
$$D = \{y \in D_{\max} : AY(a) + BY(b) = 0\}$$
where $Y = \begin{pmatrix} [y, u] \\ [y, v] \end{pmatrix}$ and A, B are 2×2 complex matrices satisfying
$$AEA^* = BEB^*, \; \operatorname{rank}(A|B) = 2, \; E = \begin{pmatrix} 0 & 1 \\ -1 & 0 \end{pmatrix}, \tag{12.2.4}$$
here A^*, B^* are the complex conjugate transposes of A, B, respectively. Then D is a self-adjoint domain and all self-adjoint domains are obtained this way.

The BC's given in part (4) can be classified into two mutually exclusive classes: separated and coupled. The former has the canonical representation:
$$\begin{array}{ll} \cos\alpha \, [y, u](a) - \sin\alpha \, [y, v](a) = 0, & \alpha \in [0, \pi), \\ \cos\beta \, [y, u](b) - \sin\beta \, [y, v](b) = 0, & \beta \in (0, \pi]; \end{array} \tag{12.2.5}$$
and the latter has the canonical representation:
$$Y(b) = e^{i\gamma} K Y(a), \tag{12.2.6}$$
where $K \in SL(2, \mathbb{R}) := \{K \in M_2(\mathbb{R}) : \det K = 1\}$ and $\gamma \in [-\pi, \pi)$. The coupled condition (12.2.6) is said to be real if $\gamma = 0$ or $\gamma = -\pi$ and is called non-real otherwise.

REMARK 12.2.2.

(i) Recall from Chapter 10 that the "BC functions" $\{u, v\}$ in Proposition 12.2.1 can be viewed as a basis in BC space and the "BC constants" α, β or K, γ, as the coordinates of a specific BC with respect to a given basis $\{u, v\}$. Given a fixed basis $\{u, v\}$, by varying the BC coordinates through their prescribed ranges, all self-adjoint domains are obtained. Thus, the same self-adjoint domain D has different representations depending on which BC basis $\{u, v\}$ is used. How do the BC coordinates change when the BC basis $\{u, v\}$ changes? This question is answered in Section 10.4.

(ii) If a is a regular endpoint, then the BC basis $\{u, v\}$ can be chosen to satisfy the initial conditions : $u(a) = 0$, $(pu')(a) = 1$; $v(a) = 1$, $(pv')(a) = 0$; in this case, the BC in case 2 of Proposition 12.2.1 reduces to the more familiar form
$$\cos\alpha \, y(a) - \sin\alpha \, (py')(a) = 0, \; \alpha \in [0, \pi).$$
Similarly for the case that b is a regular endpoint.

If both endpoints are regular, using the Naimark Patching Lemma we may construct a BC basis $\{u, v\}$ such that for $c = a$, or $c = b$,
$$u(c) = 0, \; (pu')(c) = 1; \; v(c) = 1, \; (pv')(c) = 0.$$

In this case, (12.2.5) and (12.2.6) take the more familiar forms
$$\cos\alpha\, y(a) - \sin\alpha\, (py')(a) = 0, \quad \alpha \in [0,\pi),$$
$$\cos\beta\, y(b) - \sin\beta\, (py')(b) = 0, \quad \beta \in (0,\pi],$$
and
$$Y(b) = e^{i\gamma} K\, Y(a),\ Y = \begin{pmatrix} y \\ py' \end{pmatrix};$$
respectively.

3. Existence of Eigenvalues

In this section we establish the existence of real eigenvalues for singular LD SLP's associated with equation (12.1.1).

Let S be a self-adjoint realization of (12.1.3) in H with domain D. Define functionals $\mathcal{L}$ and $\mathcal{R}$ from D to $\mathbb{R}$ by

$$\mathcal{L}f = \int_a^b (Mf)\bar{f},\ f \in D \quad \text{and} \quad \mathcal{R}f = \int_a^b |f|^2 w,\ f \in D. \tag{12.3.1}$$

We denote by U the unit sphere in the Hilbert space H.

We now state the definition of LD SLP for the general case, note that this is consistent with Definition 5.2.1 in the regular case.

DEFINITION 12.3.1. *Let (12.2.1) hold. Let U denote the unit sphere in H. Let S be a self-adjoint realization of (12.1.3) in H with domain D and let T (also with domain D) be given by (12.1.4).*

Then T is left-definite (LD) if there exists a number $c > 0$ such that

$$\mathcal{L}f \geq c \quad \text{for all } f \text{ in } D \cap U. \tag{12.3.2}$$

REMARK 12.3.1. There are various definitions of left-definiteness in the literature. Littlejohn and Wellman [449] give a general abstract definition of LD operators which are bounded below; this rules out SLP's with indefinite weight functions. Vonhoff [587], [588] - following Pleijel [517], Bennewitz and Everitt [65], Everitt [197], Niessen and Schneider [500], [499] and others, define left-definiteness for indefinite weight functions by constructing a Sobolev type Hilbert space based on the left side of equation (12.1.1) and then using the operator theory in this space rather than in the L^2 space. In [587] the self-adjoint realizations in the Sobolev type Hilbert space are characterized, in general, by λ-dependent BC's. Note that Definition (12.3.1) does not involve spectral parameter dependent boundary conditions.

In this section, under the assumption (12.2.1) we define the left-definiteness only for problems with indefinite weight functions. This is done solely for convenience in stating the results. Clearly, Definition (12.3.1) can also be applied to problems with positive weight functions.

LEMMA 12.3.1. *Let (12.2.1) hold. Let S be a self-adjoint realization of (12.1.3) in H with domain D and let T be given by (12.1.4). For each $\lambda \in \mathbb{R}$, let $S(\lambda)$ be given by (12.2.3). Then λ is in the spectrum of T if and only if $0 \in \sigma(S(\lambda))$.*

PROOF. Note that the linear operator S is self-adjoint and hence closed. So, $S(\lambda)$ and

$$\frac{1}{w}M = (sgnw)\frac{1}{|w|}M = (sgnw)S : D \to H$$

are also closed. Thus, by (12.2.3),

$$0 \notin \sigma(S(\lambda)) \iff S(\lambda) = \frac{1}{|w|}M - \lambda\, sgnw : D \to H \text{ is } 1-1 \text{ and onto}$$

$$\iff \frac{1}{w}M - \lambda = (sgnw)S(\lambda) : D \to H \text{ is } 1-1 \text{ and onto}$$

$$\iff \lambda \text{ is not in the spectrum of } \frac{1}{w}M : D \to H.$$

$\square$

LEMMA 12.3.2. *Let (12.2.1) hold. Let S be a self-adjoint realization of (12.1.3) in H with domain D and let T be given by (12.1.4). Assume T is left-definite. Then the spectrum of T is real and 0 is not in the spectrum.*

PROOF. For $f \in D$, $(Sf, f) = \mathcal{L}f$. Thus, (12.3.2) implies that c is a lower bound for the self-adjoint operator S, and hence is also a lower bound for the spectrum $\sigma(S)$. Since $c > 0$, $0 \notin \sigma(S)$. By Lemma 12.3.1 with $\lambda = 0$, 0 is not in the spectrum of T. Hence, Theorem 11.1.2 implies that the spectrum of T is real. $\square$

DEFINITION 12.3.2. *Let (12.2.1) hold. Let U denote the unit sphere in H. Let $S(\lambda), \lambda \in \mathbb{R}$, be a self-adjoint realization of (12.1.2) given by (12.2.3) with domain D. Then for each $n \in \mathbb{N}_0$*

$$\xi_n(\lambda) = \inf\{\sup\{(S(\lambda)f, f) : f \in F \cap U\} : F \in G_{n+1}(D)\} \qquad (12.3.3)$$

is defined for all $\lambda \in \mathbb{R}$. We call $\xi = \xi_n(\lambda), \lambda \in \mathbb{R}$, the n-th spectral-curve of S.

If for each $\lambda \in \mathbb{R}$, $\xi_n(\lambda)$ is an eigenvalue of $S(\lambda)$, then $\xi = \xi_n(\lambda), \lambda \in \mathbb{R}$, is called an eigencurve of $S(\lambda)$.

REMARK 12.3.2. By Lemma (12.2.1), if S is bounded below, then $\xi_n(\lambda)$ is finite for each $n \in \mathbb{N}_0$ and every $\lambda \in \mathbb{R}$. In general, a spectral-curve may contain eigenvalues for some values of λ and may equal the infimum of the essential spectrum for other values of λ. (The infimum $\inf \sigma_e(S)$ may or may not be an eigenvalue.) Thus, in general, $\xi_n(\lambda)$ may or may not be an eigencurve; and if it is not an eigencurve some part of it may consist of eigenvalues. For instance, $\xi_n(\lambda)$ may be an eigenvalue for each $\lambda > 0$ and not for $\lambda < 0$ or the other way around.

We observe that for the operator $S(\lambda)$ defined by (12.2.3) and for any $f \in U$

$$(S(\lambda)f, f) = \mathcal{L}f - \lambda\mathcal{R}f.$$

Hence (12.3.3) becomes

$$\xi_n(\lambda) = \inf\{\sup\{\mathcal{L}f - \lambda\mathcal{R}f : f \in F \cap U\} : F \in G_{n+1}(D)\}. \qquad (12.3.4)$$

Note that λ *is an eigenvalue of* (12.1.1) *if and only if* $\xi = 0$ *is an eigenvalue of* (12.1.2) *for the same boundary condition; in this case, these two eigenvalues* λ *and* 0 *have the same eigenspace.*

LEMMA 12.3.3. *Let (12.2.1) hold. Let S be a self-adjoint realization of (12.1.3) in H with domain D and let T be given by (12.1.4). Then T is left-definite if and only $\xi_0(0) > 0$.*

PROOF. From (12.3.4) we have that

$$\xi_0(0) = \inf\{\mathcal{L}f : f \in D \cap U\}.$$

Then (12.3.2) holds for some $c > 0$ if and only if $\xi_0(0) > 0$. $\square$

3. EXISTENCE OF EIGENVALUES

We can now state our main existence result, its proof will be given in Section 12.4.

THEOREM 12.3.1. *Let (12.2.1) hold. Let $S(\lambda)$, $\lambda \in \mathbb{R}$, be a self-adjoint realization of equation (12.1.2) defined by (12.2.3) with domain D and let T be given by (12.1.4), let $S = S(0)$. Let $\xi_n(\lambda)$ be defined by (12.3.3). Assume T is LD. Then for each $n \in \mathbb{N}_0$, the equation $\xi_n(\lambda) = 0$ has exactly one positive root $\lambda = \lambda_n$ and exactly one negative root $\lambda = \lambda_{-n}$, and these roots satisfy $|\lambda_{\pm n}| \geq \xi_n(0)$.*

(i) Suppose that for some $n \in \mathbb{N}_0$, there exists $\lambda_n > 0$ such that $\xi_n(\lambda_n) = 0$ and 0 is an eigenvalue of $S(\lambda_n)$. Then T has positive eigenvalues λ_j, $j = 0, \ldots, n$, which can be indexed to satisfy

$$0 < \lambda_0 \leq \lambda_1 \leq \cdots \leq \lambda_n. \qquad (12.3.5)$$

Moreover, these eigenvalues are the only points of the spectrum of T in $[0, \lambda_n]$.

(ii) Suppose that for some $m \in \mathbb{N}_0$, there exists $\lambda_{-m} < 0$ such that $\xi_m(\lambda_{-m}) = 0$ and 0 is an eigenvalue of $S(\lambda_{-m})$. Then T has negative eigenvalues λ_{-j}, $j = 0, \ldots, m$, which can be indexed to satisfy

$$\lambda_{-m} \leq \ldots \lambda_{-1} \leq \lambda_{-0} < 0. \qquad (12.3.6)$$

Moreover, these eigenvalues are the only points of the spectrum of T in $[\lambda_{-m}, 0]$.

(iii) Suppose that for some $m, n \in \mathbb{N}_0$, there exist $\lambda_{-m} < 0$ and $\lambda_n > 0$ such that $\xi_m(\lambda_{-m}) = 0 = \xi_n(\lambda_n)$ and 0 is an eigenvalue of both $S(\lambda_{-m})$ and $S(\lambda_n)$. Then $\{\lambda_j : j = -m, \cdots, -1, -0, 0, 1, \cdots, n\}$ are eigenvalues of T and they can be indexed to satisfy

$$\lambda_{-m} \leq \cdots \leq \lambda_{-1} \leq \lambda_{-0} < 0 < \lambda_0 \leq \lambda_1 \leq \cdots \leq \lambda_n. \qquad (12.3.7)$$

Moreover, these eigenvalues are the only points of the spectrum of T in $[\lambda_{-m}, \lambda_n]$.

In (12.3.5)-(12.3.7) only geometrically double eigenvalues appear twice. Strict inequalities hold throughout (12.3.5)-(12.3.7) whenever the BC is separated or non-real coupled, or at least one endpoint is LP.

PROOF. This will be given in Section 12.4 below. □

The next Corollary is an immediate consequence of Theorem 12.3.1.

COROLLARY 12.3.1. *Let the hypotheses and notation of Theorem 12.3.1 hold. Suppose that for each $n \in \mathbb{N}_0$, there exists an $m \in \mathbb{N}_0$ such that $m \geq n$ and $\xi_m(\lambda_{-m}) = 0 = \xi_m(\lambda_m) = 0$ and 0 is an eigenvalue of both operators $S(\lambda_{\pm m})$. Then T has an infinite number of positive and an infinite number of negative eigenvalues $\{\lambda_j : j \in \mathbb{Z}^*\}$ which can be indexed to satisfy*

$$\cdots \leq \lambda_{-2} \leq \lambda_{-1} \leq \lambda_{-0} < 0 < \lambda_0 \leq \lambda_1 \leq \lambda_2 \leq \cdots . \qquad (12.3.8)$$

and for each $n \in \mathbb{N}_0$, λ_n and λ_{-n} are the unique positive and negative roots of the equation $\xi_n(\lambda) = 0$, respectively. Moreover, these eigenvalues are the only points of the spectrum of T in (λ_-, λ_+), where $\lambda_{\pm} = \lim_{n \to \pm \infty} \lambda_n$. In (12.3.8) only geometrically double eigenvalues appear twice. Strict inequalities hold throughout (12.3.8) whenever the BC is separated or non-real coupled, or at least one endpoint is LP.

REMARK 12.3.3. From Lemma 12.2.2 we see that

$$\xi_n(\lambda) \leq \inf \sigma_e(S(\lambda)) \quad \lambda \in \mathbb{R}, \ n \in \mathbb{N}_0.$$

Thus if for some $n \in \mathbb{N}_0$ there exists $\lambda \in \mathbb{R}$ such that $\xi_n(\lambda) < \inf \sigma_e(S(\lambda))$, then $\xi_n(\lambda)$ is an eigenvalue of $S(\lambda)$. Therefore, the assumption of Theorem 11.3.1, (i), is satisfied if for some $n \in \mathbb{N}_0$, there exists $\lambda_n > 0$ such that $\xi_n(\lambda_n) = 0 < \inf \sigma_e(S(\lambda_n))$. Similarly, the assumption of Theorem 11.3.1, (ii), is satisfied if for some $m \in \mathbb{N}_0$, there exists $\lambda_{-m} < 0$ such that $\xi_m(\lambda_{-m}) = 0 < \inf \sigma_e(S(\lambda_{-m}))$. Furthermore, the assumption of Corollary 12.3.1 is satisfied if for each $n \in \mathbb{N}_0$, there exist $m \geq n$ and $\lambda_{-m} < 0$, $\lambda_m > 0$ such that $\xi_m(\lambda_{\pm m}) = 0 < \inf \sigma_e(S(\lambda_{\pm m}))$, respectively.

THEOREM 12.3.2. *Let the hypotheses and notation of Theorem 12.3.1 hold.*
(i) Suppose that for some $n \in \mathbb{N}_0$, $\xi = \xi_n(\lambda)$, $\lambda \in \mathbb{R}$, *is an eigencurve of S. Then the conclusion of Theorem 11.3.1, (iii) holds with* $m = n$.
(ii) Suppose that for each $n \in \mathbb{N}_0$, $\xi = \xi_n(\lambda)$, $\lambda \in \mathbb{R}$, *is an eigencurve of S. Then the conclusion of Corollary 12.3.1 holds.*

PROOF. This follows directly from Theorem 11.3.1. □

Now we state a result on the numbers of zeros of eigenfunctions for the LP case. A parallel result for the LC nonoscillatory case will be given in Section 12.5.

COROLLARY 12.3.2. *Let the hypotheses and notation of Theorem 12.3.1 hold. Suppose that at least one endpoint is LP, and suppose that* λ_n, $n \in \mathbb{Z}^*$, *is the eigenvalue of T with index n ensured by either Theorem 11.3.1, Corollary 12.3.1, or Theorem 12.3.2. Then any eigenfunction of λ_n has exactly $|n|$ zeros in the open interval J.*

PROOF. Note that λ_n is the eigenvalue of T with index n if and only if 0 is the $|n|$-th eigenvalue of $S(\lambda_n)$ and the two eigenvalues have the same eigenspace. Hence the conclusion follows from Theorem 10.12.1; see also Theorem 14.10 in [**600**]. □

REMARK 12.3.4. Whenever one or more eigencurves $\xi_n(\lambda)$ exist, Theorem 12.3.2 yields an algorithm for the numerical computation of the eigenvalues of general singular LD problems with separated or coupled boundary conditions: Use the code SLEIGN2, see [**43**], to compute an eigencurve $\xi_n(\lambda)$, then use a root finder to locate its unique positive root λ_n and its unique negative root λ_{-n}. Note that, see Remark 12.3.3 above, if the eigencurve $\xi_n(\lambda)$ exists for some $n \in \mathbb{N}_0$, then all eigencurves $\xi_j(\lambda)$ exist for $j = 0, 1, ..., n$. The asymptotic behavior and the monotonicity properties of the eigencurves, see Lemmas 12.4.2 and 12.4.4 below, guarantee an efficient root finding scheme. In particular, when neither endpoint is LP, then all eigencurves $\xi_n(\lambda)$, $n \in \mathbb{N}_0$, exist. This case is studied in detail in Section 12.5 below.

THEOREM 12.3.3. *Let (12.2.1) hold. Let S be a self-adjoint realization of equation (12.1.3) with domain D and let T be given by (12.1.4). Assume that $\sigma(S)$ is discrete and has a positive lower bound. Then T is LD, its spectrum consists of only real eigenvalues, the eigenvalues are unbounded from below and above, and can be ordered to satisfy*

$$\cdots \leq \lambda_{-2} \leq \lambda_{-1} \leq \lambda_{-0} < 0 < \lambda_0 \leq \lambda_1 \leq \lambda_2 \leq \cdots$$

with only geometrically double eigenvalues appearing twice.

PROOF. By definition T is LD. From Lemma 12.3.2 we then deduce that its spectrum is real. By Theorem 11.1.4 $\sigma_e(T) = \emptyset$ since $\sigma_e(S) = \emptyset$. Hence the spectrum of T consists of real eigenvalues only. From Theorem 12.3.2 Part (ii) we know that T has an infinite number of positive and an infinite number of negative eigenvalues. Their multiplicities are either 1 or 2, and hence can be ordered to satisfy the inequalities of the theorem. Since the essential spectrum of the problem is empty, we have that $\lambda_{\pm n} \to \pm\infty$ as $n \to +\infty$, or equivalently in this case, the eigenvalues are unbounded from below and from above. □

The next result illustrates how Theorem 12.3.3 can be used to establish the existence of eigenvalues of LD problems with an LP endpoint. It uses the well known Molchanov criterion for the discreteness of the spectrum of RD problems.

Consider the SLP consisting of the differential equation

$$-y'' + qy = \lambda wy \quad \text{on } J = (0, \infty) \tag{12.3.9}$$

and the BC

$$\cos\alpha\, y(0) - \sin\alpha\, y'(0) = 0, \ \alpha \in [0, \pi). \tag{12.3.10}$$

THEOREM 12.3.4. *Let (12.3.9), (12.3.10) hold. Assume*

(1) $q, w \in L_{loc}([0, \infty), \mathbb{R})$, *and w changes sign on J;*
(2) *there exist $k, h \in \mathbb{R}$, $k > 0$, $h > 0$, such that $q(t) \geq k$ a.e. on J and $\int_t^{t+h} q \to \infty$ as $t \to \infty$;*
(3) *there exist $k_1 > 0$, $k_2 > 0$, such that $k_1 \leq |w(t)| \leq k_2$ a.e on J.*

Then for each $\alpha \in [0, \pi/2]$, the spectrum of the SLP (12.3.9), (12.3.10), consists of only real eigenvalues. These are unbounded from below and from above and can be ordered to satisfy

$$\cdots < \lambda_{-2} < \lambda_{-1} < \lambda_{-0} < 0 < \lambda_0 < \lambda_1 < \lambda_2 < \cdots.$$

Moreover, for every $n \in \mathbb{Z}^*$, each eigenfunction for λ_n has exactly $|n|$ zeros in the open interval J.

PROOF. Consider the associated family of RD problems consisting of the equation

$$-y'' + (q - \lambda w)y = \xi|w|y \quad \text{on } J \tag{12.3.11}$$

and BC (12.3.10) with $\alpha \in [0, \pi)$. First note that (12.3.9) is regular at 0 and LP at infinity for each $\lambda \in \mathbb{R}$. The fact that 0 is a regular endpoint is clear from the hypotheses. The LP at infinity classification is well known for $|w| = 1$, see [**141**], and can be shown to hold for w bounded. In fact, the proof of Lemma 12.2.1 is still valid when the constant ξ in equation (12.1.3) is replaced by a bounded function. Let $S(\lambda, \alpha)$ denote the self-adjoint operator realization of the SLP (12.3.11), (12.3.10) in $L^2(J, |w|)$ and denote its spectrum by $\sigma(S(\lambda, \alpha))$. Then $\sigma(S(\lambda, \alpha))$ is bounded below and discrete for each $\lambda \in \mathbb{R}$ and each $\alpha \in [0, \pi)$ by the extension of the well known Molchanov criterion to bounded weight functions, see Theorem 4 in Kwong and Zettl [**415**] (where f can be chosen as 1). Hence $\xi_n(\lambda)$ is an eigencurve for each $n \in \mathbb{N}_0$ and every $\alpha \in [0, \pi)$. Fix an $\alpha \in [0, \pi/2]$. Let g be a real eigenfunction of $\xi_0(0)$ normalized to satisfy $\int_0^\infty g^2|w| = 1$. By Theorem 1 of Everitt [**198**] we have that g' and $q^{1/2}g$ are in $L^2(J,1)$ and $g(t)g'(t) \to 0$ as $t \to \infty$. This and an integration by parts show that for $\alpha \in [0, \pi/2]$,

$$\xi_0(0) = \int_0^\infty g[-g'' + qg] = g(0)g'(0) + \int_0^\infty g'^2 + \int_0^\infty qg^2 > g(0)g'(0) \geq 0.$$

Thus, $\sigma(S(0,\alpha))$ has a positive lower bound. By Theorem 12.3.3, the spectrum of the SLP (12.3.9), (12.3.10) consists of only real eigenvalues, which are unbounded from below and above, and can be ordered to satisfy the inequalities of Theorem 12.3.3. Since the eigenvalues all have multiplicity 1, these inequalities are all strict. The last conclusion of the theorem follows from Corollary 12.3.2. $\square$

4. Lemmas and Proofs

In this section we present several technical lemmas about the spectral-curves $\xi_n(\lambda)$, $n \in \mathbb{N}_0$, defined by (12.3.3), and then use these lemmas to prove Theorem 11.3.1.

LEMMA 12.4.1. *Let the hypotheses and notation of Theorem 12.3.1 hold. For any $\lambda_1, \lambda_2 \in \mathbb{R}$ we have that*

$$|\xi_n(\lambda_1) - \xi_n(\lambda_2)| \leq |\lambda_1 - \lambda_2|. \tag{12.4.1}$$

Hence $\xi_n(\lambda)$ is continuous on $\mathbb{R}$.

PROOF. Note that for any $f \in U$, the unit sphere in H, we have

$$\mathcal{R}f = \int_a^b |f|^2 w \in [-1, 1].$$

Then (12.4.1) follows from (12.3.4). This implies that $\xi_n(\lambda)$ is continuous on $\mathbb{R}$. $\square$

From Lemma 12.4.1 we obtain an interesting result on RD problems.

COROLLARY 12.4.1. *Let the hypotheses and notation of Theorem 12.3.1 hold. Let $S(\lambda)$ be a self-adjoint realization of (12.1.2) with domain D. Assume $\sigma(S(\lambda))$ is bounded below and discrete for some $\lambda \in \mathbb{R}$. Then $\sigma(S(\lambda))$ is bounded below and discrete for all $\lambda \in \mathbb{R}$.*

PROOF. Let $\sigma(S(\lambda))$ be bounded below and discrete for $\lambda = \lambda^*$. Then $\sigma(S(\lambda))$ is bounded below for all $\lambda \in \mathbb{R}$ by Lemma 12.2.1. Note that $\sigma_e(S(\lambda^*)) = \emptyset$ means that $\inf \sigma_e(S(\lambda^*)) = \infty$. Lemma 12.2.2 implies that for each $\lambda \in \mathbb{R}$,

$$\inf \sigma_e(S(\lambda)) = \lim_{n \to \infty} \xi_n(\lambda).$$

Thus $\lim_{n \to \infty} \xi_n(\lambda^*) = \infty$. From (12.4.1) we have for each $\lambda \in \mathbb{R}$, $\lim_{n \to \infty} \xi_n(\lambda) = \infty$. Therefore, $\inf \sigma_e(S(\lambda)) = \infty$, i.e., $\sigma_e(S(\lambda)) = \emptyset$. $\square$

LEMMA 12.4.2. *Let the hypotheses and notation of Theorem 12.3.1 hold. For $n \in \mathbb{N}_0$ we have that*

$$\lim_{\lambda \to \infty} \frac{\xi_n(\lambda)}{\lambda} = -1 \quad \text{and} \quad \lim_{\lambda \to -\infty} \frac{\xi_n(\lambda)}{\lambda} = 1. \tag{12.4.2}$$

4. LEMMAS AND PROOFS

PROOF. Note that the definition of $\xi_0(0)$ implies that $\mathcal{L}f \geq \xi_0(0)$ for all $f \in D \cap U$. Thus, from (12.3.4) we obtain

$$\frac{\xi_n(\lambda)}{\lambda} \geq \frac{\xi_0(0)}{\lambda} - \sup_{f \in D \cap U} \int_a^b |f|^2 w \geq \frac{\xi_0(0)}{\lambda} - \sup_{f \in D \cap U} \int_a^b |f|^2 |w|$$

$$= \frac{\xi_0(0)}{\lambda} - 1 \longrightarrow -1 \text{ as } \lambda \longrightarrow \infty.$$

Since w changes sign and $|w| > 0$ a.e. on J, there exist subsets J_1 and J_2 of J such that $w > 0$ on J_1, $w < 0$ on J_2, $m(J_j) > 0$, $j = 1, 2$, and $m(J) = m(J_1) + m(J_2)$, where m denotes Lebesgue measure. Let $\varepsilon > 0$. Since D is dense in H, we can choose an $(n+1)$-dimensional linear subspace F of D such that for all $f \in F \cap U$ we have

$$\int_{J_1} |f|^2 w > 1 - \varepsilon/2.$$

We observe that

$$1 = \int_J |f|^2 |w| = \int_{J_1} |f|^2 w - \int_{J_2} |f|^2 w.$$

Thus

$$\int_{J_2} |f|^2 w = \int_{J_1} |f|^2 w - 1 > -\varepsilon/2.$$

It follows that

$$\int_J |f|^2 w = \int_{J_1} |f|^2 w + \int_{J_2} |f|^2 w > 1 - \varepsilon.$$

Since F is finite dimensional, there exists an upper bound c for the set $\{\mathcal{L}f : f \in F \cap U\}$. Then from (12.3.4) we have

$$\frac{\xi_n(\lambda)}{\lambda} \leq \frac{c}{\lambda} - (1 - \varepsilon) = \frac{c}{\lambda} - 1 + \varepsilon.$$

The first part of (12.4.2) follows from this. The proof of the second part is similar and hence omitted. □

REMARK 12.4.1. We comment on the relationship between Lemma 12.4.2 and Theorem 2.2 in Binding and Volkmer [**79**]. For regular problems consisting of equation (12.1.2) and separated BC's with w bounded and $|w|$ replaced by 1, Theorem 2.2 in [**79**] gives an asymptotic result similar to (12.4.2) but with $-1, +1$ replaced by $-\sup w$ and $-\inf w$. Note that Lemma 12.4.2 does not require that w be bounded. Such an assumption is a relatively mild restriction in the regular case but a significant one for singular problems. Also, Lemma 12.4.2 applies to regular and singular equations, separated and coupled self-adjoint BC's. This illustrates one of the advantages in using $|w|$ instead of 1 on the right-hand side of equation (12.1.2). Moreover, (12.4.2) holds not only for eigencurves but also for spectral-curves.

LEMMA 12.4.3. *Let the hypotheses and notation of Theorem 12.3.1 hold. Let $n \in \mathbb{N}_0$ and $\lambda \in \mathbb{R}$. Assume $\xi_0(0) > 0$ and $\xi_n(\lambda) < \xi_0(0)$. Let $\varepsilon \in [0, \xi_0(0) - \xi_n(\lambda))$. Then the set*

$$D_{\lambda,\varepsilon} := \{f \in D \cap U : \mathcal{L}f - \lambda \mathcal{R}f < \xi_0(0) - \varepsilon\} \qquad (12.4.3)$$

is nonempty and for any $f \in D_{\lambda,\varepsilon}$ we have

$$\lambda \mathcal{R}f > \varepsilon. \qquad (12.4.4)$$

PROOF. Since $\xi_n(\lambda) < \xi_0(0) - \varepsilon$, $D_{\lambda,\varepsilon}$ is not empty by (12.3.4). Note that the definition of $\xi_0(0)$ implies that $\xi_0(0) \leq \mathcal{L}f$ for any $f \in D \cap U$. Thus for any $f \in D_{\lambda,\varepsilon}$
$$\mathcal{L}f - \lambda \mathcal{R}f < \mathcal{L}f - \varepsilon$$
and (12.4.4) follows. $\square$

LEMMA 12.4.4. *Let the hypotheses and notation of Theorem* 12.3.1 *hold. For* $n \in \mathbb{N}_0$, $\xi_n(\lambda)$ *is strictly decreasing in the region*
$$E_1 = \{(\lambda, \xi) : \lambda > 0, \, \xi < \xi_0(0)\}$$
and strictly increasing in the region
$$E_2 = \{(\lambda, \xi) : \lambda < 0, \, \xi < \xi_0(0)\}.$$

PROOF. Let λ be such that $\xi_n(\lambda) < \xi_0(0)$, and let $\varepsilon \in [0, \xi_0(0) - \xi_n(\lambda))$. From (12.3.4)
$$\xi_n(\lambda) = \inf\{\sup\{\mathcal{L}f - \lambda \mathcal{R}f : f \in F \cap D_{\lambda,\varepsilon}\} : F \in G_{n+1}(D)\},$$
where $D_{\lambda,\varepsilon}$ is defined by (12.4.3). Let λ_1 and λ_2 be such that $0 < \lambda_1 < \lambda_2$ and $(\lambda_1, \xi_n(\lambda_1))$ and $(\lambda_2, \xi_n(\lambda_2))$ are in E_1. Then from (12.4.3) and (12.4.4) we see that for $\varepsilon \in [0, \xi_0(0) - \xi_n(\lambda_1))$,
$$D_{\lambda_1,\varepsilon} \subset D_{\lambda_2,\varepsilon} \subset D \cap U.$$
Thus, for $i = 1$ and 2,
$$\xi_n(\lambda_i) = \inf\{\sup\{\mathcal{L}f - \lambda_i \mathcal{R}f : f \in F \cap D_{\lambda_2,\varepsilon}\} : F \in G_{n+1}(D)\}.$$
For any $f \in D_{\lambda_2,\varepsilon}$, by Lemma 12.4.3 we have that
$$\mathcal{R}f = \int_a^b |f|^2 w > \frac{\varepsilon}{\lambda_2},$$
and hence
$$(\mathcal{L}f - \lambda_2 \mathcal{R}f) - (\mathcal{L}f - \lambda_1 \mathcal{R}f) = (\lambda_1 - \lambda_2)\mathcal{R}f < \frac{\lambda_1 - \lambda_2}{\lambda_2}\varepsilon.$$
This implies that
$$\xi_n(\lambda_2) - \xi_n(\lambda_1) \leq \frac{\lambda_1 - \lambda_2}{\lambda_2}\varepsilon < 0,$$
and thus $\xi_n(\lambda_2) < \xi_n(\lambda_1)$.

The proof of the second part is similar and hence omitted. $\square$

PROOF OF THEOREM 12.3.1. Let $n \in \mathbb{N}_0$. From Lemma 12.3.3 we have that $\xi_n \geq \xi_0(0) > 0$. By Lemmas 12.4.1, 12.4.2 and 12.4.4, the n-th spectral-curve $\xi = \xi_n(\lambda)$ is continuous on $\mathbb{R}$, strictly decreasing in E_1, strictly increasing in E_2, and $\lim_{\lambda \to \pm\infty} \xi_n(\lambda) = -\infty$. Therefore, the equation $\xi_n(\lambda) = 0$ has exactly one positive root $\lambda = \lambda_n$ and exactly one negative root $\lambda = \lambda_{-n}$. Note that $\xi_n(\lambda) > 0$ for $\lambda \in (\lambda_{-n}, \lambda_n)$. Letting $\lambda_* = \lambda_{\pm n}$ and $\lambda_\# = 0$ in (12.4.1), we obtain the inequality $|\lambda_{\pm n}| \geq \xi_n(0)$.

(i) Since $0 = \xi_n(\lambda_n)$ is an eigenvalue of $S(\lambda_n)$ by assumption, we know that λ_n is an eigenvalue of T.

If $\lambda_j = \lambda_n$ for some $j \in \{0, 1, \cdots, n-1\}$, then $\lambda_j = \lambda_n$ is an eigenvalue of T; if $\lambda_j < \lambda_n$ for some $j \in \{0, 1, \cdots, n-1\}$, then
$$\xi_j(\lambda_j) = 0 < \xi_n(\lambda_j) \leq \inf \sigma_e(S(\lambda_j)),$$

which, together with Lemma 12.2.3, implies that $0 = \xi_j(\lambda_j)$ is an eigenvalue of $S(\lambda_j)$, and hence λ_j is also an eigenvalue of T. Clearly (12.3.5) holds.

Let $\lambda_* \in (0, \lambda_n) \setminus \{\lambda_0, \lambda_1, \cdots, \lambda_{n-1}\}$. Then $0 \neq \xi_i(\lambda_*)$ for $i = 0, 1, .., n-1$. Since $0 < \xi_n(\lambda_*) \leq \inf \sigma_e(S(\lambda_*))$, Lemma 12.2.3 implies that $0 \notin \sigma(S(\lambda_*))$. Thus, by Lemma 12.3.1, λ_* is not in the spectrum of T. Hence, $\lambda_0, \lambda_1, ..., \lambda_n$ are the only points from the interval $[0, \lambda_n]$ which are in the spectrum of T.

Since $\lambda_j = \lambda_{j+1}$ for some $j \in \{0, 1, \cdots, n-1\}$ if and only if $0 = \xi_j(\lambda_j) = \xi_{j+1}(\lambda_{j+1})$ is an eigenvalue of $S(\lambda_j) = S(\lambda_{j+1})$ with multiplicity 2, this happens if and only if $\lambda_j = \lambda_{j+1}$ is an eigenvalue of T with multiplicity 2. Hence only double eigenvalues appear twice in (12.3.5).

(ii) The proof is similar to (i) and hence omitted.

(iii) This is an immediate consequence of Parts (i) and (ii).

REMARK 12.4.2. Let (12.2.1) hold. Let S be a self-adjoint realization of equation (12.1.3) with domain D and let T be given by (12.1.4). Let $\xi_n(\lambda)$ be defined by (12.3.3). Assume that T is LD. Recall that for each $n \in \mathbb{N}_0$, λ_n and λ_{-n} are the unique positive and negative roots of the equation $\xi_n(\lambda) = 0$, respectively. Then the sequence $\{\lambda_n\}_{n=0}^{+\infty}$ is nondecreasing and $\{\lambda_{-n}\}_{n=0}^{+\infty}$ is nonincreasing. Set

$$\delta_\pm = \lim_{n \to \pm\infty} \lambda_n.$$

We have that

$$-\infty \leq \delta_- < 0 < \delta_+ \leq +\infty.$$

Theorem 12.3.1 implies that (δ_-, δ_+) does not intersect the essential spectrum of T, and all the eigenvalues in (δ_-, δ_+) can be obtained from spectral-curves. If δ_- is finite and is an eigenvalue, and the problem only has finitely many eigenvalues in $(\delta_-, 0)$, then δ_- can also be obtained from spectral-curves. There is a similar statement for δ_+.

5. LC Non-Oscillatory Problems

In this section we specialize to the case where (12.1.1) is LCNO. First, we clarify the concept of LCNO for LD problems, and show that every singular LD problem associated with an LCNO equation can be transformed into a regular LD problem with the same spectrum. This extends results in Niessen and Zettl [503] for singular LCNO problems with positive w. Based on it and the results of Chapter 5 on regular LD problems, we obtain further results on the existence of eigenvalues and the numbers of zeros of eigenfunctions.

LEMMA 12.5.1. *Let (12.2.1) hold. In equation (12.1.1) assume that a is LC. If (12.1.1) is NO at a for some $\lambda = \lambda_0 \in \mathbb{R}$, then equation (12.1.2) is NO at a for every $\{\lambda, \xi\} \in \mathbb{R}^2$. A similar result holds at the endpoint b.*

PROOF. This follows from Theorem 7.3.1. Note that w is not assumed to be positive in Theorem 7.3.1. Consider the case $\xi = 0$ first. By Theorem 7.3.1 equation (12.1.1) is NO at a for every $\lambda \in \mathbb{R}$. For any fixed $\lambda \in \mathbb{R}$, since (12.1.2) is NO at a for $\xi = 0$ it is NO at a for every $\xi \in \mathbb{R}$, again by Theorem 7.3.1. The proof for the endpoint b is similar. □

REMARK 12.5.1. By Lemma 12.5.1, LCNO is a property of equation (12.1.1) independent of λ even when w is indefinite. So we can simply say that (12.1.1) is LCNO if (12.1.1) is LCNO for some (and hence for all) $\lambda \in \mathbb{R}$.

LEMMA 12.5.2. *Let (12.2.1) hold and assume that equation (12.1.1) is LCNO at each endpoint. Then there exist real-valued functions u and v in the maximal domain $D_{\max}$ satisfying the following conditions:*

a) $v > 0$ on $J = (a, b)$;

b) for some $\xi_1 \in \mathbb{R}$ and $\{\lambda, \xi\} = \{\lambda_0, \xi_1\}$, u is a principal solution of equation (12.1.2) at a and v is a non-principal solution of Eq. (12.1.2) at a;

c) for some $\xi_2 \in \mathbb{R}$ and $\{\lambda, \xi\} = \{\lambda_0, \xi_2\}$, u is a principal solution of (12.1.2) at b and v is a non-principal solution of (12.1.2) at b;

d) $[u, v](a) = [u, v](b) = -1$.

PROOF. We observe that (12.1.1) is LCNO for $\lambda = \lambda_0$ means that (12.1.2) is LCNO for $\{\lambda, \xi\} = \{\lambda_0, 0\}$. Note from Lemma 12.1.1 that $D_{\max}$ is the maximum domain of (12.1.2) with $\lambda = \lambda_0$. Then the conclusions follow from the corresponding results for RD problems in Chapter 10, see Theorem 10.6.5 and Remark 10.6.1. See also [**503**], pages 564-566. □

The function v in Lemma 12.5.2 is called a *regularizing function* for (12.1.1), and $\{u, v\}$ is called a *BC basis* for (12.1.1).

COROLLARY 12.5.1. *Let (12.2.1) hold. Let equation (12.1.1) be LCNO at both endpoints. Let $\{u, v\}$ be a BC basis for (12.1.1) defined as in Lemma 12.5.2. Then for any function $y \in D_{\max}$*

$$\frac{y}{v}(a) = [y, u](a), \quad \frac{y}{v}(b) = [y, u](b).$$

PROOF. For any $y \in D_{\max}$

$$[y, u] = y(pu') - u(py')$$
$$= \frac{y}{v}[v(pu') - u(pv')] - \frac{u}{v}[v(py') - y(pv')]$$
$$= \frac{y}{v}[v, u] - \frac{u}{v}[v, y].$$

Let $t \to a^+$, and noting that $\lim_{t \to a^+}(u/v)(t) = 0$ and $[v, u](a) = 1$, we obtain the equality for a. Similarly, we obtain the equality for b. (Recall that $[y, u](a)$ and $[y, u](b)$ are finite by the Lagrange identity.) □

LEMMA 12.5.3. *Let (12.2.1) hold. Assume (12.1.1) is LCNO and let v be a regularizing function for (12.1.1). Consider the equation*

$$-(Pz')' + Qz = \lambda W z \quad \text{on } J, \tag{12.5.1}$$

where

$$P = pv^2, \quad Q = v[-(pv')' + qv], \quad W = wv^2. \tag{12.5.2}$$

Then $P > 0, W \neq 0$ a.e. on J, $1/P, Q, W \in L^1(J, \mathbb{R})$. Hence (12.5.1) is regular on J. Furthermore, y is a solution of (12.1.1) if and only if $z = y/v$ is a solution of (12.5.1), and in this case, $Pz' = -[y, v]$.

PROOF. This is a minor modification of Theorem 4.3 in [**386**] for the case with indefinite w and has a similar proof. □

REMARK 12.5.2. Lemma 12.5.1 implies that if (12.1.1) is LCO for some $\lambda \in \mathbb{R}$, then (12.1.3) is LCO for all $\lambda, \xi \in \mathbb{R}$. In this case, the spectrum of every self-adjoint realization of (12.1.3) is unbounded above and below, see Niessen and Zettl [**503**]. Therefore, there are no LD problems associated with (12.1.1) in the LCO case.

We observe from Proposition 12.2.1, (4), that all BC's determining self-adjoint domains for (12.1.1) are of the form

$$AY(a) + BY(b) = 0, \ Y = \begin{pmatrix} [y,u] \\ [y,v] \end{pmatrix}, \tag{12.5.3}$$

where A, B satisfy (12.2.4). Recall that these boundary conditions can be categorized into two mutually exclusive classes : the separated conditions and the coupled ones and the separated and coupled BC's have the canonical representations (12.2.5) and (12.2.6), respectively.

LEMMA 12.5.4. *Let (12.2.1) hold and assume that equation (12.1.1) is LCNO at both endpoints. Let $\{u,v\}$ be a BC basis for (12.1.1) defined as in Lemma 12.5.2. Then the singular SLP (12.1.1), (12.5.3) has discrete spectrum consisting of exactly the same eigenvalues as those of the regular SLP consisting of equation (12.5.1) and the BC*

$$AZ(a) + BZ(b) = 0, \ Z = \begin{pmatrix} z \\ Pz' \end{pmatrix}, \tag{12.5.4}$$

and the eigenfunctions are related by $y_n = vz_n$, $n \in \mathbb{Z}^*$. In particular, the eigenfunctions of the singular problem (12.1.1), (12.5.3) have exactly the same zeros as the eigenfunctions of the corresponding regular problem on J.

PROOF. It is well known that the spectrum is discrete whenever each endpoint is either regular or LC. (In this case the associated resolvent operator is compact.) From Lemmas 12.5.2 and 12.5.3, the transformation $z = y/v$ transforms the problem (12.1.1), (12.5.3) into the problem (12.5.1), (12.5.4) and vice versa. Thus, the two problems have the same set of eigenvalues. Note that for any eigenvalue λ, y is an eigenfunction of (12.1.1), (12.5.3) if and only if $z = y/v$ is an eigenfunction of (12.5.1), (12.5.4) for the same λ. Hence y and z have exactly the same zeros in J since $v > 0$ on J. □

Next we investigate the relationship between the left-definiteness of singular and the corresponding regular problems.

THEOREM 12.5.1. *Let the hypotheses and notation of Lemma 12.5.4 hold. Then the singular SLP (12.1.1), (12.5.3) is LD if and only if the corresponding regular SLP (12.5.1), (12.5.4) is LD.*

PROOF. Let P, Q, W be given in (12.5.2). Define U and $\tilde{U}$ to be the unit spheres in $L^2(J,|w|)$ and $L^2(J,|W|)$, respectively; and D and $\tilde{D}$ the self-adjoint domains in $L^2(J,|w|)$ and $L^2(J,|W|)$ characterized by (12.5.3) and (12.5.4), respectively. Let the functional $\mathcal{L}$ on D be defined as in (12.3.1), and the functional $\tilde{\mathcal{L}}$ on $\tilde{D}$ be defined by

$$\tilde{\mathcal{L}}g = \int_a^b [-(Pg')'\bar{g} + Q|g|^2].$$

It is easy to see that $f \in D \bigcap U$ if and only if $g = f/v \in \tilde{D} \bigcap \tilde{U}$, and for $f \in D \bigcap U$ and $g = f/v$ we have

$$-(Pg')'\bar{g} + Q|g|^2 = -(pf')'\bar{f} + q|f|^2.$$

By the definition of left-definiteness of regular SLP's (see [**387**]), SLP (12.5.1), (12.5.4) is LD if and only if there exists $c > 0$ such that $\tilde{\mathcal{L}}g \geq c > 0$ for all $g \in \tilde{D} \bigcap \tilde{U}$, hence if and only if $\mathcal{L}f \geq c > 0$ for all $f \in D \bigcap U$. This means that the left-definiteness of the regular SLP (12.5.1), (12.5.4) is equivalent to that of the singular SLP (12.1.1), (12.5.3). □

COROLLARY 12.5.2. *Let the hypotheses and notation of Lemma 12.5.4 hold.*

(1) *Assume there exists a regularizing function v given in Lemma 12.5.2 such that Q defined in (12.5.2) satisfies $Q(t) \geq 0$ and $Q(t) \not\equiv 0$ a.e. on J. Then*
   (a) *the separated SLP (12.1.1), (12.2.5) is LD if $\pi/2 \leq \alpha \leq \pi$ and $0 \leq \beta \leq \pi/2$;*
   (b) *the coupled SLP (12.1.1), (12.2.6) is LD if*
$$K = \begin{pmatrix} c & 0 \\ 0 & 1/c \end{pmatrix}$$
for some real number $c \neq 0$.

(2) *Assume there exists a regularizing function v given in Lemma 12.5.2 such that Q defined in (12.5.2) satisfies $Q = 0$ a.e. on J. Then*
   (a) *the separated SLP (12.1.1), (12.2.5) is LD if $\pi/2 \leq \alpha \leq \pi$, $0 \leq \beta \leq \pi/2$ and and $(\alpha, \beta) \notin \{(0, \pi/2), (0, \pi), (\pi/2, \pi/2), (\pi/2, \pi)\}$;*
   (b) *the coupled SLP (12.1.1), (12.2.6) is LD if*
$$K = \begin{pmatrix} c & 0 \\ 0 & 1/c \end{pmatrix}$$
for some real number $c \neq 0$ and $ce^{i\gamma} \neq 1$.

PROOF. The first part follows directly from Theorem 12.5.1 and Corollary 5.2.1. The second part can be justified by a slight modification of the proof of the first part. □

REMARK 12.5.3. Since the regularizing function v is not explicitly given, the hypotheses on Q in Corollary 12.5.1 may be difficult to verify. However, we observe the following: if there exists an $r > 0$ such that equation (12.1.2) with $\{\lambda, \xi\} = \{0, r\}$ has a positive solution v which is non-principal at both endpoints a and b, then v can be chosen as a regularizing function. In this case, $Q = vMv = r|w|v^2 \geq 0$ and $Q \not\equiv 0$ on J. If (12.1.2) with $\{\lambda, \xi\} = \{0, 0\}$ has a positive solution v which is non-principal at both endpoints a and b, then v can be chosen as a regularizing function. In this case, $Q = vMv = 0$.

EXAMPLE 12.5.1. *Consider the equation*
$$-y'' + \frac{\nu^2 - 1/4}{t^2} y = \lambda w y \text{ on } J = (0,1), \tag{12.5.5}$$
where $0 < \nu < 1, \nu \neq 1/2$, $w \in L(J; \mathbb{R})$, w changes sign and $|w| = 1$ a.e. on J.

It is easy to see that (12.5.5) is LCNO at 0 and regular at 1. The right-definite equation associated with (12.5.5) is the Bessel equation
$$-y'' + \frac{\nu^2 - 1/4}{t^2} y = \xi y \text{ on } J. \tag{12.5.6}$$

For $\xi = 0$,
$$u_1(t) = -\frac{1}{2\nu} t^{1/2+\nu} \quad \text{and} \quad v(t) = t^{1/2-\nu}$$

are principal and non-principal solutions, respectively, of (12.5.6) at 0; and $u_2(t) := u_1(t) + \frac{1}{2\nu}v(t)$ and $v(t)$ are principal and non-principal solutions, respectively, of (12.5.6) at 1. Let $u \in D_{max}$ be defined by u_1 on $(0, c]$, u_2 on $[d, 1)$, and use the Naimark Patching Lemma to construct $u \in D_{max}$ such that $u = u_1$ on $(0, c]$ and $u = u_2$ on $[d, 1)$, where $0 < c < d < 1$. Then u and v satisfy the assumptions of Lemma 12.5.2. Using $\{u, v\}$ as a BC basis, we obtain the separated BC

$$\begin{aligned}\cos\alpha\,[y,u](0) - \sin\alpha\,[y,v](0) &= 0, \\ \cos\beta\,[y,u](1) - \sin\beta\,[y,v](1) &= 0.\end{aligned} \quad (12.5.7)$$

We claim that SLP (12.5.5), (12.5.7) is LD provided $\pi/2 \leq \alpha \leq \pi$, $0 \leq \beta \leq \pi/2$, and $(\alpha, \beta) \notin \{(0, \pi/2), (0, \pi), (\pi/2, \pi/2), (\pi/2, \pi)\}$. In fact, Let $Q = vMv$. Then $Q \equiv 0$ by Remark 5.1, and hence the conclusion follows from Corollary 12.5.1.

Similarly, for the coupled BC

$$Y(1) = e^{i\gamma}KY(0), \ Y = \begin{pmatrix} [y,u] \\ [y,v] \end{pmatrix}, \quad (12.5.8)$$

where $K = \begin{pmatrix} c & 0 \\ 0 & 1/c \end{pmatrix}$ for some real number $c \neq 0$, $\gamma \in [0, \pi)$, SLP (12.5.5), (12.5.8) is LD if $ce^{i\gamma} \neq 1$.

The next result gives more information on the existence of eigenvalues and the numbers of zeros of eigenfunctions for the LCNO case.

THEOREM 12.5.2. *Let the hypotheses and notation of Lemma(12.5.4) hold. Assume SLP (12.1.1), (12.5.3) is LD. Then its spectrum consists of only real eigenvalues, the eigenvalues are unbounded from below and above, and can be ordered to satisfy the inequalities*

$$\cdots \leq \lambda_{-2} \leq \lambda_{-1} \leq \lambda_{-0} < 0 < \lambda_0 \leq \lambda_1 \leq \lambda_2 \leq \cdots \quad (12.5.9)$$

with only geometrically double eigenvalues appearing twice. Furthermore, we have the following:

(i) If the BC (12.5.3) is separated then all inequalities in (12.5.9) are strict. For $n \in \mathbb{Z}^*$, any eigenfunction u_n of λ_n has exactly $|n|$ zeros in the open interval J.

(ii) If the BC (12.5.3) is a real coupled self-adjoint BC, i.e., (12.5.9) with $\gamma = 0$, then equalities in (12.5.9) may occur. For $n \in \mathbb{Z}^*$, Let u_n be a real-valued eigenfunction of λ_n, then the number of zeros of u_n in the interval $[a, b)$ is 0 or 1 if $n = \pm 0$ or ± 1 and is $|n| - 1$ or $|n|$ or $|n| + 1$ when $|n| > 1$.

(iii) If the BC (12.5.3) is a non-real coupled self-adjoint BC, i.e., (12.5.9) with $\gamma \neq 0$, then strict inequality holds throughout (12.5.9).

For $n \in \mathbb{Z}^*$, let u_n be an eigenfunction of λ_n, then the number of zeros of $\operatorname{Re} u_n$ and of $\operatorname{Im} u_n$ in the interval $[a, b)$ is 0 or 1 if $n = \pm 0$ or ± 1 and is $|n| - 1$ or $|n|$ or $|n| + 1$ when $|n| > 1$, and u_n itself has no zero in $[a, b]$.

PROOF. From Theorem 12.5.1 and Lemma 12.5.4, we see that SLP (12.1.1), (12.5.3) is LD implies that SLP (12.5.1), SLP (12.5.4) is LD, the two problems have exactly the same eigenvalues, and the corresponding eigenfunctions have the same zeros in J. Then the conclusions follow from Theorem 5.3.1; see also Theorems 3.1 and 3.2 in [**387**]. □

EXAMPLE 12.5.2. *Consider the equation*
$$-((1-t^2)y')' + y = \lambda w y \text{ on } J = (-1, 1), \qquad (12.5.10)$$
where $|w| = 1$ and w changes sign on J. The RD equation associated with (12.5.10) is the classical Legendre equation
$$-((1-t^2)y')' + y = \lambda y \text{ on } J = (-1, 1). \qquad (12.5.11)$$
Note that both endpoints are LCNO and $u \equiv 1$ is a principal solution of Eq. (12.5.11) for $\lambda = 1$ at both endpoints of J. Then
$$[y, 1](-1) = 0 = [y, 1](1) \qquad (12.5.12)$$
defines a self-adjoint separated BC and SLP (12.5.11), (12.5.12) is the classical self-adjoint Legendre SLP. Its eigenvalues are known to be
$$\lambda_n = 1 + n(n+1), \; n \in \mathbb{N}_0,$$
and the eigenfunctions are the classical Legendre polynomials. Since $\lambda_0 = 1 > 0$, the problem consisting of equation (12.5.10) and the same BC (12.5.12) is LD. Hence by Theorem 12.5.2, eigenvalues of the problem (12.5.10), (12.5.12) are all real, unbounded from below and above, and satisfy inequalities (12.5.9) with strict inequality holding everywhere; and if u_n is an eigenfunction of λ_n for some $n \in \mathbb{Z}^*$, then u_n has exactly $|n|$ zeros in the open interval J.

6. Further Eigenvalue Properties in the LCNO Case

In this section we present results on eigenvalue inequalities, asymptotic behavior and ranges of eigenvalues for singular LD problems with regular or LCNO endpoints. These results are analogues of theorems in Chapter 5 for regular LD problems and of theorems from Chapter 4 for regular RD problems. The proofs follow from the regular case, Lemma 12.5.4 and Theorem 12.5.1, we omit the details.

Assume a and b are LCNO endpoints and $\{u, v\}$ is a boundary condition basis.. For $K = (k_{ij}) \in SL(2, \mathbb{R})$, denote by $\{\mu_n : n \in \mathbb{Z}^*\}$ and $\{\nu_n : n \in \mathbb{Z}^*\}$ the eigenvalues of equation (12.1.1) and the separated BC's
$$[y, u](a) = 0, \; k_{22}[y, u](b) - k_{12}[y, v](b) = 0 \qquad (12.6.1)$$
and
$$[y, v](a) = 0, \; k_{21}[y, u](b) - k_{11}[y, v](b) = 0, \qquad (12.6.2)$$
respectively, and by $\{\lambda_n(e^{i\gamma}K) : n \in \mathbb{Z}^*\}$ the eigenvalues of (12.1.1) and the coupled BC
$$Y(b) = e^{i\gamma} K Y(b), \; Y = \begin{pmatrix} [y, u] \\ [y, v] \end{pmatrix}. \qquad (12.6.3)$$

THEOREM 12.6.1. *Let the hypotheses and notation of Lemma 12.5.4 hold. Let $K = (k_{ij}) \in SL(2, \mathbb{R})$.*

(a) Assume that $k_{11} > 0$, $k_{12} \leq 0$, and SLP (12.1.1), (12.6.2) is LD. Then, both problem (12.1.1), (12.6.1) and problem (12.1.1), (12.6.3) with any $\gamma \in (-\pi, \pi]$ are LD. Furthermore, $\lambda_{\pm 0}(K)$ are geometrically simple, and for each $\gamma \in (-\pi, \pi)$, $\gamma \neq 0$, we have
$$\begin{aligned}
\nu_0 &\leq \lambda_0(K) < \lambda_0(e^{i\gamma}K) < \lambda_0(-K) \leq \{\mu_0, \nu_1\} \\
&\leq \lambda_1(-K) < \lambda_1(e^{i\gamma}K) < \lambda_1(K) \leq \{\mu_1, \nu_2\} \\
&\leq \lambda_2(K) < \lambda_2(e^{i\gamma}K) < \lambda_2(-K) \leq \{\mu_2, \nu_3\} \\
&\leq \lambda_3(-K) < \lambda_3(e^{i\gamma}K) < \lambda_3(K) \leq \{\mu_3, \nu_4\} \leq \cdots
\end{aligned} \qquad (12.6.4)$$

and another set of inequalities obtained by replacing λ_n, μ_n, ν_n, $<$ and $\leq$ in (12.6.4) by λ_{-n}, μ_{-n}, ν_{-n}, $>$ and $\geq$, respectively.

(b) Assume that $k_{11} \leq 0$, $k_{12} < 0$, and SLP (12.1.1), (12.6.3) with $\gamma = 0$ is LD. Then problem (12.1.1), (12.6.3) with any other $\gamma \in (-\pi, \pi]$, problem (12.1.1), (12.6.1), and problem (12.1.1), (12.6.2) are all LD. Furthermore, both $\lambda_0(K)$ and $\lambda_{-0}(K)$ are geometrically simple, and for each $\gamma \in (-\pi, \pi)$, $\gamma \neq 0$, we have

$$\begin{aligned}
\lambda_0(K) < \lambda_0(e^{i\gamma}K) < \lambda_0(-K) \leq \{\mu_0, \nu_0\} \leq \\
\lambda_1(-K) < \lambda_1(e^{i\gamma}K) < \lambda_1(K) \leq \{\mu_1, \nu_1\} \leq \\
\lambda_2(K) < \lambda_2(e^{i\gamma}K) < \lambda_2(-K) \leq \{\mu_2, \nu_2\} \leq \\
\lambda_3(-K) < \lambda_3(e^{i\gamma}K) < \lambda_3(K) \leq \{\mu_3, \nu_3\} \leq \cdots
\end{aligned} \quad (12.6.5)$$

and another set of inequalities obtained by replacing λ_n, μ_n, ν_n, $<$ and $\leq$ in (12.6.5) by $\lambda_{-n}, \mu_{-n}, \nu_{-n}$, $>$ and $\geq$, respectively.

(c) If neither case (a) nor case (b) applies to K, then either case (a) or case (b) applies to $-K$.

THEOREM 12.6.2. *Let the hypotheses and notation of Lemma 12.5.4 hold. Assume SLP (12.1.1), (12.5.3) is LD, and denote its eigenvalues by $\{\lambda_n : n \in \mathbb{Z}^*\}$. Then*

$$\lambda_{\pm n} \sim \pm \frac{n^2 \pi^2}{\left[\int_a^b \sqrt{\frac{w_\pm(t)}{p(t)}} dt\right]^2}, \quad n \to \infty,$$

where w_+ and w_- denote the positive and negative parts of w, respectively.

PROOF. In Section 12.5 we showed that there exists a regular problem which has exactly the same spectrum. The conclusion the follows from Theorem 5.3.1. □

The eigenvalues for the BC

$$[y, u](a) = 0 = [y, u](b) \quad (12.6.6)$$

play a special role in determining the upper and lower bounds of the eigenvalues of SLP (12.1.1), (12.5.3) for all self-adjoint BC's. In the RD case (12.6.6) when u is a principal solution at both endpoints, is called the Friedrichs boundary condition since it determines the so called Friedrichs extension. In the LD case there is no Friedrichs extension since the spectrum is not bounded below. Nevertheless, in analogy with the RD case, we denote the eigenvalues for the boundary condition (12.6.6) by $\{\lambda_n^F : n \in \mathbb{Z}^*\}$.

THEOREM 12.6.3. *Let the hypotheses and notation of Lemma 12.5.4 hold. Assume SLP (12.1.1), (12.5.3) is LD, and denote its eigenvalues by $\{\lambda_n : n \in \mathbb{Z}^*\}$. Then SLP (12.1.1), (12.6.6) is LD, and*

$$\lambda_n \in (0, \lambda_n^F], \ \lambda_{-n} \in [\lambda_{-n}^F, 0) \ for \ n = 0, 1,$$
$$\lambda_n \in (\lambda_{n-2}^F, \lambda_n^F], \ \lambda_{-n} \in [\lambda_{-n}^F, \lambda_{-n+2}^F) \ for \ n = 2, 3, 4, \ldots.$$

PROOF. This follows from Theorem 5.5.2. □

7. Approximating a Singular Problem with Regular Problems

In Section 10.9 we studied the approximation of a given singular right-definite problem with a sequence a regular RD problems. In this section we undertake such a study for the left-definite case.

7.1. Notation and Basic Results. For the convenience of the reader we give the notation and review the basic results needed in Section 12.7 below. We study spectral approximations of singular Sturm-Liouville problems associated with the equation

$$-(py')' + qy = \lambda w\, y \quad \text{on } J = (a,b),\ -\infty \leq a < b \leq \infty, \tag{12.7.1}$$

and suitable boundary conditions, where p, q and w satisfy

$$1/p, q, w \in L^1_{loc}(J, \mathbb{R}),\ p > 0,\ |w| > 0 \text{ a.e. on } J,\ w \text{ changes sign on } J. \tag{12.7.2}$$

By 'w changes sign on J' we mean that each of the sets $\{t \in J : w(t) > 0\}$, $\{t \in J : w(t) < 0\}$ has positive or infinite Lebesgue measure.

As in Chapter 5 and previous sections of this chapter, we base our approach on the one parameter family of equations

$$-(py')' + (q - \lambda w)y = \xi\,|w|\,y \quad \text{on } J = (a,b). \tag{12.7.3}$$

In (12.7.3) ξ is the spectral parameter and $\lambda \in \mathbb{R}$ defines a family of equations. Let

$$H = L^2(J, |w|). \tag{12.7.4}$$

For $\lambda = 0$ let $S = S(0)$ denote a self-adjoint realization of (12.7.3) in H with domain $S = D$, i.e.

$$S_{\min} \subset S = S^* \subset S_{\max},\ Sf = \frac{1}{|w|}[-(pf')' + qf],\ f \in D,$$

where $S_{\min}$ and $S_{\max}$ denote the minimal and maximal operators of (12.7.3) with $\lambda = 0$ and S^* is the adjoint of S. Then [**387**] for each $\lambda \in \mathbb{R}$

$$S(\lambda)\,f = \frac{1}{|w|}[-(pf')' + (q - \lambda w)\,f],\ f \in D(\lambda) = D$$

is a self-adjoint realization of (12.2.3) in H.

REMARK 12.7.1. The self-adjoint domain D can be characterized in terms of boundary conditions; see Section 10.4. These depend on the classification of the endpoints in the space H, as limit-point (LP) or limit-circle (LC). (If both endpoints are in the limit-point case then there are no boundary conditions and the minimal operator $S_{\min}$ is self-adjoint with no proper self-adjoint extensions in H.) Note that, although the domains $D(\lambda)$ of the operators $S(\lambda)$ are independent of $\lambda \in \mathbb{R}$, the operators $S(\lambda)$ depend on λ and this dependence gives rise to the spectral curves which play a critical role in our analysis. The LP/LC classification is independent of λ and ξ; see Lemma 12.2.1.

We remind the reader of the definition of the operator T which represents SLP consisting of equation (12.7.1) with appropriate boundary conditions when the weight function is indefinite.

DEFINITION 12.7.1. *Let D be a self-adjoint domain in H. Then we define an operator realization T of (12.7.1) in H by*

$$Tf = \frac{1}{w}[-(pf')' + qf],\ f \in D.$$

We say the operator T, or, equivalently, the boundary value problem consisting of (12.7.1) with boundary conditions that determine D, is left-definite (LD) if $S = S(0)$ is a positive operator in H. (See Chapter 11)

7. APPROXIMATING A SINGULAR PROBLEM WITH REGULAR PROBLEMS

Recall from Chapter 11 that the operator T maps D into H but is not selfadjoint or even symmetric *in the space* H. However, considered as an operator in the Krein space $K = L^2(J, w)$, see [133], T is self-adjoint. Its spectrum is the same regardless of whether it is considered as an operator in H or in K. As in Chapter 11 and the previous sections of this chapter, we identify the spectrum of T as the spectrum of the boundary value problem consisting of equation (12.7.1) together with the boundary conditions that determine the self-adjoint domain D. Note that this spectrum may consist of eigenvalues as well as essential spectrum. We remind the reader that the spectral theory of self-adjoint operators in Krein space is much more complex than that in Hilbert space; in particular the spectrum of a self-adjoint operator in Krein space is not necessarily real.

For the convenience of the reader we recall some results here from Chapter 11 and the previous sections of this chapter. These results will be used below in the subsections of this section. Note that $S = S(0)$ is not bounded below when p changes sign or when an endpoint is LCO (even with $p > 0$); thus the assumption that T is LD rules out both of these cases.

THEOREM 12.7.1. *The classification of each of the endpoints of the interval J, in the space H, for the differential equation (12.7.3) as limit point (LP), limit circle oscillatory (LCO) or limit circle non-oscillatory (LCNO) is independent of $\lambda, \xi \in \mathbb{R}$.*

PROOF. See Lemma 12.2.1 □

We continue our investigation of left-definite problems. These have real spectrum. The results summarized in the next theorem are from Chapter 11.

THEOREM 12.7.2. *Let (12.2.1) hold. Let S be a self-adjoint realization of (12.1.3) in H with domain D and let T be given by (12.1.4). Assume T is left-definite. Suppose there exist nondegenerate subintervals J_+ and J_- of J such that w is positive on J_+ and negative on J_-. Then*

(1) *The spectrum of T is real.*
(2) *If the essential spectrum of S is empty, so is the essential spectrum of T.*

PROOF. Part (1) follows from Theorem 11.1.1. By Theorem 11.1.4, the essential spectrum of T is either empty or consists of the entire complex plane. But the latter is ruled out by (1) and hence (2) holds. □

The Lemma repeats two earlier results and is stated here for the convenience of the reader.

LEMMA 12.7.1. *Let the hypotheses and notation of Theorem 12.7.2 hold. Let $S(\lambda)$ be defined by (12.2.3). Then*

(1) *The spectrum of T is precisely the set*
$$\{\lambda \in \mathbb{R} \mid 0 \in \sigma(S(\lambda))\}.$$
(2) *For each $\lambda \in \mathbb{R}$, $S(\lambda)$ is bounded below.*

PROOF. Part (1) is Lemma 12.3.1; part (2) is contained in Lemma 12.2.2. □

REMARK 12.7.2. Recall the spectral curves $\xi_n(\lambda)$ defined by (12.3.3). If $S(\lambda)$ has no essential spectrum, then the curves $\xi_n(\lambda)$ are precisely the eigenvalues of $S(\lambda)$. Otherwise it is well known that, for each fixed $\lambda \in \mathbb{R}$, the following hold:

If $\xi_n(\lambda) < \xi_{n+1}(\lambda)$ for some n, then $\xi_n(\lambda)$ is an eigenvalue, as are all $\xi_j(\lambda)$, for $0 \leq j \leq n$.

If there is some integer N such that $\xi_N(\lambda) = \xi_{N+1}(\lambda)$ and at least one endpoint of J is limit point (so that there are no double eigenvalues), then

$$\xi_N(\lambda) = \xi_{N+1}(\lambda) = \xi_{N+2}(\lambda) = \xi_{N+3}(\lambda) = \cdots = \inf \sigma_{ess}(S(\lambda)).$$

We note the following consequence of Theorem 12.3.1.

COROLLARY 12.7.1. *Let the hypotheses and notation of Theorem 12.7.2 hold.*

(1) *If for some $N \in \mathbb{N}_0$ and all $\lambda > 0$, $\xi_N(\lambda)$ is an eigencurve, then T has positive eigenvalues $\lambda_0, \lambda_1, \ldots, \lambda_N$ which can be indexed to satisfy*

$$\lambda_0 \leq \lambda_1 \leq \cdots \leq \lambda_N$$

and for each $j \in \{0, 1, \ldots N\}$, λ_j is the unique positive root of the equation $\xi_j(\lambda) = 0$. Moreover, these eigenvalues are the only points of the spectrum of T in the interval $[0, \lambda_N]$.

(2) *If for some $M \in \mathbb{N}_0$ and all $\lambda < 0$, $\xi_M(\lambda)$ is an eigencurve, then T has negative eigenvalues $\lambda_{-0}, \lambda_{-1}, \ldots, \lambda_{-M}$ which can be indexed to satisfy*

$$\lambda_{-0} \geq \lambda_{-1} \geq \cdots \geq \lambda_{-M}$$

and for each $j \in \{0, 1, \ldots M\}$, λ_{-j} is the unique negative root of the equation $\xi_j(\lambda) = 0$. Moreover, these eigenvalues are the only points of the spectrum of T in the interval $[\lambda_{-M}, 0]$.

(3) *If for some $n \in \mathbb{N}_0$ and all $\lambda \in \mathbb{R}$, $\xi_n(\lambda)$ is an eigencurve, then (1) and (2) hold with $N = M = n$.*

PROOF. This follows from Theorem 12.3.1. $\square$

For the work which follows we also require the following result from Section 12.4.

THEOREM 12.7.3. *Let the hypotheses and notation of Theorem 12.7.2 hold. Let the spectral curves $\xi_n(\lambda)$ be defined by (12.3.3). The functions $\xi_n(\lambda)$ have the following properties.*

i: $\xi_n(0) \geq \xi_0(0) > 0$.
ii: $\xi_n(\lambda) \to -\infty$ as $\lambda \to \pm\infty$.
iii: $\xi_n(\lambda)$ *is monotone decreasing on any interval $[\mu, +\infty)$ where $\mu > 0$ is any number such that $\xi_n(\mu) \leq \xi_0(0)$ (the existence of such μ is guaranteed by Property ii). In particular, ξ_n is monotone decreasing in a neighbourhood of λ_n.*
iv: $\xi_n(\lambda)$ *is monotone increasing on any interval $(-\infty, \mu]$ where $\mu < 0$ is any number such that $\xi_n(\mu) \leq \xi_0(0)$ (the existence of such μ is guaranteed by Property ii). In particular, ξ_n is monotone increasing in a neighbourhood of λ_{-n}.*

PROOF. Part (i) follows from the definition of left-definite; the other parts follow from Lemma 12.4.2. $\square$

7. APPROXIMATING A SINGULAR PROBLEM WITH REGULAR PROBLEMS

7.2. Spectral Exactness. In this subsection we consider the effects of interval truncation on the spectrum of the operator T. Let $\{a_r : r \in \mathbb{N}\}$ be a decreasing sequence converging to a and $\{b_r : r \in \mathbb{N}\}$ an increasing sequence converging to b such that
$$J_r = (a_r, b_r), \quad -\infty \leq a < a_r < b_r < b \leq \infty.$$

Let D be a domain determined by self-adjoint boundary conditions on the interval $J = (a, b)$. For each $\lambda \in \mathbb{R}$ let $S(\lambda)$ be the self-adjoint operator realization of (12.7.3) in the Hilbert space $H = L^2(J, |w|)$ determined by the domain D and note that D is λ-independent.

Assume, for the remainder of this section, that $S = S(0)$ is a positive operator and let T be the corresponding left-definite operator in H as constructed in Section 11.1; see (12.7.1).

Associated with each interval J_r, let D_r be determined by the 'inherited' regular self-adjoint boundary conditions on J_r constructed as in Section 10.9. Let $S_r(\lambda)$ denote the self-adjoint realization of (12.7.3) with $J = J_r$ in the Hilbert space $H_r = L^2(J_r, |w|)$ determined by the domain D_r and note that D_r is independent of λ. (In H_r, $|w|$ denotes the restriction of $|w|$ to J_r.) Let T and T_r be the operators defined as in Section 11.1 on the intervals J, J_r, respectively, and with the indefinite weight function w.

Do the eigenvalues of T_r approximate the spectrum of T? If so, how?

These are the questions we study in this subsection. We start by introducing some additional notation. The eigenvalues $\xi_n(\lambda)$ and λ_n of Section 12.2 are now denoted by $\xi_n(\lambda, J)$ and $\lambda_n(J)$, respectively. Similarly the eigenvalues of $S_r(\lambda)$ and of T_r will be denoted by $\xi_n(\lambda, J_r)$ and $\lambda_n(J_r)$, respectively.

THEOREM 12.7.4. *Let the hypotheses and notation of Theorem 12.3.1 hold. Then*

(1) *For all sufficiently large r, T_r is left-definite and has an infinite number of positive and negative eigenvalues $\lambda_n(J_r)$ satisfying:*

$$... \leq \lambda_{-2}(J_r) \leq \lambda_{-1}(J_r) \leq \lambda_{-0}(J_r) < 0 < \lambda_0(J_r) \leq \lambda_1(J_r) \leq \lambda_2(J_r) \leq ...$$

(2)
$$\lim_{r \to \infty} \lambda_n(J_r) = \lambda_n(J), \quad (12.7.5)$$
for $n = -M, -M+1, \ldots, -0, +0, 1, \ldots, N$.

PROOF. Since $\xi_0(J) > 0$, and $\xi_0(J_r) \to \xi_0(J)$ by Theorem 10.9.2 (see also [**38**]), it follows that T_r is left-definite for r sufficiently large. Since T_r is regular it then follows from Theorem 5.3.1 that T_r has eigenvalues which can be ordered to satisfy (1). Consider first the case $\lambda_n(J) > 0$. By Theorem 12.3.1 $\lambda_n(J)$ is the unique positive zero of the function $\xi_n(\lambda, J)$, $\lambda \in \mathbb{R}$. Given $\epsilon > 0$, let
$$\lambda_- = \lambda_n(J) - \epsilon/2, \quad \lambda_+ = \lambda_n(J) + \epsilon/2.$$

By Theorem 5.3.2 the function $\xi_n(\lambda, J)$ is strictly decreasing at $\lambda_n(J)$. Hence $\xi_n(\lambda_-, J) > 0$ and $\xi_n(\lambda_+, J) < 0$. From the spectral exactness results for right-definite problems in Section 10.9 applied to the operators $S(\lambda_-)$ and $S(\lambda_+)$, there exists $R \in \mathbb{N}$ such that for all $r > R$,
$$\xi_n(\lambda_-, J_r) > \frac{1}{2}\xi_n(\lambda_-, J) > 0,$$

$$\xi_n(\lambda_+, J_r) < \frac{1}{2}\xi_n(\lambda_+, J) < 0.$$

By Lemma 12.4.1 the eigenvalues $\xi_n(\lambda, J_r)$ are continuous functions of λ and so by the intermediate value theorem, for all $r > R$, there exists a point in (λ_-, λ_+) which is a zero of $\xi_n(\lambda, J_r)$. By the theory for regular left-definite problems in Chapter 5 this point is precisely the eigenvalue $\lambda_k(J_r)$ of T_r. Thus we have, for all $r > R$, the inequality $|\lambda_n(J_r) - \lambda_n(J)| < \epsilon$, which completes the proof for positive eigenvalues.

For negative eigenvalues the proof is similar, the primary difference being that $\xi_n(\lambda, J)$ is strictly *increasing* in a neighborhood of $\lambda_{-n}(J)$, rather than strictly decreasing. □

REMARK 12.7.3. Note that the results in Chapter 10 (see also [**38**]) for right-definite problems do not guarantee uniformity in λ of the convergence

$$\lim_{r \to \infty} \xi_n(\lambda, J_r) = \xi_n(\lambda, J), \qquad (12.7.6)$$

and hence (12.7.5) is not an immediate consequence of (12.7.6).

7.3. Spectral Inclusion. We continue to use the notation of the previous sections of this chapter. The results of the previous subsection give spectral exactness for eigenvalues in the gap in the essential spectrum $\sigma_e(T)$, of the operator T, containing 0. If there is no essential spectrum, then the whole spectrum of T consists of eigenvalues, and there is nothing further to prove. In particular this is the case when neither endpoint is LP.

In this subsection we show that points of the essential spectrum of T can also be approximated by eigenvalues of the operators T_r. In the right-definite case, spectral inclusion is a consequence of strong resolvent convergence in Hilbert space of the operators S_r to the operator S on a core of S. In the left-definite case considered here, a different approach is needed since the theory of resolvent convergence in Hilbert space is not available.

THEOREM 12.7.5. *Let the operators T, T_r be defined as in Subsection 12.7.2 and assume that T is left-definite. Then T_r is left-definite for all sufficiently large r and the spectrum of T_r is discrete consisting of eigenvalues $\lambda_n(J_r)$ which can be ordered as in (1) of Theorem 12.7.4:*

$$\ldots \leq \lambda_{-2}(J_r) \leq \lambda_{-1}(J_r) \leq \lambda_{-0}(J_r) < 0 < \lambda_0(J_r) \leq \lambda_1(J_r) \leq \lambda_2(J_r) \leq \ldots$$

(a): *Suppose λ is in the essential spectrum of T and $\lambda > 0$. Then for any $\varepsilon > 0$ there exists an $R > 0$ such that for any $r > R$ there exists a positive index n_r, depending on r, such that*

$$|\lambda - \lambda_{n_r}(J_r)| < \varepsilon. \qquad (12.7.7)$$

(b): *Suppose λ is in the essential spectrum of T and $\lambda < 0$. Then for any $\varepsilon > 0$ there exists an $R > 0$ such that for any $r > R$ there exists a positive index m_r, depending on r, such that*

$$|\lambda - \lambda_{-m_r}(J_r)| < \varepsilon. \qquad (12.7.8)$$

PROOF. The proof of Theorem 12.7.5 will be given below following some lemmas. □

LEMMA 12.7.2. *The essential spectrum of T is precisely the set of $\lambda \in \mathbb{R}$ such that 0 lies in the essential spectrum of $S(\lambda)$.*

PROOF. Elementary manipulations show that
$$T - \lambda I = sgn(w)S(\lambda).$$
The essential spectrum of T is the set of λ for which $T - \lambda I$ is not Fredholm. This happens if and only if $S(\lambda)$ is not Fredholm, which happens if and only if 0 lies in the essential spectrum of $S(\lambda)$. □

The left-definiteness of T depends on its boundary conditions. Clearly the eigenvalues of T depend on its boundary conditions. The next lemma shows that, as in the right-definite case, the essential spectrum of T is independent of its boundary conditions *as long as left-definiteness is preserved.*

LEMMA 12.7.3. *Suppose that T_1 and T_2 are left-definite and determined by the same differential expression but different boundary conditions. Then the essential spectrum of T_1 is the same as the essential spectrum of T_2.*

PROOF. This follows directly from Lemma 12.7.2 since the essential spectrum of $S(\lambda)$ is independent of its boundary conditions. □

LEMMA 12.7.4. *Let $\xi_n(\lambda, J_r)$ be the nth eigenvalue of $S_r(\lambda)$, for some fixed J_r, for $n = 0, 1, 2, \ldots$ and assume that $\xi_n(\lambda, J_r)$ is simple. Then for $\lambda > 0$,*
$$\dot{\xi}_n(\lambda, J_r) \leq \frac{\xi_n(\lambda, J_r) - \xi_0(0, J_r)}{\lambda}, \qquad (12.7.9)$$
where dot denotes differentiation with respect to λ.

PROOF. Let $y_n(\cdot, \lambda)$ be the eigenfunction of $S_r(\lambda)$ with eigenvalue $\xi_n(\lambda)$, normalized so that
$$(|w|y_n, y_n) = 1. \qquad (12.7.10)$$
Here $(\cdot, \cdot)$ denotes the $L^2(J_r)$ inner product: $(f, g) = \int_{J_r} f\bar{g}$. By Theorem 4.2 of [387] $\dot{\xi}_n(\lambda, J_r)$ exists and is given by
$$\dot{\xi}_n(\lambda, J_r) = -(wy_n, y_n). \qquad (12.7.11)$$
Since $\xi_n(\lambda, J_r)$ is the nth eigenvalue of $S_r(\lambda)$ we have
$$-(py_n')' + (q - \lambda w)y_n = \xi_n(\lambda, J_r)|w|y_n \quad \text{on } J_r \qquad (12.7.12)$$
which we can write as
$$|w|S_r(0)y_n - \lambda w y_n = \xi_n(\lambda, J_r)|w|y_n. \qquad (12.7.13)$$
Taking the inner product of both sides with y_n, we obtain
$$(|w|S_r(0)y_n, y_n) - \lambda(wy_n, y_n) = \xi_n(\lambda, J_r). \qquad (12.7.14)$$
Substituting (12.7.14) into (12.7.11) yields
$$\dot{\xi}_n(\lambda, J_r) = \frac{\xi_n(\lambda, J_r) - (|w|S_r(0)y_n, y_n)}{\lambda}. \qquad (12.7.15)$$
From the variational characterization of $\xi_0(0, J_r)$ it follows that
$$(|w|S_r(0)y_n, y_n) \geq \xi_0(0, J_r)$$
and the conclusion follows. □

PROOF OF THEOREM 12.7.5. Since the essential spectrum is independent of the boundary conditions we may assume, without loss of generality, that for all sufficiently large r the eigenvalues of T_r are simple and hence by Theorem 4.2 of [**387**] $\dot{\xi}_n(\lambda, J_r)$ exists and is given by (12.7.11).

We now use Lemmas 12.7.4 and 12.7.2, together with the spectral inclusion results in [**38**], to prove spectral inclusion of the sequence $(T_r)_{r \in \mathbb{N}}$ for the operator T.

Suppose that $\lambda > 0$ lies in the essential spectrum of T. Then by Lemma 12.7.2, the point 0 lies in the essential spectrum of $S(\lambda)$. By spectral inclusion for $S(\lambda)$ [**38**] we know that for every $\epsilon > 0$, there exists $R \in \mathbb{N}$ such that for all $r \geq R$, there exists an index n_r such that

$$|\xi_{n_r}(\lambda, J_r)| < \epsilon. \tag{12.7.16}$$

From Lemma 12.4.1 we also know that $|\dot{\xi}_{n_r}(\cdot, J_r)| \leq 1$ for all r. This implies continuity of the $\xi_{n_r}(\cdot, J_r)$ which is uniform in both arguments, at least for r sufficiently large. Combining this with (12.7.16), we can guarantee that, provided ϵ is sufficiently small,

$$|\xi_{n_r}(\mu, J_r)| < \frac{1}{2}\xi_0(0, J) \tag{12.7.17}$$

for all $\mu \in I := (\lambda - \xi_0(0, J)/4, \lambda + \xi_0(0, J)/4) \cap \mathbb{R}^+$. Now by Lemma 12.7.4 we have

$$\begin{aligned} \dot{\xi}_{n_r}(\mu, J_r) &\leq (\xi_{n_r}(\mu, J_r) - \xi_0(0, J_r))/\mu \\ &\leq -\xi_0(0, J_r)/(2\mu) \\ &\leq -\xi_0(0, J)/(4\mu) \end{aligned} \tag{12.7.18}$$

where the last inequality in (12.7.18) is valid for all sufficiently large r as a consequence of the spectrally exact convergence $\xi_0(0, J_r) \to \xi_0(0, J)$ as $r \to \infty$ for the right-definite operators $S_r(0)$ and $S(0)$. The importance of this inequality is that it guarantees that $\dot{\xi}_{n_r}(\cdot, J_r)$ is bounded away from zero on the interval I. That is,

$$\inf_I [-\dot{\xi}_{n_r}(\cdot, J_r)] \geq M > 0, \tag{12.7.19}$$

where M is independent of ϵ and r for all sufficiently small ϵ and sufficiently large r.

Now, provided ϵ is sufficiently small, the set of all μ satisfying

$$|\lambda - \mu| \leq \frac{\epsilon}{M}$$

is contained in the ϵ-independent interval I. By an elementary mean value theorem, the smallness of $|\xi_{n_r}(\lambda, J_r)|$ given in (12.7.16), together with the bound away from zero of $\dot{\xi}_{n_r}(\mu, J_r)$ given in (12.7.19), guarantee that there is a zero of $\xi_{n_r}(\lambda, J_r)$ at a point $\lambda_{n_r}(J_r)$ satisfying

$$|\lambda - \lambda_{n_r}(J_r)| \leq \frac{\epsilon}{M}.$$

Thus λ is approximated by an eigenvalue of T_r. This completes the proof of Theorem 12.7.5. □

8. Floquet Theory of Left-Definite Problems

For Sturm-Liouville problems with periodic coefficients and a positive weight function it is well known [**153**], [**600**] that the essential spectrum has a band structure and that the endpoints of these bands are eigenvalues of periodic and semi-periodic boundary conditions over an interval whose length is the fundamental period. In this Section we extend this result to problems with a weight function which changes sign. In contrast to the positive case, in this case the spectral bands go off to infinity in both directions. Their endpoints, as in the positive case, are periodic and semi-periodic eigenvalues of the 'left-definite' problem over a fundamental period.

In Chapter 5 the eigenvalues of left-definite problems are characterized in terms of an associated family of 'right-definite' problems. This characterization plays an important role in this Section.

We continue to study spectral properties of the equation

$$-(py')' + qy = \lambda w y \text{ on } J \tag{12.8.1}$$

with w indefinite by applying the well established spectral theory in the Hilbert space $H = L^2(J, |w|)$ to the one parameter family of equations

$$-(py')' + (q - \lambda w)y = \xi |w| y \text{ on } J. \tag{12.8.2}$$

The coefficients satisfy the basic conditions

$$1/p, q, w \in L_{loc}(J, \mathbb{R}), \ p > 0, \ |w| > 0 \text{ a.e. on } J. \tag{12.8.3}$$

Clearly, for any given boundary condition, $\xi = 0$ is an eigenvalue of (12.8.2) with eigenfunction y if and only if λ is an eigenvalue of (12.8.1) for the same boundary condition and the same eigenfunction y. What is the essential spectrum of equation (12.8.1)?

THEOREM 12.8.1. *Let $J = [a, \infty)$, $-\infty < a < \infty$, and assume that p, q, w are periodic with fundamental period r, $0 < r < \infty$,*

$$p(t+r) = p(t), \ q(t+r) = q(t), \ w(t+r) = w(t), \ t \in J. \tag{12.8.4}$$

Define an operator T as follows: For A_1, A_2 in $\mathbb{R}$, not both 0, let

$$D = \{f \in H = L^2(J, |w|) : f, pf' \in AC_{loc}(J),$$
$$\frac{1}{|w|}[-(pf')' + qf] \in H, \ A_1 f(a) + A_2(pf')(a) = 0.\} \tag{12.8.5}$$

$$Tf = \frac{1}{w}[-(pf')' + qf], \ f \in D \tag{12.8.6}$$

For each $\lambda \in \mathbb{R}$, let $\xi_n^P(\lambda)$, $\xi_n^S(\lambda)$, $n \in \mathbb{N}_0 = \{0, 1, 2, 3, ...\}$ denote the periodic and semi-periodic eigenvalues of (12.8.2) on the fundamental interval $[a, a + r]$, respectively, and assume that $\xi_0^P(0) > 0$. Then:

1. *$\xi_0^S(0) > 0$.*

2. *For each $n \in \mathbb{N}_0$ the equation $\xi_n^P(\lambda) = 0$ has exactly one positive root λ_n^P and exactly one negative root λ_{-n}^P; and the equation $\xi_n^S(\lambda) = 0$ has exactly one positive root λ_n^S and exactly one negative root λ_{-n}^S. Furthermore each of $\xi_n^P(\lambda)$, $\xi_n^S(\lambda)$ is continuous at each point $\lambda \in \mathbb{R}$ and strictly monotone in a neighborhood of each of its roots.*

3.

$$\sigma_e(T) = \cup_{n \in Z^*} I_n \text{ where } Z^* = \{..., -2, -1, -0, 0, 1, 2, ...\},$$
$$I_0 = [\lambda_0^P, \lambda_0^S], \ I_1 = [\lambda_1^S, \lambda_1^P], \ I_2 = [\lambda_2^P, \lambda_2^S],$$
$$I_3 = [\lambda_3^S, \lambda_3^P], \ I_4 = [\lambda_4^P, \lambda_4^S], \ I_5 = [\lambda_5^S, \lambda_5^P]...$$
$$I_{-0} = [\lambda_{-0}^S, \lambda_{-0}^P], \ I_1 = [\lambda_{-1}^P, \lambda_{-1}^S], \ I_{-2} = [\lambda_{-2}^S, \lambda_{-2}^P],$$
$$I_{-3} = [\lambda_{-3}^P, \lambda_{-3}^S], \ I_{-4} = [\lambda_{-4}^S, \lambda_{-4}^P], \ I_{-5} = [\lambda_{-5}^P, \lambda_{-5}^S], ... \quad (12.8.7)$$

PROOF. The hypothesis $\xi_0^P(0) > 0$ implies that the SLP consisting of equation (12.8.2) with $\lambda = 0$ and periodic boundary conditions on the interval $[a, a+r]$ is left-definite. The inequality $\xi_0^S(0) > \xi_0^P(0)$ follows Section 4.8 and hence equation (12.8.2) with $\lambda = 0$ and semi-periodic boundary conditions on the interval $[a, a+r]$ is also left-definite. Thus (2) follows from results of Chapter 5 and the references cited there, particularly [**387**]. To prove part (3) let $S(\lambda)$ denote the self-adjoint realization of equation (12.8.2) on J determined by the boundary condition $A_1 f(a) + A_2 (p f')(a) = 0$ of (12.8.5) for each $\lambda \in \mathbb{R}$ and note that

$$\lambda \in \sigma_e(T) \iff 0 \in \sigma_e(S(\lambda)). \quad (12.8.8)$$

This follows from Lemma 12.3.1 and the observation that λ is an eigenvalue of (12.8.1) if and only if $\xi = 0$ is an eigenvalue of (12.8.2) with the same boundary condition. Suppose $\lambda_0^P \leq v \leq \lambda_0^S$. By results from Chapter 5 the eigencurve $\{\xi_0^P(\lambda) : \lambda \in \mathbb{R}\}$ is a continuous function of λ, is zero when $\lambda = \lambda_0^P$, is strictly decreasing in a neighborhood of this point λ_0^P and has no other positive zero. Hence $\xi_0^P(v) \leq 0$. Similarly $\xi_0^S(v) \geq 0$. Therefore $0 \in [\xi_0^P(v), \xi_0^S(v)] \subset \sigma_e(S(v))$ by the classical Floquet theory applied to (12.8.2) in the Hilbert space H. From (12.8.8) we may conclude that $v \in \sigma_e(T)$. More generally, if $\lambda_n^P \leq v \leq \lambda_n^S$, $n = 0, 2, 4, ...$ we have, by a similar argument, that $\xi_n^P(v) \leq 0$ and $\xi_n^S(v) \geq 0$. This implies that $0 \in [\xi_n^P(v), \xi_n^S(v)] \subset \sigma_e(S(v))$ and thus $v \in \sigma_e(T)$. Similarly, $\lambda_n^S \leq v \leq \lambda_n^P$, $n = 1, 3, 5, ...$ implies $\xi_n^S(v) \leq 0$ and $\xi_n^P(v) \geq 0$, hence we have $0 \in [\xi_n^P(v), \xi_n^S(v)] \subset \sigma_e(S(v))$ and thus $v \in \sigma_e(T)$. Thus we have shown that all the intervals I_n with $n \geq 0$ of (12.8.7) are in $\sigma_e(T)$.

To prove that all the intervals I_n with $n \leq 0$ of (12.8.7) also lie in $\sigma_e(T)$, suppose that $\lambda_{-n}^S \leq v \leq \lambda_{-n}^P$, $n = 0, 2, 4, ..$; then arguing as before but noting that the eigencurves $\xi_{-n}^P(\lambda), \xi_{-n}^S(\lambda)$ are increasing in a neighborhood of each zero $\lambda_{-n}^P, \lambda_{-n}^S$ we see that $\xi_{-n}^P(v) \leq 0$, $\xi_{-n}^S(v) \geq 0$ and thus $0 \in [\xi_{-n}^P(v), \xi_{-n}^S(v)] \subset \sigma_e(S(v))$ and therefore $v \in \sigma_e(T)$. Similarly $v \in [\lambda_{-n}^P, \lambda_{-n}^S]$ implies that $\xi_{-n}^P(v) \geq 0$, $\xi_{-n}^S(v) \leq 0$ and thus $0 \in [\xi_{-n}^P(v), \xi_{-n}^S(v)] \subset \sigma_e(S(v))$ and consequently $v \in \sigma_e(T)$. This concludes the proof showing that all intervals (12.8.7) lie in $\sigma_e(T)$.

To finish the proof of the Theorem we need to show that no other points of $\mathbb{R}$ are in $\sigma_e(T)$. Suppose $v \in (\lambda_{-0}^P, \lambda_0^P)$. Then $\xi_0^P(v) > 0$. This implies that 0 is below the beginning of the essential spectrum: $\xi_0^P(v) = \inf \sigma_e(S(v))$. Then (12.8.8) implies that v is not in $\sigma_e(T)$. Note that if $\lambda_n^P = \lambda_{n+1}^P = v$ for some n, then $\xi_n^P(v) = 0 = \xi_{n+1}^P(v)$ and $0 \in \sigma_e(S(v))$ implying that $v \in \sigma_e(T)$. In this case the 'gap' $(\lambda_n^P, \lambda_{n+1}^P)$ is missing i.e. the two adjacent spectral bands combine. Similarly if $\lambda_n^S = \lambda_{n+1}^S = v$ for some n, then $v \in \sigma_e(T)$ and the 'gap' $(\lambda_n^S, \lambda_{n+1}^S)$ is missing. Suppose $\lambda_n^P < v < \lambda_{n+1}^P$ for some $n = 1, 2, 3, ...$ Then $\xi_n^P(v) < 0$ and $\xi_{n+1}^P(v) > 0$ implying that 0 lies in the gap $(\xi_n^P(v), \xi_{n+1}^P(v))$ and thus not in $\sigma_e(S(v))$. Then

(12.8.8) implies that v is not in $\sigma_e(T)$. The proofs of the other cases are all similar and hence omitted. $\square$

There is an analogous result for the whole line case.

THEOREM 12.8.2. *Let $J = (-\infty, \infty)$, and assume that p, q, w are periodic with fundamental period r, $0 < r < \infty$,*

$$p(t+r) = p(t),\ q(t+r) = q(t),\ w(t+r) = w(t),\ t \in J. \tag{12.8.9}$$

Define an operator T from H to H as follows:

$$D = \{f \in H = L^2(J, |w|) : f,\ pf' \in AC_{loc}(J),\ \frac{1}{|w|}[-(pf')' + qf] \in H,\}$$

$$Tf = \frac{1}{w}[-(pf')' + qf],\ f \in D \tag{12.8.10}$$

For each $\lambda \in \mathbb{R}$, let $\xi_n^P(\lambda)$, $\xi_n^S(\lambda)$, $n \in \mathbb{N}_0 = \{0, 1, 2, 3, ...\}$ denote the periodic and semi-periodic eigenvalues of (12.8.2) on the fundamental interval $[a, a+r]$, respectively, and assume that $\xi_0^P(0) > 0$. Then

1. $\xi_0^S(0) > 0$.
2. *For each $n \in \mathbb{N}_0$ the equation $\xi_n^P(\lambda) = 0$ has exactly one positive root λ_n^P and exactly one negative root λ_{-n}^P; and the equation $\xi_n^S(\lambda) = 0$ has exactly one positive root λ_n^S and exactly one negative root λ_{-n}^S. Furthermore each of $\xi_n^P(\lambda)$, $\xi_n^S(\lambda)$ is continuous at each point $\lambda \in \mathbb{R}$ and monotone in a neighborhood of each of its roots.*
3.
    $$\sigma_e(T) = \cup_{n \in Z^*} I_n \text{ where } Z^* = \{..., -2, -1, -0, 0, 1, 2, ...\},$$
    $$I_0 = [\lambda_0^P, \lambda_0^S],\ I_1 = [\lambda_1^S, \lambda_1^P],\ I_2 = [\lambda_2^P, \lambda_2^S],$$
    $$I_3 = [\lambda_3^S, \lambda_3^P],\ I_4 = [\lambda_4^P, \lambda_4^S],\ I_5 = [\lambda_5^S, \lambda_5^P]...$$
    $$I_{-0} = [\lambda_{-0}^S, \lambda_{-0}^P],\ I_{-1} = [\lambda_{-1}^P, \lambda_{-1}^S],\ I_{-2} = [\lambda_{-2}^S, \lambda_{-2}^P],$$
    $$I_{-3} = [\lambda_{-3}^P, \lambda_{-3}^S],\ I_{-4} = [\lambda_{-4}^S, \lambda_{-4}^P],\ I_{-5} = [\lambda_{-5}^P, \lambda_{-5}^S],...$$

PROOF. This proof is similar to the proof of Theorem 12.8.1 and hence omitted. $\square$

9. Comments

The results of this chapter are based, in large part, on Kong, Wu and Zettl [389], Kong, Möller, Wu and Zettl [382], and Marletta and Zettl [466], [464]. There has been considerable interest in left-definite Sturm-Liouville problems: see, for instance, Atkinson and Jabon [28], Atkinson and Mingarelli [29], Bennewitz and Everitt [65], Binding and Browne [72], Binding and Volkmer [80], Constantin [128], Čurgus and Langer [133], Daho and Langer [135], Kong, Wu and Zettl [387], Kong, Möller, Wu and Zettl [382] and Kong, Wu and Zettl [389]. The last two of these specifically address the question of the characterization of the spectrum of a singular left-definite problem in terms of the spectra of a related one parameter family of right-definite problems.

A special class of singular LD problems with limit-circle non-oscillatory endpoints and a specific singular BC was studied by Kaper, Kwong, and Zettl [365].

For a "LD Hilbert space" approach to the study of singular boundary value problems with an indefinite weight function see Vonhoff [**587**], [**588**], Bennewitz and Everitt [**65**], and the references cited therein. For a Krein space approach to LD and indefinite problems see Curgus and Langer [**133**], [**134**] and the reference cited there. Also see the seminal paper of Weyl [**607**].

(1) This section introduces the left-definite operators T and the associated family of right-definite operators upon which the study of T is based.

(2) Basic results are given in some detail here since we believe this section and this chapter will form the starting point for the investigation of singular left-definite problems. Also some of the basic notation used in this chapter is introduced.

(3) When w is positive the existence of eigenvalues and other parts of the spectrum can be established using the theory of self-adjoint operators in the Hilbert space $H = L^2(J, w)$. When w changes sign many authors have tried to establish the existence of eigenvalues and other parts of the spectrum by using operator theory in a Krein space or in a different Hilbert space, one constructed from the left side of the equation i.e. a left-definite Hilbert space. In the opinion of this writer these efforts have had only very limited success. It is therefore interesting to note that the proof of the existence of eigenvalues and other parts of the spectrum of left-definite problems can be based on the right-definite theory *for a one parameter family of problems*.

(4) The important Lemma 12.4.2 was inspired by the work of Binding and Volkmer [**79**] but is more general, and we believe, more 'natural' than their result partly because it does not require any additional boundedness assumption on w as in [**79**]. Eigencurves for regular problems on a bounded interval J with separated boundary conditions, and with $|w|$ replaced by 1, have been investigated by other authors, see Binding and Volkmer [**79**], Binding and Browne [**69**], [**70**], [**72**]. While we have been strongly influenced by the work of these authors, our approach differs from theirs in several important respects: the boundary conditions can be coupled as well as separated, the problems can be singular with limit-point and/or limit-circle endpoints, and the weight function in equation (12.1.2) is $|w|$ rather than 1. The "interplay" between w and $|w|$ in (12.1.2) has significant advantages, particularly for singular problems. See Lemma 12.4.2 and Remark 12.4.1.

(5) In many respects the results of this section are analogues of the results of Niessen and Zettl [**503**] for $w > 0$ and the proofs are similar.

(6) This is a continuation of Section 12.5 and includes a study of the variation of the eigenvalues with the boundary conditions analogous to the eigenvalue inequalities in Eastham, Kong, Wu and Zettl [**145**] for $w > 0$.

(7) In this section the results in Bailey, Everitt, Weidmann and Zettl [**38**] for $w > 0$ are extended to the case when w is indefinite. The results for the right-definite case, in particular spectral exactness and spectral inclusion, are obtained from the theory of norm resolvent and strong resolvent convergence of self-adjoint operators in Hilbert space. In contrast with Section 12.5 were the proofs from the $w > 0$ case are adapted to get similar results for w indefinite, here new methods seem to be needed. As in

Section 12.4 these methods are based on using the right-definite results for a one parameter family of related operators.

The results of Section 12.7 cover (a) spectrally exact approximation of eigenvalues in the spectral gap of T containing 0; (b) spectral inclusion for all other spectral points of T. Since $\lambda = 0$ is not in the spectrum of the left-definite operator T, the spectrum of T has a 'gap' containing 0. This gap is the analogue of the first spectral gap $(-\infty, \inf \sigma_e)$ in the right-definite case i.e. when $w > 0$ a.e. on J and the spectrum is bounded below. If the spectrum is discrete (in both cases), this 'gap' is the whole real line. In this case (12.7.5) gives spectral convergence for all the eigenvalues of T. In the presence of essential spectrum (12.7.5) gives only convergence for the eigenvalues in the spectral gap of T which contains 0. This gap may be a bounded interval or a half-line. Thus Theorem 12.7.4 is an analogue of the results in [**38**] which establish spectral convergence for the eigenvalues in the gap $(-\infty, \inf \sigma_e)$ for the right-definite case.

For $w > 0$ Stolz and Weidmann [**567**] have shown that spectral convergence can be achieved in each gap with an astute choice of boundary conditions. But these conditions depend on the particular gap. This approach does not seem to lend itself to the left-definite problems studied here since changing the boundary conditions may not preserve the positivity of the operator $S(0)$ i.e. the left-definiteness of T. The latter is used in the characterization of the eigenvalues given by Kong, Wu and Zettl [**389**], see Section 12.3, which plays a critical role here.

The problem of spectral approximation of singular differential operators in terms of regular operators is a topic of interest in numerical analysis, see Markowich [**460**], Abramov, Balla and Konyukhova [**1**], Brown and Marletta [**106**].

(8) This Section is based on Marletta and Zettl [**464**]. Note that the spectral gap $(\lambda^P_{-0}, \lambda^P_0)$ is always present since $\lambda^P_{-0} < 0 < \lambda^P_0$. As already mentioned in connection with Section 12.7, in the left-definite theory it plays the role that the 'first' gap $(-\infty, inf\sigma_e)$ plays in the right definite theory. In principle, each of the other gaps may be present or missing. There are examples to show that all other gaps are present and that all other gaps are missing. But we say 'in principle' because we don't know of an example where, say $\lambda^P_k = \lambda^P_{k+1}$ for $k = 7, 9, 14$ and for no other k. This comment applies to left-definite and to right definite problems.

The hypothesis $p > 0$ a.e. on J is needed in (12.8.3) since without it the left-definite hypothesis $\xi^P_0(0) > 0$ could not be fulfilled. This hypothesis plays an important role in our proof. Do Theorems 12.8.1 and 12.8.2 hold without it? If not, what are the corresponding results without this hypothesis? For the classical Mathieu equation

$$-y'' + (\sin t)y = \xi y \qquad (12.9.1)$$

the lowest periodic eigenvalue on the interval $[0, 2\pi]$ is approximately $\xi^P_0(0) \sim -0.378$. Hence the above results do not apply directly to (12.9.1). But they do apply when the spectral parameter is shifted by replacing $\sin t$

by $\sin t + 1$ and the weight function w is taken to be a function of type:
$$w(t) = \begin{cases} 1 \text{ on } J \setminus I, \\ -1 \text{ on } I, \end{cases}$$
for $J = [a, \infty)$, $-\infty < a < \infty$ or $J = (-\infty, \infty)$ and I a proper subinterval of J.

We believe this section is a starting point for the development of Floquet theory for left-definite problems of second order and higher order.

In this chapter we investigated only the real coefficient case. Results for a periodic complex potential q, when $p = 1 = w$, were obtained by Rofe-Beketov in [**551**].

Part 5

Examples and other Topics

Euler, Leonhard (1707 - 1783)
> [upon losing the use of his right eye]
> Now I will have less distraction.

In H. Eves In Mathematical Circles, Boston: Prindle, Weber and Schmidt, 1969.

To illustrate some of the behavior of the spectrum of SLP we discuss a number of examples including the 32 examples from the SLEIGN2 package. These examples have been chosen to illustrate some of the main features of the spectrum of SLP.

If there is a singularity in the *interior* of the underlying interval then neither the results from Part I nor Part II apply. In fact by Theorem 1.2.3 it is always the case that either not all solutions or not all quasi-derivatives of solutions can be continuously extended through the singular point. The case of one interior singularity is discussed in Chapter 10; this extends in a completely straightforward way to any finite number of such singularities. The case of infinitely many interior singularities involves some new features; in particular the self-adjoint realizations may be infinite dimensional extensions of the minimal operator. This case is studied in Everitt and Zettl [**238**].

The results of Chapter 13 also apply and are interesting when the interior "singular" points are in fact regular. For instance it is well known, see Zettl [**622**], [**620**], that in the classical theory there are no three point self-adjoint boundary value problems. That is, if all solutions and their quasi-derivatives are required to be continuous at every interior point then every self-adjoint extension of the minimal operator can be characterized in terms of *two point* boundary conditions.

What if the solutions or their quasi-derivatives are allowed to have finite jump discontinuities at some regular interior point? Is it possible to give conditions on the jumps to get a self-adjoint problem? The answer is yes. For example, if the solution is required to be continuous but its quasi-derivative is allowed to have a jump discontinuity at an interior point then the problem is self-adjoint provided the jump is proportional to the value of the solution at the point and the proportionality constant c is real. This is called a "point interaction of strength c" in the Physics literature; see pp. 20, 21 in [**268**]. The roles of the solution and its quasi-derivative can be reversed in the above statement.

Similar results apply at an interior singularity; here there may be infinite jumps or wild oscillations, but these are controlled with the Lagrange form $[y, u]$ which is finite.

CHAPTER 13

Two Intervals

Descartes, René (1596-1650)
> Each problem that I solved became a rule which served afterwards to solve other problems.

Discours de la Méthode, 1637.
> It is not enough to have a good mind. The main thing is to use it well.

Discours de la Méthode, 1637.

1. Introduction

In Parts II and IV we studied SLP on one interval. In this chapter we study Sturm-Liouville boundary value problems on two intervals. As in the one interval case we take these intervals to be open to allow for the possibility of singularities at the endpoints, but otherwise the intervals are arbitrary: they may be identical, disjoint, overlap, abut; each interval may be the whole real line. On each interval we have an SL expression. In the case of identical intervals these expressions may be the same or different. There are four endpoints: the left and right endpoints of each interval. If the intervals abut, then the right endpoint of the left interval is the same as the left endpoint of the right interval and these "two" endpoints are counted among the four.

In [**243**], partly motivated by problems from the applied literature (see the Comments section at the end of this chapter for illustrations), Everitt and Zettl embarked on a systematic rigorous study of two interval SLP in the framework of a direct sum of Hilbert spaces. A primary goal of this study was to characterize *all* self-adjoint operators in terms of boundary conditions as explicitly as possible.

Let $J_1 = (a_1, b_1)$, $J_2 = (a_2, b_2)$, let M_1 be an expression on J_1 with weight function w_1 and M_2 on J_2 with w_2. A simple way of getting self-adjoint operators S in the direct sum space $H = H_1 + H_2$, $H_k = L^2(J_k, w_k)$, $k = 1, 2$ is to take the direct sum of self-adjoint operators from H_1 and H_2. If these were all the self-adjoint operator realization from the two intervals there would be no need for a "two interval" theory. In fact there are many more in general. These "new" self-adjoint operators involve interactions between the two intervals. Characterizing these interactions as explicitly as possible is our main goal in this chapter. At regular endpoints this characterization can be given as conditions on the values of solutions and their quasi-derivatives at the endpoints. In the singular case, since solutions and their quasi-derivatives do not exist in general, the Lagrange sesquilinear form involving solutions and certain fixed maximal domain functions can be used as replacements for the values of the solutions and their quasi-derivatives. As a special case of the regular case we get the so called point interactions discussed in the Physics literature [**268**].

2. Notation and Basic Assumptions

Let $-\infty \leq a_r < b_r \leq \infty$, let $J_r = (a_r, b_r)$, and let

$$p_r^{-1}, q_r, w_r \in L_{loc}(J_r, \mathbb{R}), \quad w_r > 0, \quad r = 1, 2. \tag{13.2.1}$$

Define differential expressions M_r by

$$M_r y = -(p_r y')' + q_r y \quad \text{on} \quad J_r, \quad r = 1, 2. \tag{13.2.2}$$

Let $H_r = L^2(J_r, w_r)$. Below we use the notation from Part I with a subscript r to denote the $r-th$ interval. The subscript r is sometimes omitted when it is clear from the context.

The two interval maximal and minimal domains and operators are simply the direct sums of the corresponding one interval domains and operators:

$$D_{\max} = D_{1\max} + D_{2\max}; \quad D_{\min} = D_{1\min} + D_{2\min}, \tag{13.2.3}$$

$$S_{\max} = S_{1\max} + S_{2\max}; \quad S_{\min} = S_{1\min} + S_{2\min}. \tag{13.2.4}$$

Elements of $H = H_1 + H_2$ will be denoted in bold face type: $\mathbf{f} = \{f_1, f_2\}$ with $f_1 \in H_1$, $f_2 \in H_2$. As usual the inner product in H is defined by

$$(\mathbf{f}, \mathbf{g}) = (f_1, g_1)_1 + (f_2, g_2)_2$$

where $(\cdot, \cdot)_r$ is the usual inner product in H_r. The subscript r may be omitted when it is clear from the context. As in the one interval case the Lagrange sesquilinear form plays an important role. It is defined by

$$[\mathbf{f}, \mathbf{g}] = [f_1, g_1]_1(b_1) - [f_1, g_1]_1(a_1) + [f_2, g_2]_2(b_2) - [f_2, g_2]_2(a_2). \tag{13.2.5}$$

Note that the two interval form $[\mathbf{f}, \mathbf{g}]$ connects all four endpoints with each other. It is through this form that the two intervals are connected to each other in both the regular and singular cases even when the solutions blow up or oscillate wildly at one or more of the endpoints.

3. Characterization of all Self-Adjoint Extensions

All self-adjoint extensions of the two interval minimal operator are characterized in terms of boundary conditions. First some preliminary lemmas.

LEMMA 13.3.1. *We have*
1. $S_{\min}^* = S_{1\min}^* + S_{2\min}^* = S_{1\max} + S_{2\max} = S_{\max}$; $S_{\max}^* = S_{1\max}^* + S_{2\max}^* = S_{1\min} + S_{2\min} = S_{\min}$.
   In particular, $D_{\max} = D(S_{\max}) = D(S_{1\max}) + D(S_{2\max})$; $D_{\min} = D(S_{\min}) = D(S_{1\min}) + D(S_{2\min})$.
2. *The minimal operator $S_{\min}$ is a closed, symmetric, densely defined operator in the Hilbert space H with deficiency index $d = d^+ = d^-$ given by*

$$d = d_1 + d_2. \tag{13.3.1}$$

PROOF. See [**243**] for a proof as well as a definition and discussion of deficiency indices. Since the coefficients and the weight function are all real the upper and lower deficiency indices are equal and the common value is denoted by d in the two interval case and by d_1, d_2 for intervals 1 and 2. □

3. CHARACTERIZATION OF ALL SELF-ADJOINT EXTENSIONS

We start with the general characterization of the domains of self-adjoint extensions of the two interval minimal operator. A set of functions $\psi_1, \psi_2, ..., \psi_d$ from the maximal domain is said to be linearly independent modulo the minimal domain if no nontrivial linear combination of them is in the minimal domain.

LEMMA 13.3.2. *Let (13.2.1), (13.2.2) hold and let the two interval minimal and maximal domains $D_{\min}, D_{\max}$ and operators $S_{\min}$ and $S_{\max}$ be defined as above. Let the Lagrange form $[\cdot, \cdot]$ be given by (10.2.3). If the operator S with domain $D(S)$, $D_{\min} \subset D(S) \subset D_{\max}$ is a self-adjoint extension of the minimal operator $S_{\min}$ with deficiency index d, then there exists $\mathbf{g}_1, \mathbf{g}_2, ..., \mathbf{g}_d \in D(S) \subset D_{\max}$ satisfying the following conditions:*

(1) $\mathbf{g}_1, \mathbf{g}_2, ..., \mathbf{g}_d$ *are linearly independent modulo $D_{\min}$;*
(2) $[\mathbf{g}_j, \mathbf{g}_k] = 0, \quad j, k = 1, 2, ..., d;$
(3) $D(S) = \{\mathbf{f} \in D_{\max} : [\mathbf{f}, \mathbf{g}_j] = 0, \quad j = 1, 2, ..., d.\}$

Conversely, given $\mathbf{g}_1, \mathbf{g}_2, ..., \mathbf{g}_d \in D_{\max}$ satisfying conditions (1) and (2), the set $D(S)$ defined by (3) is a self-adjoint domain.

PROOF. These results are extensions of Theorem 10.4.1 to the multi interval case. See Theorem 3.1 and Corollary 3.3 in Everitt and Zettl [**238**]. □

REMARK 13.3.1. When $d = 0$ conditions (1), (2) and (3) of Lemma 13.3.2 are vacuous. In this case the minimal operator $S_{\min}$ is itself self-adjoint and has no proper self-adjoint extension: $S_{\min} = S_{\max}$ in this case. This case occurs if and only if all endpoints are LP.

Conditions (3) are "boundary conditions" and conditions (1) and (2) are the conditions on the "boundary conditions" which determine self-adjoint domains. Condition (1) specifies the number of boundary conditions needed for self-adjointness and condition (2) characterizes the types of conditions which are self-adjoint.

REMARK 13.3.2. All three of the conditions of Lemma 13.3.2 depend on the maximal domain functions $\mathbf{g}_j$; these depend on the coefficients p, q and on the weight function w. This dependence is implicit and complicated. Next we endeavor to make the conditions (2) and (3) more explicit. In the case of a regular endpoint c this can be done in terms of the values $y(c), y^{[1]}(c)$ of maximal domain functions y. These values do not exist, in general, at a singular endpoint. We will see below that at singular LC endpoints there is a parallel theory with $y(c), y^{[1]}(c)$ replaced by the Lagrange form $[y, u]_j(c)$ and $[y, v]_j(c)$ where u, v are fixed, independent of λ, normalized, maximal domain functions which are linearly independent modulo the minimal domain. Recall that at an LP endpoint c, $[y, u]_j(c) = 0$ for any maximal domain functions y, u and that at an LC endpoint, say c, $[y, u](c)$ exists as a finite limit and $[y, u](c) = W(y, u)(c)$ if u is real valued in a neighborhood of c.

THEOREM 13.3.1. *Let (13.2.1), (13.2.2) hold and let the two interval minimal and maximal domains $D_{\min}, D_{\max}$ and operators $S_{\min}$ and $S_{\max}$ be defined as above. Let the Lagrange form $[\cdot, \cdot]$ be given by (10.2.3). Let d be the deficiency index of the (two interval) minimal operator $S_{\min}$. Then $0 \leq d \leq 4$. For $1 \leq d \leq 4$, let $\mathbf{u} = \{u_1, u_2\}, \mathbf{v} = \{v_1, v_2\}$ be real-valued maximal domain functions such that u_j, v_j are linearly independent modulo $D_{j\min}$, and are normalized to satisfy $[u_j, v_j](c) = 1$ at each non LP endpoint c of J_j, $j = 1, 2$. Then all self-adjoint extensions of the minimal operator $S_{\min}$ can be characterized as follows:*

Case 1. $d = 0$. In this case $S_{\min} = S_{\max}$, and $S_{\min}$ is itself a self-adjoint operator with no proper self-adjoint extension. This case occurs if and only if all four endpoints are LP. In particular there are no boundary conditions required or allowed to determine self-adjoint extensions. Also note that for all $\mathbf{f} = \{f_1, f_2\}$, $\mathbf{g} = \{g_1, g_2\} \in D_{\max}$ we have

$$[\mathbf{f}, \mathbf{g}] = [f_1, g_1]_1(b_1) - [f_1, g_1]_1(a_1) + [f_2, g_2](b_2) - [f_2, g_2](b_2) = 0 - 0 + 0 - 0 = 0.$$

Case 2. $d = 1$. This case occurs if and only if exactly three endpoints are LP; the other, say c, is either regular or LC. In this case condition (3) of Lemma 13.3.2 reduces to

$$c_{11}[y, u](c) + c_{12}[y, v](c) = 0, \quad c_{11}, c_{12} \in \mathbb{R}, \ (c_{11}, c_{12}) \neq (0, 0). \tag{13.3.2}$$

If c is regular, then (13.3.2) can be further reduced to the more familiar form

$$c_{11} y(c) + c_{12} y^{[1]}(c) = 0, \quad c_{11}, c_{12} \in \mathbb{R}, \ (c_{11}, c_{12}) \neq (0, 0). \tag{13.3.3}$$

In this case the symmetry conditions (2) of Lemma 13.3.2 reduce to

$$c_{11}\overline{c}_{12} - \overline{c}_{11} c_{12} = 0. \tag{13.3.4}$$

Multiplying (13.3.4) by $\overline{c}_{11}$ we may assume that c_{11} is real; (13.3.4) then implies that c_{12} is real. From condition (1) it follows that not both of c_{11} and c_{12} can be zero.

To summarize this case we can say that all self-adjoint extensions of the minimal operator are determined by boundary conditions of the form (13.3.2) at the non LP endpoint c

where both of c_{11} and c_{12} are real and not both are zero.

Case 3. $d = 2$. This case occurs when exactly two of the four endpoints are LP. Let c and d denote the two non LP endpoints. Each of these may be R or LC. This case is identical with the one interval theory on an interval with endpoints c and d (even though c and d may be any two of the four endpoints, e.g. c may be the left endpoint of the first interval and d the right endpoint of the second interval). Thus there are separated and coupled self-adjoint boundary conditions at c and d. Conditions (3) of Lemma 13.3.2 reduce to

$$c_{11}[y, u](c) + c_{12}[y, v](c) + d_{11}[y, u](d) + d_{12}[y, v](d) = 0, \tag{13.3.5}$$
$$c_{21}[y, u](c) + c_{22}[y, v](c) + d_{21}[y, u](d) + d_{22}[y, v](d) = 0, \tag{13.3.6}$$

where $c_{ij}, d_{ij} \in \mathbb{C}$ and $\mathbf{u} = \{u_1, u_2\}, \mathbf{v} = \{v_1, v_2\}$ are maximal domain functions which are normalized to satisfy $[u, v](c) = 1 = [u, v](d)$.

Conditions (2) of Lemma 13.3.2 reduce to

$$c_{11}\overline{c_{22}} - c_{12}\overline{c_{21}} = d_{11}\overline{d_{22}} - d_{12}\overline{d_{21}}, \tag{13.3.7}$$
$$c_{11}\overline{c_{12}} - c_{12}\overline{c_{11}} = d_{11}\overline{d_{12}} - d_{12}\overline{d_{11}}, \tag{13.3.8}$$
$$c_{21}\overline{c_{22}} - c_{22}\overline{c_{21}} = d_{21}\overline{d_{22}} - d_{22}\overline{d_{21}}. \tag{13.3.9}$$

Condition (1) of Lemma 13.3.2 reduces to

$$\mathrm{rank}(C, D) = 2, \quad C = (c_{ij}), D = (d_{ij}), \tag{13.3.10}$$

where (C, D) denotes the 2 by 4 matrix whose first two columns are C and whose last two columns are D.

Note that conditions (13.3.9) and (13.3.8) hold whenever the matrices $C = (c_{ij})$ and $D = (d_{ij})$ are both real and (13.3.7) reduces to $\det C = \det D$ in this case. The special case of this special case when $\det C = 0 = \det D$ is equivalent to the separated boundary condition case:

$$c_{11}[y,u](c) + c_{12}[y,v](c) = 0, \quad c_{11}, c_{12} \in \mathbb{R}, \quad (c_{11}, c_{12}) \neq (0,0), \qquad (13.3.11)$$

$$d_{11}[y,u](d) + d_{12}[y,v](d) = 0, \quad d_{11}, d_{12} \in \mathbb{R}, \quad (d_{11}, d_{12}) \neq (0,0). \qquad (13.3.12)$$

The boundary conditions (13.3.5), (13.3.6) together with the self-adjointness conditions (13.3.7), (13.3.8), (13.3.9), (13.3.10) have the canonical representation

$$Y(d) = \exp(i\gamma) \, K \, Y(c), \quad -\pi < \gamma \leq \pi, \quad K \in SL_2(\mathbb{R}), \qquad (13.3.13)$$

where

$$Y = \begin{bmatrix} [y,u] \\ [y,v] \end{bmatrix}. \qquad (13.3.14)$$

At a regular endpoint, say c, u and v can be chosen so that (12.3.14) reduces to the familiar form

$$Y(c) = \begin{bmatrix} y(c) \\ y^{[1]}(c) \end{bmatrix}. \qquad (13.3.15)$$

If the two non-LP endpoints are from the same interval then all the two interval self-adjoint extensions are obtained simply as direct sums of the self-adjoint operators from the interval with the two non LP endpoints with the minimal operator from the other interval. However if the two non LP endpoints are from the two different intervals (or the same interval but for different differential expressions M) then there is "mixing" between the two intervals, i.e. the canonical projection of the two interval self-adjoint operators to the one interval Hilbert space is not self-adjoint.

Case 4. $d = 3$. In this case there is exactly one LP endpoint. Assume that b_2 is LP and, to avoid subscripts, let $a = a_1, b = b_1, c = a_2$. The other cases are similar but there is a sign difference if two of the three non LP endpoints are right, rather than left, endpoints due to the difference in sign of the Lagrange form $[\cdot,\cdot]$. This difference will be pointed out below.

Condition (3) of Lemma 13.3.2 reduces to

$$\begin{aligned} &a_{11}[y,u](a) + a_{12}[y,v](a) + b_{11}[y,u](b) + \\ &b_{12}[y,v](b) + c_{11}[y,u](c) + c_{12}[y,v](c) = 0, \\ &a_{21}[y,u](a) + a_{22}[y,v](a) + b_{21}[y,u](b) + \\ &b_{22}[y,v](b) + c_{21}[y,u](c) + c_{22}[y,v](c) = 0, \\ &a_{31}[y,u](a) + a_{32}[y,v](a) + b_{31}[y,u](b) + \\ &b_{32}[y,v](b) + c_{31}[y,u](c) + c_{32}[y,v](c) = 0. \end{aligned} \qquad (13.3.16)$$

Let $A = (a_{ij})$, $B = (b_{ij})$, $C = (c_{ij})$ be the 3 by 2 matrices with entries as indicated. Condition (1) of Lemma 13.3.2 implies that

$$\operatorname{rank}(A, B, C) = 3, \qquad (13.3.17)$$

i.e. the 3 by 6 matrix constructed by putting B to the right of A and then C to the right of B has full rank.

The symmetry condition (2) of Lemma 13.3.2 reduces to the following conditions on the matrix (A, B, C):

$$(a_{j1}\bar{a}_{k2} - a_{j2}\bar{a}_{k1}) - (b_{j1}\bar{b}_{k2} - b_{j2}\bar{b}_{k1}) + (c_{j1}\bar{c}_{k2} - c_{j2}\bar{c}_{k1}) = 0, \quad j,k = 1,2,3. \quad (13.3.18)$$

Note that only 6 of these conditions are independent since the conditions are symmetric in j, k.

Although, for the sake of definiteness, we assumed that b_2 is the LP endpoint the conditions (13.3.18) hold in general provided that there is a plus sign in front of every group of terms corresponding to a left endpoint and a minus sign for each right endpoint. These groups of terms have been grouped with parentheses in (13.3.18).

Case 5. $d = 4$. *This is the case when there is no LP endpoint, i.e. each endpoint is either R or LC. Let $A = (a_{ij})$, $B = (b_{ij})$, $C = (c_{ij})$, $D = (d_{ij})$ be 4 by 2 matrices with complex entries and let (A, B, C, D) be the 4 by 8 matrix whose first two columns are those of A, etc. To simplify the notation let $a = a_1$, $b = b_1$, $c = a_2$, $d = b_2$. Then the three conditions of Lemma 11.3.4 are, respectively, equivalent to the following:*

(1) *The matrix (A, B, C, D) has full rank.*
(2) *The boundary conditions are*

$$a_{11}[y,u](a) + a_{12}[y,v](a) + b_{11}[y,u](b) + b_{12}[y,v](b)+$$
$$c_{11}[y,u](c) + c_{12}[y,v](c) + d_{11}[y,u](d) + d_{12}[y,v](d) = 0,$$
$$a_{21}[y,u](a) + a_{22}[y,v](a) + b_{21}[y,u](b) + b_{22}[y,v](b)+$$
$$c_{21}[y,u](c) + c_{22}[y,v](c) + d_{21}[y,u](d) + d_{22}[y,v](d) = 0,$$
$$a_{31}[y,u](a) + a_{32}[y,v](a) + b_{31}[y,u](b) + b_{32}[y,v](b)+$$
$$+c_{31}[y,u](c) + c_{32}[y,v](c) + d_{31}[y,u](d) + d_{32}[y,v](d) = 0,$$
$$a_{41}[y,u](a) + a_{42}[y,v](a) + b_{41}[y,u](b) + b_{42}[y,v](b)+$$
$$+c_{41}[y,u](c) + c_{42}[y,v](c) + d_{41}[y,u](d) + d_{42}[y,v](d) = 0. \tag{13.3.19}$$

(3) *Given the rank condition (1), necessary and sufficient conditions on the boundary conditions for self-adjointness are:*

$$(a_{j1}\bar{a}_{k2} - a_{j2}\bar{a}_{k1}) - (b_{j1}\bar{b}_{k2} - b_{j2}\bar{b}_{k1}) + (c_{j1}\bar{c}_{k2} - c_{j2}\bar{c}_{k1})$$
$$- (d_{j1}\bar{d}_{k2} - d_{j2}\bar{d}_{k1}) = 0, \quad j,k = 1,2,3,4. \tag{13.3.20}$$

These conditions can be expressed more compactly as

$$AEA^* - BEB^* + CEC^* - DED^* = 0, \quad E = \begin{bmatrix} 0 & -1 \\ 1 & 0 \end{bmatrix}. \tag{13.3.21}$$

REMARK 13.3.3. At a regular endpoint, say c, u and v can be chosen with the initial conditions:

$$u(c) = 0, \quad u^{[1]}(c) = 1, \quad v(c) = -1, \quad v^{[1]}(c) = 0.$$

With this choice we have

$$[y,u](c) = W(y,u)(c) = y(c), \quad [y,v](c) = W(y,v)(c) = y^{[1]}(c),$$

and the boundary conditions reduce to the familiar form for the regular case, see Chapter 4.

REMARK 13.3.4. At an LCNO endpoint u can be chosen to be a principal solution and v can be chosen to be a positive non-principal solution; see Chapter 10.

REMARK 13.3.5. Theorem 13.3.1 characterizes all self-adjoint extensions of the two interval minimal operator or, equivalently, all self-adjoint restrictions of the two interval maximal operator. If the minimal operator is replaced by another symmetric restriction S_{sym} of the maximal operator then one can ask the question: What are all the self-adjoint extensions of S_{sym}? Consider the case when $d = 4$. Determine S_{sym} by fixing a "self-adjoint interaction" through two of the four endpoints. Then all self-adjoint extensions of S_{sym} are obtained from the general self-adjoint boundary conditions at the other two endpoints. This remark takes on added significance when there are infinitely many intervals. In that case the maximal domain is an infinite dimensional extension of the minimal domain and an additional condition has to be added to the three of Lemma 13.3.2; see Everitt and Zettl [**238**]. By choosing fixed "self-adjoint interactions" at all but two of the non LP endpoints the usual two-point regular or singular self-adjointness conditions can be applied at the remaining two endpoints. In many applications to Physics, particularly quantum mechanics, the two "remaining" endpoints are the "outside" ones i.e. all the others are "interior" singularities. These remarks can be illustrated in terms of the self-adjointness conditions (13.3.21): First take conditions at two endpoints, say a, d which are independent of those at c, b; this can be done by choosing the matrices A, D to have two rows of zeros each and then choose B, C to have zeros in the complementary rows but such that (A, B, C, D) has full rank. Now choose the non zero entries of A, D such that

$$AEA^* = DED^*$$

and keep A, D fixed. Let $S(A, D)$ denote the restriction of the maximal operator determined by A, D at the endpoints a, d. Then every pair of matrices B, C satisfying (13.3.21) and the rank condition $rank(A, B, C, D) = 4$ determines a self-adjoint extension of $S(A, D)$. The roles of the endpoints a, b, c, d can be interchanged as long as the signs in (13.3.21) are respected.

PROOF. This is similar to the proof for the one interval case given in considerable detail in Chapter 10, so we only outline it here. The idea is to make the three conditions of Lemma 13.3.2 more concrete. It is clear that condition (1) determines the number of linearly independent boundary conditions. Let c be an LC endpoint. From Lemma 10.4.2, the Bracket Decomposition lemma, we get

$$[y, \psi](c) = [\overline{\psi}, v](c)[y, u](c) - [\overline{\psi}, u](c)[y, v](c).$$

Since at any non LP endpoint the Lagrange brackets can be assigned arbitrary values in $\mathbb{C}$ we have that

$$[\overline{\psi}, v](c) = c_{11}, \quad [\overline{\psi}, u](c) = -c_{12}$$

for arbitrary $c_{11}, c_{12} \in \mathbb{C}$. Thus the boundary conditions (3) of Lemma 13.3.2 assume the form (13.3.5), (13.3.6) at each non LP endpoint c. Now a straightforward but tedious computation shows that the symmetry conditions (2) of Lemma 13.3.2 reduce to the conditions described above for the cases for $d = 1, d = 2, d = 3$ (with a change in sign if two of the three LC endpoints are right endpoints) and $d = 4$. □

To illustrate and, hopefully, illuminate the self-adjoint boundary conditions given by Theorem 13.3.1 we give some examples. Since the conditions when $d = 0$ or 1 or 2 are the same as in the one interval case we give examples here only for

$d = 3$ and $d = 4$. (Note, however, that when $d = 2$ arbitrary self-adjoint boundary conditions can be applied at the two non LP endpoints regardless of whether these endpoints are from the same interval or not.)

EXAMPLE 13.3.1. *Assume $d = 3$. Let*

$$A = \begin{bmatrix} 1 & 0 \\ 0 & 0 \\ 0 & 0 \end{bmatrix}, B = \begin{bmatrix} 0 & 0 \\ 1 & 0 \\ 0 & 1 \end{bmatrix}, C = \begin{bmatrix} 0 & 0 \\ -1 & 0 \\ h & -1 \end{bmatrix}.$$

It is easy to check that the self-adjointness conditions (13.3.18) are satisfied for any $h \in \mathbb{R}$. When all three endpoints are regular these three matrices yield the boundary conditions:

$$y(a) = 0, \quad y(b) = y(c), \quad y^{[1]}(b) - y^{[1]}(c) = h\, y(c).$$

Thus, if $b = c$, these conditions require y to be continuous at $b = c$ but allow the quasi-derivative (py') to have a jump discontinuity at c. If this jump is proportional to the value of y at c with a real proportionality constant h ($h = 0$ is allowed and reduces to the continuous case) then the jump is self-adjoint.

For the general case of regular or singular endpoints these conditions are:

$$[y, u](a) = 0, \; [y, u](b) = [y, u](c),$$
$$[y, v](b) - [y, v](c) = h\, [y, u](c),$$

for h real. Note that, in either the regular or singular case, the condition at a is independent of the condition at c, b.

EXAMPLE 13.3.2 (10.3.2). *Suppose $d = 4$. Let*

$$A = \begin{bmatrix} 1 & 0 \\ 0 & 0 \\ 0 & 0 \\ 0 & 0 \end{bmatrix}, B = \begin{bmatrix} 0 & 0 \\ 1 & 0 \\ 0 & 1 \\ 0 & 0 \end{bmatrix}, C = \begin{bmatrix} 0 & 0 \\ -1 & 0 \\ h & -1 \\ 0 & 0 \end{bmatrix}, D = \begin{bmatrix} 0 & 0 \\ 0 & 0 \\ 0 & 0 \\ 1 & 0 \end{bmatrix}. \quad (13.3.22)$$

It is easy to check that the self-adjointness conditions (13.3.20) are satisfied for any $h \in \mathbb{R}$. When all four endpoints are regular these four matrices yield the boundary conditions:

$$y(a) = 0 = y(d), \quad y(b) = y(c), \quad y^{[1]}(b) - y^{[1]}(c) = h\, y(c). \quad (13.3.23)$$

Thus, if $b = c$, conditions (13.3.22) require y to be continuous at $b = c$ but allow the quasi-derivative to have a jump discontinuity at c. If this jump is proportional to the value of y at c with a real proportionality constant h ($h = 0$ is allowed and reduces to the continuous case) then the jump is self-adjoint. For the general case of regular or singular endpoints conditions (13.3.22) are:

$$[y, u](a) = 0 = [y, u](d), \; [y, u](b) = [y, u](c),$$
$$[y, v](b) - [y, v](c) = h\, [y, u](c), \quad (13.3.24)$$

for h real. Note that, in either the regular or singular case, the conditions at a, d are independent of those at c, b and the conditions at a, d can be replaced by any self-adjoint conditions at these two endpoints i.e. by

$$A_1 E A_1^* = D_1 E D_1^*, \quad E = \begin{bmatrix} 0 & -1 \\ 1 & 0 \end{bmatrix}, \quad \operatorname{rank}(A_1, D_1) = 2,$$

where A_1, D_1 are 2 by 2 matrices and A, D are the 4 by 2 matrices obtained by adding a row of zeros on the top and the bottom.

EXAMPLE 13.3.3. *Replacing the matrix C by*

$$C = \begin{bmatrix} 0 & 0 \\ -1 & h \\ 0 & -1 \\ 0 & 0 \end{bmatrix}$$

we get a self-adjoint problem for any real h with continuous quasi-derivatives but discontinuous solutions when $h \neq 0$.

The classical Legendre equation

$$-(py')' = \lambda y, \quad p(t) = 1 - t^2$$

is usually studied on the interval $(-1, 1)$ since it has singularities at both ± 1. (Recall that these singularities are determined, not by p, but by $1/p$.) For instance the celebrated Legendre orthogonal polynomials are the eigenfunctions of a SLP on this interval with a singular a boundary conditions at ± 1. (See Example 1 in Chapter 14 for more details of the classical one interval theory of the Legendre equation.) As an illustration of the multi-interval theory the next example will describe all self-adjoint realizations of this equation in the Hilbert space $L^2(-\infty, \infty)$. These will include interactions through the interior singular points at ± 1. To achieve this we identify $L^2(-\infty, \infty)$ with the direct sum of the L^2 spaces on the three intervals $(-\infty, -1), (-1, +1), (1, \infty)$ and apply the three interval SLP theory to these intervals. Although we have only dealt with the two interval case so far in this chapter the extensions of all the results of this chapter to the three interval case are straightforward and so we will use them for this example.

EXAMPLE 13.3.4. *Consider the classical Legendre equation*

$$-(py')' = \lambda y, \quad p(t) = 1 - t^2, \quad t \in \mathbb{R}, \quad \lambda \in \mathbb{R}, \quad \text{on} \quad J,$$
$$J = J_1 \cup J_2 \cup J_3, \quad J_1 = (-\infty, -1), \quad J_2 = (-1, 1), \quad J_3 = (1, \infty). \quad (13.3.25)$$

This equation has singular points at $-\infty$, at $+\infty$ and at the interior points -1 and $+1$ from both sides. The singularities at $-\infty$ and at $+\infty$ are due to both the leading coefficient p and the weight function $w = 1$ since neither w nor $1/p$ is integrable near $-\infty$, and $+\infty$; the singularities at -1 and at $+1$ from both sides are due to the fact that $1/p$ is not integrable near these points from either side. (Recall that the regular-singular classification, as well as the solutions, depend not on p but on $1/p$.) We make the following observations:

(1) *Both $-\infty$ and $+\infty$ are LP.*
(2) *The equation is LCNO at $-1^-, -1^+, 1^-, 1^+$;*
(3) *For $\lambda = 0$, the equation has linearly independent solutions*

$$u(t) = 1, \quad v(t) = \frac{-1}{2} \ln(|\frac{1-t}{1+t}|).$$

Note that u is a solution on all of $\mathbb{R}$ but v blows up at ± 1 from both sides. Also u is the principal solution at ± 1 from both sides and v is a non principal solution at these points.

(4) *Since u is not in the three interval maximal domain we consider the function u in point (3) of Lemma 13.3.2 as defined only on $(-2, 2)$ and define a new function still denoted by u on $\mathbb{R}$ by*

$$u(t) = \begin{cases} 1 & -1 < t < 1, \quad -2 \leq t < 1 \quad 1 < t \leq 2, \\ 0 & t \leq -3 \quad\quad\quad\quad\quad 3 \leq t, \end{cases}$$

and for $-3 < t < -2$ define u in such a way that u is continuously differentiable on $\mathbb{R}$.

Similarly we define a smooth maximal domain function which coincides with v on $[-2, 2]$ and has compact support in $[-3, 3]$. This new function is still denoted by v.

(5) *$[u, v](t) = 1$ for $-2 < t < 2$; note that $[u, v](t)$ is defined at $t = -1$ and $t = 1$ even though v does not exist at these points.*

(6) *For any maximal domain function $y = \{y_1, y_2, y_3\}$ we have*

$$[y_j, u]_j = -py'_j, \quad [y_j, v]_j = y_j - v(py'_j), \quad j = 1, 2, 3.$$

Here we have omitted the subscripts on u and v since these are all given by the same formula, see point (3) above.

(7) *The three interval Lagrange form is given by*

$$[y, z] = [y, z](-1^-) + [y, z](1^-) - [y, z](-1^+) - [y, z](1^+), \tag{13.3.26}$$

for any maximal domain functions $y = \{y_1, y_2, y_3\}$, $z = \{z_1, z_2, z_3\}$. Note that the terms involving $\pm\infty$ are missing since these are LP endpoints and consequently $[y, z](-\infty) = 0 = [y, z](\infty)$.

THEOREM 13.3.2. *Let $S_{\min}$ and $S_{\max}$ denote the three interval minimal and maximal Legendre operators, respectively. (Recall that each of these is simply the direct sum of the corresponding operators from each of the three intervals: $(-\infty, -1^-), (-1^+, 1^-), (1^+, \infty)$.) Let $a = -1^-$, $b = -1^+$, $c = 1^-$, $d = 1^+$ i.e. a denotes the right endpoint of the first interval, b the left endpoint of the second interval, c the right endpoint of the second interval and d the left endpoint of the third interval. Suppose $A = (a_{ij}), B = (b_{ij}), C = (c_{ij}), D = (d_{ij})$ are 4 by 2 complex matrices satisfying the following two conditions:*

(1)
$$\operatorname{rank}(A, B, C, D) = 4, \tag{13.3.27}$$

(2)
$$(a_{j1}\bar{a}_{k2} - a_{j2}\bar{a}_{k1}) - (b_{j1}\bar{b}_{k2} - b_{j2}\bar{b}_{k1}) + (c_{j1}\bar{c}_{k2} - c_{j2}\bar{c}_{k1})$$
$$-(d_{j1}\bar{d}_{k2} - d_{j2}\bar{d}_{k1}) = 0, \quad j, k = 1, 2, 3, 4. \tag{13.3.28}$$

These conditions can be expressed more compactly as

$$AEA^* - BEB^* + CEC^* - DED^* = 0, \quad E = \begin{bmatrix} 0 & -1 \\ 1 & 0 \end{bmatrix}. \tag{13.3.29}$$

Define D by $y \in D$ if $y \in D_{\max} = D(S_{\max})$ and y satisfies:

$$AY(a) + BY(b) + CY(c) + DY(d) = 0,$$
$$Y = \begin{bmatrix} W(y, u) \\ W(y, v) \end{bmatrix} = \begin{bmatrix} -(py') \\ y - v(py') \end{bmatrix}. \tag{13.3.30}$$

Then D is the domain of a self-adjoint operator S satisfying $S_{\min} \subset S \subset S_{\max}$.

Conversely, given a self-adjoint operator S, $S_{\min} \subset S \subset S_{\max}$, with domain $D = D(S)$ there exist complex matrices A, B, C, D satisfying the two conditions (1) and (2) such that D is given by (13.3.30).

PROOF. The proof is similar to that of Case 5 of Theorem 13.3.1 after noting that $d = d_1 + d_2 + d_3 = 1 + 2 + 1$ and $W(y, u) = -(py')$, $W(y, v) = y - v(py')$. □

4. Comments

The two interval results of this chapter are based on Everitt and Zettl [**243**] and extend readily to any finite number of intervals. The study of the two interval theory is partly motivated by the applied literature, In particular the paper of Boyd [**98**] and its references. The extension to a countable number of intervals is not routine. One significant difference is the fact that for the maximal operator may be an infinite dimensional extension of the minimal operator. See Everitt and Zettl [**238**] for an investigation of the countable interval case for second and higher order problems. For an alternative treatment based on the theory of symplectic spaces for a finite or infinite number of intervals, see the papers and monograph of Everitt and Markus [**220**].

The regular self-adjoint two interval theory has been studied by Zettl in [**620**] with a completely different approach using Green's functions.

CHAPTER 14

Examples

Halmos, Paul R.
> ...the source of all great mathematics is the special case, the concrete example. It is frequent in mathematics that every instance of a concept of seemingly great generality is in essence the same as a small and concrete special case.

I Want to be a Mathematician, Washington: MAA Spectrum, 1985.

In this chapter we give a number of examples to illustrate some of the concepts and results discussed above. These include the examples in the SLEIGN2 package. We follow the notation established above. The endpoint classifications given above are supplemented with the weakly regular (WR) classification used by SLEIGN2 : A finite regular endpoint is called weakly regular if at least one of the coefficients $1/p, q, w$ is unbounded in any neighborhood of this point.

In the first example below, the Legendre equation is given as in the SLEIGN2 code with $q(t) = 1/4$ rather than $q(t) = 0$ as in Example 13.3.4. This is done for numerical reasons, it merely shifts the eigenvalues by $1/4$.

(1) The Legendre equation on $(-1, 1)$.

$$p(t) = 1 - t^2,\ q(t) = 1/4,\ w(t) = 1,$$

$$u(t) = 1,\ v(t) = \frac{1}{2}\ln(\frac{1+t}{1-t}),\ -1 < t < 1.$$

The maximal domain functions u, v are solutions for $\lambda = 1/4$, -1 is LCNO, 1 is LCNO. The BC

$$[y, u](-1) = -(py')(-1) = 0,\ [y, u](1) = -(py)'(1) = 0,$$

determines the Friedrichs extension whose eigenvalues are given by

$$\lambda_n = \frac{1}{4} + n(n+1),\ n \in N_0;$$

and the eigenfunctions are the classical Legendre polynomials.
The BC

$$[y, v](-1) = 0,\ [y, v](1) = 0,$$

is a separated non-Friedrichs BC.

Analogues of the regular periodic and semi-periodic BC are

$$[y, u](-1) = [y, u](1),\ [y, v](-1) = [y, v](1);$$

$$[y, v](-1) = -[y, v](1),\ [y, u](-1) = -[y, u](1);$$

respectively. Note that these depend on the choice of the boundary condition functions u, v.

(2) The Liouville form of the Bessel equation on $(0,\infty)$. Also see Example 13 below - the hydrogen atom equation.
$$p = 1, \; q(t) = \frac{\nu^2 - 1/4}{t^2}, \; w = 1.$$
LP at ∞ for all ν.
At 0:
- LCNO for $-1 < \nu < 1$ but $\nu^2 \neq 1/4$
- R for $\nu^2 = 1/4$
- LP for $\nu^2 \geq 1$.

$$u(t) = t^{\nu+1/2}, \; v(t) = t^{-\nu+1/2}, \; for \; \nu \neq 0, -1/2, 1/2,$$
$$u(t) = t, \; v(t) = 1, \; for \; \nu = -1/2,$$
$$u(t) = t, \; v(t) = -1, \; for \; \nu = 1/2,$$
$$u(t) = \sqrt{t}, \; v(t) = \sqrt{t}\log(t), \; for \; \nu = 0.$$

For $\nu \geq 0$, u is the principal solution; thus $[y,u](0) = 0$ is the Friedrichs condition at 0; there is no BC at ∞ since this endpoint is LP. But for $-1/2 < \nu < 0$ note that v is the principal solution and hence $[y,v](0) = 0$ is the Friedrichs BC.

(3) The Halvorsen equation on $(0,\infty)$.
$$p(t) = 1, \; q(t) = \frac{e^{-2/t}}{t^4}, \; w(t) = 1;$$

0 is WR; ∞ is LCNO. In this case the BC vector Y has the form $Y(0) = \begin{bmatrix} y(0) \\ (py')(0) \end{bmatrix}$ at 0 and $Y(\infty) = \begin{bmatrix} [y,u](\infty) \\ [y,v](\infty) \end{bmatrix}$ where $u(t) = 1$, $v(t) = t$.

(4) The Boyd equation on $(-\infty, 0)$ and on $(0, \infty)$.
$$p = 1, \; q(t) = -1/t, \; w = 1.$$
LP at $-\infty$ and ∞; LCNO at 0^+ and 0^-.
$$u(t) = t, \; v(t) = 1 - t(\log|t|)$$

Solutions can be given in terms of Whittaker functions, see [**39**]; the given u, v are maximal domain functions which are not solutions for any λ. This equation on the interval $(-1, 1)$, *hence with an interior singularity at* 0, arose in a model in connection with the study of eddies in the atmosphere, see Boyd [**98**].

(5) The regularized Boyd equation on $(-\infty, 0)$ and on $(0, \infty)$.
$$p = r^2, \; q = -r^2(\log|t|), \; w = r^2,$$
where
$$r(t) = e^{-(t\log|t|)-t}.$$
LP at $-\infty$ and ∞; WR at 0^- and 0^+.

This is a WR form of example 4; the singularity at zero has been "regularized" using quasi-derivatives. There is a one-to-one correspondence between all self-adjoint BC of this example and example 4 considered as a "two interval" problem. Each self-adjoint realization of the Boyd equation on $(-\infty, \infty)$ is unitarily equivalent to a self-adjoint realization of the regularized Boyd equation and conversely. Thus these problems have the same spectrum. Also the eigenfunctions are closely related (but, of course not the same since one problem is singular and the other regular). For details see [**26**], [**39**], [**205**], [**503**].

(6) The Sears-Titchmarsh equation on $(0, \infty)$.
$$p(t) = t, \; q(t) = -t, \; w(t) = 1/t.$$
LP at 0; LCO at ∞.

This equation was studied by Titchmarsh [573] and by the two named authors [558].

For problems on $[1, \infty)$ the spectrum is discrete but unbounded above and below; for some numerical results see [39].
$$u(t) = \frac{\cos(t) + \sin(t)}{\sqrt{t}}, \; v(t) = \frac{\cos(t) - \sin(t)}{\sqrt{t}}.$$

(7) The BEZ equation on $(-\infty, 0)$ and on $(0, \infty)$.
$$p(t) = t, \; q(t) = 1/t, \; w(t) = 1;$$
LP at $-\infty$ and ∞; LCO at 0^- and 0^+. See example 5 in [39] for some numerical results.
$$u(t) = \cos(\ln|t|), \; v(t) = \sin(\ln|t|).$$

(8) The Laplace tidal wave equation on $(0, \infty)$.
$$p(t) = 1/t, \; q(t) = \frac{k}{t^2} + \frac{k^2}{t}, \; w = 1, \; k \in \mathbb{R}, \; k \neq 0.$$
LCNO at 0 for all k; LP at ∞ for all k. This is only a very special case of the named equation, see Homer [342] for the general equation and references to the applied literature.

Even for this special case there are no representations of solutions in terms of the well known special functions. Thus to determine boundary conditions one must use maximal domain functions. Such functions are given by
$$u(t) = t^2, \; v(t) = t - 1/k.$$

(9) The Latzko equation on $(0, 1)$.
$$p(t) = 1 - t^7, \; q = 0, \; w(t) = t^7,$$
WR at 0 (since $w(0) = 0$); LCNO at 1. The singularity at 1 requires the use of maximal domain functions such as
$$u(t) = 1, \; v(t) = -\ln(1 - t)$$
to determine the boundary condition vector $Y(1)$. This example has a long and celebrated history; see Fichera [246].

(10) A weakly regular equation on $(0, \infty)$.
$$p(t) = \sqrt{t}, \; q = 0, \; w(t) = \frac{1}{\sqrt{t}}.$$
LP at ∞ and weakly regular at zero due to both p and w.

(11) The Plum equation on $(-\infty, \infty)$.
$$p(t) = 1, \; q(t) = 100\cos^2(t), \; w(t) = 1$$
LP at $-\infty$ and ∞.

This is a form of the Mathieu equation. Since both endpoints are LP there are no boundary conditions needed or allowed; there is a unique self-adjoint realization of this equation which has no eigenvalues. It's spectrum consists of an infinite number of compact intervals, the spectral bands, separated by gaps. The first few spectral band are rather thin, the first has a width on the order of 10^{-4}.

The eigenvalues of this problem on the interval $(-b, b)$, $0 < b < \infty$, with Dirichlet boundary conditions

$$y(-b) = 0 = y(b),$$

tend to "bunch up" in the spectral bands of the whole line problem, particularly the fist band. From Theorem 10.8.4 and the well known fact that there are no eigenvalues below the essential spectrum we know that (see Section 10.8)

$$\lambda_n(b) \to \sigma_0 = \inf \sigma_e \quad as \quad b \to \infty.$$

In particular, for any positive ε and any positive integer n the first n eigenvalues differ from each other by less than ε if b is sufficiently large, e.g. one can choose b large enough so that the first 7 million eigenvalues agree to the first 1000 digits. Given any numerical code for the computation of Sturm-Liouville eigenvalues, it can be defeated simply by choosing a large enough b. All this for such a "simple" regular SLP. The singular problem on $(-\infty, \infty)$ exerts a strong influence on the behavior of the eigenvalues of the regular problems on $(-b, b)$ for $0 < b < \infty$.

(12) A Mathieu equation on $(-\infty, \infty)$ and on $(0, \infty)$.

$$p = 1, \; q(t) = \sin(t), \; w = 1.$$

LP at $-\infty$ and at ∞; R at 0.

This classical Mathieu equation has a celebrated history and voluminous literature. There are no eigenvalues for this problem on $(-\infty, \infty)$. On $(0, \infty)$ there may be one eigenvalue depending on the boundary condition at 0. The essential spectrum is the same for the whole line problem and the half line problem and consists of an infinite number of disjoint compact intervals separated by gaps. All the gaps are present. The endpoints of the spectral bands and gaps are periodic and semi-periodic eigenvalues of the problem on the interval $[0, 2\pi]$. These can be computed with SLEIGN2.

The above remarks, with some appropriate modifications, apply to the general Hill's equation. This is the SL equation with periodic coefficients p, q, w with the same fundamental period.

Of special interest is the starting point of the essential spectrum σ_0, which is finite in these cases for $p > 0$, $w > 0$. This point σ_0 is also the "oscillation number" of the equation; this means the equation is NO for $\lambda < \sigma_0$ and O for $\lambda > \sigma_0$. It may be O or NO for $\lambda = \sigma_0$. For the Mathieu equation considered here ($p = 1 = w$, $q(t) = \sin(t)$) σ_0 is not known explicitly but is approximately $\sigma_0 \sim -0.378$. This can be checked by computing the lowest eigenvalue of the periodic problem on $[0, 2\pi]$ i. e. $\sigma_0 = \lambda_0^P(0, 2\pi)$.

(13) The hydrogen atom equation on $(0, \infty)$.

$$p(t) = 1, \; q(t) = \frac{k}{t} + \frac{h}{t^2}, \; w(t) = 1, \; k, h \in \mathbb{R}.$$

This is the two parameter version of the classical one-dimensional equation for quantum theory modelling of the hydrogen atom; see [**349**], Section 10 where most of the results reported on here can be found. A few of these results can be found in the commentary file xamples.tex of the package of files comprising the SLEIGN2 package. For all h, k there are no positive eigenvalues, ∞ is LP and the essential spectrum is $[0, \infty)$. If $k = 0$ the equation reduces to a Bessel equation, see Example 2 with $h = \nu^2 - 1/4$.

LP at ∞ for all h, k.

At 0:
- R for $h = 0 = k$,
- LCNO for $h = 0$ and all $k \neq 0$
- LCNO for $-1/4 \leq h < 3/4$, but $h \neq 0$ and all k
- LCO for $h < -1/4$ and all k
- LP for $h \geq 3/4$ and all k.

Let
$$\rho = \sqrt{h + 1/4}, \text{ for } h \geq -1/4.$$

(i) For $h \geq 3/4$ and $k \geq 0$ there is at most one negative eigenvalue and $\lambda = 0$ may be an eigenvalue; for $h \geq 3/4$ and $k < 0$ there are infinitely many negative eigenvalues given by
$$\lambda_n = \frac{-k^2}{(2n + 2\rho + 1)^2}, \ n \in N_0,$$
and $\lambda = 0$ is not an eigenvalue.

(ii) For $h = 0$, $u(t) = t$, $v(t) = 1 + kt \log t$. For some computed eigenvalues see [**39**] and [**349**], Section 10.

(iii) $-1/4 < h < 3/4$, $0 < \rho < 1$ but $\rho \neq 1/2$. All the following results hold for the non-Friedrichs boundary condition : $[y, v](0) = 0$, where
$$u(t) = t^{\rho + 1/2}, \ v(t) = t^{1/2 - \rho} + \frac{k}{1 - 2\rho} t^{3/2 - \rho}.$$

(a) $k > 0$, $0 < \rho < 1/2$ there are no negative eigenvalues,

(b) $k > 0$, $1/2 < \rho < 1$ there is exactly one negative eigenvalue given by
$$\lambda_0 = \frac{-k^2}{(2\rho - 1)^2},$$

(c) if $k < 0$, $0 < \rho < 1/2$ there are infinitely many negative eigenvalues given by
$$\lambda_n = \frac{-k^2}{(2n - 2\rho + 1)^2}, \ n \in N_0,$$

(d) if $k < 0$, $1/2 < \rho < 1$ there are infinitely many negative eigenvalues given by
$$\lambda_n = \frac{-k^2}{(2n - 2\rho + 3)^2}, \ n \in N_0.$$

The next few results refer to the BC
$$A_1[y, u](0) + A_2[y, v](0) = 0.$$

(e) if $k = 0$ and $A_1 A_2 < 0$ there is exactly one negative eigenvalue given by:
$$\lambda_0 = -4 \left(\frac{-A_1 \Gamma(1 + \rho)}{A_2 \Gamma(1 - \rho)} \right)^{1/\rho}.$$

(f) Note that for $h = -1/4$, $k \in \mathbb{R}$, the LCNO classification at 0 holds. Here
$$u(t) = \sqrt{t} + kt\sqrt{t}, \; v(t) = 2\sqrt{t} + (\sqrt{t} + kt\sqrt{t})\log(t).$$

For $k = 0$ and $A_1 A_2 < 0$ there is exactly one negative eigenvalue given by
$$\lambda_0 = c\,e^{2A_1/A_2}, \; c = 4\,e^{4-2\gamma}, \; \gamma = 0.5772156649...,$$
is Eulers constant.

(g) $h < -1/4$, $k \in \mathbb{R}$, the equation is LCO at 0. For $k = 0$ this equation reduces to the Krall equation, see example 20. For $k \neq 0$ explicit formulas for the eigenvalues are not available; some qualitative properties of the spectrum are: for all k there are infinitely many negative eigenvalues going exponentially to $-\infty$ for $k > 0$ the point 0 is not an accumulation point of eigenvalues for $k \leq 0$ the eigenvalues also accumulate at 0.

(14) The Marletta equation on $(0, \infty)$.
$$p = 1, \; q(t) = \frac{3(t-31)}{4(t+1)(4+t)^2}, \; w = 1.$$

This equation is R at 0 and LP at ∞. For some boundary conditions at 0 this equation deceives both codes SLEIGN (not SLEIGN2) and SLEDGE into falsely reporting $\lambda = 0$ as an eigenvalue since there is a "near" eigenfunction there. For details see Marletta's certification report for the code SLEIGN [**461**].

(15) The harmonic oscillator equation on $(-\infty, \infty)$.
$$p = 1, \; q(t) = t^2, \; w = 1$$
LP at $-\infty$ and at ∞. Thus there is a unique self-adjoint extension; it has discrete spectrum given by
$$\lambda_n = 2n + 1, \; n \in \mathbb{N}_0.$$

See [**573**] for a classical treatment.

(16) The Jacobi equation on $(-1, 1)$.
$$p(t) = (1-t)^{\alpha+1}(1+t)^{\beta+1}, \; q = 0, \; w(t) = (1-t)^{\alpha}(1+t)^{\beta}$$

At -1 for all α:
- LP for $\beta \leq -1$ and for $\beta \geq 1$
- WR for $-1 < \beta < 0$
- LCNO for $0 \leq \beta < 1$

At +1 for all β:
- LP for $\alpha \leq -1$ and for $\alpha \geq 1$
- WR for $-1 < \alpha < 0$
- LCNO for $0 \leq \alpha < 1$.

The boundary condition functions u, v can be taken as follows:
- For $t < 0$:
  (i) if $-1 < \beta < 0$ then $u(t) = (1+t)^{-\beta}$, $v(t) = 1$,
  (ii) if $\beta = 0$ then $u(t) = 1$, $v(t) = \log\frac{1+t}{1-t}$,
  (iii) if $0 < \beta < 1$ then $u(t) = 1$, $v(t) = (1+t)^{-\beta}$.
- For $t > 0$:
  (i) if $-1 < \alpha < 0$ then $u(t) = (1-t)^{-\alpha}$, $v(t) = 1$,
  (ii) if $\alpha = 0$ then $u(t) = 1$, $v(t) = \log\frac{1+t}{1-t}$,
  (iii) if $0 < \alpha < 1$ then $u(t) = 1$, $v(t) = (1-t)^{-\alpha}$.

To get the classical Jacobi polynomials take $-1 < \alpha$, $-1 < \beta$; then note the following endpoint classifications and required boundary conditions:

At +1:
  (1) $-1 < \alpha < 0$, WR, $(py')(1) = 0$,
  (2) $0 \leq \alpha < 1$, $LCNO$, $[y,u](1) = 0$,
  (3) $1 \leq \alpha$, LP.

At -1:
  (1) $-1 < \beta < 0$, WR, $(py')(-1) = 0$,
  (2) $0 \leq \beta < 1$, $LCNO$, $[y,u](-1) = 0$,
  (3) $1 \leq \beta$, LP.

For the classical Jacobi orthogonal polynomials the eigenvalues are given by:

$$\lambda_n = n(n + \alpha + \beta + 1), \; n \in N_0.$$

It is interesting to observe that the required boundary condition for the Jacobi polynomials is the Friedrichs condition in the LCNO case but not in the WR case.
(17) The rotation Morse oscillator on $(0, \infty)$.

$$p = 1, \; w = 1, \; q(t) = \frac{2}{t^2} - 2000(2E - E^2), \; E = e^{-1.7(t-1.3)}.$$

LP at 0 and ∞. Hence there is a unique self-adjoint realization; its essential spectrum is $[0, \infty]$ and it has exactly 26 negative eigenvalues. (Ask SLEIGN2 to compute the first 28 eigenvalues and note the appearance of the 26 negative eigenvalues together with a statement that there appear to be only 26 eigenvalues and the starting point of the essential spectrum at (approximately) 0.)
(18) The Dunsch equation on $(-1, 1)$. See Dunford and Schwartz [141] chapter VIII, pp. 1510-1520 for a discussion of this problem.

$$p(t) = 1 - t^2, \; q(t) = \frac{2\alpha^2}{1+t} + \frac{2\beta^2}{1-t}, \; w = 1, \; 0 \leq \alpha, \; 0 \leq \beta.$$

At -1:
  LP for $\alpha \geq 1/2$ and all β,
  LCNO for $0 \leq \alpha < 1/2$ and all β.
At +1:
  LP for $\beta \geq 1/2$ and all α,
  LCNO for $0 \leq \beta < 1/2$ and all α.
Boundary condition functions can be obtained as follows:

$$At\;-1: u_-(t) = (1+t)^\alpha, \; v_-(t) = (1+t)^{-\alpha},$$

$$At\;+1: u_+(t) = (1+t)^\beta, \; v_+(t) = (1+t)^{-\beta}.$$

Note that u, v are maximal domain functions but not solutions. In [141] on p. 1519 it is claimed that the boundary value problem determined by the boundary conditions

$$[y, u_-](-1) = 0 = [y, u_+](1)$$

has eigenvalues given by

$$\lambda_n = (n + \alpha + \beta + 1)(n + \alpha + \beta), \; n \in N_0.$$

(19) The Donsch equation on $(-1, 1)$. This is a modification of Example 18 which illustrates an LCNO/LCO mix. Replace α in 18 by $i\gamma$. This changes the singularity at -1 from LCNO to LCO. For $\gamma > 0$ and $0 < \beta < 1/2$ we have

$$At\ -1: u(t) = \cos(\gamma \log(1+t)),\ v(t) = \sin(\gamma \log(1+t)),$$
$$At\ +1: u(t) = (1-t)^\beta,\ v(t) = (1-t)^{-\beta}.$$

These u, v are maximal domain functions which are not solutions.

(20) The Krall equation on $(0, \infty)$. This example is a special case of the Bessel equation, see example 2 above. Its solutions can be obtained in terms of modified Bessel functions. (We have followed the sleign2 package here by adding the constant 1 to q, this is done in sleign2 to facilitate numerical computations.)

$$p = 1,\ q(t) = 1 - \frac{k^2 + 1/4}{t^2},\ w = 1,\ k \in \mathbb{R},\ k \neq 0.$$

LCO at 0 and LP at ∞ for all $k \neq 0$. The essential spectrum is $[1, \infty)$.

For the boundary condition

$$[y, u](0) = 0,\ u(t) = t^{1/2} \cos(k \log(t))$$

there are an infinite number of eigenvalues which cluster at $-\infty$ and at 1. (Note that u need only be a maximal domain function on $[0, d)$ for $0 < d < \infty$. To get a maximal domain function on $(0, \infty)$ one can patch u at d appropriately.) The eigenvalues approach $-\infty$ and 1 very rapidly, see [**39**] or use sleign2 for more details; also see [**399**] for more information.

(21) The Fourier equation. See Subsection 4.3 for a description of the eigenvalues for various boundary conditions, self-adjoint and non-self-adjoint.

(22) The Laguerre equation on $(0, \infty)$.

$$p(t) = t^{\alpha+1} e^{-t},\ q = 0,\ w(t) = t^\alpha e^{-t},\ \alpha \in \mathbb{R}.$$

LP at ∞ for all α.

At 0:
- LP for $\alpha \leq -1$,
- WR for $-1 < \alpha < 0$,
- LCNO for $0 \leq \alpha < 1$,
- LP for $\alpha > 1$.

This is the classical form of the celebrated equation, which for parameter values $\alpha > 1$ produces the Laguerre polynomials as eigenfunctions; for the appropriate boundary at 0, when needed, the eigenvalues are given by

$$\lambda_n = n,\ n \in \mathbb{N}_0.$$

Remarkably, these are independent of α, see Abramovitz and Stegun [**2**], Chapter 22, Section 22.6 for more details. See the file xamples.f (this is not a typo) of the SLEIGN2 package for details of the boundary condition functions u, v.

SLEIGN2 has only very limited success with this problem; for numerical computations the Laguerre/Liouville equation, which has the same eigenvalues (for the appropriate corresponding boundary conditions) is more convenient; see Example 23 to follow.

(23) The Laguerre/Liouville equation on $(0, \infty)$.

$$p = 1,\ w = 1,\ q(t) = \frac{\alpha^2 - 1/4}{t^2} - \frac{\alpha + 1}{2} + \frac{t^2}{16},\ \alpha \in \mathbb{R}.$$

LP at ∞ for all α
LCNO for $-1 < \alpha < 1$ but $\alpha^2 \neq 1/4$
R for $\alpha^2 = 1/4$
LP for $\alpha \geq 1$

See the xamples.f file of the SLEIGN2 package for details of appropriate boundary condition functions.

(24) Jacobi/Liouville form of the Jacobi equation. See the files xamples.f and xamples.tex of the SLEIGN2 package for details.

(25) The Meissner equation on $(-\infty, \infty)$.

$$p = 1,\ q = 0,\ w = \begin{cases} 1 & for\ t < 0, \\ 9 & for\ t \geq 0. \end{cases}$$

LP at $-\infty$ and ∞.

This equation is well known in the applied literature in connection with the modelling of crystals in one dimension; see [**153**], [**341**].

Periodic boundary conditions on $(-1/2, 1/2)$. We have $\lambda_0 = 0$ and

$$\lambda_{4n+1} = (2n\pi + \alpha)^2,\ \lambda_{4n+2} = (2(n+1)\pi - \alpha)^2,$$

$$\lambda_{4n+3} = \lambda_{4n+4} = (2(n+1)\pi)^2,\ \alpha = \cos^{-1}(\frac{-7}{8}),\ n \in N_0.$$

Note that there are infinitely many simple and infinitely many double periodic eigenvalues on the interval $(-1/2, 1/2)$.

Semi-Periodic eigenvalues on $(-1/2, 1/2)$.

$$\lambda_{4n} = (2n\pi + \beta)^2,\ \lambda_{4n+1} = (2n\pi + \gamma)^2,\ \lambda_{4n+2} = (2(n+1)\pi - \gamma)^2,$$

$$\lambda_{4n+3} = (2(n+1)\pi - \beta)^2,\ \beta = \cos^{-1}(\frac{1+\sqrt{33}}{16}),\ \gamma = \cos^{-1}(\frac{1-\sqrt{33}}{16}).$$

Observe that the semi-periodic eigenvalues are all simple.

(26) The Lohner equation on $(-\infty, \infty)$.

$$p = 1,\ w = 1,\ q(t) = 1000t.$$

LP at $-\infty$ and ∞. Lohner in [**454**] computed eigenvalues of this equation on a compact interval with regular boundary conditions using interval arithmetic. He obtained good approximations with rigorously guaranteed upper and lower bounds.

(27) The Jörgens equation on $(-\infty, \infty)$. This is a remarkable example from Jörgens; see [**349**], part II, Section 10.

$$p = 1,\ w = 1,\ q(t) = \frac{1}{4}e^{2t} - k\,e^t,\ k \in \mathbb{R}.$$

LP at $-\infty$ and at ∞.

The essential spectrum starts at 0; for $k \leq 1/2$ there are no eigenvalues; for

$$h < k - 1/2 \leq h + 1,\ h = 0, 1, 2, 3, \ldots$$

there are exactly $h+1$ eigenvalues and these are all below the essential spectrum i.e. they are all negative. They can be given explicitly by

$$\lambda_n = -(k - 1/2 - n)^2,\ n = 0, 1, 2, 3, \ldots, h,\ h \in N_0.$$

(28) The Behnke-Goerisch [**59**] equation on $(-\infty, \infty)$.

$$p = 1,\ w = 1,\ q(t) = k\cos^2(t),\ k \in \mathbb{R}.$$

LP at $-\infty$ and at $+\infty$. This is a form of the Mathieu equation previously discussed, see examples 11 and 12 above. These authors computed Neumann eigenvalues on a compact interval using interval arithmetic and obtained good approximations with rigorous upper and lower bounds.

(29) The Whittaker equation on $(0, \infty)$.

$$p = 1, \ q(t) = \frac{1}{4} + \frac{k^2 - 1}{4t^2}, \ w(t) = \frac{1}{t}, \ k \in [1, \infty).$$

LP at 0 and LP at $+\infty$ for all $k \in [1, \infty)$. This equation is studied in Part II, Section 10 of [349] where it is shown that the spectrum is discrete and is explicitly given by

$$\lambda_n = n + (k+1)/2, \ n \in N_0.$$

(30) The Littlewood-McLeod equation on $(0, \infty)$:

$$p = 1, \ w = 1, \ q(t) = t\sin(t), \ t \in (0, \infty)$$

is regular at 0 and LP at ∞. All self-adjoint realizations of this equation have discrete spectrum $\sigma = \{\lambda_n : n \in \mathbb{Z}\}$ which is unbounded above and below:

$$\lambda_n \to -\infty \text{ as } n \to -\infty; \ \lambda_n \to \infty \text{ as } n \to \infty.$$

For each of these problems, every eigenfunction is oscillatory at ∞. See Littlewood [450] and McLeod [469] for details.

SLEIGN2 and other codes have difficulties computing the eigenvalues of these problems; there seems to be no algorithm for the numerical computation of eigenvalues when the spectrum is discrete and unbounded below as well as above. SLEIGN2 has a built in algorithm for every class of self-adjoint Sturm-Liouville problems with $p > 0$, $w > 0$ and both separated and coupled boundary conditions, except for this class.

(31) The Morse equation on $(-\infty, +\infty)$:

$$p = 1, \ w = 1, \ q(t) = 9e^{-2t} - 18e^{-t}, \ t \in (-\infty, \infty)$$

is LP at both endpoints. This SLP has been studied in [33]; it has exactly three negative eigenvalues

$$\lambda_n = -(n - 2.5)^2, \ n = 0, 1, 2;$$

and essential spectrum $[0, \infty)$.

(32) The Ince equation. This is a 'name' equation in Floquet theory. It has periodic coefficients and three parameters in addition to the spectral parameter. See the books [14], [344],

Here we present some recent results by Volkmer [580]. The general form of the Ince equation is

$$(1 + \alpha\cos(2t))y'' + (\beta\sin(2t))y' + (\lambda + \gamma\cos(2t))y = 0$$

and its Sturm-Liouville form

$$-(py')' + qy = \lambda w y$$

has the coefficients

$$\begin{cases} p(t) := e^{\int_0^t \frac{\beta\sin(2s)}{1+\alpha\cos(2s)}ds}, \\ q(t) = -p(t)\frac{\gamma\cos(2t)}{1+\alpha\cos(2t)}, \\ w(t) := e^{-\frac{\beta}{2}\cos(2t)}, \ if \ \alpha = 0, \\ w(t) = (1 + \alpha\cos(2t))^{-1-\beta/(2\alpha)}, \ if \ \alpha \neq 0; \end{cases}$$

with λ denoting the spectral parameter, as usual, and the parameters α, β, γ are real and $|\alpha| < 1$. Hence the equation is regular on any compact interval $[a,b]$. Let $\{\lambda_n^P(a,b) : n \in \mathbb{N}_0\}$, $\{\lambda_n^S(a,b) : n \in \mathbb{N}_0\}$ denote the periodic and semi-periodic eigenvalues on the interval (a,b), $-\infty < a < b < \infty$, respectively.

THEOREM 14.0.1 (Volkmer). *Let $\alpha, \beta, \gamma \in \mathbb{R}$ and $|\alpha| < 1$. Suppose the equation*
$$2a(v-1/2)^2 - b(v-1/2) - d/2 = 0$$

(a) *has at least one integer solution for v and let $l = \min\{|v|\}$ where v ranges over all integer solutions. Then*

(b) $\lambda_n^S(0, \pi/2)$ *is double for all $n \in \mathbb{N}_0$ if $l = 0$;*

(c) $\lambda_n^S(0, \pi/2)$ *is simple for $n = 0, 1, ..., l-1$ and $\lambda_n^S(0, \pi/2)$ is double for $n > l - 1$ if $l > 0$;*

(d) $\lambda_n^P(0, \pi)$ *is simple for $n = 0, 1, ..., l$ and $\lambda_n^P(0, \pi)$ is double for all $n > l$ if $l > 0$.*

It follows from this theorem that for any integer $l \in \mathbb{N}$ there exist parameters $\alpha, \beta, \gamma \in \mathbb{R}$ with $|\alpha| < 1$, such that the first l semi-periodic eigenvalues on the interval $(0, \pi/2)$ are simple and all the rest double and that the first l periodic eigenvalues on $(0, \pi)$ are simple and all the rest double. Recall that the lowest periodic eigenvalues $\lambda_0^P(a,b)$ on any interval is always simple by Theorem 4.8.1. So if $l = 0$ all semi-periodic eigenvalues on an interval of length half the period are double but on an interval of length the period all the periodic eigenvalues are double except for λ_0^P.

(33) The Heun equation on $(0,1)$. This is another one of the "name" equations of Applied Mathematics; see the book by Ronveaux [**553**] and the article by Lay and Slavyanov [**432**] for further discussions and some applications of the Heun equation, also see Volkmer [**581**] for other recent results. We present a number of results for the Heun equation taken from the paper of Bailey, Everitt, Hinton and Zettl [**37**].

The Sturm-Liouville form of the Heun equation is
$$My := -(p(x)y'(x))' + q(x)y(x) = \lambda w(x)y(x) \text{ for all } x \in (0,1)$$
where λ is the spectral parameter and for all $x \in (0,1)$ the coefficients p, q, w are defined by
$$\begin{cases} p(x) := x^c(1-x)^d(x+s)^e, \\ q(x) := abx^c(1-x)^{d-1}(x+s)^{e-1}, \\ w(x) := x^{c-1}(1-x)^{d-1}(x+s)^{e-1}. \end{cases}$$

The six parameters in the definition of p, q, w are restricted as follows:

(a) $a, b, e \in \mathbb{R}$, $s \in (0, \infty)$;

(b) $c, d \in (0, \infty)$ and satisfy: $a + b + 1 = c + d + e$.

We fix the four parameters a, b, e, s and study SLP for varying c, d subject to (b).

We note that the coefficients p, q, w have the properties:

$$\begin{cases} (i) & p, q, w : [0, 1] \to \mathbb{R}, \\ (ii) & q, w \in C[0,1], \text{ i.e. } q, w \in L^1(0,1), \\ (iii) & p(x), w(x) > 0 \text{ for all } x \in (0,1), \\ (iv) & 1/p \equiv p^{-1} \in L^1_{\text{loc}}(0,1), \\ (v) & p^{-1} \in L^1(0, 1/2) \text{ if } c \in (0,1), \quad p^{-1} \notin L^1(0, 1/2) \text{ if } c \in [1, \infty), \\ (vi) & p^{-1} \in L^1(1/2, 1) \text{ if } d \in (0,1), \quad p^{-1} \notin L^1(1/2, 1) \text{ if } d \in [1, \infty). \end{cases}$$

Thus the Heun equation is regular at all points of the open interval $(0,1)$; regular at endpoint 0, respectively endpoint 1, if the parameter $c \in (0,1)$, respectively $d \in (0,1)$; otherwise the endpoints 0 and 1 are singular.

We now consider the maximal domain (linear manifold) $D_{\max}(c,d)$ generated by M in $L^2((0,1); w)$; this domain depends upon all the parameters a,b,c,d,e,s but is denoted by $\Delta(c,d)$ to emphasize that the other four parameters are fixed and also that it depends critically on c,d.

$$D_{\max}(c,d) := \{f \in D(M) : f, w^{-1}M[f] \in L^2((0,1); w)\}.$$

For the purpose of defining self-adjoint differential operators in the Hilbert space, $H = L^2((0,1); w)$, generated by the differential expression M, it is necessary to classify both of the endpoints $0,1$ of M in H.

The endpoints $0,1$ of M are classified, disjointly, as either *regular* (R) or *limit-circle nonoscillatory (LCNO)* or *limit-point (LP)* in $L^2((0,1); w)$; (the LCO case does not occur for the Heun equation considered here).

Recall that the regular case classification can be regarded, without loss of generality, as a special case of the LCNO classification; the limit-point classification is distinct from the limit-circle case, and hence from the regular case.

We can now state

THEOREM 14.0.2 (Bailey, Everitt, Hinton and Zettl). *The differential expression M has the following endpoint classifications in the space $H = L^2((0,1); w)$:*

(a) *The endpoint 0 is:*
  (i) *regular for all $c \in (0,1)$,*
  (ii) *LCNO for all $c \in [1,2)$,*
  (iii) *LP for all $c \in [2, \infty)$.*
(b) *The endpoint 1 is:*
  (i) *regular for all $d \in (0,1)$,*
  (ii) *LCNO for all $d \in [1,2)$,*
  (iii) *LP for all $d \in [2, \infty)$.*

PROOF. See Theorem 6.1 in [**37**]. $\square$

Let $[\cdot,\cdot] : D_{\max}(c,d) \times D_{\max}(c,d) \times (0,1) \to \mathbb{C}$ denote the usual Lagrange bracket given by

$[f,g](x) := f(x).(p\overline{g}')(x) - (pf')(x).\overline{g}(x)$ for all $f, g \in D_{\max}(c,d)$ and all $x \in (0,1)$.

We now define the self-adjoint operator

$$T_{c,d} : D(T_{c,d}) \subset H = L^2((0,1); w) \to H.$$

DEFINITION 14.0.1. *Define the operator $T_{c,d} : D(T_{c,d}) \subset H \to H$ for all $c,d \in (0, \infty)$,*

$$\begin{cases} D(T_{c,d}) := D_{\max}(c,d) \text{ if both } c,d \in [2,\infty), \\ \quad := \{f \in D_{\max}(c,d) : \lim_{x \to 0^+} [f,1](x) = 0\} \text{ if } c \in (0,2) \text{ and } d \in [2,\infty) \\ \quad := \{f \in D_{\max}(c,d) : \lim_{x \to 1^-} [f,1](x) = 0\} \text{ if } c \in [2,\infty) \text{ and } d \in (0,2) \\ \quad := \{f \in D_{\max}(c,d) : \lim_{x \to 0^+} [f,1](x) = \lim_{x \to 1^-} [f,1](x) = 0\} \text{ if } c,d \in (0,2), \end{cases}$$

and

$$T_{c,d}f := w^{-1}M[f] \text{ for all } f \in D(T_{c,d}).$$

This definition leads to

THEOREM 14.0.3. *For all $c, d \in (0, \infty)$ the linear operator $T_{c,d}$ with domain $D(T_{c,d})$ is self-adjoint in the Hilbert function space H. Its spectrum $\sigma(T_{c,d}) = \{\lambda_n^{c,d} : n \in \mathbb{N}_0\}$ is simple, discrete and bounded below. Furthermore,*

$$\lambda_0^{c,d} \geq -K(c,d), \; K(c,d) = \sup\{|\frac{q(x)}{w(x)}| : x \in (0,1)\}.$$

PROOF. See Chapter 10 above for the self-adjointness of $T_{c,d}$ and Bailey, Everitt, Hinton and Zettl [**37**] for the rest. □

COROLLARY 14.0.1. *For all $c, d \in (0, \infty)$ the self-adjoint operator $T_{c,d}$ is bounded below in the Hilbert function space $L^2((0,1); w)$. If the non-negative number $K(c,d) \in [0, \infty)$ is defined by*

$$K(c,d) := \sup\{|w^{-1}(x)q(x)| : x \in (0,1)\},$$

then, for all $c, d \in (0, \infty)$,

$$(T_{c,d}f, f)_w \geq -K(c,d)(f,f)_w \text{ for all } f \in D(T_{c,d}).$$

If the parameters a, b of the Heun differential equation are non-negative, then

$$(T_{c,d}f, f)_w \geq 0 \text{ for all } f \in D(T_{c,d}).$$

If both the parameters a, b are positive, then

$$(T_{c,d}f, f)_w > 0 \text{ for all } f \in D(T_{c,d}) \text{ but } f \neq 0.$$

PROOF. This follows readily from the above theorems; see [**37**] for details. □

(34) The King-Cooper problem. This problem has been studied by Andy King and Richard Cooper, both in the School of Mathematics and Statistics at the University of Birmingham. The results given here are taken from the paper by Bailey, Billingham, Cooper, Everitt, King, Kong, Wu and Zettl [**36**].

Consider the Sturm-Liouville differential equation

$$M[y](x) := -(xy'(x))' - x^3 y(x) = \lambda\, x\, y(x) \text{ for all } x \in (0, b]$$

where the endpoint $b \in (0, \infty)$, and the spectral parameter $\lambda \in \mathbb{C}$.

This differential equation is to be studied in the weighted Hilbert function space $H = L^2((0, b); x)$.

The maximal domain $D_{\max}(b)$ for the differential expression $M[\cdot]$ is defined as

$$D_{\max}(b) := \{f : (0, b] \to \mathbb{C} : f, xf' \in AC_{\text{loc}}(0, b] \text{ and } f, x^{-1}M[f] \in H\}.$$

LEMMA 14.0.1. *The following endpoint classifications hold in the space H : The endpoint b is regular; the endpoint $+0$ is limit-circle non-oscillatory.*

PROOF. The classification at b is clear. Consider then the endpoint $+0$; define the boundary condition functions $u, v \in \Delta(b)$ by

$$u(x) := 1 \quad v(x) := \ln(x) \quad \text{both for all } x \in (0, b].$$

Clearly both $u, v \in L^2((0,b); x)$; a computation shows that

$$M[u](x) = -x^3 \text{ for all } x \in (0, b], \text{ so that } x^{-1}M[u] \in H$$

and

$$M[v](x) = -x^3 \ln(x) \text{ for all } x \in (0, b], \text{ so that } x^{-1}M[v] \in H.$$

Thus $u, v \in D_{\max}(b)$ and, using the standard definition of the bilinear form $[\cdot, \cdot]$: $D_{\max}(b) \times D_{\max}(b) \to \mathbb{C}$, we find

$$[u, v](x) = x\left(1.x^{-1} - 0.\ln(x)\right) = 1 \text{ for all } x \in (0, b];$$

the limit-circle non-oscillatory classification at $+0$ in H now follows. □

Now define the separated boundary value problems for this differential equation using the boundary condition functions u, v defined above, by

$$D: \quad [y, u](0^+) = 0 \text{ and } y(b) = 0,$$
$$N: \quad [y, u](0^+) = 0 \text{ and } y'(b) = 0.$$

The spectrum for both these boundary value problems is simple and discrete; let these spectra be represented by

$$\{\lambda_n^D : n \in \mathbb{N}_0\} \quad \text{and} \quad \{\lambda_n^N : n \in \mathbb{N}_0\},$$

respectively.

THEOREM 14.0.4 (Bailey, Billingham, Cooper, Everitt, King, Kong, Wu and Zettl). *We have*

$$D: \quad \lim_{b \to 0^+} \lambda_n^D(b) = +\infty \quad \text{for all} \quad n \in \mathbb{N}_0$$

and

$$N: \quad \begin{array}{l} (i) \; \lambda_0^N(b) < 0 \;\; \text{for all } b \in (0, \infty) \\ (ii) \; \lim_{b \to 0^+} \lambda_0^N(b) = 0 \\ (iii) \; \lim_{b \to 0^+} \lambda_n^N(b) = +\infty \text{ for all } n \in \mathbb{N}. \end{array}$$

PROOF. Using the transformation used by Niessen and Zettl in [**503**], see Section 10.6, the problem D is transformed to a regular problem with Dirichlet boundary conditions at both endpoints $0+$ and b. This transformation leaves the eigenvalues unchanged. The result for problem D now follows from Theorem 4.1 in [**392**].

Part (i) follows from the variational characterization of the lowest eigenvalue:

$$\lambda_0^N = \inf \frac{\int_0^b \overline{f} M f}{\int_0^b |f|^2 w}$$

where the infimum is taken over all f in the domain $D(S^N)$ of the self-adjoint operator realization S^N of the equation with the N boundary condition and w is the weight function $w(x) = x$. Since $u \in D(S^N)$ we have

$$\lambda_0^N \leq \frac{\int_0^b -x^3 \, dx}{\int_0^b x \, dx} < 0.$$

To establish properties (ii) and (iii) we consider the SLP on the interval $[\epsilon, b]$ with Neumann boundary conditions at both endpoints, i.e.

$$(xy')(\epsilon) = 0 = (xy')(b), \; 0 < \epsilon < b$$

and denote the eigenvalues of this problem by $\{\lambda_n^N(\epsilon, b) : n \in \mathbb{N}_0\}$. From Theorem 4.4 of [**392**] we have

$$\lim_{b \to \epsilon^+} \lambda_0^N(\epsilon, b) = \frac{q(\varepsilon)}{w(\varepsilon)} = -\varepsilon^2, \text{ for each } \varepsilon, \; 0 < \varepsilon < b;$$

$$\lim_{b \to \varepsilon^+} \lambda_n^N(\varepsilon, b) = +\infty, \text{ for each } n \geq 1, \text{ and for each } \varepsilon, \ 0 < \varepsilon < b.$$

We now observe that

$$[y, u](\varepsilon) = [y(xu') - u(xy')](\varepsilon) = -(xy')(\varepsilon), \ 0 < \varepsilon < b,$$

and thus - in the terminology of [**38**] - the Neumann boundary condition on the interval (ε, b) is the "inherited" boundary condition corresponding to the singular condition N. Hence from Theorem 4.1 of [**38**] it follows that

$$\lambda_n^N(\varepsilon, b) \to \lambda_n^N, \text{ as } \varepsilon \to 0, \text{ for each } n \in \mathbb{N}_0.$$

Parts (ii) and (iii) now follow. □

(35) **The Kwong-Wong equations on $[0, \infty)$ and on $[0, 2\pi]$.** These authors studied the oscillation at infinity of equations

(1) $y'' + (\sin x)y' + (\cos x)y = 0,$
(2) $y'' - (\cos x)y' + (\sin x)y = 0,$
(3) $y'' + (\cos x)y' + (\sin x)y = 0.$

The symmetric form of these equations, when written in the usual notation with spectral parameter λ, is

(4) $-(e^{-\cos x}y')' - (e^{-\cos x}\cos x)y = \lambda y,$
(5) $-(e^{-\sin x}y')' - (e^{-\sin x}\sin x)y = \lambda y,$
(6) $-(e^{\sin x}y')' - (e^{\sin x}\sin x)y = \lambda y.$

Note that the coefficients are real-valued, periodic, and the fundamental period is 2π. Consider each of equations (4), (5) and (6) on the interval $[0, 2\pi]$ with periodic boundary conditions and denote the lowest periodic eigenvalue by λ_0, μ_0, ν_0, respectively. Then $\lambda_0 = 0$ since $y(x) = e^{\cos x}$ is a solution of (4) when $\lambda = 0$ satisfying the periodic boundary conditions. Similarly $\mu_0 = 0$, since $y(x) = e^{\sin x}$ is a solution of (5) when $\lambda = 0$ satisfying the periodic boundary conditions. We know from Chapter 4 that λ_0 and μ_0 are the lowest periodic eigenvalues because their eigenfunction (which we also know from Chapter 4 is unique up to constant multiples because the lowest periodic eigenvalue is always simple) has no zero on the interval $[0, 2\pi]$. Hence $\lambda = 0$ is the oscillation number of each of equations (4) and (5) and the essential spectrum of any self-adjoint realization of equations (4) and (5) in the Hilbert space $L^2(0, \infty)$ starts at 0 and has a band structure consisting of closed subintervals of the nonnegative real line, the spectral bands. Thus both equations (4) and (5) are oscillatory at ∞ when $\lambda > 0$ and nonoscillatory at ∞ when $\lambda \leq 0$. In particular equations (1) and (2) are both NO at ∞.

In contrast to equations (4) and (5), equation (6) is oscillatory at ∞ when $\lambda = 0$. It's oscillation number ν_0 is estimated by SLEIGN2 to be:

$$\nu_0 \approx -1.228.$$

Thus equation (6) is also oscillatory for some negative values of λ. In particular equation (3) is O at ∞.

The spectral bands and gaps are also known as the stability and instability intervals of the equation. For all three equations: are all of the spectral gaps present? If not, which ones are missing? For which boundary conditions at 0 is there an eigenvalue to the left of the essential spectrum? Are there any eigenvalues in gaps?

(36) The Boumenir-Fulton equation on $[0, 2\pi]$. These authors have studied the equation:
$$-y'' + qy = \lambda y \text{ on } [0, 2\pi], \ q(x) = k^2 \sin^2 x - 3k \cos x, \ k \in \mathbb{R}.$$

For $\lambda = 1$ and any $k \in \mathbb{R}$,
$$u(x) = e^{k \cos x}(\sin x)$$

is a solution satisfying the periodic boundary conditions on $[0, 2\pi]$. From Chapter 4 we know that the lowest periodic eigenvalue $\lambda_0^P(k)$ is simple and its eigenfunction has no zero in the interval $(0, 2\pi)$. Since $u(\pi) = 0$ it follows that $1 \neq \lambda_0^P(k)$, i.e. 1 is not the lowest periodic eigenvalue. Also, since u has no other zero in the open interval $(0, 2\pi)$, from Chapter 4, Theorem 4.3.1, $\lambda_n^P(k) \neq 1$ for $n > 2$. Therefore, for each $k \in \mathbb{R}$, we must have
$$\lambda_1^P(k) = 1 \text{ or } \lambda_2^P(k) = 1.$$

Boumenir [**97**] proved that $\lambda_1^P(k) = 1$ for all 'small' k. Is this true for all $k \in \mathbb{R}$? As far as this writer knows, this is an open problem. Note that, although $\lambda_1^P(k)$ and $\lambda_2^P(k)$ are both continuous functions of k since they are continuous functions of the coefficient q, the general theory does not rule out the possibility that as k varies $\lambda_1^P(k)$ and $\lambda_2^P(k)$ change from being simple eigenvalues to double for some value of k and then 'switching identities' as k varies further.

(37) The BMZ equation. (See Remark 10.8.5) In [**107**], [**106**] Brown, McCormack and Zettl introduced a new method for proving the existence of an eigenvalue below the essential spectrum. This method uses classical analysis, functional analysis, and interval analysis. They illustrated this method by proving that the SLP
$$-y'' + qy = \lambda y, \text{ on } J = [0, \infty), \ y(0) = 0$$

has an eigenvalue below the essential spectrum when q is given by
$$q(t) = \sin(t + \frac{1}{1 + t^2}), \ t \in J.$$

This potential is an $L^1(J)$ perturbation of the potential $q(t) = \sin t$ for which this problem is known to have no eigenvalues for the same boundary condition $y(0) = 0$. Note that the problem is regular at 0 and LP at ∞.

CHAPTER 15

Notation

Russell, Bertrand (1872-1970)

> A good notation has a subtlety and suggestiveness which at times make it almost seem like a live teacher.

In J. R. Newman (ed.) The World of Mathematics, New York: Simon and Schuster, 1956.

In this Chapter we provide a brief list of the mathematical notation used.

(a, b) denotes the open interval with finite or infinite endpoints, $-\infty \leq a < b \leq \infty$

$[a, b]$ is used to denote the closed interval with finite or infinite endpoints; thus f continuous on $[a, b]$ with $a = -\infty$ means that f has a finite limit at $-\infty$.,

$[a, b)$ includes a but not b; similarly for $(a, b]$,

$\mathbb{N}_0 = \{0, 1, 2, 3, ...\}$,

$\mathbb{N} = \{1, 2, 3, ...\}$,

$\mathbb{Z} = \{..., -3, -2, -1, 0, 1, 2, 3, ...\}$,

$\mathbb{Z}^* = \{..., -3, -2, -1, -0, 0, 1, 2, 3, ...\}$,

$\mathbb{R}$ is the set of real numbers,

$\mathbb{C}$ is the set of complex numbers,

$L(J, \mathbb{C})$ is the set of Lebesgue integrable complex valued functions defined almost everywhere on a Lebesgue measurable subset J of $\mathbb{R}$.

$L(J, \mathbb{R})$ is the set of real valued Lebesgue integrable functions on a Lebesgue measurable set J.

$L(J, \mathbb{R}_+)$ is the set of positive valued Lebesgue integrable functions on a Lebesgue measurable set J.

$L_{loc}(J, \mathbb{R})$ is used to denote the set of functions y satisfying $y \in L([a, b], \mathbb{R})$ for every compact subinterval $[a, b]$ of J.

$L_{loc}(J, \mathbb{C})$ is used to denote the set of functions y satisfying $y \in L([a, b], \mathbb{C})$ for every compact subinterval $[a, b]$ of J.

$AC_{loc}(J)$ is the collection of complex valued functions which are absolutely continuous on all compact subintervals of J. (Note: $AC_{loc}(J) = AC_{loc}(J, \mathbb{C})$ but we do not need this notation since we have no need for just the real valued absolutely continuous functions on compact subintervals.)

$M_{n,m}(S)$ denotes the n by m matrices with entries from S; if $n = m$ we abbreviate this to $M_n(S)$. Also if $m = 1$ we sometimes write S^n for $M_{n,1}(S)$.

$|P|$ denotes the absolute value of P if P is a real or complex number or function. If P is a real or complex matrix constant or function then $|P|$ denotes a matrix norm. Since all matrix norms are topologically equivalent (in a finite dimensional

vector space) this matrix norm can be taken as the $1-norm$:
$$|P| = \sum |p_{ij}|.$$

$||Y||$ is denotes a norm in a vector space; this space is either specified or is clear from the context

$\{X_n : n \in N\}$ denotes the sequence $X_1, X_2, X_3, ...$

$L^2(J, w) = \{f : J \to \mathbb{C}, \int_J |f|^2 w < \infty\}$. This is the Hilbert space of square-integrable functions with weight w if $w > 0$ a.e. on J. Since we have no need for the Hilbert space of real-valued square integrable functions we do not use or need the notations $L^2(J, \mathbb{C}, w)$, $L^2(J, \mathbb{R}, w)$.

$L^2(J, w) = \{f : J \to \mathbb{C}, \int_J |f|^2 w < \infty\}$ This is the Krein space of square-integrable functions with weight w if w changes sign on J. Since we have no need for the Krein space of real-valued square integrable functions, we do not use or need the notations $L^2(J, \mathbb{C}, w)$, $L^2(J, \mathbb{R}, w)$.

$\Phi(t, u,)$ or $\Phi(t, u, P)$. This is the "primary fundamental matrix" of the system $Y' = PY$ on J. The primary fundamental matrix Φ is a matrix solution satisfying: $\Phi(u, u) = I$ for all $u \in J$. here I is the identity matrix.

$T'(x)$ the Frechet derivative in Banach spaces; see Section 1.7.

$o(|h|)$, See Section 1.7.

CHAPTER 16

Comments on Some Topics not Covered

So far as the mere imparting of information is concerned, no university has had any justification for existence since the popularization of printing in the fifteenth century.

Alfred North Whitehead, The Aims of Education.

As the title suggests, in this chapter we comment on some topics not covered and give a few references. No effort is made to make these references comprehensive or up to date.

(1) The LP/LC dichotomy. In Chapter 7 we merely give a brief introduction to this topic. It is a currently still active research area and, in this writers opinion, deserves its own monograph. See the monograph by Kauffman, Read and Zettl [**368**] for a brief introduction to LP and LC criteria. There are many sufficient conditions known for LC and also for LP and even some necessary and sufficient conditions but no necessary and sufficient conditions *which can be checked in each case*. Finding such conditions is still an open problem. A major obstacle to finding such conditions is the fact that for the LP case to hold it is enough to give conditions on a sequence of intervals, with almost no requirements on the coefficients outside these intervals, whereas essentially pointwise conditions are required for the LC case. For an entirely different approach see Zettl [**624**] for a construction of LC expressions. Also see [**22**].

(2) The O/NO alternative. This is an active research area and could use its own monograph. See Kauffman, Read and Zettl [**368**] for a brief introduction to O/NO criteria, also see the monograph of Halvorsen and Mingarelli [**289**] and its references. Many conditions, both necessary and sufficient, are known but there are no known necessary and sufficient conditions *which can be checked in each case*. As in the LC/LP situation the most general sufficient conditions for O are of "interval" type, see e.g. Kwong and Zettl [**413**], but such conditions are not appropriate for the NO case to hold. For a completely different approach i.e. a construction of *all* disconjugate (NO) equations see Zettl [**623**].

(3) Absolutely continuous spectrum. See the contributions from Last and Jitomirskaya and from Stolz in [**566**]. We have only considered the simplest division of the spectrum into its discrete and essential (continuous) parts. See the seminal paper of Gilbert and Pearson [**276**] for criteria involving subordinate solutions - an extension of the notion of principal solutions - for the absolutely continuous spectrum; see also Hinton and Shaw [**338**], [**335**], [**336**], Gesztesy, Gurarie, Holden, Klaus, Sadun, Simon, and Vogl [**266**] and the references therein. There is an extensive

literature on spectral properties of Sturm-Liouville operators by Simon and his 100 co-authors. Also see Weidmann [**599**], [**600**] for some operator theory background. Also see [**8**], [**338**], [**336**], [**335**], [**596**].

(4) There is an extensive literature on left definite problems, when the weight function w is allowed to change sign. See the monograph by Mingarelli [**472**] and its references. Such problems can be studied in the setting of Krein and Pontryagin spaces rather than Hilbert space. Although the abstract operator theory in these spaces is well developed, explicit applications to SLP are fragmentary at best. As mentioned in Chapters 5 and 12, there is no universally agreed upon definition of 'left-definite Sturm-Liouville problems'. See the Comments sections at the end of these chapters for references to other approaches to left-definite theories.

For oscillatory properties of the eigenfunctions when the weight function changes sign or is identically zero on subintervals, see Everitt, Kwong and Zettl, [**214**]. Other papers for problems with an indefinite weight function [**12**], [**27**], [**28**].

(5) Another popular topic we have not discussed is the two, or multi-parameter theory. See Volkmer [**584**] and the papers by Binding and Brown, [**73**], [**71**], [**70**].

(6) In Chapter 13 we discuss the two interval theory. The extension of this theory to any finite number of intervals is routine. The extension to an infinite number of intervals is not routine since convergence problems arise, among other things. See Everitt, Shubin, Stolz and Zettl [**236**] for an introduction to SLP problems on infinitely many intervals. Also see Geszteszy and Kirsch [**268**], [**269**], [**265**]. If an equation has an interior singularity then, in general, the solutions cannot be continuously moved through this singularity. One can study this problem on two separate intervals each of which has this singular point on the boundary. Take the direct sum of two self-adjoint operators from the two intervals and you have a two-interval self-adjoint operator which is not particularly interesting because it doesn't "connect" the two intervals together. More interesting operators are obtained by connecting solutions through the singular point, even if they blow up there, in such a way as to get a new "two interval" self-adjoint operator. Actually this construction is also of interest when the interior point is not singular but regular, obtaining what is known as "point interactions", [**620**], [**111**]. For a symplectic algebra approach to the single interval and multi-interval theory for ordinary and partial differential equations see the papers and monographs by Everitt and Markus [**220**], [**221**], [**224**], [**222**], [**223**].

(7) Discreteness criteria. See [**423**], [**484**], [**277**], [**23**], [**173**], [**329**], [**415**], [**419**], [**417**], [**420**], [**414**], [**422**], [**412**], [**445**].

(8) Inverse spectral theory. See the landmark paper of Gelfand and Levitan [**264**], the elegant exposition of Pöschel and Trubowitz [**522**] and the references therein. See the list of open problems below. See also the seminal paper of Borg [**93**], and [**263**], [**270**], [**443**], [**252**].

(9) Eigenparameter dependent boundary conditions. Fulton [**258**], [**60**], [**76**], [**77**], [**68**], [**139**], [**140**], [**258**], [**257**], [**327**], [**337**], [**591**].

(10) Expansion Theorems. For problems which are self-adjoint in a Hilbert space the spectral theorem can be applied. Most of the standard books discuss expansion theorems: [**600**], [**599**], [**574**], [**573**], [**515**], [**487**], [**444**], [**123**], [**21**], [**141**], [**6**], [**84**], [**159**], Our approach to the study of left-definite regular and singular problems does not directly yield expansion theorems. What is an appropriate expansion theory for the left-definite problems discussed in Chapters 5 and 12? Some expansion theorems for non-self-adjoint (right definite or left-definite) problems can be found in Mennicken and Möller [**471**], Locker [**451**], [**452**], Eberhard and Freiling [**159**], Eberhard, Freiling and Zettl [**160**], [**357**], [**524**].

(11) Half-Range Expansions. Perhaps these problems should be called 'half-domain expansions' since they involve expansions in terms of eigenfunctions restricted to a subinterval of the domain. [**55**], [**53**], [**363**].

(12) Integral Inequalities for derivatives. A Sturm-Liouville equation involves a function y, its derivative y' or quasi-derivative py' and y'' or $(py')'$. For the study of such equations relationships, particularly relationships involving integrals are important. For instance inequalities of the form

$$\left(\int_J |y'|^2 w\right)^2 \leq K \int_J |y|^2 w \int_J |y''|^2 w$$

play an important role in the study of discreteness conditions for the spectrum. The names of Landau, , Hardy-Littlewood, Kolmogorov, Schonberg-Cavaretta , Hadamar , Gabushin Kallman-Rota, are often associated with these kinds of inequalities and their extensions. See the monograph of Kwong and Zettl [**428**] and its references for some information on these kinds of inequalities. Also see [**13**], [**101**], [**121**], [**182**], [**185**], [**184**], [**183**], [**172**], ,[**171**], [**202**], [**208**], [**237**], [**180**], [**241**], [**178**], [**251**], [**250**], [**280**], [**421**], [**416**], [**426**], [**425**], [**427**], [**423**].

(13) Half-Linear Sturm-Liouville equations The Prüfer equation method for establishing the existence of eigenvalues can be modified for 'half-linear' equations. [**163**], [**83**], [**162**], [**379**].

(14) Non-Linear Sturm Liouville problems. This is not a well-defined area but there is a considerable literature available but this writer is not well informed on this literature. [**78**], [**99**], [**614**], [**616**], [**621**].

(15) Strong limit-point conditions. These were introduced by Everitt and have been studied by several authors. [**194**], [**198**], [**203**], [**410**].

(16) Separation of SL equations. This topic was also introduced by Everitt and has an active literature. [**204**], [**176**].

(17) Oscillation theory for nonlinear SL equations. This is also a currently active research area. An example is the Emden-Fowler equation which has received a lot of attention in the literature. [**614**], [**616**].

(18) Eigenvalues below the essential spectrum. [**105**], [**106**], [**308**], [**296**].

(19) Gaps in the essential spectrum [**27**], [**271**], [**312**], [**307**], [**311**], [**552**], [**568**], [**567**], [**603**].

(20) Essential spectrum [**297**], [**303**], [**320**], [**333**], [**511**], [**512**], [**629**], [**626**], [**646**].

(21) Bounds for the starting point of the essential spectrum. [**190**], [**174**].

(22) Perturbation theory for the spectrum. [**101**], [**100**], [**110**], [**109**], [**247**], [**265**], [**556**], [**647**].

(23) Systems and higher order linear ordinary differential equations. [1], [4], [3], [4], [52], [124], [137], [142], [143], [144], [157], [166], [196], [219], [234], [181], [177], [240], [242], [239], [178], [244], [253], [254], [259], [274], [282], [313], [339], [340], [356], [372], [465], [470], [476], [477], [475], [533], [531], [532], [544], [557], [594], [621], [635], [643], [620], [622], [637], [632], [631], [630], [634], [625], [633], [636], [627], [639], [628], [642], [624], [623], [640], [641], [638], [159].

(24) Transformation theory. Transforming one equation into another by a change of dependent or independent variable or some other way, e.g. by 'regularization'. [5], [26], [5], [195], [205], [232], [233], [364], [493], [492].

(25) Scattering theory. [15], [48], [47], [54], [87], [137], [369], [495], [494].

(26) The m-function. [25], [20], [19], [66], [64], [119], [154], [190], [190], [201], [199], [206], [207], [309], [292], [293], [291], [294], [295], [300], [299], [301], [331], [332], [361], [362].

(27) Orthogonal Polynomials. [24], [18], [212], [215], [216].

(28) Numerical Approximations[34], [32], [33], [31], [31], [35], [39], [40], [41], [42], [43], [46], [31], [51], [59], [102], [45], [44], [86], [529], [97], [105], [107], [106], [118], [261], [260], [281], [282], [453], [461], [462], [460], [463], [521], [525], [527], [526], [528], [348], [626], [10].

(29) Complex coefficients. Although we discussed the LP/LC dichotomy for complex coefficients in Chapter 7 and also established the existence of continuous eigenvalue branches in Chapter 3 we have not studied in detail spectral properties of these problem. [62], [57], [104], [175], [563].

(30) Stability theory. [95], [94], [97], [108].

(31) Asymptotic form of solutions. [305].

(32) Other topics. [112], [116], [120], [210], [211], [235], [189], [188], [193], [197], [187], [200], [186], [214], [213], [229], [245], [273], [266], [262], [269], [275], [284], [283], [287], [290], [298], [304], [310], [314], [316], [315], [323], [330], [341], [346], [347], [349], [351], [365], [360], [353], [370], [365], [373], [374], [381], [398], [383], [384], [386], [285], [387], [286], [391], [390], [114], [388], [389], [385], [380], [388], [385], [396], [395], [392], [394], [393], [399], [403], [404], [405], [400], [401], [407], [435], [434], [431], [432], [430], [436], [433], [439], [441], [440], [438], [446], [447], [450], [448], [455], [459], [467], [469], [472], [473], [479], [480], [483], [481], [478], [485], [488], [491], [496], [498], [501], [500], [499], [508], [510], [514], [516], [520], [530], [543], [548], [547], [550], [554], [558], [559], [122], [561], [560], [562], [565], [566], [569], [570], [575], [572], [576], [578], [577], [579], [582], [581], [583], [589], [598], [597], [595], [605], [644], [645], [9], [7], [6], [14], [16], [17], [30], [56], [61], [84], [90], [91], [96], [117], [129], [130], [132], [152], [153], [155], [164], [165], [167], [168], [246], [267], [278], [277], [279], [289], [319], [322], [324], [326], [334], [341], [344], [345], [350], [366], [359], [358], [371], [367], [397], [402], [406], [428], [429], [444], [451], [452], [456], [458], [471], [482], [484], [486], [487], [515], [513], [523], [522], [540], [541], [542], [538], [539], [549], [553], [555], [564], [571], [573], [574], [584], [592], [593], [599], [600], [601], [602], [606], [604], [608], [619]. [161], [158].

CHAPTER 17

Open Problems

Unfortunately what is little recognized is that the most worthwhile scientific books are those in which the author clearly indicates what he does not know; for an author most hurts his readers by concealing difficulties.
Galois, Evariste, In N. Rose (ed.) Mathematical Maxims and Minims, Raleigh NC: Rome Press Inc., 1988.

These problems are 'open' as far as the author knows at the time of this writing and are stated in random order. Some may be intractable, some accessible but challenging, others routine. I have made no effort to grade these problems by their difficulty. Some I feel can be done with a moderate amount of effort, others with considerable effort and some I have no idea how to do.

I. CHARACTERIZE OSCILLATION. Let (6.1.1) and (6.1.2) hold. At each endpoint the equations (6.1.1) fall into two disjoint classes: oscillatory (O) and nonoscillatory (NO). Characterize the O class in such a way that one can tell whether the equation is in the O or NO class directly from the coefficients p, q.

II. CHARACTERIZE THE LIMIT CIRCLE CASE. Let (7.4.1), (7.4.2) hold. At each endpoint the equations (7.4.1) fall into two disjoint classes: limit-circle (LC) or limit-point (LP). Characterize the LC class in such a way that one can tell whether the equation is in the LC or LP class directly from the coefficients p, q, w. Consider the same problem with the condition (7.4.2) replace by (1),

$$J = (a,b), \quad -\infty \leq a < b \leq \infty, \quad 1/p, q, w \in L_{loc}(J, \mathbb{R}), \quad w > 0, \text{ a.e. on } J;$$

or by (2),

$$J = (a,b), \quad -\infty \leq a < b \leq \infty, \quad 1/p, q, w \in L_{loc}(J, \mathbb{R}), \quad |w| > 0, \text{ a.e. on } J;$$

or by (3),

$$J = (a,b), \quad -\infty \leq a < b \leq \infty, \quad 1/p, q, w \in L_{loc}(J, \mathbb{C}), \quad |w| > 0, \text{ a.e. on } J.$$

Recall that when w is not positive the LC classification is defined with respect to the Hilbert space $L^2(J, |w|)$.

III. Find a theory which contains as special cases (i) Sturm-Liouville theory and (ii) the analogue of Sturm-Liouville theory for difference equations. In particular such a theory should produce difference equations analogues of every theorem in this monograph.

IV. Find a SLP whose spectrum is the primes. Given any finite set F of distinct real numbers, in [585] Volkmer and Zettle have constructed a Sturm-Liouville problem of Atkinson type whose spectrum is precisely F. Note that Example 22 in Chapter 13 above gives a class of Laguerre problems all having spectrum given by

$$\sigma = \{\lambda_n = n : n \in \mathbb{N}_0\}.$$

V. Can the technical condition (6.3.9) be omitted from the hypotheses of Theorem 6.3.4?

VI. Let (7.4.1), (7.4.2) hold and assume that one or both endpoints are LP. Find necessary and sufficient conditions for the spectrum to be bounded below and discrete. (See the Comments Section of Chapter 10 for a discussion of the well known Molchanov criterion and its extensions to the case when p and w are not necessarily equal to one.)

VII. Let (7.4.1), (7.4.2) hold and assume that one or both endpoints are LP. Find necessary and sufficient conditions for the spectrum to be discrete.

VIII. Let (7.4.1), (7.4.2) hold. Let $A, B \in M_2(\mathbb{C})$, the set of 2×2 matrices over $\mathbb{C}$ satisfying the following conditions:

$$rank(A : B) = 2;$$

$$AEA^* = BEB^*, \ E = \begin{bmatrix} 0 & -1 \\ 1 & 0 \end{bmatrix}.$$

Given any finite or infinite sequence of nonnegative integers $n_1, n_2, n_3, \cdots$ find a regular self-adjoint Sturm-Liouville problem (satisfying the stated conditions) with eigenvalues $\{\lambda_n : n \in \mathbb{N}_0\}$ such that all eigenvalues λ_n are simple except for those with indices $n_1, n_2, n_3, \cdots$ (This writer has heard claims at meetings that such problems are known - even with periodic coefficients - but does not know of any literature containing a construction for such problems.) (Recall from Chapter 4 that the algebraic and geometric multiplicities of eigenvalues for problems of this kind are equal.)

IX. Let (5.1.4) hold. By Theorem 5.8.2 the problem (5.1.1) (4.1.3) (4.1.4) has at most $2k$ nonreal eigenvalues where k is the number of negative eigenvalues of problem (5.1.2), (4.1.3), (4.1.4). Determine the value of k in terms of the coefficients and the boundary conditions.

X. Let (5.1.4) hold. By Theorem 5.8.2 the problem (5.1.1) (4.1.3) (4.1.4) has at most $2k$ nonreal eigenvalues where k is the number of negative eigenvalues of problem (5.1.2), (4.1.3), (4.1.4). Determine the number of nonreal eigenvalues. Of particular interest is the case when there are no nonreal eigenvalues.

XI. In [**522**] Pöschel and Trubowitz give a beautiful exposition of an inverse spectral theory for regular self-adjoint Sturm-Liouville problems with $J = (0, 1)$, $p = 1 = w$ on J, $q \in L^2(J, \mathbb{R})$ and Dirichlet boundary conditions: $y(0) = 0 = y(1)$. Do their results hold for $q \in L^1(J, \mathbb{R})$?

XII. Do the results in [**522**] hold for $q \in L^1(J, \mathbb{R})$, $J = (a, b)$, $-\infty \leq a < b \leq \infty$, and general self-adjoint separated or coupled regular boundary conditions?

XIII. What is the theory corresponding to that in [**522**] for regular self-adjoint Sturm-Liouville problems with $1/p, q, w \in L^1(J, \mathbb{R})$, $p > 0$, $w > 0$ on J and general self-adjoint separated or coupled boundary conditions?

XIV. What is the theory corresponding to that in [**522**] for regular self-adjoint *left-definite* Sturm-Liouville problems with $1/p, q, w \in L^1(J, \mathbb{R})$, $p > 0$, w changes sign on J?

XV. What theorem for left-definite problems corresponds to the Gelfand-Levitan inverse spectral theorem for right-definite problems?

XVI. Find the difference equations analogues of all theorems in this monograph.

XVII. Find the 'measure chain' analogues of all theorems in this monograph. This includes the previous topic since difference equations are a special case of measure chains. The theory of measure chains is a candidate for a theory which encompasses both differential and difference equation. (See Problem III above.) For an introduction to the theory of measure chains see the book by Bohner and Petersen [**92**].

XVIII. Determine the spectrum of Sturm-Liouville problems with complex valued coefficients. The methods of self-adjoint operator theory in Hilbert space do not seem to apply to the complex coefficient case. The results here are scattered, for some results see Brown, McCormack, Evans and Plum [**104**], Sims [**563**], Rofe-Beketov [**551**], Glazman [**277**], [**175**].

XIX. In Section 3.5 we construct a metric space of regular Sturm-Liouville problems and show (see Theorem 3.5.1) that, given an isolated eigenvalue λ of a problem in this space, all nearby problems have an eigenvalue close to λ. (If the given problem is self-adjoint, the nearby problems are not necessarily self-adjoint.) Is there a topology on the space of Sturm-Liouville problems each of whose endpoints is either regular or LC such that Theorem 3.5.1 holds?

XX. Many SLP have the same spectrum. What spectral information determines the regular self-adjoint problems $\omega = \omega(a, b, A, B, 1/p, q, w) \in \Omega_{s-a}$ as defined in Section 4.4 uniquely? (By spectral information we include information about the eigenfunctions, e.g. norming constants or zeros of eigenfunctions.)

XXI. What spectral information determines a singular self-adjoint Sturm-Liouville problem uniquely?

XXII. Let the operators S and T be defined as in Theorem 11.1.1. Assume that the essential spectrum of S is empty. Does it follow that the essential spectrum of T is empty?

XXIII. Can the hypotheses on w involving the subintervals J_+, J_- in Theorem 11.1.1 and $J_1, J_2, ..., J_n$ in Theorem 11.1.3 be eliminated?

Bibliography

1. A.A. Abramov, K. Balla, and N. Konyukhova, *Stable initial manifolds and singular boundary value problems for systems of ordinary differential equations*, Comput. Math. Banach Center Publications 13, 319-351 (1984).
2. A.B. Abramowitz and I.A. Stegun, *Handbook of Mathematical Functions with Formulas and Mathematical Tables*, Dover Publications Inc., New York, 1972.
3. C. Ahlbrandt, *Equivalent boundary value problems for self-adjoint differential systems*, J. Differential Eqs., 9 (1971),420-435.
4. C. D. Ahlbrandt, *Disconjugacy criteria for self-adjoint differential systems*, J. Diff. Equations 6 (1969), 271-295.
5. C. D. Ahlbrandt, D. B. Hinton, and R. T. Lewis, *The effect of variable change on oscillation and disconjugacy criteria with applications to spectral theory and asymptotic theory*, J. Math. Anal. and Appl. 81 (1981), 234-277.
6. N. I. Akhiezer and I. M. Glazman, *Theory of Linear Operators in Hilbert Space v. II*, Ungar, New York, 1963.
7. _____, *Theory of linear operators in Hilbert space, volumes I and II*, Pitman and Scottish Academic Press, London and Edinburgh, 1980.
8. I. Al-Naggar and D. B. Pearson, *A new asymptotic condition for absolutely continuous spectrum of the Sturm-Liouville operator on the half-line*, Helv. Physics Acta 67, (1994), 144-166.
9. S. Albeverio, F. Gesztesy, R. Hoegh-Krohn, and H. Holden, *Solvable models in quantum mechanics*, Springer-Verlag, Heidelberg and New York, 1988.
10. G. Alefeld and J. Herzberger, *Introduction to interval computations*, Academic Press, New York/London, 1983.
11. W. Allegretto and A. Mingarelli, *Boundary problems of the second order with an indefinite weight function*, J. Reine Angew. Math. 398 (1989), 1-24.
12. _____, *On the nonexistence of positive solutions for a Schroedinger equation with an indefinite weight function*, C. R. Math. Rep. Acad. Sci. Canada 8 (1986), 69-73.
13. T. G. Anderson and D. B. Hinton, *Relative boundedness and compactness theory for 2nd order differential operators*, pre-print.
14. F.M. Arscott, *Periodic differential equations*, Pergamon, 1964.
15. A.A. Arsen, *Resonances in the scattering problem for the Sturm-Liouville operator*, Mat. Sb. **191** (2000), no. 3, 3–12.
16. M. Ashbaugh, R. Brown, and D. Hinton, *Interpolation inequalities and nonoscillatory differential equations, p. 243-255*, International Series of Numerical Mathematics, 103, Birkhauser, Basel, 1992.
17. R.A. Askey, T. H. Koornwinder, and Editors W. Schempp, *Special functions: group theoretical aspects and applications*, D. Reidel Publishing Co., Dordrecht, 1984.
18. F. V. Atkinson, *Estimation of an eigenvalue occurring in a stability problem*, Math. Zeit.,68 (1957), 82-99.
19. _____, *On the asymptotic behavior of the Titchmarsh-Weyl m-coefficient and the spectral function for the scalar second-order differential expressions*, Lecture Notes in Mathematics, vol. 964 (1982), 1-27.
20. _____, *On the location of the Weyl circles*, Proc. R. Soc. Edinburgh A88 (1981), 345-356.
21. _____, *Discrete and continuous boundary value problems*, Academic Press, New York /London, 1964.
22. F. V. Atkinson, M. S. P. Eastham, and J. B. McLeod, *The limit-point, limit-circle nature of rapidly oscillating potentials*, Proc. Royal Soc. Edinburgh (A), to appear.

23. F. V. Atkinson and W. N. Everitt, *Bounds for the point spectrum for a Sturm-Liouville equation*, Proc. Royal Soc. Edinburgh (A) 80 (1978), 57-66.
24. _____, *Orthogonal polynomials which satisfy second-order differential equations*, Proceedings of the Christoffel Symposium, (1979), 11-19. Birkhauser-Verlag; Basel, 1981.
25. F. V. Atkinson, W. N. Everitt, and K. S. Ong, *On the m-coefficient of Weyl for a differential equation with an indefinite weight function*, Proc. London Math. Soc. (3) 29 (1974),368-384.
26. F. V. Atkinson, W. N. Everitt, and A. Zettl, *Regularisation of a Sturm-Liouville problem with an interior singularity using quasi-derivatives*, Differential and Integral Equations 1 (1988), 213-221.
27. F. V. Atkinson and D. Jabon, *Indefinite Sturm-Liouville problems*, Argonne Reports ANL-87-27, v. I, edited by Kaper, Kwong and Zettl, (1987), 31-45.
28. _____, *Indefinite Sturm-Liouville problems,pp. 31-45 of spectral theory of Sturm-Liouville differential operators ed. H. G. Kaper and A. Zettl*, Argonne Reports ANL-84-73,Argonne National Laboratory (1984).
29. F. V. Atkinson and A. B. Mingarelli, *Asymptotics of the number of zeros and of the eigenvalues of general weighted Sturm-Liouville problems*, J. Reine Angew. Math. 375/376 (1987), 380-393.
30. T. Ya. Azizov and I. S. Iohvidov, *Linear operators in space with an indefinite metric*, Wiley, 1989.
31. P. B. Bailey, *On the approximation of eigenvalues of singular Sturm-Liouville problems by those of suitably chosen regular problems*, Spectral theory and computational methods of Sturm-Liouville problems. (Knoxville, TN, 1996), 171-182, Lecture Notes in Pure and Appl. Math., 191, Dekker, New York, 1997.
32. _____, *Sleign: an eigenvalue-eigenfunction code for Sturm-Liouville problems*, Report SAND77-2044 (Sandia National Laboratory, Albuquerque, New Mexico, USA: 1976.
33. _____, *A slightly modified Prufer transformation useful for calculating Sturm-Liouville eigenvalues*, J. Comput. Phys. 29 (1978), 306-310.
34. _____, *Sturm-Liouville eigenvalues via a phase function*, SIAM J. Appl. Math 14 (1966), 242-249.
35. _____, *Sturm-Liouville problems: Coupled and Separated Boundary Conditions*, private communication.
36. P. B. Bailey, J. Billingham, R. J. Cooper, W. N. Everitt, A. C. King, Q. Kong, H. Wu, and A. Zettl, *Eigenvalue problems in fuel cell dynamics*, Proc. Roy. Soc. London (A) 459 (2003), 241-261.
37. P. B. Bailey, W. N. Everitt, D. B. Hinton, and A. Zettl, *Some spectral properties of the Heun differential equation*, Operator Theory: Advances and Applications 132 (2002), 87-110.
38. P. B. Bailey, W. N. Everitt, J. Weidmann, and A. Zettl, *Regular approximations of singular Sturm-Liouville problems*, Results in Mathematics, v. 23 (1993), 3-22.
39. P. B. Bailey, W. N. Everitt, and A. Zettl, *Computing eigenvalues of singular Sturm-Liouville problems*, Results in Mathematics, v. 20 (1991), 391-423.
40. _____, *On the numerical computation of the spectrum of singular Sturm-Liouville problems*, NSF Final Report for Grant DMS-9106470; 1995.
41. _____, *Regular and singular Sturm-Liouville problems with coupled boundary conditions*, Proc. Royal Soc. of Edinburgh (A) 126 (1996), 505-514.
42. _____, *SLEIGN2: a Fortran code for the numerical approximation of eigenvalues, eigenfunctions and continuous spectrum of Sturm-Liouville problems: 1997. available on the WWW at www.math.niu.edu./pub/papers/Zettl/SL2*.
43. _____, *The SLEIGN2 Sturm-Liouville code*, ACM TOMS, ACM Trans. Math. Software 21, (2001), 143-192.; presently available as the LaTeX file bez3.tex on the web site (http://www.math.niu.edu/ zettl/SL2).
44. P. B. Bailey, B. S. Garbow, H. Kaper, and A. Zettl, *Algorithm 700, A FORTRAN software package for Sturm-Liouville problems*, ACM TOMS v. 17 (1991), 500-501.
45. _____, *Eigenvalue and eigenfunction computations for Sturm-Liouville problems*, ACM TOMS v. 17(1991), 491-499.
46. P. B. Bailey, M. K. Gordon, and L. F. Shampine, *Solving Sturm-Liouville eigenvalues problem*, Report SAND76-0560 Sandia National Laboratory, Albuquerque, New Mexico, USA; 1976.

47. V. Bargmann, *On the connection between phase shifts and scattering potential*, Rev. Mod. Phys. 21 (1949), 488-493.
48. _____, *Remarks on the determination of a central field of force from the elastic scattering phase shifts*, Phys. Rev. 75 (1949), 301-303.
49. J. H. Barrett, *Disconjugacy of second order linear equations with non-negative coefficients*, Proc. Amer. Math. Soc. 10, (1959), 552-561.
50. _____, *Oscillation theory of ordinary linear differential equations*, Advances in Math., 3, (1969), 415-509.
51. J. V. Baxley, *Eigenvalues of singular differential operators by finite difference methods I*, J. Math. Anal. Appl. 37 (1972), 244-254.
52. R. Beals, *An abstract treatment of some forward-backward problems of transport and scattering*, J. Funct. Anal. 34 (1979), 1-20.
53. _____, *Indefinite Sturm-Liouville problems and half-range completeness*, J. Differential Equations 56 (1985), no. 3, 391-407.
54. _____, *On an equation of mixed type from electron scattering*, J. Math. Anal. Appl. 58 (1977), 32-45.
55. _____, *Partial-range completeness and existence of solutions to two-way diffusion equations*, J. Math. Phys. 22 (1981), 954-960: erratum, J. Math. Phys. 24 (1983), 1932.
56. P. R. Beesack, *Gronwall inequalities*, Carleton University, Dept. of Mathematics, Carleton Mathematical Lecture Notes, no. 11, Ottawa, 1975.
57. P.R. Beesack, *Nonoscillation and disconjugacy in the complex domain*, Trans. Amer. Math. Soc., 81 (1956), 211=242.
58. H. Behncke and H. Focke, *Deficiency indices of singular Schroedinger operators*, Math. Z. 158 (1978), 87-98.
59. H. Behnke and F. Goerisch, *Inclusions for eigenvalues of self-adjoint problems*, J. Herzberger (Hrsg.): Topics in Validated Computation, North Holland Elsevier, Amsterdam, usw. (1994).
60. B. P. Belinskiy and J. P. Dauer, *On a regular Sturm-Liouville problem on a finite interval with the eigenvalue parameter appearing linearly in the boundary condition*, Spectral theory and computational methods of Sturm-Liouville problems, Marcel Dekker, (1997), 183-196.
61. W.W. Bell, *Special functions for scientists and engineers*, Van Nostrand, London, 1968.
62. Carl M. Bender, Stefan Boettcher, and Van M. Savage, *Conjecture on the interlacing of zeros in complex Sturm-Liouville problems*, J. Math. Phys. **41** (2000), no. 9, 6381–6387.
63. C. Bennewitz, *A generalization of Niessen's Limit-Circle Criterion*, Proc. Roy. Soc. of Edinburgh, Vol. 78A (1977), 81-90.
64. _____, *A note on the Titchmarsh-Weyl m-function*, 1987 Symposium, Argonne National Laboratory, Argonne, IL., Report ANL-87-26, vol. 2, 105-111.
65. C. Bennewitz and W. N. Everitt, *On the second-order left-definite boundary value problems*, Lecture Notes in Mathematics 1032 (1983), 31-67, Springer-Verlag, Heidelberg.
66. _____, *Some remarks on the Titchmarsh-Weyl m-coefficient*, In Tribute to Ake Pleijel, Mathematics Department, University of Uppsala, Sweden, (1980), 49-108.
67. A. Bielecki, *Une remarque sur la methode de Banach-Cacciopoli-Tikhonov dans la theorie des equations differentielles ordinaires*, Bull. Acad. Polon. Sci. Cl. III 4(1956), 261-264.
68. P. Binding, *A hierarchy of Sturm-Liouville problems*, Math. Methods in the Appl. Sci., 26 (2003), 349-357.
69. P. A. Binding and P. Browne, *Asymptotics of eigencurves for second order ordinary differential equations, I.*, J. Differential Equations, 88 (1990), 30-45.
70. _____, *Spectral properties of two parameter eigenvalue problems II.*, Proc. Roy. Soc. Edinburgh 106A (1987),39-51. Corrigend (with R. H. Picard) ibid. 115A (1990), 87-90.
71. P. A. Binding and P. J. Browne, *Applications of two parameter spectral theory to symmetric generalized eigenvalue problems*, J. Applied Analysis 29, (1988),107-142.
72. _____, *Left-definite Sturm-Liouville problems with eigenparameter dependent boundary conditions*, J. Differential and Integral Equations 12 (1999), 167-182.
73. _____, *Multiparameter Sturm theory*, Proc. Roy. Soc. Edinburgh, 99A, (1984), 173-184.
74. _____, *Oscillations theory for indefinite Sturm-Liouville problems with eigenparameter-dependent boundary conditions*, Proc. Roy. Soc. 127A (1997), 1123-1136.
75. P. A. Binding and P. J. Browne, *Sturm-Liouville problems with non-separated eigenvalue dependent boundary conditions*, Proc. Roy. Soc. Edinburgh Sect. A **130** (2000), no. 2, 239–247.

76. P. A. Binding, P. J. Browne, and K. Seddihi, *Sturm-Liouville problems with eigenparameter dependent boundary conditions*, Proc. Edinburgh Math. Soc. (2) 37 (1994), no. 1, 161-182.
77. P. A. Binding, P. J. Browne, and B. A. Watson, *Inverse spectral problems for Sturm-Liouville equations with eigenparameter dependent boundary conditions*, J. London Math. Soc. (2) 62 (2000), no. 1, 161-182.
78. P. A. Binding and Y.-X. Huang, *Existence and nonexistence for positive eigenfunctions for the p-Laplacian*, Proc. Amer. Math. Soc. 123, 1833-1838.
79. P. A. Binding and H. Volkmer, *Eigencurves for two parameter Sturm-Liouville equations*, SIAM Review 38 (1996),no. 1, 27-48.
80. _____, *Oscillation theory for Sturm-Liouville problems with indefinite coefficients*, Proc. Royal Soc. Edinburgh, 131 A (2001), 989-1002.
81. _____, *Pruefer Angle Asymptotics for Atkinson's semi-definite Sturm-Liouville Eigenvalue Problem*, pre-print.
82. P. A. Binding and Q. Ye, *Variational principles for indefinite eigenvalue problems*, Linear Algebra Appl. 218 (1995), 251-262.
83. Paul A. Binding and Bryan P. Rynne, *Half-eigenvalues of periodic Sturm-Liouville problems*, J. Differential Equations 206 (2004), 280-305.
84. G. Birkhoff and G-C. Rota, *Ordinary differential equations*, Ginn and Company, New York, 1962.
85. G..D. Birkhoff, *Existence and oscillation theorem for a certain boundary value problem*, Trans. Amer. Math. Soc. 10 (1909), 259-270.
86. S. Blackford and J. Dongarra, *LAPACK working Note 81*, Version 3.0, June 30, 1999. Available from www.netlib.org/lapack.
87. F. Bloch, *Ueber die Quantenmechanik der Elektron in Kristallgittern*, Zeit. fuer Physik, 52 (1928), 555-600.
88. L.E. Blumenson, *On the eigenvalues of Hill's equation*, Comm. Pure Appl. Math., 16 (1963), 261-266.
89. M. Boecher, *The theorems of oscillation of Sturm and Klein*, Bull. Amer. Math. Soc. 4 (1897-1898), 295-313,365-376.
90. _____, *Lecons sur les methodes de Sturm dans la theorie des equations differentielles lineaires, et leurs developpements modernes*, Gauthier-Villars, Paris, 1917.
91. J. Bognar, *Indefinite inner product spaces*, Springer, 1974.
92. M. Bohner and A. Peterson, *Dynamic Equations on Time Scales*, Birkhauser, Boston, MA., 2001.
93. G. Borg, *Eine Umkehrung der Sturm-Liouville Eigenwertaufgabe. Bestimmung der Differentialgleichung durch die Eigenwerte*, Acta. Math., 78 (1946), 1-96.
94. _____, *On a Liapounoff criterion of stability*, Amer.J. Math., 71 (1949), 67-70.
95. _____, *Ueber die stabilitaet gewisser Klassen von Linearen Differentialgleichungen*, Ark. Mat. Astr. Fys.,31 (1944), 1-31.
96. O. Boruvka, *Linear differential transformations of the second order*, English Universities Press, London, 1971; translated from the German edition of 1967.
97. A. Boumenir, *A rigorous verification of a numerically computed eigenvalue*, Comput. Math. Appl.,V.38, (1999), pages 39-41.
98. J. P. Boyd, *Sturm-Liouville eigenvalue problems with an interior pole*, J. Math. Physics 22 (1981), 1575-1590.
99. J. S. Bradley, *Comparison theorems for the square integrability of solutions of $(r(t)y')' + q(t)y=f(t,y,)$*, Glasgow Mathematical Journal, Vol. 13, part 1, (1972), 75-79.
100. I. Brinck, *Self-adjointness and spectra of Sturm-Liouville operators*, Math. Scand. 7 (1959), 219-239.
101. B. M. Brown and M. Eastham, *The Hurwitz theorem for Bessel functions and antibound states*, Proc. Roy. Soc. London, to appear.
102. B. M. Brown, W. D. Evans, V. G. Kirby, and M. Plum, *Safe numerical bounds for the Titchmarsh - Weyl $m(\lambda)$ function*, Math. Proc. Cambridge Phil. Soc. 113, (1993), 583-599.
103. B. M. Brown and M. Marletta, *Spectral inclusion and spectral exactness of elliptic PDEs on exterior domains*, To appear in IMA J. Numer. Analysis (2004).
104. B. M. Brown, D. K. R. McCormack, W. D. Evans, and M. Plum, *On the spectrum of second order differential operators with complex coefficients*, Proc. Roy. Soc. of London, A, (1999), 455, 1234-1257.

105. B. M. Brown, D. K. R. McCormack, and M. Marletta, *On computing enclosures for the eigenvalues of Sturm-Liouville problems*, pre-print.
106. B. M. Brown, D. K. R. McCormick, and A. Zettl, *On a computer assisted proof of the existence of eigenvalues below the essential spectrum of the Sturm-Liouville problem*, Journal of Computational and Applied Mathematics 125, (2000), 385-393.
107. _____, *On the existence of an eigenvalue below the essential spectrum*, Proc. R. Soc. London, A, (1999), 2229-2234.
108. R. Brown, S. Clark, and D. Hinton, *Some function space inequalities and their application to oscillation and stability problems in differential equations, pp.19-41*, Analysis and Applications, Narosa Pub. House, New Delhi, 2002.
109. M. Burnat, *Die Spektraldarstellung einiger Differentialoperatoren mit periodischen Koeffizienten im Raume der fastperiodischen Funktionen*, Studia Math., 25 (1964/5), 33-64.
110. _____, *The stability of eigenfunctions and the spectrum of Hill's equation*, Bull. Acad. Polon. Ser. Math. Astr. Phys., 9 (1961), 795-797. (Russian).
111. D. Buschmann, G. Stolz, and J. Weidmann, *One-dimensional Sschroedinger operators with local point interactions*, J. Reine Angew. Math. 467 (1995), 169-186.
112. P. L. Butzer, W. N. Everitt, and G. Schoettler, *Sturm-Liouville boundary value problems and Legrange interpolations series*, Rend. di Mat. (VII) 14 (1994), 87-126.
113. R. Byers, B. Harris, and M. K. Kwong, *Weighted means and oscillation conditions for second-order matrix differential equations*, J. Diff. Equations, 61, 2, (1986), 164-177.
114. Xifang Cao, Q. Kong, H. Wu, and A. Zettl, *Geometric aspects of Sturm-Liouville problems III. Level surfaces of the n-th Eigenvalue*, preprint.
115. Xifang Cao, Qingkai Kong, Hongyou Wu, and Anton Zettl, *Sturm-Liouville problems Whose Leading Coefficient Function Changes Sign*, Canad. J. Math. 55 (2003), 724-749.
116. R. Carmona, *One-dimensional Schroedinger operators with random or deterministic potentials: new spectral types*, J. Functional Analysis, 51 (1983), 229-258.
117. K. Chadan and P.C. Sabatier, *Inverse Problems in Quantum Scattering Theory, 2nd ed.*, Springer-Verlag, New York, 1989.
118. B. Chanane, *Computing eigenvalues of regular Sturm-Liouville problems*, Appl. Math. Lett. **12** (1999), no. 7, 119–125.
119. Jyoti Chaudhuri and W. N. Everitt, *On the spectrum of ordinary second-order differential operators*, Proc. Royal Soc. Edinburgh (A) 68(1969), 95-119.
120. _____, *On the square of a formally self-adjoint differential expression*, J. London Math. Soc. (2) 1 (1969), 661-673.
121. R. S. Chisholm and W. N. Everitt, *On bounded integral operators in the space of integrable-square functions*, Proc. Royal Soc. Edinburgh (A) 69 (1971), 199-204.
122. C. Shubin Christ and G. Stolz, *Spectral theory of one-dimensional Schroedinger operators with point interactions*, J. Math. Anal. Appl. 184 (1994), 491-516.
123. E. A. Coddington and N. Levinson, *Theory of ordinary differential equations*, McGraw-Hill, New York /London/Toronto, 1955.
124. E. A. Coddington and A. Zettl, *Hermitian and anti-Hermitian properties of Green's matrices*, Pacific J. Math. 18 (1966), 451-454.
125. W. J. Coles, *An oscillation criterion for second order linear differential equations*, Proc. Amer. Math. Soc. 19 (1968), 755-759.
126. _____, *A simple proof of a well-known oscillation theorem*, Proc. Amer. Math. Soc. 19 (1968), 507.
127. W.J. Coles and D. Willett, *Summability criteria for oscillation of second order linear differential equations*, Annl. Mat. Pura App. 79 (1968), 391-398.
128. A. Constantin, *A general-weighted Sturm-Liouville problem*, Scuola Normale Superiore Pisa 24, (1997), 767-782.
129. J. B. Conway, *Functions of One Complex Variable*, Springer-Verlag, Berlin, Heidelberg, New York, 1973.
130. W. A. Coppel, *Stability and asymptotic behavior of differential equations*, D. C. Heath, Boston, 1965.
131. _____, *Disconjugacy*, Lecture Notes in Mathematics No. 220, Springer-Verlag, New York Berlin, 1971.
132. E.T. Copson, *Theory of functions of a complex variable*, Oxford University Press, Oxford, 1946.

133. B. Curgus and H. Langer, *A Krein space approach to symmetric ordinary differential operators with an indefinite weight function*, J. Differential Equations 79 (1989), 31-61.
134. K. Daho and H. Langer, *Sturm-Liouville Operators with an indefinite weight function*, Proc. Roy. Soc. Edinburgh 78A (1997), 161-191.
135. M. Dauge and B. Helffer, *Eigenvalues Variation. I. Neumann Problem for Sturm-Liouville Operators*, J. Diff. Equations, v. 104 (1993), 243-262.
136. _____, *Eigenvalues Variation. II. Neumann Problem for Sturm-Liouville Operators*, J. Diff. Equations, v. 104 (1993), 263-297.
137. P. Deift and E. Trubowitz, *Inverse scattering on the line*, Comm. Pure Appl. Math.32 (1979), 121-251.
138. J. Dieudonne, *Foundations of Modern Analysis*, Academic Press, New York/London, 1969.
139. A. Dijksma, *Eigenfunction expansion for a class of J-selfadjoint ordinary differential operators with boundary conditions containing the eigenvalue parameter*, Proc. Roy. Soc. Edinburgh, 86A (1980), 1-27.
140. A. Dijksma, H. Langer, and H. de Snoo, *Symmetric Sturm-Liouville operators with eigenvalue depending boundary conditions*, CMS Conference Proceeding Vol. 8: Oscillation, Bifurcation, Chaos (1987), 87-116.
141. N. Dunford and J. T. Schwartz, *Linear operators , vol.II*, Wiley, New York, 1963.
142. H. I. Dwyer, *Eigenvalues of matrix Sturm-Liouville problems with separated or coupled boundary conditions*, Doctoral Thesis, Northern Illinois University, 1993.
143. H. I. Dwyer and A. Zettl, *Computing eigenvalues of regular Sturm-Liouville problems*, Electronic J. Diff. Equations (EJDE), Directory Volumes/1994/06- Dwyer-Zettl. (Latex file Dwyer-Zettl-tex 24KB; Postscript file Dwyer-Zettl.ps 377 Kb).
144. _____, *Eigenvalue computations for regular matrix Sturm-Liouville problems*, Electronic J. Diff. Equ, EDJE.
145. E.S.P. Eastham, Q. Kong, H. Wu, and A. Zettl, *Inequalities among eigenvalues of Sturm-Liouville problems*, J. of Inequalities and Applications, 3, (1999), 25-43.
146. M. S. P. Eastham, *The first stability interval of the periodic Schroedinger equation*, J. London Math. Soc. (2), 4 (1972), 587-592.
147. _____, *Limit-circle differential expressions of the second-order with an oscillating coefficient*, Quart. J. Math. Oxford (2) 24 (1973).
148. _____, *On a limit-point method of Hartman*, Bull. London Math. Soc. 4 (1972), 340-344.
149. _____, *The periodic Schroedinger equation and the Brillouin zone*, J. London Math. Soc. (2), 5 (1972), 240-242.
150. _____, *The Schroedinger equation with a periodic potential*, Proc. Royal Soc. Edinburgh, 69, (1971), 125-131.
151. _____, *Semi-bounded second-order differential operators*, Proc. Royal Soc. Edinburgh (A), 72, 2, (1973), 9-16.
152. _____, *Theory of ordinary differential equations*, Van Nostrand Reinhold Company, London, 1970.
153. _____, *The spectral theory of periodic differential equations*, Scottish Academic Press, Edinburgh/London, 1973.
154. M. S. P. Eastham, *A connection formula for Sturm-Liouville spectral functions*, Proc. Roy. Soc. Edinburgh Sect. A **130** (2000), no. 4, 789–791. MR1 776 676
155. M. S. P. Eastham and H. Kalf, *Schroedinger-type operators with continuous spectrum*, Pitman, London, 1982.
156. M. S. P. Eastham and M. L. Thompson, *On the limit point, limit-circle classification of second-order ordinary differential expressions*, Quart. J. Math. Oxford Ser. (2) (1973), 531-535.
157. M. S. P. Eastham and A. Zettl, *Second order differential expressions whose squares are limit-3*, Proc. Royal Soc. Edinburgh, 16(1977), 223-238.
158. W. Eberhard and G. Freiling, *The distribution of the eigenvalues for second order eigenvalue problems in the presence of an arbitrary number of turning points*, Results Math. 21 (1992), no. 1-2, 24-41.
159. W. Eberhard and G. Freiling, *Stone-reguläre Eigenwertprobleme*, Math. Z. 160 (1978), 139-161.
160. W. Eberhard, G. Freiling, and A. Zettl, *Sturm-Liouville Problems with Singular Non-Self-Adjoint Boundary Conditions*, Math. Nachrichten (to appear).

161. Walter Eberhard, *Irregular eigenvalue-problems on the half-axis*, Results Math. 14 (1-2)(1988), 48-63.,.
162. Walter Eberhard and Árpad Elbert, *Half-linear eigenvalue problems*, Math. Nachr. 183 (1997), 55-72.
163. Walter Eberhard and Árpád Elbert, *On the eigenvalues of half-linear boundary value problems*, Math. Nachr. 213 (2000), 57-76.
164. D.E. Edmunds and W.D. Evans, *Spectral Theory and Differential Operators*, Oxford University Press, New York, 1987.
165. U. Elias, *Oscillation Theory of Two-Term Differential Equations*, Kluwer, Dordrecht, 1997.
166. L. Erbe, Q. Kong, and S. Ruan, *Kamenev type theorems for second order matrix differential systems*, Proc. Amer. Math. Soc., 117 (1993), 957-962.
167. A. Erdelyi, *Higher transcendental functions: I, II, and III*, McGraw-Hill, New York, 1953.
168. _____, *Asymptotic Expansions*, Dover Publ., New York, 1956.
169. W. D. Evans, *On the limit-point and Dirichlet-type results for second-order differential expressions*, Proceedings of the 1976 Dundee Conference on Differential Equations, Lecture Notes in Mathematics, Springer-Verlag.
170. _____, *On the limit-point, limit-circle classification of a second-order differential equation with a complex coefficient*, J. London Math. Soc. (2), 4 (1971), 245-256.
171. W. D. Evans and W. N. Everitt, *On an inequality of Hardy-Littlewood type: I.*, Proc. Roy. Soc. Edinburgh (A) 101 (1985), 131-140.
172. _____, *A return to the Hardy-Littlewood integral inequality*, Proc. R. Soc. London A 380 (1982), 447-486.
173. W. D. Evans, M. K. Kwong, and A. Zettl, *Lower bounds for the spectrum of ordinary differential operators*, J. of Diff. Equ., 48(1983), 123-155.
174. _____, *Lower bounds for the spectrum of ordinary differential operators*, J. of Diff. Equations, 48 (1983), 123-155.
175. W. D. Evans, R.L. Lewis, and A. Zettl, *On the spectrum of nonselfadjoint differential operators*, Lecture Notes in Mathematics, Springer-Verlag, (1982).
176. W. D. Evans and A. Zettl, *Dirichlet and separation results for Schroedinger type operators*, Proc. Royal Soc. Edinburgh, 80A (1978), 151-162.
177. _____, *Interval limit-point criteria for differential expressions and their powers*, J. London Math. Soc. (2) 15,no. 1, (1977), 119-133.
178. _____, *Interval limit-point criteria for differential expressions and their powers*, J. London Math. Soc. (to appear).
179. _____, *Levinson's limit-point criterion and powers*, J. Functional Analysis and Applications, V. 62 (1978), 629-639.
180. _____, *Norm inequalities involving derivatives*, Proc. Royal Soc. Edinburgh, A. (1978), 51-70.
181. _____, *On the deficiency indices of powers of real 2nth order symmetric differential expressions*, J. London Math. Soc. (2), 13 (1976), 543-556.
182. W.D. Evans, W.N. Everitt, W.K. Hayman, and D.S. Jones, *Five integral inequalities: an inheritance from Hardy and Littlewood*, Journal of Inequalities and Applications 2 (1998), 1-36.
183. W. N. Everitt, *A general integral inequality associated with certain ordinary differential operators*, Quaestiones Mathematicae 2 (1978), 479-494.
184. _____, *Integral inequalities and spectral theory*, Proceedings of the Symposium on Spectral Theory and Differential Equations, Dundee, 1974. Lecture Notes in Mathematics 448 (1974), 148-166; Springer-Verlag, Heidelberg.
185. _____, *Integral inequalities and the Liouville transformation*, Proceedings of the Conference on the Theory of Ordinary and Partial Differential Equations, Dundee, 1974. Lecture Notes in Mathematics 415 (1974), 338-352.
186. _____, *Legendre polynomials and singular differential operators*, Proceedings of the Conference on Ordinary and Partial Differential Equations, Dundee, 1978l Lecture Notes in Mathematics 827 (1980), 83-106; Springer-Verlag, Heidelberg.
187. _____, *A note on the Dirichlet condition for second-order differential expressions*, Canadian J. Math. XXVIII (1976), 312-320.
188. _____, *A note on the self-adjoint domains of second-order differential equations*, Quart. J. of Math. (Oxford) (2) 14 (1963), 41-45.

189. _____, *On a generalization of Bessel functions and a resulting class of Fourier kernels*, Quart. J. of Math. (Oxford) (2) 10 (1959), 270-279.
190. _____, *On a property of the m-coefficient of a second-order linear differential equation*, J. Lond. Math. Soc. (2), 4 (1972), 443-457.
191. _____, *On the limit-circle classification of second-order differential expressions*, Quart. J. Math. (Oxford) (2) 23 (1972), 193-196.
192. _____, *On the limit-point classification of second-order differential operators*, J. London Math. Soc. 41 (1966), 531-534.
193. _____, *On the spectrum of a second-order linear differential equation with a p-integrable coefficient*, Applicable Analysis 2 (1972), 143-160.
194. _____, *On the strong limit- point condition of second-order differential expressions*, Proceedings of the International Conference on Differential Equations, Los Angelos, (1974), 287-307. Academic Press, Inc., New York.
195. _____, *On the transformation theory of ordinary second-order linear symmetric differential equations*, Czechoslovak Mathematical Journal 32 (1982), 275-306.
196. _____, *Self-adjoint boundary value problems on finite intervals*, J. London Math. Soc. 37 (1962), 372-384.
197. _____, *Some remarks on a differential expression with an indefinite weight function*, Proceedings of the Conference in Spectral Theory and Asymptotics of Differential Equations, Scheveningen, 1973. North Holland Mathematics Studies 13 (1974), 13-28. Edited by E. M. de Jager.
198. _____, *Some remarks on the strong limit-point condition of second-order linear differential expressions*, Casopsis Pro Pestovani Matematiky 111 (1986), 137-145.
199. _____, *Some remarks on the Titchmarsh-Weyl m-coefficient and associated differential operators*, Differential Equations, Dynamical Systems and Control Science; A festschrift in Honor of Lawrence Markus; 33-53. Lecture Notes in Pure and Applied Mathematics, volume 152; Marcel Dekker, Inc., New York (1994); edited by K. D. Elworthy, W. N. Everitt, E. B. Lee.
200. _____, *Spectral theory of the Wirtinger inequality*, Proceedings of the Conference on Ordinary and Partial Differential Equations, Dundee, 1976. Lecture Notes in Mathematics 564 (1976), 93-105; Springer-Verlag, Heidelberg.
201. _____, *The Titchmarsh- Weyl m-coefficient for second-order linear ordinary differential equations: a short survey of properties and applications*, Proceedings of the 5th Spanish National Congress on Differential Equations and Applications, (1985), 249-265. Published by the University of La Laguna, Tenerife.
202. W. N. Everitt and W. D. Evans, *HELP inequalities for limit-circle and regular problems*, Proc. Royal Soc. London (A) 432 (1991), 367-390.
203. W. N. Everitt, M. Giertz, and J. B. McLeod, *On the strong and weak limit-point classification of second- order differential expressions*, Proc. London Math. Soc. (3) 29 (1974), 142-158.
204. W. N. Everitt, M. Giertz, and J. Weidmann, *Some remarks on a separation and limit-point criterion of second-order ordinary differential expressions*, Math. Ann. 200 (1973), 335-346.
205. W. N. Everitt, J. Gunson, and A. Zettl, *Some comments on Sturm-Liouville eigenvalue problems with interior singularities*, Zeitschrift fuer Angewandte Mathematik und Physik (ZAMP) v. 38 (1987), 813-838.
206. W. N. Everitt and S. G. Halvorsen, *On the asymptotic form of the Titchmarsh-Weyl m-coefficient*, Applicable Analysis 8 (1978), 153-169.
207. W. N. Everitt, D. B. Hinton, and J.K. Shaw, *The asymptotic form of the Titchmarsh-Weyl m-coefficient for Dirac systems*, J. London Math. Soc. (2) 27 (1983), 465-476.
208. W. N. Everitt and D. S. Jones, *On a integral inequality*, Proc. Royal Soc. London (A) 357 (1977), 271-288.
209. W. N. Everitt, I. W. Knowles, and T. T. Read, *Limit-point and limit-circle criteria for Sturm-Liouville equations with intermittently negative coefficients*, Proc. Royal Soc. Edinburgh (A) 103 (1986), 215-218.
210. W. N. Everitt, A. M. Krall, L. L. Littlejohn, and V. P. Onyango-Otieno, *Differential operators and the Laguerre type polynomials*, SIAM J. Math. Anal. 23 (1992), 722-736.
211. _____, *The laguerre Type Operator on a Left Definite Hilbert Space*, Journal of Math. Anal. and Applications, Vol. 192 (1995), pp. 460-468.

212. W. N. Everitt, K. H. Kwon, L. L. Littlejohn, and R. Wellman, *On the spectral analysis of the Laguerre Polynomials for positive integers k*, Spectral theory and computational methods of Sturm-Liouville problems, Marcel Dekker, (1997), 251-283.
213. W. N. Everitt, M. K. Kwong, and A. Zettl, *Differential operators and quadratic inequalities with a degenerate weight*, J. Math. Anal. and Appl., 98 (1984), 378-399.
214. _____, *Oscillation of eigenfunctions of weighted regular Sturm-Liouville problems*, J. London Math. Soc. 27(1983), 106-120.
215. W. N. Everitt and L. L. Littlejohn, *Differential Operators and the Legendre Type Polynomials*, Differential and Integral Equations, Vol. 1 (1988), pp. 97-116.
216. _____, *Orthogonal Polynomials and Spectral Theory: A Survey*, Orthogonal Polynomials and their Applications, C. Berzinski, L. Gori and A. Ronveaux (Editors), Proceedings of the third International Symposium held in Erice, Italy 1990 (1991), pp. 21-55.
217. W. N. Everitt, L. L. Littlejohn, and S. M. Loveland, *The operator domains for the Legendre differential expressions*, Research Report 3/92/57, Department of Mathematics and Statistics, Utah State University, Logan, Utah, USA; February, 1993.
218. W. N. Everitt, L. L. Littlejohn, and S. C. Williams, *The left-definite Legendre type boundary problem*, Constructive Approximation 7 (1991), 485-500.
219. W. N. Everitt and C. Markett, *On a generalization of Bessel functions satisfying higher-order differential equations*, Jour. Comp. Appl. Math. 54 (1994), 325-349.
220. W. N. Everitt and L. Markus, *Boundary value problems and symplectic algebra for ordinary and quasi-differential operators*, Mathematical Surveys and Monographs, 61, American Mathematical Society, RI, USA, 1999.
221. W. N. Everitt and L. Markus, *Complex symplectic geometry with applications to ordinary differential operators*, Trans. Amer. Math. Soc. 351 (12) (1999), 4905-4945.
222. _____, *Elliptic partial differential operators and symplectic algebra*, Mem. Amer. Math. Soc. 162 (770) (2003), x+111.
223. _____, *Infinite dimensional complex symplectic spaces*, Mem. Amer. Math. Soc. 171 (810) (2004), x+76.
224. _____, *Multi-interval linear ordinary boundary value problems and complex symplectic algebra*, Mem. Amer. Math. Soc. 151 (715) (2001), viii+64.
225. W. N. Everitt, M. Marletta, and A. Zettl, *Inequalities and eigenvalues of Sturm-Liouville problems near a singular boundary*, J. Inequalities and Applications, 6, (2001), 405-413.
226. _____, *Inequalities and eigenvalues of Sturm-Liouville problems near a singular boundary*, J. Inequalities and Applications 6 (2001), 405-413.
227. W. N. Everitt, M. Moeller, and A. Zettl, *Discontinuous dependence of the n-th Sturm-Liouville eigenvalue*, Birkhauser, General Inequalities 7, v. 123, (1997), 145-150. Proceedings of 7th International Conference at Oberwolfach, Nov. 13-19, 1995.
228. _____, *Sturm-Liouville problems and discontinuous eigenvalues*, Proc. Roy. Soc. Edinburgh, 129A, (1999), 707-716.
229. W. N. Everitt and G. Nasri-Roudsari, *Sturm-Liouville problems with coupled boundary conditions and Lagrange interpolation series: II.* Rend. Mat. Ser. VII 20, (2000), 199-236.
230. W. N. Everitt and F. Neuman, *A concept of adjointness and symmetry of ordinary differential expressions based on the generalized Lagrange identity and Green's formula*, Lecture Note in Math. 1032 , Springer Verlag, Berlin (1983), 161-169.
231. W. N. Everitt and D. Race, *On necessary and sufficient conditions for the existence of Caratheodory solutions of ordinary differential equations*, Quaestiones Math. 3, no.2, (1976), 507-512.
232. _____, *The regular representation of singular second-order differential expression using quasi-derivatives*, Proc. London Math. Soc. (3) 64 (1992), 383-404.
233. _____, *The regularization of singular second-order differential expressions using Frentzen-type quasi-derivatives*, Quart. J. Math. Oxford (2) (1993), 301-313.
234. _____, *Some remarks on linear ordinary quasi-differential expressions*, Proc. London Math. Soc. v.3, no.54 (1987), 300-320.
235. W. N. Everitt, G. Schoettler, and P. L. Butzer, *Sturm-Liouville boundary-value problems and Lagrange interpolation series*, Rend. Mat. Appl. (7) 14 (1994), 87-126.
236. W. N. Everitt, C. Shubin, G. Stolz, and A. Zettl, *Sturm-Liouville problems with an infinite number of interior singularities*, Marcel Dekker, v. 191, (1997), 211-249. Proceedings of the 1996 Knoxville Barrett Conference, edited by D. B. Hinton and P. Schaefer.

237. W. N. Everitt and S. D. Wray, *On quadratic integral inequalities associated with second-order symmetric differential expressions*, Lecture Notes in Mathematics 1032 (1983), 170-223, Springer-Verlag, Heidelberg.
238. W. N. Everitt and A. Zettl, *Differential operators generated by a countable number of quasi-differential expressions on the line*, Proc. London Math. Soc.(3) v.64 (1992), 524-544.
239. _____, *Generalized symmetric ordinary differential expressions I. The basic theory*, Nieuw Arch. voor Wisk. (3), XXVII (1979), 363-397.
240. _____, *The number of integrable-square solutions of products of differential expressions*, Proc. Royal Soc. Edinburgh, 76A, (1977), 215-226.
241. _____, *On a class of integral inequalities*, J. London Math. Soc., 17 (1978), 291-303.
242. _____, *Products of differential expressions without smoothness assumptions*, Quaestiones Mathematicae, 3 (1978), 67-82.
243. _____, *Sturm-Liouville differential operators in direct sum spaces*, Rocky Mountain J. of Mathematics, (1986), 497-516.
244. W.N. Everitt and A. Zettl, *The Kallman-Rota inequality; a survey*, Gian-Carlo Rota on Analysis and Probability; Selected papers and Commentaries. Chapter 5; Pages 227-250. Birkhauser, Boston; 2003.
245. W. Karwowski F. Gesztesy and Z. Zhao, *Limits of soliton solutions*, Duke Math. J. 68 (1992), 101-150.
246. G. Fichera, *Numerical and quantitative analysis*, Pitman Press, London, 1978.
247. N. E. Firsova, *Levinson formula for perturbed Hill operator*, Theoret. Math. Phys. 62 (1985), 130-139.
248. W.B. Fite, *Concerning the zeros of the solutions of certain differential equations*, Trans. Amer. Math. Soc. 19 (1980), 341-352.
249. A. Fleige, *Non-semibounded sesquilinear forms and left-indefinite Sturm-Liouville problems*, Int. Equ. Op. Theory 33 (1999), 20-33.
250. Z. M. Franco, H. G. Kaper, M. K. Kwong, and A. Zettl, *Best constants in norm inequalities for derivatives on a half-line*, Proc. Royal Soc. Edinburgh, 100A(1985), 67-84.
251. _____, *Bounds for the best constants in Landau's inequality on the line*, Proc. Royal Soc. Edinburgh, 95A(1983), 257-262.
252. G. Freiling and V. Yurko, *Inverse Sturm-Liouville Problems and Their Applications*, Nova Science, Huntington, New York, 2000.
253. H. Frentzen, *Equivalents, adjoints and symmetry of quasi-differential expressions with matrix valued coefficients and polynomials in them*, Proc. Roy. Soc. Edinburgh A 92 (1982), 123-146.
254. H. Frentzen, D. Race, and A. Zettl, *On the commutativity of certain quasi-differential expression II*, Proc. Royal Soc. Edinburgh, 123A, (1993), 27-43.
255. K. O. Friedrichs, *Spectraltheorie halbbeschraenkter Operatoren und Anwendungen auf die Spektralzerlegung von Differentialoperatoren. I,II.*, Math. Ann.109 (1933/34), 465-487. Corrections, ibid. 110 (1934/35), 777-779.
256. _____, *Ueber die ausgezeichnete Randbedingung in der Spektraltheorie der halbbeschraenkten gewoehnlichen Differentialoperatoren zweiter Ordnung*, Math. Ann. 112 (1935/36), 1-23.
257. C. Fulton, *Singular eigenvalue problems with eigenvalue parameter contained in the boundary conditions*, Proc. Roy. Soc. Edinburgh 87 (1980/81), no. 1, 1-34.
258. _____, *Two-point boundary value problems with eigenvalue parameter contained in the boundary conditions*, Proc. Roy. Soc. Edinburgh 77A (1977), 293-308.
259. C. Fulton and A. Krall, *Self-adjoint 4th-order boundary value problems in the limit-4 case*, Proceedings of Dundee Conference 1982, Springer-Verlag. Lecture Notes in Mathematics. 1032, pp. 240-256.
260. C.T. Fulton and S. Pruess, *Eigenvalue and eigenfunction asymptotics for regular Sturm-Liouville problems*, J. Math. Anal. Appl., 188, (1994), 297-340.
261. _____, *Mathematical software for Sturm-Liouville problems*, ACM. Trans. Math. Software 19 (1993), 360-376.
262. C.S. Gardner, J.M. Greene, M.D. Kruskal, and R.M. Miura, *Korteweg-deVries equation and generalizations, VI. Methods for exact solution*, Comm. Pure Appl. Math. 27 (1974), 97-133.
263. Z. M. Gasimov, *Inverse singular periodic problem of Sturm-Liouville*, Trans. Acad. Sci. Azerb. Ser. Phys.-Tech. Math. Sci. **XIX** (1999), no. 5, Math. Mech., 27–31 (2000), Translated by V. K. Panarina. MR1 784 880

264. J. M. Gelfand and B. M. Levitan, *On the determination of a differential equation from its spectral function*, English transl., AMS transl. (2), 1 (1955), 253-304.
265. F. Gesztesy, *On the one-dimensional Coulomb Hamiltonian*, J. Physics A 13 (1980), 867-875.
266. F. Gesztesy, D. Gurarie, H. Holden, M. Klaus, L. Sadun, B. Simon, and P. Vogel, *Trapping and cascading of Eigenvalues in the Large Coupling Limit*, CMP 118 (88), 597-634.
267. F. Gesztesy and H. Holden, *Soliton equations and their algebro-geometric solutions. vol. I: (1+1)-dimensional continuous models*, Cambridge Studies in Advanced Mathematics, Vol. 79, Cambridge University Press, 2003.
268. F. Gesztesy and W. Kirsch, *One-dimensional Schroedinger operators with interactions on a discrete set*, J. fuer Mathematik 362 (1985), 28-50.
269. F. Gesztesy, C. Macedo, and L. Streit, *An exactly solvable periodic Schroedinger operator*, J. Phys. A: Math. Gen. 18 (1985), 503-507.
270. F. Gesztesy and B. Simon, *A new approach to inverse spectral theory, II. general real potentials and the connection to the spectral measure*, Ann. Math. 152 (2000), 593-643.
271. F. Gesztesy, B. Simon, and G. Teschl, *Zeros of the Wronskian and renormalized oscillation theory*, Amer. J. Math., 118 (1996), 571-594.
272. F. Gesztesy and M. Unal, *Perturbative oscillation criteria and Hardy-type inequalities*, Math. Nach. 189 (1998), 121-144.
273. F. Gesztesy and R. Weikard, *Elliptic algebro-geometric solutions of the Kdv and AKNS hierarchies-an analytic approach*, Bull. Amer. Math. Soc. 35 (1998), 271-317.
274. M. Giertz, M. K. Kwong, and A. Zettl, *Commuting linear differential expressions*, Proc. Royal Soc. Edinburgh, 87A(1981), 331-347.
275. D. J. Gilbert and B. J. Harris, *Connection formulae for spectral functions associated with singular Sturm-Liouville equations*, Proc. Roy. Soc. Edinburgh Sect. A 130 (2000) (1), 25-34.
276. D. J. Gilbert and D. B. Pearson, *On subordinacy and analysis of the spectrum of one-dimensional Schroedinger operators*, J. Math. Anal. Appl. 128 (1987), 30-56.
277. I. M. Glazman, *Direct Methods for the qualitative spectral analysis of singular differential operators*, Eng. transl., Israel Program for Scientific Translations, Jerusalem, 1965.
278. I. C. Gohberg and M. G. Krein, *Theory and Application of Volterra Operators in Hilbert Spaces*, Trans. Math. Monographs 24 AMS, Providence, RI, 1970.
279. S. Goldberg, *Unbounded linear operators*, McGraw-Hill, New York /London, 1966.
280. J. Goldstein, M. K. Kwong, and A. Zettl, *Weighted Landau inequalities*, J. Math. Anal. and Appl., 95 (1983), 20-28.
281. L. Greenberg, *A Pruefer method for calculating eigenvalues of self-adjoint systems of ordinary differential equations*, Parts 1 and 2, University of Maryland Technical Report TR91-24.
282. L. Greenberg and M. Marletta, *The code SLEUTH for solving fourth order Sturm-Liouville problems*, ACM Trans. Math. Software 23 (1997), no. 4., 453-493.
283. J. Gunson, *Perturbation theory for a Sturm-Liouville problem with an interior singularity*, Proc. Royal Soc. London Ser. A. 414 (1987), 255-269.
284. I. M. Guseĭnov, A. A. Nabiev, and R. T. Pashaev, *Transformation operators and asymptotic formulas for the eigenvalues of a polynomial pencil of Sturm-Liouville operators*, Sibirsk. Mat. Zh. 41 (3) (2000), 554-566, ii.
285. K. Haertzen, Q. Kong, H. Wu, and A. Zettl, *Geometric aspects of Sturm-Liouville problems, I. Structures on spaces of boundary conditions*, Proc. Roy Soc. Edinburgh 130A, 561-589, (2000), 561-589.
286. ———, *Geometric aspects of Sturm-Liouville problems, II. Spaces of boundary conditions for left-definiteness*, Trans. Amer.Math. Soc. Vol. 356, 1, (2004),no. 1, 135-157.
287. S. G. Halvorsen, *A function-theoretic property of solution of the equation $x'' + (lw - q)x = 0$*, Quart. J. Math. Oxford (2), 38 (1987), 73-76.
288. ———, *On the quadratic integrability of solutions of $x'' + fx = 0$*, Math. Scand. 14 (1964), 111-119.
289. S. G. Halvorsen and A. B. Mingarelli, *Non-Oscillation Domains of Differential Equations with Two Parameters*, Lecture Notes in Mathematics no 1338, Springer-Verlag, Berlin, New York, 1988.
290. R. J. Hangelbroek, *A functional-analytic approach to the linear transport equation*, Transport Theory Statist. Phys. 5 (1976), 1-85.
291. B. J. Harris, *The asymptotic form of the spectral functions associated with a class of Sturm-Liouville equations*, Proc. Royal Soc. Edinburgh, 100A, (1985), 343-360.

292. _____, *The asymptotic form of the Titchmarsh-Weyl m-function*, J. London Math. Soc. 2, 30 (1984), 110-118.
293. _____, *The asymptotic form of the Titchmarsh-Weyl m-function associated with a Dirac system*, J. London Math. Soc.(2), 31 (1985), 321-330.
294. _____, *The asymptotic form of the Titchmarsh-Weyl m-function associated with a second-order differential equation with locally integrable coefficient*, Proc. Royal Soc. Edinburgh, 102A, (1986), 243-251.
295. _____, *The asymptotic form of the Titchmarsh-Weyl m-function for second-order linear differential equations with analytic coefficient*, J. Diff Equations, 65, 2, (1986), 219-234.
296. _____, *Bounds for the point spectra of Sturm-Liouville equations*, J. London Math. Soc. 25 (1), (1982), 145-162.
297. _____, *Criteria for Sturm-Liouville equations to have continuous spectra*, J. Math. Anal. Appl., (93), (1983), 235-249.
298. _____, *Eigenvalue problems for second-order differential equations*, Proc. 1986-87 Focused Research Program on Spectral Theory and Boundary Value Problems, ANL 87-26.
299. _____, *An exact method for the calculation of certain Tischmarsh-Weyl m-functions*, Proc. Royal Soc. Edinburgh 106A (1987), 137-142.
300. _____, *The form of the spectral functions associated with a class of Sturm-Liouville equations with integrable coefficient*, Proc. Royal Soc. Edinburgh, 105A (1987), 215-227.
301. _____, *An inverse problem involving the Titchmarsh-Weyl m-functions*, Proc. Royal Soc. Edinburgh 110A (1988), 63-71.
302. _____, *Limit circle criteria for Sturm-Liouville equations*, Quart. J. Math. (2), 35, (1984), 415-427.
303. _____, *Lower bounds for the spectrum of second-order linear differential equation with a coefficient whose negative part is p-integrable*, Proc. Royal Soc Edinburgh, 97A,(1984), 105-107.
304. _____, *A note on a paper of Atkinson concerning the asymptotics of an eigenvalue problem with interior singularity*, Proc. Royal Soc. Edinburgh 110A (1988), 63-71.
305. _____, *On the asymptotic properties of linear differential equations*, Mathematika 34 (1987), 187-198.
306. _____, *On the oscillation of solutions of linear differential equations*, Mathematika 21, (1984), 214-226.
307. _____, *On the spectra and stability of periodic differential equations*, Proc. London Math. Soc. (3)40(1980), 161-192.
308. _____, *On the spectra of self-adjoint operators*, Proc. Royal Soc. Edinburgh 86A,(1980), 261-274.
309. _____, *On the Titchmarsh-Weyl m-function*, Proc. Royal Soc. Edinburgh, 95A, (1983), 223-237.
310. _____, *A series of certain Ricatti equations with applications to Sturm-Liouville problems*, Journal of Math Analysis and Applications, 137, 2 (1989), 462-470.
311. _____, *Some spectral gap results*, Proceedings of 1980 Dundee Differential Equations Conference, Springer-verlag 846.
312. _____, *A systematic method of estimating gaps in the essential spectrum of self-adjoint operators*, J. London Math. Soc.(2) 18, (1978), 115-132.
313. P. Hartman, *Comparison theorems for selfadjoint second order systems and uniqueness of eigenvalues of scalar boundary value problems, pp. 1-22*, Contributions to analysis and geometry, Johns Hopkins Univ. Press, Baltimore, 1980.
314. _____, *Differential equations with non-oscillatory eigenfunctions*, Duke Math. J., 15 (1948), 697-709.
315. _____, *On an ordinary differential equation involving a convex function*, Amer. Math. Soc. 146 (1969),179-202.
316. _____, *On linear second order differential equations with small coefficients*, Amer. J. Math. 73 (1951), 955-962.
317. _____, *On nonoscillatory linear differential equations of second order*, Amer. J. Math. 74 (1952), 389-400.
318. _____, *On the number of L2-solutions of $x'' + q(t)x = lx$*, Amer. J. Math. 73 (1951), 635-645.

319. _____, *Ordinary differential equations*, Wiley and Sons, Inc., New York /London/Sydney, 1964.
320. P. Hartman and A. Wintner, *A separation theorem for continuous spectra*, Amer. J. Math. 71 (1949), 650-662.
321. O. Haupt, *Ueber eine Methode zum Beweis von Oscillationstheoreme*, Mat. Ann. 76 (1915), 67-104.
322. _____, *Untersuchungen uber Oszillationtheoreme*, Teubner, Leipzig, 1911.
323. S.W. Hawking and R. Penrose, *The singularities of gravity collapse and cosmology*, Proc. Roy. Soc. London Ser. A, 314 (1970), 529-548-.
324. D. Hilbert, *Grundzuge einer allgemeinen Theorie der linearen Integralgleichungen,pp. 195-204*, Leipzig and Berlin, 1912.
325. E. Hille, *Nonoscillation theorems*, Trans. Amer. Math. Soc. 64 (1948), 234-252.
326. _____, *Lectures on ordinary differential equations*, Addison-Wesley, London, 1969.
327. D. B. Hinton, *An expansion theorem for an eigenvalue problem with eigenvalue parameter in the boundary condition*, Quart. J. Math. Oxford (2), 41 (1990), 189-224.
328. _____, *Limit point-limit circle criteria for ....*, Lecture Notes in Mathematics 415 (1974), 173-183. (Springer Verlag; Heidelberg, 1974;edited by I.M. Michael and B.D. Sleeman).
329. _____, *On the location of the least point of the essential spectrum*, J. Comput. Appl. Math. 148 (2002), no. 1, 77-89.
330. _____, *Sturm's 1836 Oscillation Results: Evolution of the Theory*, preprint.
331. D. B. Hinton, M. Klaus, and J.K. Shaw, *Levinson's theorem and Titchmarsh-Weyl theory of Dirac systems*, Proc. Royal Soc. Edinburgh. 109A (1988), 173-186.
332. _____, *On the Titchmarsh-Weyl function for the perturbed periodic Hill's equation*, Quart. J. Math. Oxford (2), (1979), 33.
333. D. B. Hinton and R.T. Lewis, *Singular differential operators with spectra discrete and bounded below.*, Proc. Roy. Soc. Edinburgh A 84 (1979), 117-134.
334. D. B. Hinton and P. W. Schaefer, *Lecture notes in Pure and Applied Mathematics, v. 191*, Marcel Dekker, New York, 1997.
335. D. B. Hinton and J. K. Shaw, *Absolutely continuous spectra of perturbed periodic Hamiltonian systems*, Rocky Mountain J. Math. 17, no.4, (1987), 727-748.
336. _____, *Absolutely continuous spectra of second-order differential operators with long and short range potentials*, SIAM J. Math. Anal. 17 (1986), 182-196.
337. _____, *Differential operators with spectral parameter incompletely in the boundary conditions*, Funkcial. Ekvac. 33 (1990), 363.
338. _____, *On the absolutely continuous spectrum of the perturbed Hill's equations*, Proc. London Math. Soc. (3) 50 (1985), 175-192.
339. _____, *On the spectrum of a singular Hamiltonian system*, Quaes. Math. 5 (1982), 29-81.
340. _____, *Spectrum of a Hamiltonian system with spectral parameter in a boundary condition*, Canad. Math. Soc. Proc. 8 (1987), 171.
341. H. Hochstadt, *A special Hill's equation with discontinuous coefficients*, Amer. Math. Monthly, 70 (1963), 18-26.
342. M. S. Homer, *Boundary value problems for the Laplace tidal wave equation*, Proc. Royal Soc. London Ser. A 428 (1990), 157-180.
343. A. Hurwitz, *Ueber die Nullstellen der Bessel'schen Funktion*, Math. Annalen, 33 (1889), 246-266.
344. E. Ince, *Ordinary Differential Equations*, Dover Publications, New York, 1956.
345. I. S. Iohvidov, M. G. Krein, and H. Langer, *Introduction to the spectral theory of operators in spaces with an indefinite metric. Mathematische Forschung, vol. 9*, Berlin, Akademie, 1982.
346. R. S. Ismagilov, *On the self-adjointness of the Sturm-Liouville operator*, Uspehi. Mat. Nauk. 18 (1963), 161-166.
347. C. Jacobi, *Zur Theorie der Variationsrechnung und die Differentialgleichungen*, J. Fur die reine und angewantdte Mathematik, 17 (1837), 68-82.
348. J.D.Pryce, *A test package for Sturm-Liouville problems*, ACM Trans. Math. Software 25 (1999), 21-57.
349. K. Joergens, *Spectral theory of second order ordinary differential operators*, Lectures delivered at Aarhus Universitet 1962/63, Matimatisk Institut Aarhus 1964.

350. K. Jorgens and F. Rellich, *Eigenwerttheorie gewoehnlicher Differentialgleichungen*, Springer-Verlag, Heidelberg, 1976.
351. I. S. Kac, *Power-Asymptotic estimates for spectral functions of generalized boundary value problems of second- order*, Sov. Math. Dokl. 13 (2) (1972), 453-457.
352. H. Kalf, *A characterization of the Friedrichs extension of Sturm-Liouville operators*, J. London Math. Soc. (2) 17 (1978), 511-521.
353. _____, *Remarks on some Dirichlet-type results for semi-bounded Sturm-Liouville operators*, Math. Ann. 210 (1974), 197-205.
354. I. Kamenev, *Integral criterion for oscillations of linear differential equations of second order*, Mat.Zametki, 23 (1978), 249-251.
355. E. Kamke, *A new proof of Sturm's comparison theorem*, Amer. Math. Monthly, 46 (1939), 417-421.
356. _____, *Ueber die definiten selbstadjungierten Eigenwertaufgaben bei gewoehnlichen linearen Differentialgleichungen, I. und II.,*, Math. Z. 46 (1940), 231-250 and 251-286. (In German).
357. _____, *Zum Entwicklungssatz bei polaren Eigenwertaufgaben*, Math. Z. 45 (1939), 706-718.
358. _____, *Differentialgleichungen, Loesungsmethoden und Loesungen: Gewoehliche Differentialgleichungen, 3rd edition*, Chelsea Publishing Company, New York, 1948.
359. _____, *Differentialgleichungen, Loesungsmethoden und Loesungen*, Chelsea, New York, 1971.
360. H. G. Kaper, M. K.Kwong, and A. Zettl, *Characterization of the Friedrichs extensions of singular Sturm-Liouville expressions*, SIAM J. Math. Anal., 7(1986), 772-777.
361. H. G. Kaper and M. K. Kwong, *Asymptotics for the Titchmarsh-Weyl m-coefficient for integrable potentials*, Proc. Royal Soc. Edinburgh 103A (1986), 347-358.
362. _____, *Asymptotics for the Titchmarsh-Weyl m-coefficient for integrable potentials II*, Springer L.N.M. vol. 1285, Proc. Conf. Diff. Eq. and Math. Phys., Birmingham, Alabama, (1986), 222-229.
363. H. G. Kaper, M. K. Kwong, C. G. Lekkerkerker, and A. Zettl, *Full and partial-range eigenfunction expansions for Sturm-Liouville problems with indefinite weights*, Proc. Royal Soc. Edinburgh, 98A(1984), 69-88.
364. H. G. Kaper, M. K. Kwong, and A. Zettl, *Regularizing transformations for certain singular Sturm-Liouville boundary value problems*, SIAM J. Math. Anal., V. 15(1984), 957-963.
365. _____, *Singular Sturm-L problems with nonnegative and indefinite weights*, Monatsheft fuer Mathematik, 97(1984), 177-189.
366. _____, *Proceedings of the focused research program on spectral theory and boundary value problems*, Argonne Reports ANL-87-26, vol 1, vol 2, vol 3, and vol 4. (Edited by Kaper, Kwong and Zettl), 1988 and 1989.
367. T. Kato, *Perturbation theory for linear operators, 2nd edn.*, Springer, Heidelberg, 1980.
368. R. M. Kauffman, T. T. Read, and A. Zettl, *The deficiency index problem for powers of ordinary differential expressions*, Lecture Notes in Mathematics 621, Springer-Verlag, 1-127, 1977.
369. I. Kay and H.E. Moses, *Reflectionless transmission through dielectrics and scattering potentials*, J. Appl. Phys. 27 (1956),1503-1508.
370. M. V. Keldysh, *On the characteristic values and characteristic functions of certain classes of non-selfadjoint equations*, Dokl. Akad. Nauk SSr (in Russian) 77 (1951), 11.
371. H.B. Keller, *Numerical methods for two-point boundary value problems*, Ginn-Blaisdell, Waltham, Mass., 1968.
372. V. I. Khrabustovski, *The discrete spectrum of perturbed differential operators of arbitrary order with periodic matrix coefficients*, Math. Notes 21 (1977), 467-472.
373. M. Klaus, *On the variation- diminishing property of Schroedinger operators*, Proc. Canadian Math. Soc.: 1986 Seminar on Oscillation, Bifurcation and Chaos, American Math. Soc., Providence, 1987.
374. M. Klaus and J. Shaw, *On the eigenvalues of Zakharov-Shabot systems*, SIAM J. Math. Analysis, 34 (2003), 759-773.
375. A. Kneser, *Untersungen ueber die reelen Nullstellen der Integral linearer Differentialgleichungen*, Math. Ann. 42 (1893), 409-435.
376. I. Knowles, *A limit-point criterion for a second-order linear differential operator*, J. London Math. Soc. 8 (1974), 719-727.
377. _____, *Note on a limit-point criterion*, Proc. Amer. Math. Soc. 41 (1973), 117-119.

378. _____, *On second-order differential operators of limit-circle type*, Lecture Notes in Mathematics, no. 415, Springer-Verlag, (1974), 184-187.
379. L. Kong and Q. Kong, *Right-definite Half-linear Sturm-Liouville problems*, pre-print.
380. L. Kong, Q. Kong, H. Wu, and A. Zettl, *Regular Approximations of Singular Sturm-Liouville problems with limit-circle endpoints*, Results Math. 45 (2004), no. 1-3, 274-292.
381. Q. Kong, Q. Lin, H. Wu, and A. Zettl, *A new proof of the inequalities among Sturm-Liouville eigenvalues*, PanAmerican Math. J., 10, (2000), 1-11.
382. Q. Kong, M. Moeller, H. Wu, and A. Zettl, *Indefinite Sturm-Liouville problems*, Proc. Roy. Soc. Edinburgh, Sect. A, 133, (2003), 639-652.
383. Q. Kong, H. Wu, and A. Zettl, *Dependence of eigenvalues on the problem*, Mathematische Nachrichten, 188, (1997), 173-201.
384. _____, *Dependence of n-th Sturm-Liouville eigenvalue on the problem*, J. Differential Equations, 156, (1999), 328-354.
385. _____, *Eigenvalue of Sturm-Liouville Problems when the domain shrinks to a point*, pre-print.
386. _____, *Inequalities among eigenvalues of singular Sturm-Liouville problems*, Dynamical Systems and Applications, 8, (1999), 517-531.
387. _____, *Left-definite Sturm-Liouville problems*, J.Differential Equations, I77, (2001), 1-26.
388. _____, *Multiplicity of Sturm-Liouville eigenvalues*, J. of Computational and Applied Mathematics 171 (2004),no. 1-2, 291-309.
389. _____, *Singular Left-Definite Sturm-Liouville Problems*, J. Differential Equations,(2005), 1-29.
390. _____, *Sturm-Liouville problems whose leading Coefficient Function Changes sign*, Canadian J. Math. v. 55 (4), 2003, 724-749.
391. _____, *Sturm-Liouville problems with Finite Spectrum*, J. Math. Anal. and Appl. (2001), 1-15.
392. Q. Kong and A. Zettl, *Dependence of eigenvalues of Sturm-Liouville problems on the boundary*, J. Differential Equations, vol.126,no.2, (1996), 389-407.
393. _____, *The derivative of the matrix exponential function*, Cubo Matematica Educacional, V. 3., (2001), 121-124.
394. _____, *Eigenvalues of regular Sturm-Liouville problems*, J. Differential Equations, vol. 131, no.1, (1996), 1-19.
395. _____, *Interval oscillation conditions for difference equations*, SIAM J. Math. Analysis, 26, no.4 (1995), 1047-1060.
396. _____, *Linear ordinary differential equations*, WSSIAA v.3, (1994), (Special volume dedicated to W.Walter), Inequalities and Applications, edited by R. P. Agarwal, 381-397.
397. T.H. Koornwinder, *Jacobi functions and analysis on noncompact semisimple Lie groups, special functions: group theoretical aspects and applications, 1-85*, D. Reidel Publishing Co., Dordrecht, 1984; Edited by R.A. Askey, T.H. Koornwinder and W. Schempp.
398. S. Kotani, *Lyapunov exponents and spectra for one-dimensional random Schroedinger-operators*, Comtemp. Math. 50 (1986), 277-286.
399. A. M. Krall, *Boundary value problems for an eigenvalue problem with a singular potential*, J. Diff. Equations, 45 (1982), 128-138.
400. A. M. Krall and A. Zettl, *Singular self-adjoint Sturm-Liouville problems*, Differential and Integral Equations v. 1 (1988), 423-432.
401. _____, *Singular Sturm-Liouville problems II., interior singular points*, SIAM J. Math. Anal., (1988), 1135-1141.
402. W. Kratz, *Quadratic Functionals in Variational Analysis and Control Theory*, Akademic Verlag, Berlin, 1995.
403. M. G. Krein, *The theory of self-adjoint extensions of semi-bounded Hermitian transformations and its applications i*, Mat. Sb. 20 (1947), 431-495.
404. _____, *The theory of self-adjoint extensions of semi-bounded Hermitian transformations and its applications ii.*, Mat. Sb. 21 (1947), 365-404. (In Russian.).
405. K. Kreith, *PDE generalization of the Sturm comparison theorem*, Memoirs of the American Mathematical Society, 48 (1984), 31-45.
406. _____, *Oscillation Theory*, Lecture Notes in Mathematics, 324, Springer-Verlag, Berlin, 1973.

407. N. P. Kupcov, *Conditions of non-self adjointness of a second-order linear differential operator*, Dokl. Akad. Nauk. SSSR 138 (1961), 767-770.
408. H. Kurss, *A limit-point criterion for non-oscillatory Sturm-Liouville differential operators*, Proc. Amer. Math. Soc. 18 (1967), 445-449.
409. M. K. Kwong, *Lp-perturbations of second-order linear differential equations*, Math. Am. 215 (1975), 23-34.
410. _____, *Note on a strong limit-point condition of the second-order differential expressions*, Oxford Ser. (2) 28 (1977), no. 110, 201-208.
411. _____, *On boundedness of solutions of second-order differential equations in the limit-circle case*, Proc. Amer. Math. Soc. 52 (1975), 242-246.
412. M. K. Kwong and A. Zettl, *An alternate proof of Kato's inequality*, Evolution Equations: Proceedings in honor of J.S. Goldstein's 60th Birthday, Marcel Dekker, New York.
413. _____, *Asymptotically constant functions and second-order linear oscillation*, J. of Math. Anal. and Applications, 93 (1983), 475-494.
414. _____, *Best constants for discrete Kolmogorov inequalities*, Houston J. Math., (1989), 99-119.
415. _____, *Discreteness conditions for the spectrum of ordinary differential operators*, J. Diff. Equations v. 40 n. 1 (1981), 53-70.
416. _____, *An extension of the Hardy-Littlewood inequality*, Proc. Amer. Math. Soc., 77(1979), 117-118.
417. _____, *Extremals in Landau's inequality for the difference operator*, Proc. Royal Soc. Edinburgh, v. 107(1987), 299-311.
418. _____, *Integral inequalities, and second order linear oscillation*, J. Diff. Equations, 45(1982), 16-33.
419. _____, *Landau's inequality*, Proc. Royal Soc. Edinburgh, 97a(1984), 161-163.
420. _____, *Landau's inequality for the difference operator*, Proc. Amer. Math. Soc., v. 104 (1988), 201-206.
421. _____, *Norm inequalities for dissipative operators on inner product spaces*, Houston J. Math. v.5(1979), 543-557.
422. _____, *Norm inequalities for the powers of a matrix*, American Math. Monthly, v.98,no.6(1991), 533-538.
423. _____, *Norm inequalities of product form is weighted Lp spaces*, Proc. Royal Soc. Edinburgh, 89A(1981), 293-307.
424. _____, *On the limit point classification of second order differential operators*, Mathematische Zeitschrift, (1973), 297-304.
425. _____, *Ramifications of Landau's inequality*, Proc. Royal Soc. Edinburgh, 86A(1980), 175-212.
426. _____, *Remarks on best constants for norm inequalities among powers of an operator*, J. Approx. Theory, 26(1979), 248-258.
427. _____, *Weighted norm inequalities of sum form involving derivatives*, Proc. Royal Soc. Edinburgh, A88 (1981), 121-134.
428. _____, *Norm inequalities for derivatives and differences*, Lecture Notes in Mathematics 1536, Springer-Verlag, 1-150, 1993.
429. V. Lakshmikantham and S. Leela, *Differential and Integral Inequalities; Theory and Applications*, Academic Press, New York, 1969.
430. P. Lancaster, A. Shkalikov, and Q. Ye, *Strongly definitizable linear pencils in Hilbert space*, Int. Equ. Oper. Theory 17 (1993), no. 3, 338-360.
431. H. Langer and A. Schneider, *On spectral properties of regular quasidefinite pencils $F - \lambda G$*, Res. Math. 19 (1991), 89-109.
432. W. Lay and S. Yu Slavyanov, *Heun's equation with nearby singularities*, Proc. Roy. Soc. London A 455 (1999), 4347-4361.
433. Andrien-Marie Legendre, *Suite des Recherches sur la Figure des Planetes*, Histoire de l'Academie Royale des Sciences avec les Memoires de Mathematique et de Physique',Paris 1789 (1793), pp. 372-454.
434. W. Leighton, *Comparison theorems for linear differential equations of second order*, Proc. Amer. Math. Soc., 13 (1962), 603-610.
435. _____, *On self-adjoint differential equations of second order*, J. London Math. Soc. 27 (1952), 37-47.

436. _____, *Principal quadratic functionals*, Trans. Amer. Math. Soc. 67 (1949), 253-274.
437. W. Leighton and Z. Nehari, *On the oscillation of solutions of self-adjoint linear differential equations of the fourth order*, Trans. Amer. Math. Soc., 89 (1958), 325-377.
438. A. Ju. Levin, *Classification of nonoscillatory cases for the equation .... where q(t) is of constant sign*, Dokl. Akad. Nauk SSSR 171 (1966), 1037-1040= Soviet Math. Dokl. 7 (1966), 1599-1602.
439. _____, *A comparison principle for second-order differential equations*, Soviet Math. Dokl. 1 (1960), 1313-1316.
440. _____, *Distribution of the zeros of a linear differential equation*, Soviet Math. Dokl., 5 (1964), 818-822.
441. _____, *Some properties bearing on the oscillation of linear differential equations*, Soviet Math. Dokl., 4 (1963), 121-124.
442. N. Levinson, *Criteria for the limit-point case for the second-order linear differential operators*, Casopis Pest. Mat. Fys. 74 (1949), 17-20.
443. _____, *On the uniqueness of the potential in a Schroedinger equation for a given asymptotic phase*, Mat.-Fys. Medd. Danske Vid. Selsk. 25 (1949), 1-29.
444. B. M. Levitan and I. Sargsjan, *Introduction to Spectral Theory: Selfadjoint Ordinary Differential Operators*, American Math. Soc. Translations of Mathematical Monographs 39, Providence, 1975.
445. R. Lewis, *The discreteness of the spectrum of self-adjoint, even order, one-term, differential operators*, Proc. Amer. Math. Soc., 42 (1974), 480-482.
446. A. Liapunov, *Probleme General de la Stabilite du Mouvement*, (French translation of a Russian paper dated 1893),Ann. Fac. Sci. Univ. Toulouse, 2 (1907), 27-247; reprinted as Ann. Math. Studies, No. 17, Princeton, 1949.
447. J. Liouville and J. C. F. Sturm, *Estrait d'un memoire sur le developpement des functions en serie*, Jour. Math. Pures et Appl. de Liouville II (1837), 220-223.
448. L. Littlejohn and A. Zettl, *Variants of the Fourier Expansion*, preprint.
449. L.L. Littlejohn and R. Wellman, *A general left-definite theory for certain self-adjoint operators with applications to differential equations*, J. Differential Equations 181 (2002), no. 2, 280-339.
450. J. E. Littlewood, *On linear differential equations of the second order with a strongly oscillating coefficient of y*, J. London Math. Soc. 41, (1966), 627-638.
451. John Locker, *Functional analysis and two-point differential operators*, Pitman Research Notes in Mathematics, vol. 144, Longmans, Harlow, Essex, 1986.
452. _____, *Spectral Theory of Non-Self-Adjoint Two-Point Differential Operators*, American Mathematical Society, Mathematical Surveys and Monographs, vol.73, Providence, Rhode Island, 2000.
453. R. J. Lohner, *Einschliesung der Gewoehnlicher Anfangs und Randwertaufgaben und Anwendungen*, PhD thesis University of Karlsruhe, 1988.
454. _____, *Verified solution of eigenvalue problems in ordinary differential equations*, private communication, 1995.
455. J. Lutzen, *Sturm and Liouville's work on ordinary linear differential equations. The emergence of Sturm-Liouville theory*, Arch. Hist. Exact Sci. 29 (1984), 309-376.
456. _____, *Joseph Liouville 1809-1882: Master of Pure and Applied Mathematics*, Springer-Verlag, Berlin, 1990.
457. J. W. Macki and J. S. W. Wong, *Oscillation theorems for linear second order differential equations*, Proc. Amer. Math. Soc. 20 (1969), 67-72.
458. V. A. Marchenko, *Sturm-Liouville operators and applications*, Birkhauser, Basel, 1986.
459. L. Marcus and R. A. Moore, *Oscillation theorems for linear second order differential with almost periodic coefficients*, Acta. Math. 96 (1956), 99-123.
460. P. A. Markovich, *Eigenvalue problems on infinite intervals*, Mathematics of Computation 39, 421-444 (1982).
461. M. Marletta, *Certification of algorithm 700; Numerical tests of the SLEIGN software for Sturm-Liouville problems*, ACM TOMS, v17 (1991), 501-503.
462. _____, *Numerical solution of eigenvalue problems for Hamiltonian systems*, Advances in Computational Mathematics 2, (1994), 155-184.
463. M. Marletta and A. Zettl, *Counting and computing Eigenvalues of Left-Definite Sturm-Liouville Problems*, J. Computational and Applied Mathematics, 148 (2002), 65-75.

464. _____, *Floquet theory for Left-Definite Sturm-Liouville Problems*, J. Math. Analysis Appl. 305, no. 2, 15 May 2005, 477-482.
465. _____, *The Friedrichs extension of singular differential operators*, J. Differential Equations 160, (2000), 404-421.
466. _____, *Spectral exactness and spectral inclusion for singular left-definite Sturm-Liouville problems*, Results in Mathematics, 45, (2004), 299-308.
467. J. B. McLeod, *On the spectrum of wildly oscillating functions*, J. London Math. Soc. 39 (1964), 623-634.
468. _____, *On the spectrum of wildly oscillating functions (ii)*, J. London Math. Soc. 40 (1965), 655-661.
469. _____, *Some examples of wildly oscillating potentials*, J. London Math. Soc. 43 (1968), 647-654.
470. F. Meng and A. Mingarelli, *Oscillation of linear Hamiltonian systems*, Proc. Amer. Math. Soc., 131 (2003), 897-904.
471. R. Mennicken and M. Moeller, *Nonselfadjoint boundary eigenvalue problems*, Elsevier Publishing Co., North Holland, 2003.
472. A. B. Mingarelli, *A survey of the regular weighted Sturm-Liouville Problem-the non-definite case*, World Scientific Publishing, Singapore, (1986), 109-137.
473. M. Moeller, *On the essential spectrum of a class of operators in Hilbert space*, Math. Nachr. 194 (1998), 185-196.
474. _____, *On the unboundedness below of the Sturm-Liouville operator*, Proc. Royal Soc. Edinburgh 129A (1999), 1011-1015.
475. M. Moeller and A. Zettl, *Differentiable dependence of eigenvalues of operators in Banach spaces*, J. Operator Theory, 36(1996), 335-355.
476. _____, *Semi-boundedness of ordinary differential operators*, J. Differential Equations, v.115, no.1, (1995), 24-49.
477. _____, *Symmetric differential operators and their Friedrichs extension*, J. Differential Equations, v.115, no.1, (1995), 50-69.
478. _____, *Weighted norm inequalities for quasi-derivatives*, Results in Mathematics 24 (1993), 153-160.
479. R. A. Moore, *The behavior of solutions of a linear differential equation of second order*, Pacific J. Math. 5 (1955), 125-145.
480. _____, *The least eigenvalue of Hill's equation*, J. Analyse Math. 5 (1956/57), 183-196.
481. M. Morse, *A generalization of Sturm separation and comparison theorems*, Math. Annalen, 103 (1930), 52-69.
482. _____, *Variational Analysis: Critical Extremals and Sturmian Extensions*, Wiley-Interscience, New York, 1973.
483. P. M. Morse, *Diatomic molecules according to the wave mechanics; II: Vibration levels.*, Phys. Rev. 34 (1920), 57-61.
484. E. Mueller-Pfeiffer, *Spectral Theory of Ordinary Differential Equations*, Ellis Horwood Limited, Chichester, 1981.
485. I. M. Nabiev, *Multiplicity and relative position of the eigenvalues of a quadratic pencil of Sturm-Liouville operators*, Mat. Zametki 67 (3) (2000), 369-381.
486. M. A. Naimark, *Lineare Differentialoperatoren. Mathematische Lehrbuecher und Monographien*, Akadmie-Verlag, Berlin, 1960.
487. _____, *Linear Differential Operators: II*, Ungar Publishing Company, New York, 1968.
488. H. Narnhofer, *Quantum theory for 1/r2-potentials'*, Acta Phys. Austriaca 40 (1974), 306-322.
489. Z. Nehari, *On the zeros of solutions of second order linear differential equations*, Amer. J. Math., 76 (1954), 689-697.
490. _____, *Oscillation criteria for second order linear differential equations*, Trans. Amer. Math. Soc. 85 (1957), 428-445.
491. J. W. Neuberger, *Concerning boundary value problems*, Pacific J. Math. 10 (1960), 1385-1392.
492. F. Neuman, *On a problem of transformations between limit-point and limit-circle differential equations*, Proc. Royal Soc. Edinburgh 72 (1973/74), 187-193.
493. _____, *On the Liouville transformation*, Rendiconti di Matematica 3 (1970), 132-139.

494. R. Newton, *Inverse scattering by a local impurity a periodic potential in one dimension II*, J. Math. Phys. 26 (1885), 311-316.
495. _____, *Inverse scattering by a local impurity in a periodic potential in one dimension*, J. Math. Phys. 24 (1983), 2152-2162.
496. _____, *The Marchenko and Gelfand-Levitan methods in the inverse scattering problem in one and three dimensions*, In Conference on Inverse Scattering: Theory and Application, Society for Industrial and Applied Mathematics, Philadelphia, 1983.
497. H.-D. Niessen, *A Necessary and Sufficient Limit-Circle Criterion for Left-Definite Eigenvalue Problems*, Lecture Notes in Mathematics 415 (1974), pp. 205-210.
498. _____, *Zum verallgemeinerten zweiten Weylschen Satz*, Archiv der Mathematik, Vol.2 (1971), 648-656.
499. H.-D. Niessen and A. Schneider, *Linksdefinite singulaere kanische Eigenwertprobleme II*, Journal fur die reine und angewandte Mathematik, Vol. 289 (1977), 62-84.
500. _____, *Linksdefinite singulaere kanonische Eigenwertproblems I*, Journal fur die reine und angewandte Mathematik, Vol. 281 (1976), 13-52.
501. _____, *Spectral Theory for Left-Definite Singular Systems of Differential Equations I and II*, Mathematical Studies 13 (1974), 29-56 (= North-Holland Publishing Company).
502. H.-D. Niessen and A. Zettl, *The Friedrichs extension of regular ordinary differential operators*, Proc. Roy. Soc. Edinburgh, 114A, (1990), 229-236.
503. _____, *Singular Sturm-Liouville problems: The Friedrichs extension and comparison of eigenvalues*, Proc. London Math. Soc. v.64, (1992), 545-578.
504. C. Olech, Z. Opial, and T. Wazewski, *Sur le probleme d'oscillation des integrales de l'equation..............=0*, Bull. Acad. Polon. Sci. Cl. III 5 (1957), 621-626.
505. K. S. Ong, *On the limit-point and limit-circle theory of a second-order differential equation*, Proc. Royal Soc. Edinburgh 72 (19975), 245-256.
506. Z. Opial, *Sur eine critere d'oscillation des integrales de l'equation differentielle...........*, Ann. Polon. Math. 4 (1959-1960), 99-104.
507. _____, *Sur les integrales oscillantes de l'equation differentielle ....*, Ann. Polon. Math. 4 (1958), 308-313.
508. M. Otelbaev, *On summability with a weight of a solution of the Sturm-Liouville equation*, Mat. Zametki 16 (1974), 969-980.
509. W. T. Patula and P. Waltman, *Limit-point classification of second-order linear differential equations*, J. London Math. Soc. 8 (1974), 209-216.
510. W. T. Patula and J. S. W. Wong, *An Lp-analogue of the Weyl alternative*, Math. Ann. 197 (1972), 9-28.
511. D. B. Pearson, *Value distribution and spectral analysis of differential operators*, J. Physics A26, (1993), 4067-4080.
512. _____, *Value distribution and spectral theory*, Proc. London Math. Soc.(3) 68, (1994), 127-144.
513. _____, *Quantum scattering and spectral theory*, Academic Press, London, 1988.
514. W. Peng, M. Racovitan, and H. Wu, *Geometric aspects of Sturm-Liouville Problems V. Natural Loops of Boundary conditions for Monotonicity of Eigenvalues and their Applications*, pre-print.
515. I. G. Petrovski, *Ordinary differential equations*, Prentice Hall, Inc., London, 1966.
516. M. Picone, *Sui valori eccezionali di un parametroducui dipende un equazioni differenziale lineare ordinaria del second ordine*, Ann. Scuola Norm. Pisa, 11 (1909), 1-141.
517. A. Pleijel, *Complementary remarks about the limit-point and limit-circle theory*, ibid. 8 (1971), 45-47.
518. _____, *Generalized Weyl circles*, Lecture Notes in Mathematics 415 (1974), 211-226.(Springer Verlag; Heidelberg, 1974; edited by I.M. Michael and B. D. Sleeman).
519. _____, *Some remarks about the limit-circle theory*, Arkiv. Mat. 7 (1969), 543-550.
520. _____, *A Survey of Spectral Theory for Pairs of Ordinary Differential Operators*, Lecture Notes in Mathematics 448 (1975), 256-270.
521. M. Plum, *Eigenvalue inclusions for second-order ordinary differential operators by a homotopy method*, J. Applied Mathematics and Physics (ZAMP), 41, (1990), 205-226.
522. J. Poeschel and E. Trubowitz, *Inverse Spectral Theory*, Academic Press, New York /London/Sydney, 1987.

523. E.G.P. Poole, *Introduction to the theory of linear differential equations*, Oxford University Press, Oxford, 1936.
524. H. Pruefer, *Neue Herleitung der Sturm-Liouvilleschen Reihenentwicklung stetiger Funktionen*, Math. Ann., 95 (1926), 499-518.
525. S. Pruess, *Estimating the eigenvalues of Sturm-Liouville problems by approximating the differential equation*, SIAM J. Numer. Anal., 10, (1973), 55-68.
526. _____, *High order approximations to Sturm-Liouville eigenvalues*, Numer. Math. 24, (1975), 241-247.
527. _____, *Solving linear boundary value problems by approximating the coefficients*, Math. Comp., 27, (1973), 551-561.
528. S. Pruess and C. Fulton, *An asymptotic numerical method for a class of singular Sturm-Liouville problems*, SIAM J. Numer. Anal., 32, (1995), 1658-1676.
529. J. D. Pryce, *Numerical solution of Sturm-Liouville problems*, Oxford University Press, Oxford, U. K., 1993.
530. C. R. Putnam, *On the spectra of certain boundary value problems*, Amer. J. Math. 71 (1949), 109-111.
531. D. Race and A. Zettl, *Characterization of the factors of quasi-differential expressions*, Proc. Royal Soc. Edinburgh 120A (1992), 297-312.
532. _____, *Nullspaces, representations and factorizations of quasi-differential expressions*, J. Differential and Integral Equations v.6, no. 4, (1993), 949-960.
533. _____, *On the commutativity of certain quasi-differential expressions*, J. London Math. Society, v.42(1990), 489-504.
534. T. T. Read, *A limit-point criterion for $-(py')' + qy$*, Lecture notes in Mathematics, Springer-Verlag, 1976.
535. _____, *A limit-point criterion for expressions with intermittently positive coefficients*, J. London Math. Soc. (2) 15 (1997), no. 2, 271-276.
536. _____, *A limit-point criterion for expressions with oscillatory coefficients*, Pac. J. Math. 66 (1976), 243-255.
537. _____, *On the limit-point condition for polynomials in second-order differential expressions*, J. London Math. Soc. 10 (1975), 357-366.
538. M. Reed and B. Simon, *Methods of Modern Mathematical Physics, vol.I*, Academic Press, New York, 1972.
539. _____, *Methods of Modern Mathematical Physics, vol. 4*, Academic Press, New York, 1978.
540. W. T. Reid, *Ordinary differential equations*, Wiley and Sons, Inc., New York, 1971.
541. _____, *Riccati Differential Equations*, Academic Press, New York, 1972.
542. _____, *Sturmian Theory for Ordinary Differential Equations*, Applied Mathematical Sciences 31, Springer-Verlag, Berlin, 1980.
543. F. Rellich, *Die zulassingen Randbedingungen bei den Eigenvertproblemen der mathematischen Physik, (Gewoehnliche Differentialgleichungen zweiter Ordnung)*, Math. Z. 49 (1944), 702-723.
544. _____, *Halbbeschraenkte gewoehnliche Diffferentialoperatoren zweiter Ordnung*, Math. Ann. 122 (1950/51), 343-368.
545. R. Richardson, *Theorems of oscillation for two linear differential equations of the second order with two parameters*, Trans. Amer. Math. Soc., 13 (1912), 22-34.
546. R. G. D. Richardson, *Contributions to the study of oscillatory properties of the solutions of linear differential equations of the second-order*, American J. Math. 40 (1918), 283-316.
547. J. Ridenhour and A. Zettl, *Construction of Singular Green's Functions of Sturm-Liouville Problems*, preprint.
548. _____, *Singular Green's functions*, preprint.
549. B. Riemann and H. Weber, *Die Partiellen Differentialgleichungen der Mathematischen Physik, II,5th ed.*, Braunschweig, 1912.
550. F. S. Rofe-Beketov, *Non-semibounded differential operators*, Teor. Funkcional. Anal. i Prilozen Vyp. 2 (1966),178-184.
551. _____, *The spectrum of non-selfadjoint differential operators with periodic coefficients*, Sov. Math. Dokl. 4 (1963), 1563-1566.
552. _____, *A test for the finiteness of the number of discrete levels introduced into the gaps of a continuous spectrum by perturbations of a periodic potential*, Sov. Math. Dokl. 5 (1964), 689-692.

553. A. Ronveaux, *Heun's differential equations*, Oxford University Press (1995).
554. R. Rosenberger, *Charakterisierung der Friedrichsfortsetzung von halbbeschraenkten Sturm-Liouville Opatoren*, Dissertation, Technische Hochschule Darmstadt, 1984.
555. H. L. Royden, *Real Analysis, third edition*, Macmillan, New York/London, 1988.
556. D. Schenk and M. A. Shubin, *Asymptotic expansion of the state density and the spectral function of a Hill operator*, Math. USSR Sbornik 56 (1987), 473-490.
557. Bernd Schultze, *Green's Matrix and the Formula of Titchmarsh-Kondaira for Singular Left-Definite Canonical Eigenvalue-Problems*, Proceedings of the Royal Society of Edinburgh, Vol. 83A (1979), 147-183.
558. D. B. Sears and E. C. Titchmarsh, *Some eigenvalue formulae*, Quart. J. Math. Oxford (2) 1 (1950), 165-175.
559. J. K. Shaw, A.P. Baronavski, and H.D. Ladouceur, *Applications of the Walker method, Spectral Theory and Computational Methods of Sturm-Liouville problems*, Lecture Notes in Pure and Applied Mathematics 191 (1997), 377-395. (Marcel Dekker, Inc., New York: 1997).
560. B. Simon, *Resonances in one dimension and Fredholm determinants*, J. Functional Anal., 178 (2000), 396-420.
561. _____, *Schroedinger semigroups*, Bull. Am. Math. Soc. 7 (1982), 447-526.
562. B. Simon and T. Wolff, *Singular continuous spectrum under rank one perturbations and localisation for random Hamiltonians*, Comm. Pure Appl. Math. 39, (1986), 75-90.
563. A. R. Sims, *Secondary conditions for linear differential operators of the second order*, J. Math. and Mechanics, 6, 247-285, 1957.
564. S. Yu. Slavyanov and W. Lay, *Special functions: a unified theory based on singularities*, Oxford University Press, Oxford, 2000.
565. H. M. Srivastava, Vu Kim Tuan, and S. B. Yakubovich, *The Cherry transform and its relationship with a singular Sturm-Liouville problem*, Q. J. Math.
566. G. Stolz, *Pure point spectra for Wannier-Stark Hamiltonians*, pre-print.
567. G. Stolz and J. Weidmann, *Approximation of isolated eigevalues of general singular ordinary differential operators*, Results in Mathematics 28, 345-358 (1995).
568. _____, *Approximation of isolated eigevalues of ordinary differential operators*, J. Reine Angew. Math. 445 (1993), 31-44.
569. C. Sturm, *Sur les equations differentielles lineaires du second ordre*, J. Math. Pures Appl. 1 (1836), 106-186.
570. _____, *Sur une classe d'equations a differences partielles*, J. Math. Pures Appl., 1 (1936), 373-444.
571. C. A. Swanson, *Comparison and oscillation theory of linear differential equations*, Academic Press, New York, 1968.
572. F.J. Tipler, *General relativity and conjugate ordinary differential equations*, J. Differential Eqs. 30 (1978), 165-174.
573. E. C. Titchmarsh, *Eigenfunction Expansions Associated with Second Order Differential Equations, part II, (2nd ed)*, Clarendon Press, Oxford, 1962.
574. _____, *Eigenfunction expansions I. sec. ed.*, Oxford Univ. Press, 1962.
575. A. L. Treskunov, *Sharp two-sided estimates for eigenvalues of some class of Sturm-Liouville problems*, J. Math. Sci. (New York).
576. G. M. Tuynman, *The derivation of the exponential map of matrices*, The American Mathematical Monthly v.102 (1995), 818-820.
577. E. J. M. Veling, *Asymptotic analysis of a singular Sturm-Liouville boundary value problem*, Integral Equations Operator Theory 7 (1984), no. 4, 561-587.
578. M. Venkatesulu and Pallav Kumar Baruah, *A classical approach to eigenvalue problems associated with a pair of mixed regular Sturm-Liouville equations. I*, J. Appl. Math. Stochastic Anal. 13 (3) (2000), 303-312.
579. H. Volkmer, *Sturm-Liouville problems with indefinite weights and Everitt's inequality*, Proc. Royal Soc. Edinburgh, A126, (1996), 1097-1112.
580. Hans Volkmer, *Coexistence of periodic solutions of Ince's equation*, Analysis (Munich) 23 (2003), no. 1, 97-105.
581. _____, *The coexistence problem for Heun's Differential Equation*, preprint.
582. Hans Volkmer, *Convergence radii for eigenvalues of two-parameter Sturm-Liouville problems*, Analysis (Munich) 20 (3) (2000), 225-236.
583. Hans Volkmer, *Eigenvalues associated with Borel sets*, pre-print.

584. _____ , *Multiparameter Eigenvalue Problems and Expansion Theorems*, Lecture Notes in Mathematics no 1356, Springer-Verlag, Berlin, New York, 1988.
585. Hans Volkmer and A. Zettl, *An inverse spectral theory for Sturm-Liouville Problems with finite spectrum*, preprint.
586. R. Vonhoff, *Eigenvalue Problems of Left-Definite Differential Equations*, preprint.
587. _____ , *A Left-Definite study of Legendre's Differential Equations and of the Fourth-Order Legendre Type Differential Equation*, Results. Math. 37 (2000), 155-196.
588. _____ , *Spektraltheoretische Untersuchung linksdefiniter Differentialgleichungen im singulaeren Fall*, Dissertation, Universitat Dortmund 1995.
589. S. Wallach, *The spectra of periodic potentials*, Am. J. Math. 70 (1948), 842-848.
590. J. Walter, *Bemerkungen zu dem Grenzpunktfallkriterium von N. Levinson*, Math. Z. 105 (1968), 345-350.
591. _____ , *Regular eigenvalue problems with eigenvalue parameter in the boundary conditions*, Math. Z. 133 (1973), 301-312.
592. Wolfgang Walter, *Differential and Integral Inequalities*, Springer-Verlag, Berlin/New York, 1970.
593. G. N. Watson, *A Treatise on the Theory of Bessel Functions,sec. ed.*, Cambridge U. P., Cambridge, 1962.
594. J. Weidmann, *Oszillationsmethoden fuer systeme gewoehnlicher Differentialgleichungen*, Math. Zeit. 119 (1971), 349-373.
595. _____ , *Spectral theory of Sturm-Liouville Operators; Approximation by Regular Problems*, preprint.
596. _____ , *Uniform Nonsubordinacy and the Absolutely Continuous Spectrum*, Analysis 16 (1996), 98-99.
597. _____ , *Verteilung der Eigenwerte for eine Klasse von Integraloperatoren of L2(a,b)*, J. reine angew. Math. 276 (1975), 213-220.
598. _____ , *Zur Spektraltheorie von Sturm-Liouville-Operatoren*, Math. Zeit. 98 (1967), 268-302.
599. _____ , *Linear Operators in Hilbert Spaces*, Graduate Texts in Mathematics, vol. 68, Springer Verlag, Berlin, 1980.
600. _____ , *Spectral Theory of Ordinary Differential Operators*, Lecture Notes in Mathematics 1258, Springer Verlag, Berlin, 1987.
601. _____ , *Lineare Operatoren in Hilbertraeumen, Teil I Grundlagen*, Teubner, Stuttgart/Leipiz/Wiesbaden, 2000.
602. _____ , *Lineare Operatoren in Hilbertraeumen, Teil II Anwendungen*, Teubner, Stuttgart/Leipiz/Wiesbaden, 2003.
603. J. Weidmann and G. Stolz, *Approximation of isolated eigenvalues of ordinary differential operators*, preprint.
604. H. Weinberger, *Variational Methods for Eigenvalue Approximation*, CBMS-NSF Regional Conference Series in Applied Mathematics 15. SIAM, Philadelphia, 1974.
605. A. Weinstein, *On the Sturm-Liouville theory and the eigenvalues of intermediate problems*, Numer. Math. 5 (1963), 238-245.
606. R. Weinstock, *Calculus of variations*, McGraw-Hill, New York, 1952.
607. H. Weyl, *Ueber gewoehnliche Differentialgleichungen mit Singularitaeten und die zugehoerigen Entwicklungen willkuerlicher Funktionen*, Math. Ann. 68 (1910), 220-269.
608. E.T. Whittaker and G. N. Watson, *Modern analysis*, Cambridge University Press, Cambridge, 1950.
609. D. Willett, *Classification of second order linear differential equations with respect to oscillation*, Advanc. Math. 3 (1969), 594-623.
610. _____ , *On the oscillatory behavior of the solution of second order linear differential equations*, Ann. Polon. Math 21 (1969), 175-194.
611. A. Wintner, *A criteria of oscillatory stability*, Quant. Appl. Math. 7 (1949), 115-117.
612. _____ , *On the comparison theorem of Kneser-Hille*, Math. Scand. 5 (1957), 255-260.
613. _____ , *On the nonexistence of conjugate points*, Amer. J. Math. 73 (1951), 368-380.
614. J. S. W. Wong, *On second order nonlinear oscillation*, Math. Research Ctr., Univ. of Wisconsin, Technical Report No. 836, also Funkcial. Ekvac. 11 (1969), 207-234.
615. _____ , *Oscillation and nonoscillation of solutions of second order linear differential equations with integrable coefficients*, Trans. Amer. Math. Soc. 144 (1969), 197-215.

616. _____, *Oscillation theorems for second order nonlinear differential equations*, Bull. Inst. Math. Acad. Sinica 3 (1975), 283-309.
617. _____, *Second order linear oscillation with integrable coefficients*, Bull. Amer. Math. Soc. 74(1968), 909-911.
618. J. S. W. Wong and A. Zettl, *On the limit-point classification of second-order differential equations*, Math. Z. 132 (1973), 297-304.
619. Y. Yakubovich and V. Starzhinskii, *Linear Differential Equations with Periodic Coefficients*, vol. I, II, John Wiley and Sons, New York, 1975.
620. A. Zettl, *Adjoint and self-adjoint problems with interface conditions*, SIAM J. Applied Math. 16, (1968), 851-859.
621. _____, *Adjoint linear differential operators*, Proc. Amer. Math. Soc. 16, (1965), 1239-1241.
622. _____, *Adjointness in non-adjoint boundary value problems*, SIAM J. Applied Math. 17, (1969), 1268-1279.
623. _____, *An algorithm for the construction of all disconjugate operators*, Proc. Royal Soc. Edinburgh, 75A, 4, (1976).
624. _____, *An algorithm for the construction of limit circle expressions*, Proc. Royal Soc. Edinburgh, 75A, 1, (1976).
625. _____, *A characterization of the factors for ordinary linear differential operators*, Bulletin Amer. Math. Soc., V. 80, No. 3 (1974), 498-500.
626. _____, *Computing continuous spectrum*, Proc. International Symposium on Trends and developments in ordinary differential equations, Yousef Alavi and Po-Fang Hsieh, editors, World Scientific,(1994), 393-406.
627. _____, *A constructive characterization of disconjugacy*, Bull. Amer. Math. Soc., Vol 81, No. 1 (1975).
628. _____, *Deficiency indices of polynomials in symmetric differential expressions II*, Proc. Royal Soc. Edinburgh, 73A, 20(1974/75).
629. _____, *The essential spectrum of nonselfadjoint ordinary differential operators*, University of Missouri at Rolla Press; edited by J. L. Henderson (1985), 152-168. Proceedings of the Twelfth and Thirteenth Midwest Conference.
630. _____, *Explicit conditions for the factorization of nth order differential operators*, Transactions American Math. Soc., 41, No.1,(1973), 137-145.
631. _____, *Factorization and disconjugacy of third order differential equations*, Proc. Amer. Math. Soc., 31(1972), 203-208.
632. _____, *Factorization of differential operators*, Proc. Amer. Math. Soc., 27(1970), 425-426.
633. _____, *Formally self-adjoint quasi-differential operators*, Rocky Mountain J. Math. 5(1975), 453-474.
634. _____, *General theory of the factorization of nth order linear differential operators*, Transactions American Math. Soc., V. 197 (1974), 31-353.
635. _____, *The lack of self-adjointness in three point boundary value problems*, Proc. Amer. Math. Soc., 17(1966), 368-371.
636. _____, *The limit point and limit circle cases for polynomials in a differential operator*, Proc. Royal Society of Edinburgh, (1975), 219-224.
637. _____, *A note on square integrable solutions of linear differential equations*, Proc. Amer. Math. Soc., 21(1969), 671-672.
638. _____, *On the Friedrichs extension of singular differential operators*, Communications in Applied Analysis 2 (1998), no.1, 31-36.
639. _____, *Perturbation of the limit circle case*, Quart. J. Math. Oxford, (1975).
640. _____, *Perturbation theory of deficiency indices of differential operators*, J. London Math. Soc., (2), 12(1976).
641. _____, *Powers of symmetric differential expressions without smoothness assumptions*, Quaestiones Mathematicae, 1(1976), 83-94.
642. _____, *Separation for differential expressions and the Lp spaces*, Proc. Amer. Math. Soc., V. 55, 1, (1976).
643. _____, *Some identities related to Polya's property W for linear differential equations*, Proc. Amer. Math. Soc., 18(1967), 992-994.
644. _____, *Square integrable solutions of Ly=f(t,y)*, Proc. Amer. Math. Soc.,(1970), 635-639.
645. _____, *Sturm-Liouville problems*, Marcel Dekker, v. 191, (1997), 1-104. Proceedings of the 1996 Knoxville Barrett Conference, edited by D. B. Hinton and P. Schaefer.

646. V. A. Zheludev, *Eigenvalues of the perturbed Schroedinger operator with a periodic potential*, Topics in Math. Phys. 2 (1968), 87-101.
647. _____, *Perturbation of the spectrum of one-dimensional self-adjoint Schroedinger operator with a periodic potential*, Topics in Math. Phys. 4 (1971), 55-75.
648. M. Zlamal, *Oscillation criterions*, Cas. Pest. Mat. a Fis. 75 (1950), 213-217.

Index

Abel's formula, 7
Absolutely continuous spectrum, 295
Adjoint identity for fundamental matrices, 20
adjoint systems, 20
Atkinson type problems, 57

BC basis, 182, 184
Bielecki norm, 4
boundary value problem space, 53
bounds of solutions, 10
bracket decomposition, 175

Canonical Form of Self-Adjoint Boundary Conditions, 179
canonical form of singular coupled self-adjoint boundary conditions, 191
Canonical forms of self-adjoint boundary conditions, 71
Change of BC bases theorem, 185
changes sign, 69, 232
Characteristic function
    Transcendental function
        Transcendental characterization of eigenvalues, 44
characteristic function, 44
Complex coefficients, 298
Continuity of zeros of an analytic function, 54
Continuous dependence of solutions of initial value problems on the problem, 12
Continuous dependence of solutions of scalar initial value problems on the problem, 27
continuous eigenvalue branch, 55
continuous eigenvalue branches, 56
continuous extension of solutions, 10
Continuous extension of solutions to endpoints, 11
Continuous extensions to the endpoints, 27
Coupled boundary conditions
    Real coupled
        Complex coupled, 71

coupled boundary conditions, 46

deficiency index, 175
Derivative of the matrix exponential function, 18
derivatives of eigenvalues, 56
Differentiable dependence of solutions, 28
Differentiable dependence of solutions of initial value problems on all the parameters of the problem, 14
Disconjugacy, 140
discrete spectrum, 174
Discreteness criteria, 296

eigencurves, 113
eigenfunction, 44
Eigenparameter dependent boundary conditions, 296
eigenvalue, 44
eigenvalue index for left-definite problems, 113
Eigenvalues in gaps, 297
Embedded eigenvalues, 297
essential spectrum, 174
    continuous spectrum, 203
Expansion theorems, 297
exponential law, 15

Finite spectrum, 58
Floquet theory, 35
Frechet derivative, 14
fundamental matrix, 8

Gabushin inequalities, 297
geometric multiplicity of an eigenvalue, 44
GKN Glazman, Krein, Naimark, 71
Green's Function, 61
Green's Kernel Operator, 98
Gronwall inequality, 8

Hadamar inequalities, 297
Half-linear equations, 297
Half-Range expansions, 297
Hardy-Littlewood inequalities, 297

indefinite, 232
Inherited boundary conditions
    Inherited operators, 202
Initial value problem, 4
Inverse initial value problems, 20
Inverse spectral theory, 296
isolated zeros, 32

Jump set of boundary conditions, 78

Kallman-Rota inequality, 297
Kolmogorov inequalities, 297

Lagrange bracket decomposition
    bracket decomposition, 175
Lagrange form, 172
Landau inequalities, 297
LC/LP dichotomy
    Two parameter LC/LP dichotomy, 230
Left-Definite, 109
limit circle
    LC, 145
limit-circle nonoscillatory
    LCNO, 145
limit-circle oscillatory
    LCO, 145
limit-point
    LP, 145

Maximal domain
    Maximal operator, 164
Maximal operator, 173
metric, 53
Minimal domain
    Minimal operator, 164
Minimal operator, 173
Multi-parameter theory, 296

Naimark Patching Lemma, 175
Non-Principal solution, 131
Nonlinear oscillation, 297
Nonlinear problems, 297
nonoscillatory
    NO, 145
norm inequalities, 297
normalized eigenfunction, 70
Numerical approximations, 298

Order 1/2, 31
oscillation (O), 33
oscillatory
    (O), 145

Patching Lemma, 175
periodic coefficients, 35
Picone Identity, 34
primary fundamental matrix, 8
Principal solution, 131
Product formula for solutions, 15
Pruefer transformation, 81

quasi-derivative, 25

Rank invariance, 6, 7
Rank invariance of solutions at endpoints, 11
regular endpoint, 27, 145
resolvent set, 174
Ricatti equation, 140

Schonberg-Cavaretta inequalities, 297
self-adjoint boundary conditions, 70
self-adjoint extension, 173
self-adjoint realization, 173
self-adjoint restriction, 173
separated boundary conditions, 46
Separated self-adjoint boundary conditions, 71
separated singular self-adjoint boundary conditions, 191
Separation, 297
simple eigenvalue, 44
singular characteristic function, 188
singular endpoint, 27
singular initial conditions, 130
singular self-adjoint boundary conditions, 245
solution, 25
solution of a system, 3
spectral bands, 210
spectral gaps, 210
spectral parameter, 26
spectrum, 174
Strong limit-point conditions, 297
Sturm Comparison Theorem, 32
Sturm Separation Theorem, 32
Successive approximations, 5
symmetric differential equation, 70
symmetric equation, 69

The Friedrichs extension, 193
trivial solution, 69

variation of parameters formula, 8

weight function, 26

Titles in This Series

121 **Anton Zettl,** Sturm-Liouville Theory, 2005
120 **Barry Simon,** Trace Ideals and Their Applications, 2005
119 **Tian Ma and Shouhong Wang,** Geometric theory of incompressible flows with applications to fluid dynamics, 2005
118 **Alexandru Buium,** Arithmetic differential equations, 2005
117 **Volodymyr Nekrashevych,** Self-similar groups, 2005
116 **Alexander Koldobsky,** Fourier analysis in convex geometry, 2005
115 **Carlos Julio Moreno,** Advanced analytic number theory: L-functions, 2005
114 **Gregory F. Lawler,** Conformally invariant processes in the plane, 2005
113 **William G. Dwyer, Philip S. Hirschhorn, Daniel M. Kan, and Jeffrey H. Smith,** Homotopy limit functors on model categories and homotopical categories, 2004
112 **Michael Aschbacher and Stephen D. Smith,** The classification of quasithin groups II. Main theorems: The classification of simple QTKE-groups, 2004
111 **Michael Aschbacher and Stephen D. Smith,** The classification of quasithin groups I. Structure of strongly quasithin K-groups, 2004
110 **Bennett Chow and Dan Knopf,** The Ricci flow: An introduction, 2004
109 **Goro Shimura,** Arithmetic and analytic theories of quadratic forms and Clifford groups, 2004
108 **Michael Farber,** Topology of closed one-forms, 2004
107 **Jens Carsten Jantzen,** Representations of algebraic groups, 2003
106 **Hiroyuki Yoshida,** Absolute CM-periods, 2003
105 **Charalambos D. Aliprantis and Owen Burkinshaw,** Locally solid Riesz spaces with applications to economics, second edition, 2003
104 **Graham Everest, Alf van der Poorten, Igor Shparlinski, and Thomas Ward,** Recurrence sequences, 2003
103 **Octav Cornea, Gregory Lupton, John Oprea, and Daniel Tanré,** Lusternik-Schnirelmann category, 2003
102 **Linda Rass and John Radcliffe,** Spatial deterministic epidemics, 2003
101 **Eli Glasner,** Ergodic theory via joinings, 2003
100 **Peter Duren and Alexander Schuster,** Bergman spaces, 2004
99 **Philip S. Hirschhorn,** Model categories and their localizations, 2003
98 **Victor Guillemin, Viktor Ginzburg, and Yael Karshon,** Moment maps, cobordisms, and Hamiltonian group actions, 2002
97 **V. A. Vassiliev,** Applied Picard-Lefschetz theory, 2002
96 **Martin Markl, Steve Shnider, and Jim Stasheff,** Operads in algebra, topology and physics, 2002
95 **Seiichi Kamada,** Braid and knot theory in dimension four, 2002
94 **Mara D. Neusel and Larry Smith,** Invariant theory of finite groups, 2002
93 **Nikolai K. Nikolski,** Operators, functions, and systems: An easy reading. Volume 2: Model operators and systems, 2002
92 **Nikolai K. Nikolski,** Operators, functions, and systems: An easy reading. Volume 1: Hardy, Hankel, and Toeplitz, 2002
91 **Richard Montgomery,** A tour of subriemannian geometries, their geodesics and applications, 2002
90 **Christian Gérard and Izabella Łaba,** Multiparticle quantum scattering in constant magnetic fields, 2002
89 **Michel Ledoux,** The concentration of measure phenomenon, 2001

TITLES IN THIS SERIES

88 **Edward Frenkel and David Ben-Zvi,** Vertex algebras and algebraic curves, second edition, 2004
87 **Bruno Poizat,** Stable groups, 2001
86 **Stanley N. Burris,** Number theoretic density and logical limit laws, 2001
85 **V. A. Kozlov, V. G. Maz'ya, and J. Rossmann,** Spectral problems associated with corner singularities of solutions to elliptic equations, 2001
84 **László Fuchs and Luigi Salce,** Modules over non-Noetherian domains, 2001
83 **Sigurdur Helgason,** Groups and geometric analysis: Integral geometry, invariant differential operators, and spherical functions, 2000
82 **Goro Shimura,** Arithmeticity in the theory of automorphic forms, 2000
81 **Michael E. Taylor,** Tools for PDE: Pseudodifferential operators, paradifferential operators, and layer potentials, 2000
80 **Lindsay N. Childs,** Taming wild extensions: Hopf algebras and local Galois module theory, 2000
79 **Joseph A. Cima and William T. Ross,** The backward shift on the Hardy space, 2000
78 **Boris A. Kupershmidt,** KP or mKP: Noncommutative mathematics of Lagrangian, Hamiltonian, and integrable systems, 2000
77 **Fumio Hiai and Dénes Petz,** The semicircle law, free random variables and entropy, 2000
76 **Frederick P. Gardiner and Nikola Lakic,** Quasiconformal Teichmüller theory, 2000
75 **Greg Hjorth,** Classification and orbit equivalence relations, 2000
74 **Daniel W. Stroock,** An introduction to the analysis of paths on a Riemannian manifold, 2000
73 **John Locker,** Spectral theory of non-self-adjoint two-point differential operators, 2000
72 **Gerald Teschl,** Jacobi operators and completely integrable nonlinear lattices, 1999
71 **Lajos Pukánszky,** Characters of connected Lie groups, 1999
70 **Carmen Chicone and Yuri Latushkin,** Evolution semigroups in dynamical systems and differential equations, 1999
69 **C. T. C. Wall (A. A. Ranicki, Editor),** Surgery on compact manifolds, second edition, 1999
68 **David A. Cox and Sheldon Katz,** Mirror symmetry and algebraic geometry, 1999
67 **A. Borel and N. Wallach,** Continuous cohomology, discrete subgroups, and representations of reductive groups, second edition, 2000
66 **Yu. Ilyashenko and Weigu Li,** Nonlocal bifurcations, 1999
65 **Carl Faith,** Rings and things and a fine array of twentieth century associative algebra, 1999
64 **Rene A. Carmona and Boris Rozovskii, Editors,** Stochastic partial differential equations: Six perspectives, 1999
63 **Mark Hovey,** Model categories, 1999
62 **Vladimir I. Bogachev,** Gaussian measures, 1998
61 **W. Norrie Everitt and Lawrence Markus,** Boundary value problems and symplectic algebra for ordinary differential and quasi-differential operators, 1999
60 **Iain Raeburn and Dana P. Williams,** Morita equivalence and continuous-trace C^*-algebras, 1998
59 **Paul Howard and Jean E. Rubin,** Consequences of the axiom of choice, 1998

For a complete list of titles in this series, visit the
AMS Bookstore at **www.ams.org/bookstore/**.